Springer Series in

MATERIALS SCIENCE 150

Springer Series in

MATERIALS SCIENCE

Editors: R. Hull C. Jagadish R.M. Osgood, Jr. J. Parisi Z. Wang

The Springer Series in Materials Science covers the complete spectrum of materials physics, including fundamental principles, physical properties, materials theory and design. Recognizing the increasing importance of materials science in future device technologies, the book titles in this series reflect the state-of-the-art in understanding and controlling the structure and properties of all important classes of materials.

Please view available titles in *Springer Series in Materials Science* on series homepage http://www.springer.com/series/856

Amalia Patanè
Naci Balkan
Editors

Semiconductor Research

Experimental Techniques

With 192 Figures

Springer

Editors

Amalia Patanè
The University of Nottingham, School of Physics and Astronomy
University Park, NG7 2RD Nottingham, UK
E-mail: Amalia.Patane@nottingham.ac.uk

Naci Balkan
The University of Essex, School of Computer Science and Electronic Engineering
Essex, UK
E-mail: balkan@essex.ac.uk

Springer Series in Materials Science ISSN 0933-033X
ISBN 978-3-642-23350-0 e-ISBN 978-3-642-23351-7
DOI 10.1007/978-3-642-23351-7
Springer Heidelberg Dordrecht London New York

Library of Congress Control Number: 2012935492

Printed on acid-free paper

Springer is part of Springer Science+Business Media (www.springer.com)

Preface

This book intends to provide its readers with the fundamentals and applications of experimental techniques commonly used in semiconductor research. Each chapter describes the physics concepts underlying a specific technique and its latest developments in the investigation of novel semiconductor materials and heterostructures, including InN, dilute nitride III-N-V alloys, InAs/GaSb heterostructures and self-assembled quantum dots.

Chapter 1 focuses on the investigation of semiconductor surfaces by reflection high-energy electron diffraction (RHEED) and low-energy electron diffraction (LEED). These diffraction-based techniques give access to the structural properties of crystalline layers and junctions, which represent the core of modern devices. Several examples of LEED and RHEED patterns are described and provide the reader with the basic tools for interpreting RHEED and LEED data. *Chapter 2* describes transmission electron microscopy (TEM) and high resolution electron microscopy techniques. These are based on the analysis of a transmitted electron beam through an electron-transparent sample. Operation principles of TEM and examples of spatial mapping of composition and strain at the nanoscale and atomic resolution are the focus of this chapter. *Chapter 3* reviews common techniques used to investigate the energy and momentum relaxation rates of hot carriers in semiconductors where the carrier heating is achieved by either the application of an electrical field or by an optical excitation. This condition is frequently met in optical and electronic device. Also, this chapter reviews steady-state spectral and transient measurement techniques. *Chapter 4* describes the principles, experimental setups, and theoretical approaches used in optical modulation spectroscopy studies. Particular attention is dedicated to contactless electroreflectance (CER) and photoreflectance (PR). These are non-destructive techniques, which are widely applied to study the band structure properties of semiconductor materials and devices. The complementary technique of photoluminescence (PL) and its relation to absorption and photoluminescence excitation (PLE) spectroscopy are described in *Chap. 5*. Typical experimental setups for optical studies, with and without an applied magnetic field, are discussed and examples of their application to study electronic properties, disorder effects and carrier thermalization in III–V semiconductor alloys and heterostructures are provided. High pressure is one of the most valuable characterization tools available in semiconductor research. *Chapter 6* describes how it can be used to investigate several important physical phenomena in semiconductor materials and devices such as the Gunn effect and avalanche breakdown. It also considers laser devices and the interesting changes that take place when a semiconductor material is subject to high pressures. *Chapter 7* describes the principles and instrumentation of techniques used in spatially resolved spectroscopy, including micro-photoluminescence (μ-PL), scanning near-field optical microscopy (SNOM), and spatially resolved cathodoluminescence (CL). Since the spatial resolution is often limited not only by instrumental capabilities but also by the spreading of the photoexcited carriers outside the photoexcited volume, the mechanisms of diffusion, photon recycling, phonon wind, and Fermi pressure are reviewed. The availability of ultrashort laser pulses has offered a new investigation tool to the optical spectroscopy field, giving access not only to the spectral features of luminous phenomena, but also directly to their dynamics. *Chapter 8* presents an overview of the most popular ultra-fast time resolved optical spectroscopy techniques used in semiconductor physics in the picosecond and sub-picosecond time range. For each of these techniques, which include time resolved PL (TRPL), pump and probe time resolved spectroscopy, and time-correlated single photon counting (TCSPC), the operating principles and the fundamental concepts are introduced together with typical experimental setups and applications. The development of a new generation of Raman spectroscopy systems in recent years has contributed to the widespread of Raman spectroscopy in materials science. The aim of *Chap. 9* is to offer an up-to-date overview of the fundamentals and use of Raman spectroscopy. The chapter is restricted to standard spontaneous Raman spectroscopy. It discusses the basic concepts of inelastic light scattering and its application to study phonon and impurity modes, crystal quality, and strain effects in semiconductors. High magnetic fields have played a key role in elucidating the electronic properties of semiconductors. The cyclotron resonance (CR) and the Hall effects are perhaps the most celebrated examples of the use of magnetic fields in semiconductor physics. Cyclotron resonance is the focus of *Chap. 10*, which describes the basic theory of CR and discusses experimental setups and applications of CR to unravel the determination of the electron mass in various materials. *Chapter 11* examines instead the physics of electron motion in the presence of a magnetic field, with particular reference to recent applications in which high magnetic fields have been used to elucidate the electronic and quantum properties of novel heterostructures and nanostructures. Also, it describes how magneto-tunnelling spectroscopy (MTS) can be used to measure the band structure of semiconductors and to investigate and manipulate the energy eigenvalues and eigenfunctions of electrons confined in low-dimensional systems. Finally, photoconductivity (PC), photo-induced transient spectroscopy (PITS), and deep level transient spectroscopy (DLTS) are presented in *Chap. 12*. These techniques provide powerful tools to investigate the intrinsic and extrinsic energy levels of a semiconductor, time constants associated with carrier recombination, activation energies, carrier capture cross-section and densities of energy traps.

Several colleagues have contributed to this book by kindly making available their expertise and knowledge. We are very grateful for their help in completing this book, which we hope will assist students and researchers in their work, thus contributing to novel and exciting research in semiconductor physics.

Nottingham, UK *Amalia Patanè*
Essex, UK *Naci Balkan*

Contents

Contributors

A. Adams Advanced Technology Institute, University of Surrey, Guildford, Surrey, GU2 7XH, UK, Alf.Adams@surrey.ac.uk

T. Amand Physics Department - LPCNO, Université de Toulouse, INSA-CNRS-UPS, LPCNO, 135, Av. de Rangueil, 31077 Toulouse, France

M.Ç. Arikan Science Faculty, Department of Physics, Istanbul University, 34134 Vezneciler, Istanbul, Turkey, arikan@istanbul.edu.tr

N. Balkan School of Computer Science and Electronic Engineering, University of Essex, Essex, UK, balkan@essex.ac.uk

A. Balocchi Physics Department - LPCNO, Université de Toulouse, INSA-CNRS-UPS, LPCNO, 135, Av. de Rangueil, 31077 Toulouse, France

M. Beltrán Departamento de Ciencia de los Materiales e I.M. y Q.I., Facultad de Ciencias, Universidad de Cádiz, Campus Río San Pedro, s/n, 11510 Puerto Real, Cádiz, Spain

T. Ben Departamento de Ciencia de los Materiales e I.M. y Q.I., Facultad de Ciencias, Universidad de Cádiz, Campus Río San Pedro, s/n, 11510 Puerto Real, Cádiz, Spain

M. Capizzi CNISM-Dipartimento di Fisica, Sapienza Università di Roma, P.le A. Moro 2, 00185 Roma, Italy, mario.capizzi@roma1.infn.it

R. Cuscó Institut Jaume Almera, Consell Superior d'Investigacions Científiques (CSIC), Lluís Solé i Sabarís s/n, 08028 Barcelona, Catalonia, Spain, rcusco@ictja.csic.es

O. Drachenko Institute of Ion Beam Physics and Materials Research, Helmholtz-Zentrum Dresden-Rossendorf, P.O. Box 510119, 01314 Dresden, Germany, o.drachenko@hzdr.de

L. Eaves School of Physics and Astronomy, The University of Nottingham, Nottingham NG7 2RD, UK

A. Erol Science Faculty, Department of Physics, Istanbul University, 34134 Vezneciler, Istanbul, Turkey, ayseerol@istanbul.edu.tr

P.L. Galindo Departamento de Lenguajes y Sistemas Informáticos, Universidad de Cádiz, Campus Río San Pedro, 11510 Puerto Real, Cádiz, Spain

R. García Departamento de Ciencia de los Materiales e I.M. y Q.I., Facultad de Ciencias, Universidad de Cádiz, Campus Río San Pedro, s/n, 11510 Puerto Real, Cádiz, Spain

D. González Departamento de Ciencia de los Materiales e I.M. y Q.I., Facultad de Ciencias, Universidad de Cádiz, Campus Río San Pedro, s/n, 11510 Puerto Real, Cádiz, Spain

M. Guina Optoelectronics Research Centre, Tampere University of Technology, FI-33101 Tampere, Finland

M. Helm Institute of Ion Beam Physics and Materials Research, Helmholtz-Zentrum Dresden-Rossendorf, P.O. Box 510119, 01314 Dresden, Germany, m.helm@hzdr.de

M. Herrera Departamento de Ciencia de los Materiales e I.M. y Q.I., Facultad de Ciencias, Universidad de Cádiz, Campus Río San Pedro, s/n, 11510 Puerto Real, Cádiz, Spain

J. Ibáñez Insa Institut Jaume Almera, Consell Superior d'Investigacions Científiques (CSIC), Lluís Solé i Sabarís s/n, 08028 Barcelona, Catalonia, Spain, jibanez@ictja.csic.es

R. Kudrawiec Institute of Physics, Wrocław University of Technology, Wybrzeże Wyspiańskiego 27, 50-370 Wrocław, Poland, robert.kudrawiec@pwr.wroc.pl

P. Laukkanen Department of Physics and Astronomy, University of Turku, FI-20014 Turku, Finland, peklau@utu.fi

Optoelectronics Research Centre, Tampere University of Technology, FI-33101 Tampere, Finland

J.G. Lozano Departamento de Ciencia de los Materiales e I.M. y Q.I., Facultad de Ciencias, Universidad de Cádiz, Campus Río San Pedro, s/n, 11510 Puerto Real, Cádiz, Spain

J.M. Mánuel Departamento de Ciencia de los Materiales e I.M. y Q.I., Facultad de Ciencias, Universidad de Cádiz, Campus Río San Pedro, s/n, 11510 Puerto Real, Cádiz, Spain

X. Marie Physics Department - LPCNO, Université de Toulouse, INSA-CNRS-UPS, LPCNO, 135, Av. de Rangueil, 31077 Toulouse, France, marie@insa-toulouse.fr

J. Misiewicz Institute of Physics, Wrocław University of Technology, Wybrzeże Wyspiańskiego 27, 50-370 Wrocław, Poland, jan.misiewicz@pwr.wroc.pl

S.I. Molina Departamento de Ciencia de los Materiales e I.M. y Q.I., Facultad de Ciencias, Universidad de Cádiz, Campus Río San Pedro, s/n, 11510 Puerto Real, Cádiz, Spain, sergio.molina@uca.es

F.M. Morales Departamento de Ciencia de los Materiales e I.M. y Q.I., Facultad de Ciencias, Universidad de Cádiz, Campus Río San Pedro, s/n, 11510 Puerto Real, Cádiz, Spain

A. Patanè School of Physics and Astronomy, The University of Nottingham, Nottingham NG7 2RD, UK, amalia.patane@nottingham.ac.uk

M. Paz Guerrero-Lebrero Departamento de Lenguajes y Sistemas Informáticos, Universidad de Cádiz, Campus Río San Pedro, 11510 Puerto Real, Cádiz, Spain

G. Pettinari CNISM-Dipartimento di Fisica, Sapienza Università di Roma, P.le A. Moro 2, 00185 Roma, Italy

High Field Magnet Laboratory, Institute for Molecules and Materials, Radboud University Nijmegen, Toernooiveld 7, 6525 ED Nijmegen, The Netherlands

School of Physics and Astronomy, The University of Nottingham, NG7 2RD Nottingham, UK

J. Pizarro Departamento de Lenguajes y Sistemas Informáticos, Universidad de Cádiz, Campus Río San Pedro, 11510 Puerto Real, Cádiz, Spain

A. Polimeni CNISM-Dipartimento di Fisica, Sapienza Università di Roma, P.le A. Moro 2, 00185 Roma, Italy

A. Prins Advanced Technology Institute, University of Surrey, Guildford, Surrey, GU2 7XH, UK

J. Sadowski MAX-lab, Lund University, SE-221 00 Lund, Sweden

Institute of Physics, Polish Academy of Sciences, al. Lotnikow 32/46, 02-668 Warszawa, Poland

D.L. Sales Departamento de Ciencia de los Materiales e I.M. y Q.I., Facultad de Ciencias, Universidad de Cádiz, Campus Río San Pedro, s/n, 11510 Puerto Real, Cádiz, Spain, david.sales@uca.es

M. Sánchez Departamento de Ciencia de los Materiales e I.M. y Q.I., Facultad de Ciencias, Universidad de Cádiz, Campus Río San Pedro, s/n, 11510 Puerto Real, Cádiz, Spain

S. Sweeney Advanced Technology Institute, University of Surrey, Guildford, Surrey, GU2 7XH, UK

G. Tamulaitis Semiconductor Physics Department and Institute of Applied Physics, Vilnius University, Vilnius, Sauletekio al. 9-III, Lithuania, gintautas.tamulaitis@ff.vu.lt

Chapter 1
Surface Studies by Low-Energy Electron Diffraction and Reflection High-Energy-Electron Diffraction

P. Laukkanen, J. Sadowski, and M. Guina

Abstract In this chapter, we present the basic concepts of the low-energy electron diffraction (LEED) and reflection high-energy electron diffraction (RHEED) experiments. The main goal is to provide an overview of the exploitation of these instrumental methods for analyzing the surfaces of technologically important III–V compound semiconductors. In particular, the interpretation of LEED and RHEED patterns is discussed for the most representative reconstructions of GaAs(100), GaInAsN(100), and Bi-stabilized III–V(100) surfaces. Other application examples concern the use of RHEED for optimizing the growth conditions and growth rates used in molecular beam epitaxy of III–V device heterostructures.

1.1 Basics of RHEED and LEED

The ability to synthesize novel semiconductor compounds and to study their crystalline properties and interfaces is essential for the development of new electronic and optoelectronic devices. Nobel laureate Herbert Kroemer has described the importance of device interfaces as follows: "Often, it may be said that the interface is the device" [1]. However, the controlled fabrication of high-quality interfaces is not straightforward because the surface of a crystal forming the interface is usually the most defective part of the material. In addition, the atomic structures of semiconductor surfaces often differ from the structures of the corresponding planes deeper in the crystal (bulk) because on many semiconductor surfaces, the bulk-plane structure is not energetically favored. This indeed affects the properties of the semiconductor surfaces and the interface formation. In general, any improvement in

P. Laukkanen (✉)
Department of Physics and Astronomy, University of Turku, FI-20014 Turku, Finland
Optoelectronics Research Centre, Tampere University of Technology, FI-33101 Tampere, Finland
e-mail: peklau@utu.fi

A. Patanè and N. Balkan (eds.), *Semiconductor Research*, Springer Series in Materials Science 150, DOI 10.1007/978-3-642-23351-7_1,

the crystal ordering at the interface or in the film will decrease the density of harmful interface defects. Crystalline interfaces and thin films are also essential for investigating the fundamental properties of these materials because the interpretation of measured data from an amorphous (not well-defined) material is often challenging. The diffraction-based analysis techniques are very useful for understanding and engineering the properties of crystalline layers and junctions. This chapter deals with the characterization of several technologically relevant semiconductor surface layers by means of two surface sensitive measurements: reflection high-energy electron diffraction (RHEED) and low-energy electron diffraction (LEED), which visualize the reciprocal lattice of a surface layer studied. RHEED is the standard equipment in the epitaxial growth chambers providing a great opportunity to monitor surface properties during the growth process. This real-time in-situ method makes it possible to control the growth with atomic layer precision. In contrast, the LEED characterization is usually done in a separate vacuum chamber that is connected to an epitaxial growth chamber. LEED measurements are particularly useful in determining the surface geometry of the starting substrate used for interface growth, such as an insulator–semiconductor interface. Both RHEED and LEED analyses are also excellent methods to solve detailed surface atomic structures via the comparison of angle or electron-energy dependent diffraction intensity with the corresponding intensity curves calculated and refined with the potential atomic models. While discussing several examples of LEED and RHEED patterns, our goal is to provide the reader with the basic tools to interpret own RHEED and LEED data. For an in-depth study of the instrumental and physical concepts, we recommend several excellent books and review articles, for example [2–10] given here.

Generally speaking, the instrumental part of LEED and RHEED consists of an electron gun and a luminescence screen that detects the electrons diffracted by the sample under study. Since its discovery in 1927 by Davisson and Germer [11, 12], LEED has become a widely used method for analyzing surface structures. In a typical LEED setup, the electrons are accelerated with voltage V of 10–500 V. Then the de Broglie wavelength of the electrons, $\lambda_E = h/(2m_E Ve)^{1/2}$, ranges from 0.87 to 2.75 Å; these values are small enough for electrons to experience diffraction from semiconductor crystals. Here h is the Planck constant, m_E is electron rest mass, and e is the elementary charge. The LEED electron beam impinges on the surface along the surface-normal direction, and the elastically backscattered electrons give an intensity pattern which visualizes (or maps) the reciprocal lattice of the surface layer. On the basis of this information about the reciprocal lattice, the real-space surface lattice can be constructed. The mathematical relation between the lattice and reciprocal-lattice vectors can be found for example in [2–5]. The diffraction is described by the Laue condition: $\boldsymbol{k}_{\|} = \boldsymbol{G}$. Here $\boldsymbol{G}$ is the surface reciprocal lattice vector, and $\boldsymbol{k}_{\|}$ is the parallel component of the change $(\boldsymbol{K}_f - \boldsymbol{K}_i)$ in which $\boldsymbol{K}_i$ and $\boldsymbol{K}_f$ are the incident and scattered wave vectors of the electron, as represented schematically in Fig. 1.1. The Laue equation also provides a physical meaning for the reciprocal vectors: every diffraction beam corresponds to the reciprocal vector. Because LEED operates in a low-energy regime, which is close to the minimum of

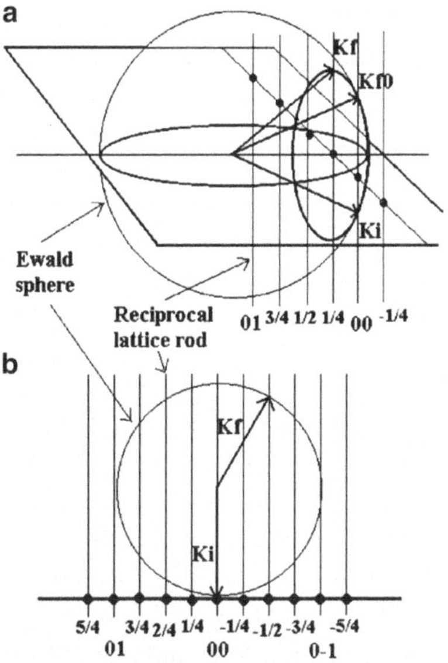

Fig. 1.1 The reciprocal Ewald-sphere constructions in (**a**) RHEED and (**b**) LEED for the ideal single plane of atoms, where the plane distance is infinite in the surface-normal direction in real space, corresponding to the infinitely small distance in reciprocal space; in other words, the reciprocal rods are formed along the surface normal. These rods intersect at the reciprocal lattice points (*black spots* in the plane) determined by the surface reciprocal vector $\boldsymbol{G}$. The fractional order spots are due to a surface reconstruction. $\boldsymbol{K}_{\mathbf{i}}$ and $\boldsymbol{K}_{\mathbf{f}}$ are the incident and diffracted wave vectors of electrons; $\boldsymbol{K}_{\mathbf{f}}^{0}$ represents the specular reflection. The magnitude of the $\boldsymbol{K}$ vector is $2\pi/\lambda_{\mathrm{E}}$ for the incident and elastically diffracted electrons. The intersections of the sphere and rods provide the diffraction directions. Data adapted from [2, 6]

the "universal" curve of electrons mean-free path in solids, the elastic scattering occurs mainly on the topmost layers. However, we should note that a LEED diffraction pattern would also include some signal from the bulk planes beneath the surface, i.e., the intensity for the (1×1) reference spots.

In RHEED, electrons are accelerated at much higher voltages (i.e., from about 5 to 50 kV). For a voltage of 10 kV, the de Broglie wavelength of the electrons is 0.123 Å. The electrons hit the target at a small angle of 1–8° relative to the surface plane making the formation of a RHEED pattern extremely sensitive to the surface geometry. Nevertheless, the incident electron beam penetrates slightly into the crystal and hence, a RHEED pattern includes also some signal from the (1×1) bulk planes. The diffraction is described by the Laue condition as for LEED. In addition to the change in the electron momentum perpendicular to a surface, the diffraction process leads to a change of the momentum parallel to the surface. This gives rise to a set of diffracted beams on either side of the specular reflected beam. If the incident electron beam is parallel to the rows of surface atoms that are separated by the lattice vector $\boldsymbol{d}_{\mathrm{S}}$ in the direction perpendicular to the beam direction, then the diffracted electron beams contribute to constructive interference. The angle of diffracted electrons relative to the incident direction, θ, fulfills the well-known formula: $d_{\mathrm{S}} \sin\theta = n\lambda_{\mathrm{E}}$, where n is an integer.

For fabrication of high-quality semiconductor devices, it is of utmost importance to start with a substrate having a surface which is structurally well-defined (with a good crystal structure) and is free from amorphous oxides and carbon contaminants. LEED and RHEED are very useful tools for checking whether the substrate surface is clean and well ordered. Nevertheless, it is well known that a substrate surface exhibiting reasonable LEED and RHEED patterns could still have contaminated areas with poor crystalline quality. This is because the LEED and RHEED electrons have a short coherence length, and hence the diameter of the maximum surface area from which emitted electrons add coherently to the intensity is about 10–50 nm for LEED and 50–100 nm for RHEED. Based on this estimation, we could also say that RHEED is somewhat more sensitive to surface defects than LEED. The coherence length of electrons should not be mixed with macroscopic probe areas of electron beams which are of an order of a millimeter in diameter. We should note here that X-ray photoelectron spectroscopy (XPS) and scanning tunneling microscopy (STM) are routinely used to provide complementary information on the surface quality. Although LEED and RHEED do not provide chemical information about surfaces, they are faster to use than, for example STM and XPS, for assessing the surface quality. The absence of the diffraction spots indicates that a surface layer is amorphous (e.g., badly oxidized). If the bulk-plane related (1×1) pattern appears with a strong background intensity, the surface must be cleaned to obtain sharp spots with a low background intensity. Furthermore, clean and well-defined semiconductor surfaces are usually reconstructed. Thus, the LEED or RHEED pattern from such a well-ordered substrate surface should include extra "superstructure" intensity spots in addition to the (1×1) spots. In RHEED, narrow and intensive streaks indicate a smooth surface. It is still worth noting that monitoring LEED or RHEED during the surface cleaning is also helpful to determine the integer (1×1) spots (or streaks) for the pattern analysis described below.

In terms of application, RHEED has certain advantages over LEED: it provides an opportunity for real-time in-situ monitoring of the surface structure during a layer growth, and can be used to determine the growth rate of epitaxial layers. We should mention that LEED can also be utilized in real-time in-situ manner for studying the surface under a heat treatment and/or short time exposure to an adsorbate flux. LEED observations have been increasingly performed also at low temperatures, however one should be aware of possible electron beam-induced changes in surface structures at temperatures below 40 K [13]. The normal features of RHEED and LEED are summarized in Table 1.1.

1.2 Analysis of LEED and RHEED Patterns

The left column of Fig. 1.2 illustrates the real space lattice cells for an unreconstructed cubic bulk (100) plane and some of the reconstructed surfaces whose *superlattices* are simply related to the $(100)(1 \times 1)$ plane structure. The right column of Fig. 1.2 illustrates the corresponding LEED patterns that visualize the

Table 1.1 Comparison of the normal features of RHEED and LEED

	RHEED	LEED
Electron energy	5–50 keV	10–500 eV
Electron beam diameter	~1 mm	~1 mm
Coherence length	50–100 nm	10–50 nm
Electron incidence angle	1–8° from surface plane	About perpendicular to surface
Vacuum condition	$<10^{-3}$ mbar	$<10^{-7}$ mbar
Applicable to monitor epitaxy in situ	Yes; MBE chamber is usually equipped with RHEED	No
Use	To monitor epitaxy, to determine surface lattice, to elucidate surface atomic structures by comparing measured and calculated diffraction intensity curves	To determine surface lattice, to elucidate surface atomic structures by comparing measured and calculated diffraction intensity curves
Strength	Easy and quick probe of surface lattice, can be employed in situ in MBE, enables atomic structure determination even with accuracy of 1/100 of atom radius	Easy and quick probe of the surface lattice from a single pattern, different ordered reconstructions coexisting on the surface can be readily concluded, enables atomic structure determination even with accuracy of 1/100 of atom radius
Weakness	Does not give a 2D picture of lattice, solution of the atomic structure requires extensive calculations	Not sensitive to local surface defects, solution of the atomic structure requires extensive calculations

reciprocal lattices. On the basis of such information about the reciprocal lattices, one can determine the lattice vectors and cell of the surface studied by LEED, which further is essential to solving a structural unit cell of the surface (e.g., those in Figs. 1.3 and 1.4). The relative change in the reciprocal-lattice vector length (and cell area) for a surface layer, as illustrated by LEED, is inversely proportional to the corresponding change in the surface lattice dimensions, as compared to the bulk plane. The surfaces are often expressed in terms of the Wood notation [2–5]: $S(hkl)(n \times m)R\theta - A$, where $S(hkl)$ gives the crystal plane of the material with the chemical composition of S, n and m are the proportional lengths of the vectors defining the real-space surface lattice compared to the reference vectors of the (1×1) bulk plane, θ is the angle of possible rotation between the lattices contributing to the reconstruction, and A is the possible adsorbate that induces the reconstruction. The $R\theta - A$ term is not marked if the rotation angle is zero and the surface layer consists of only the substrate elements. The surface lattice may partially match with the bulk plane lattice; in this case, the surface layer is called *commensurate*. If the surface lattice does not match with the bulk structure at all, the surface is termed *incommensurate*. An *incommensurate* structure may form, for example, on adsorbate-covered surfaces in which the adsorbate layer is weakly

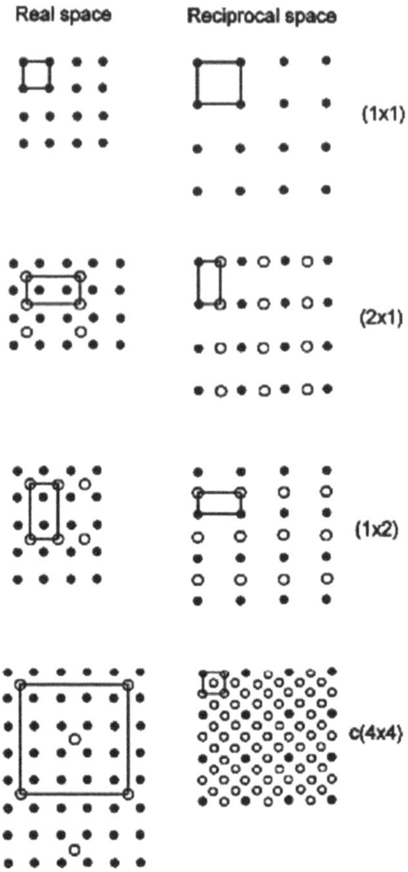

Fig. 1.2 Schematic lattices (*left column*) and reciprocal lattices (*right column*), as visualized by LEED, with cells for a cubic (100)(1 × 1) bulk-plane terminated surface and the reconstructed (100)(2 × 1), −(1 × 2), and −c(4 × 4) surfaces. The *black dots* describe the bulk-plane lattice [or represent the integer (1 × 1) LEED spots], and *open circles* describe the surface lattice (or represent the additional reconstruction-related LEED spots)

bonded to a substrate. The Wood's notation is not always applicable, and a matrix notation must be used for the surface lattice description [2–5]. The letter "c" is used in front of the Wood $(n \times m)$ term if the surface has a centered lattice cell. A further note is that it is possible to describe a surface with different cells; equally well, a III–V(100)c(4 × 4) surface (Fig. 1.2) may be described with a primitive $(2\sqrt{2} \times 2\sqrt{2})R45°$ cell of which area is smaller than the c(4 × 4) area.

Figure 1.3 shows examples of RHEED and LEED patterns corresponding to (2 × 1)-reconstructed and bulk-terminated (1 × 1) surfaces of a III–V(100) substrate. It also shows possible atomic structures causing the patterns. It can be observed as an essential difference between LEED and RHEED: a single LEED pattern visualizes the two-dimensional reciprocal lattice, while a RHEED pattern gives the reciprocal lattice periodicity only in one crystal direction. Therefore, in order to study the lattice periodicity in the orthogonal direction using RHEED, the sample must be rotated by 90°. This particular feature of RHEED makes it sometimes difficult to decide whether there are two or more surface reconstructions coexisting on a substrate. The RHEED pattern shown in Fig. 1.3 illustrates the reciprocal lattice

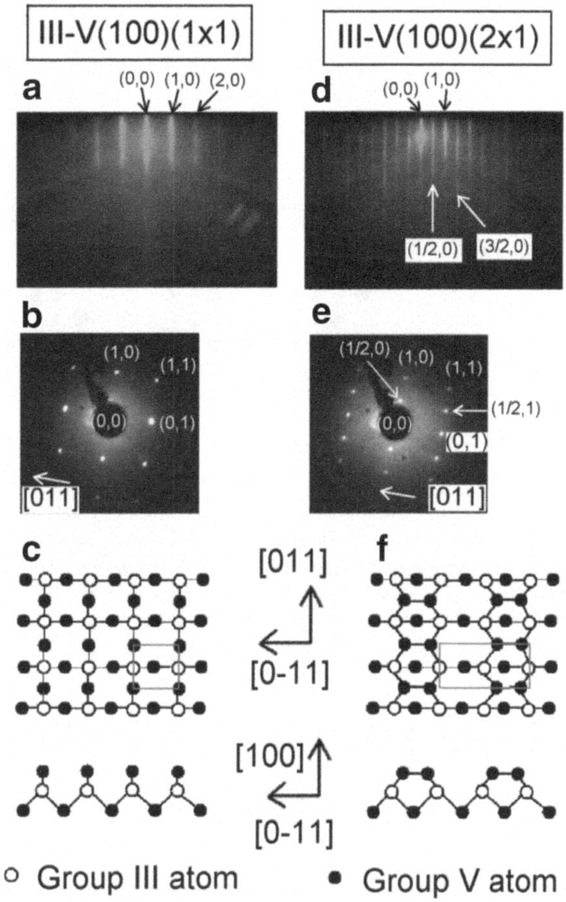

Fig. 1.3 RHEED and LEED examples for III–V(100)(1 × 1) and −(2 × 1) surfaces, with possible atomic arrangements. The reciprocal lattice of an unreconstructed III–V(100)(1 × 1) surface is identical to that of the bulk plane. The vectors, defined by the (00) and (10) spots and the (00) and (01) spots, represent the reciprocal lattice vectors for the bulk plane, as visualized by LEED. For a reconstructed III–V(100)(2 × 1) surface, the two vectors defined by the (00) and (1/2 0) spots and the (00) and (01) spots determine the reciprocal lattice of a surface layer. The *gray square* and *rectangle* represent the surface lattice cells

in the [0-11] direction, while the electron beam lies in the [011] direction. If the electron beam lies in the [0-11] direction, then RHEED patterns from the (1 × 1) and (2 × 1) surfaces resemble that of Fig. 1.3a. For the (2 × 1)-reconstructed surface, both the RHEED and LEED patterns reveal the 1/2 fraction spots (or streaks) between the integer (1 × 1) spots; that is, the length of the reciprocal lattice vector of a surface layer, defined by the (00) and (1/2 0) spots, is half of the length of the reciprocal vector for the bulk plane in the [0-11] direction. This means doubling of the lattice cell length for the surface layer in the same [0-11] direction, as compared to the bulk (1 × 1) lattice. It should be noted that for the square and rectangle lattices, the directions of the lattice vectors are identical to the corresponding reciprocal vector directions [2–5]. In the case of III–V(100) surfaces, the doubling usually means that surface atoms form new bonds creating new pairs, so-called dimers. All the surface layers considered in this chapter are simply related to the bulk plane lattices (i.e., are the *superstructures*).

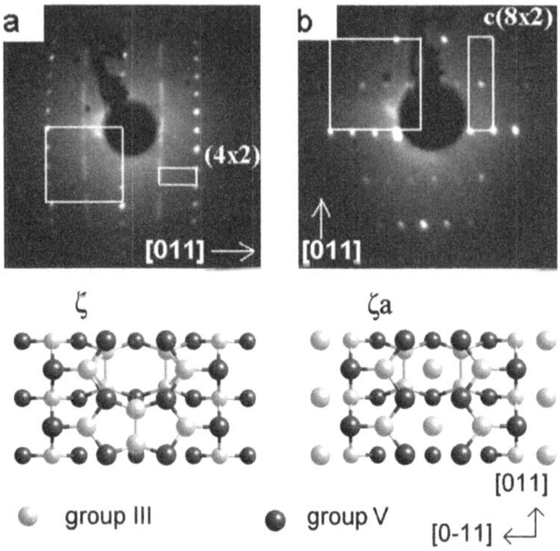

Fig. 1.4 The LEED patterns from (**a**) InAs(100)(4×2) and (**b**) InSb(100)c(8×2) substrate surfaces (please see the text also). The electron energies were (**a**) 55 eV and (**b**) 44 eV. The reciprocal cell of the bulk plane is illustrated by the *white square* and the surface cell by the *white rectangle*. The InAs and InSb substrate pieces were rotated in the different positions relative to the LEED equipment. Two most likely (4 × 2) structural unit cells, which describe both the InAs(100)(4 × 2) and InSb(100)c(8 × 2) surfaces, are also shown

1.3 Using LEED to Study III–V Surfaces

1.3.1 The c(8 × 2) Surfaces of InAs(100) and InSb(100)

Figure 1.4 shows typical LEED patterns from InAs(100) and InSb(100) surfaces. The substrate surfaces were cleaned by argon-ion sputtering (1.5 keV, 10 mA, for 0.5 h) heating the substrate at a temperature of about 300°C and by subsequent heating for 0.5 h to 470 and 440°C, respectively. This cycle was repeated six times before the patterns shown in Fig. 1.4 were observed. The surface cleanness was confirmed by STM and XPS measurements.

For the interpretation of LEED patterns, it is helpful to determine the integer intensity spots (00), (10), (01), and (11) that set and visualize the reciprocal lattice vectors and cell for a bulk plane, i.e., the reference (1 × 1) reciprocal cell. Often the intensity of the integer spots is higher than that of other possible spots, because the integer spots include also the signal from the bulk planes, and does not depend on the electron energy as strongly as the intensity of the additional spots. The cubic zinc-blende III–V(100) substrates, predominantly used for fabrication optoelectronics devices, have the square lattice as well as reciprocal lattice for the bulk plane as shown in Fig. 1.2. The square areas indicated in Fig. 1.4 represent the reciprocal cells for the InAs(100) and InSb(100) planes. The side lengths of these squares can be readily determined with the (1 × 1) spots along the [011] direction.

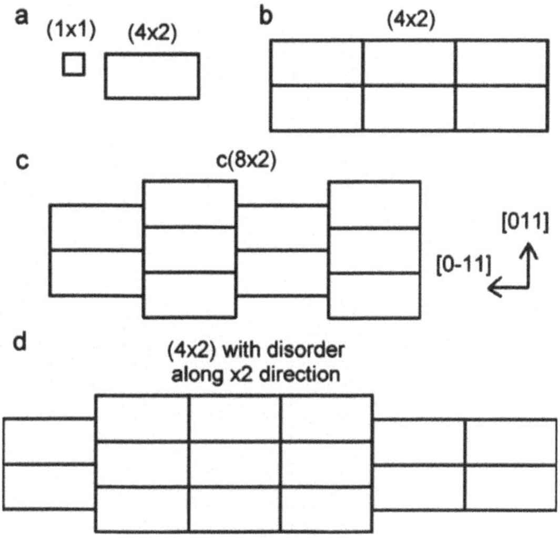

Fig. 1.5 (**a**) The III–V(100)(1 × 1) and –(4 × 2) real-space lattice cells. (**b**) The arrangement of the (4 × 2) cells that would cause a clear (4 × 2) LEED pattern. (**c**) The (4 × 2)-cell arrangement for the InSb(100)c(8 × 2) LEED in Fig. 1.4. (**d**) A possible (4 × 2)-cell arrangement which causes the ×2 streaks instead of the ×2 spots in the LEED pattern of the InAs(100)(4 × 2) in Fig. 1.4

The other, fraction LEED spots in Fig. 1.4, evidence the reconstruction and the change in the reciprocal lattice (and further in the lattice) of a surface layer for both the InAs(100) and InSb(100) substrates. The reciprocal cells for these InAs(100)(4× 2) and InSb(100)c(8 × 2) surfaces are shown in Fig. 1.4. Two most-likely structural unit cells for the reconstructions are also shown in Fig. 1.4 indicating a significant rearrangement of the surface atoms. These atomic models were solved by combining STM, diffraction measurements, and ab initio calculations [14–17]. The (4 × 2) lattice cell arrangement that causes the c(8 × 2) periodicity is illustrated in Fig. 1.5; both surfaces can be described by the (4 × 2) cells shown in Fig. 1.4. It appears that the ×2 order is more visible on the InSb surface than on the InAs surface (Fig. 1.4); this can be linked to some stacking disorder in the ×2 lattice direction on the InAs(100)(4 × 2) surface. On the basis of the shown LEED pattern, it is not clear whether the InAs(100) surface layer has the (4×2) or c(8×2) lattice. A possible reason for the lattice disorder is schematically shown in Fig. 1.5. We also note that the In-V(100)c(8 × 2)/(4 × 2) surface is potentially useful as a starting substrate for producing high-quality III–V heterointerfaces [18] and insulator—III–V interfaces [19, 20]. These types of insulator interfaces are required for the development of future metal-oxide-semiconductor field effect transistors (MOSFETs) with reduced power consumption and prolonged lifetime.

1.3.2 The GaAs(100) Reconstructions

The GaAs(100) is one of the most studied semiconductor surfaces [21–40]. LEED patterns from the common GaAs(100) reconstructions are shown in Fig. 1.6. The reconstructions depend on the substrate treatment and are often related to the amount of arsenic on the surface and the substrate temperature. For example, the

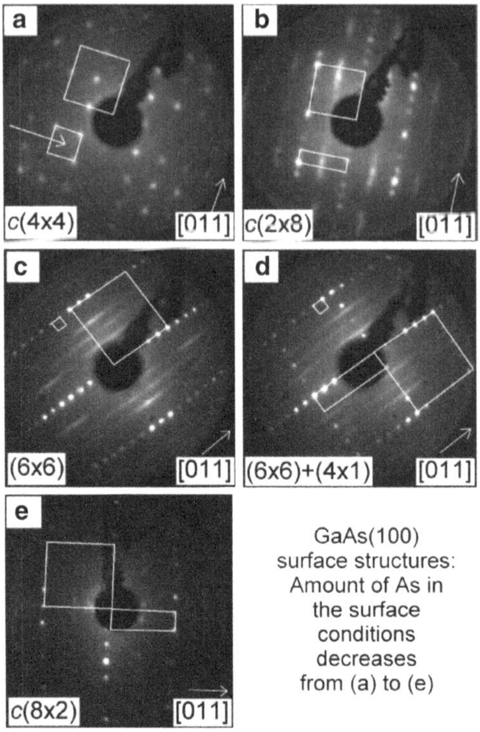

Fig. 1.6 LEED from the GaAs(100) surface reconstructions with decreasing the As amount in the surface conditions from (**a**) to (**e**). The *white squares* represent the (1 × 1) bulk-plane reciprocal cells and the *white rectangles* represent the reciprocal lattice cells of the surface layers. In pattern (**a**), the *smaller white square* is the c(4 × 4) cell, in which the *white arrow* shows the missing spot of the c(4 × 4). This spot was seen with different electron energies. The 6× streaks in (**c**) and (**d**) arise most likely from the same type of the disorder as described for the InAs(100)(4 × 2) in Fig. 1.5. In pattern (**d**), the (4 × 1) cell is drawn instead of the (4 × 2) because the ×2 streak or spots were not seen properly. In contrast, in the pattern (**e**) those ×2 spots are clearly seen

GaAs(100)c(4 × 4) surface, of which LEED is shown in Fig. 1.6a, was obtained by heating an arsenic-capped GaAs(100) substrate in UHV overnight at 320°C. The As-cap layer was deposited by MBE onto a 100-nm GaAs buffer layer that was grown on n-type GaAs(100). We should note that it is not easy to determine the absolute temperature of a sample accurately; the errors may be as large as ±25°C, partly due to variations of the temperature across the substrate. Anyway, an increase of the substrate temperature to 530°C changed the GaAs(100)c(4 × 4) surface to the GaAs(100)c(2 × 8) or −(2 × 4) one (Fig. 1.6b). The most likely atomic structures of the III-As(100)c(2 × 8) or (2 × 4) surfaces, which often appear during the epitaxial growth of III-As alloys, are the β2 and α2 [26, 32, 34, 38, 39]. When the GaAs(100) substrate is further heated to 580°C, a (6 × 6) LEED pattern appears as shown in Fig. 1.6c. Prolonged heating leads to the coexistence of two different lattices: (6 × 6) and 4× spots as reveled in Fig. 1.6d. Sometimes, this phenomenon complicates the interpretation of LEED patterns. STM graphs confirmed the presence of areas exhibiting the (6×6) reconstruction as well as of areas with the 4× row structure. We note here that the areas with a ×2 order might be too small to produce the (4 × 2) or c(8 × 2) LEED periodicity in Fig. 1.6d. Further heating did not yield the pure (4 × 2) or c(8 × 2) pattern; this might be due to the presence of residual arsenic in the chamber and partial adsorption of arsenic back to the GaAs surface during its cooling. In a separate experiment, the sputtering cleaning and subsequent heating to 580°C resulted in a pure c(8 × 2) pattern as shown in Fig. 1.6e. The c(8 × 2) pattern corresponds to the atomic structure labeled ξ in Fig. 1.4 [14]. According to ab initio calculations [40], the GaAs(100)(6 × 6) surface is energetically unstable and represents an exceptional case since all the other phases, i.e., c(4 × 4), (2 × 4), and c(8 × 2) are stable.

The information gained by analyzing the series of patterns (LEED or RHEED) can be used to identify the crystal orientations. For example, one might not readily conclude whether a pattern is (2 × 4) or (4 × 2) when measuring a substrate piece cut from the full wafer, but the surface directions can be unambiguously determined by monitoring the change of the patterns. For the InAs(100), the reconstructions follow a change from c(4 × 4) to (2 × 4)/c(2 × 8) and then c(8 × 2) when decreasing the amount of As on the surface [28, 31, 33, 41–46]. For the InSb(100), the pattern changes from (4 × 4) through c(2 × 6) and (1 × 3) to c(8 × 2) when the Sb content on the surface decreases [46–50]. The GaSb(100) exhibits c(2 × 10), c(2 × 6), and (1 × 3) reconstructions [50–55] while the InP(100) surfaces are characterized by the c(4 × 4) and (2 × 4) reconstructions [56–64]. Thus, the GaSb(100) and InP(100) do not show the intrinsic c(8 × 2) and (4 × 2) patterns. It is worth noticing that the InP(100) produces the (2 × 4) pattern for a large range of the surface conditions, making a RHEED optimization of the InP growth with this reconstruction difficult. The occurrence of the c(8 × 2) pattern for the InP(100) substrate usually indicates that a thin InAs(100) layer has been formed during arsenic exposure.

1.3.3 The Bi-Induced Reconstructions on III–V(100)

Understanding the Bi-induced surface reconstructions is essential for advancing the epitaxy of relatively unexplored III–V–Bi compounds and the Bi-surfactant-mediated epitaxy. This section is focused on discussing the Bi-induced reconstructions observed so far.

As a surfactant, Bi floats on the growth front with little incorporation into the crystal. The Bi surfactant has been reported to smoothen the growth of an III–V overlayer [65–69] and to improve the photoluminescence (PL) of dilute nitride GaInAsN quantum wells (QWs) [66]. It also enhances nitrogen incorporation into GaAsN [68, 70] and removes a CuPt-like order of GaInP layers [71, 72]. Sometimes, it is hard to detect the presence of Bi on a substrate surface by LEED or RHEED because Bi induces the reconstructions quite similar to those induced by the other group-V elements. Nevertheless, there are "fingerprint patterns" associated with

Bi. Such an example is a Bi-induced (1 × 3) on the GaAs(100) [65–67]. This reconstruction can be observed, for example, when depositing Bi by MBE onto the GaAs(100)(2 × 4) substrate while keeping the arsenic valve closed and decreasing the substrate temperature. When further decreasing the substrate temperature after observing the (1 × 3) pattern, the GaAs(100)c(4 × 4)-Bi surface is formed under a Bi flux. Another "fingerprint" of Bi is an appearance of a (2 × 1) pattern [67, 73]. This GaAs(100)(2 × 1)-Bi reconstruction is unusual due to its metallic character [73, 74] and might improve the atomic diffusion on the growth front [73]. The (2 × 1) pattern corresponds to the less As-rich surface [73,74] in contrast to the GaAs(100)(1×3)-Bi surface which is more As rich. Apparently, there is no report so far in the literature concerning the structural study of the GaAs(100)(1 × 3)-Bi.

Other fingerprint reconstructions induced by Bi include a (2 × 8) phase on the InAs and InP(100) substrates, a c(2 × 10) pattern on the InSb(100) substrate, and a very interesting c(2×12) structure on the InAs(100) surface. For the latter, the LEED pattern indicates the dimer-row separation in the [011] direction to be $6 \times a_{(1\times1)}$, where $a_{(1\times1)} = 4.3$ Å is the length of the (1 × 1) square lattice of the InAs(100) bulk plane. On the other hand, STM reveals that the dimer row separation is $10 \times a_{(1\times1)}$ and that twisting of the dimer rows leads to a c(2 × 12) pattern [75]. Indeed, it was observed that the heating of the c(2 × 12) surface straightened the dimer rows and led to a (2 × 10) pattern. This large lattice cell causes a very dense row of the LEED spots and complicates the determination of the spot separation.

The Bi-induced reconstructions found so far can be summarized as follows. The addition of Bi on the GaAs(100)c(8 × 2) surface and the subsequent heating induce the (2 × 4) pattern, if the maximum Bi coverage is 0.5 ML, and the (2 × 1) pattern for the Bi coverage between 0.5 and 1.0 ML. If the substrate has the GaAs(100)(2 × 4) or −c(4 × 4) starting surface, Bi induces the (1 × 3) pattern, which changes to the c(4 × 4) one at lower temperatures. Bismuth desorbs strongly from the GaAs(100) at about 450°C. The corresponding desorption temperature for the InAs(100) is about 350°C. When starting with the InAs(100)c(8×2) surface, Bi produces (2×4), (2×8), and (2 × 1) patterns, for up to 1 ML coverage, and further the (2 × 10) and c(2 × 12) reconstructions for a coverage of 1–1.5 MLs. Except for the c(2 × 12) and (2 × 10), the same series of reconstructions has been observed on the InP(100)(2 × 4); here, Bi desorbs at about 400°C. When starting with the InSb(100)c(8 × 2) surface and increasing the Bi coverage to 1 ML, the reconstruction evolves from (1×3) to c(2×6) and finally to (2 × 5); the strong desorption takes place at about 300°C.

1.4 Using RHEED to Study III–Vs

1.4.1 Optimizing the Growth Conditions of GaAs/AlAs Heterostructures

Although the GaAs/AlAs interfaces have been studied for several decades, it has remained somewhat unclear what it is the optimum As/group-III flux ratio for the growth of low-dimensional structures, such as multiquantum wells (MQWs). This issue was recently addressed by combining RHEED observations and photoluminescence measurements for GaAs/AlAs MQWs [76]. The results show that an optimum flux ratio is near the conditions corresponding to a change in the GaAs(100) reconstruction from (2 × 4) to (4 × 2). The main findings of this study are presented in Fig. 1.7 and serve as an example of how RHEED can be utilized to control and understand the epitaxial growth. The RHEED patterns were obtained during the growth of the GaAs and AlAs layers at different substrate temperatures while decreasing the As flux. The (2 × 4) pattern changed first to a (1 × 1) and then to (4 × 2). The growth time in the (1 × 1) and (4 × 2) conditions was minimized because they are known to result in poor quality of the crystal.

1.4.2 The GaAs(100) Reconstructions

Figure 1.8 shows the RHEED patterns for three different reconstructions on the GaAs(100) substrate. The more intense streaks of the GaAs(100)(4 × 6) are the integer diffractions, and their distance provides the reference reciprocal vector length of the bulk plane while the less intense streaks are the fraction diffractions. The streaky and sharp patterns indicate smooth GaAs(100) surfaces. The (4 × 6) and (2 × 4) patterns were obtained at the substrate temperature of about 550°C, i.e.,

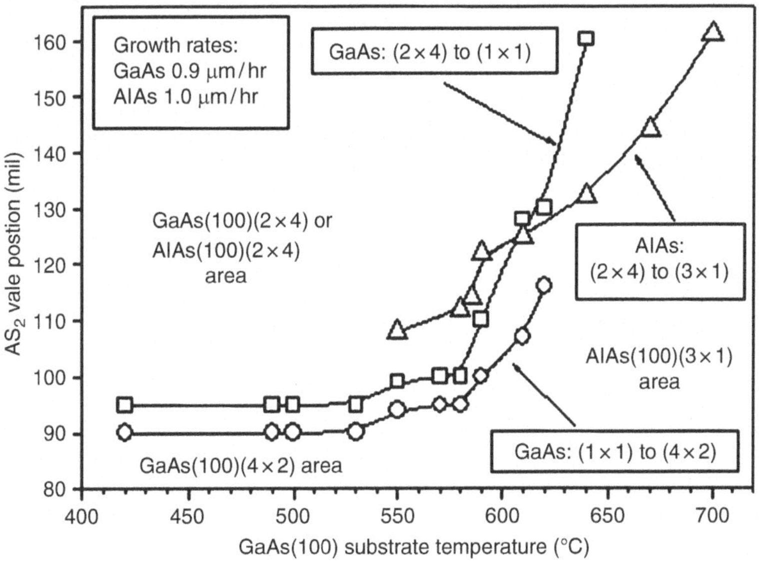

Fig. 1.7 A diagram showing how the GaAs(100) and AlAs(100) surface structures change by tuning the substrate temperature and As flux during the epitaxial growth. The 100-mil As flux corresponds to the As/Ga flux ratio of 5 and to the As/Al flux ratio of 18 approximately

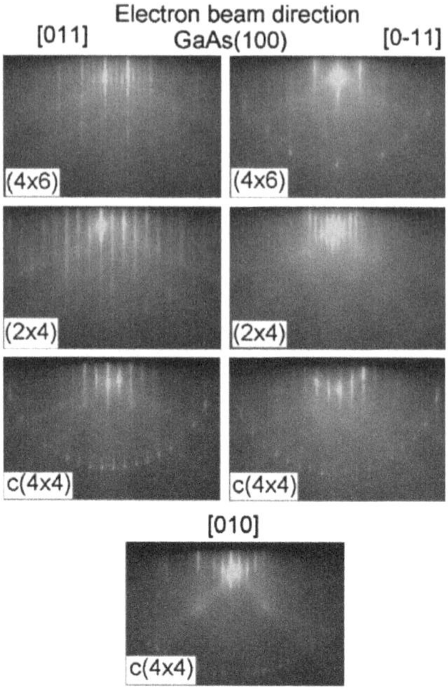

Fig. 1.8 RHEED patterns for different GaAs(100) reconstructions in different surface directions. The more intense streaks of the GaAs(100)(4 × 6) are the integer diffractions and visualize the (1 × 1) reciprocal cell side while the less intense streaks are the fraction diffractions. For the GaAs(100)c(4 × 4), it is also shown the diffraction along the [010] direction, i.e., the 45° rotation relative to the [011]. Its integer streak distance is smaller than that along the [011] or [0-11] direction and is divided in the four equal parts (some of the fraction streaks are weaker than the other ones), please see also Fig. 1.2

30°C higher than the temperature at which the transition from (2 × 4) to c(4 × 4) occurs under the 1×10^{-7} mbar As flux. The (4 × 6) pattern was obtained by closing the As shutter and exposing the surface to Ga flux. Most likely, this RHEED pattern corresponds to a Ga-rich surface as reported recently [37], but it cannot be excluded from the possibility that the surface included both the (4 × 1) and (6 × 6) areas, as shown by the LEED images of Fig. 1.6d. The c(4 × 4) pattern was observed at the substrate temperature of 480°C for an As_2 flux corresponding to a partial pressure of 1×10^{-7} mbar. This c(4 × 4) reconstruction remained stable after the As flux was closed and the substrate temperature was decreased.

1.4.3 The GaAs(111) Reconstructions

The GaAs(111)A substrate surface exhibits only (2 × 2) reconstruction while GaAs(111)B surface reveals also a more complicated $(\sqrt{19} \times \sqrt{19})R23.4°$ reconstruction [77–81]. The (2 × 2) and $(\sqrt{19} \times \sqrt{19})R23.4°$ patterns were found on the GaAs(111)B substrate at the temperatures of 480 and 520°C, respectively and an As pressure of about 1×10^{-7} mbar. The GaAs(111)A surface exhibits a (2 × 2) pattern for a wide range of the surface conditions. For example, for the same As pressure of about 1×10^{-7} mbar, the (2 × 2) RHEED pattern is maintained for substrate temperatures ranging from 350 to 590°C, and it does not change visually at all. Therefore, it might be difficult to optimize the growth parameters of these surfaces using with the (2×2) RHEED. It is worth noting that the $(\sqrt{19} \times \sqrt{19})R23.4°$ reconstruction on the GaAs(111)B surface is an exception because it appears to be a metallic surface with partially filled dangling bonds. The metallic character of the III–V surface might be a useful growth front [82].

1.4.4 Probing Surface Reconstructions in GaInAsN(100)

In this section, we discuss the reconstructions of dilute nitride (GaInAsN) lattice matched to the GaAs(100) and having a N content of about 3%. The RHEED patterns were monitored as a function of the substrate temperature and the As flux. For the growth temperatures within 300–400°C, only a (1×3) reconstruction was observed (Fig. 1.9). Heating the GaInAsN film and supplying As flux modified the reconstruction from a $c(4 \times 4)$ pattern at temperatures below 300°C to the (1×3) pattern for temperatures within 300–450°C. Further change to a (2×4) pattern occurred for a temperature range of 500–570°C. Closing the As flux at 570°C produced a (3×1) pattern. The GaInAsN(100)(1×3) reconstruction is interesting because usually it is not seen on the III-As(100) surfaces. The observation agrees with recent calculations proving that the (4×3) reconstructions are almost stable on the GaAs(100) and are likely important in kinetically limited growth conditions as well as for strain-stabilized III–V alloy surfaces [83]. Thus the (1×3) RHEED observation provides an experimental evidence for the calculated results; the surfaces producing a (1×3) pattern are, in fact, composed of (4×3) building blocks [50, 55]. We would like to remind the reader that depending on the measurement by which the lattice periodicity has been determined (e.g., RHEED or STM), the same surface may be called (1×3) or $c(2 \times 6)$. Actually, more recent measurements and calculations have revealed that the real unit cell is (4×3) and not (1×3), although it is sometimes referred as the (1×3) reconstruction. The (1×3), $c(2 \times 6)$, and (4×3) reconstructions are similar to each other because these surfaces consist of dimer rows with the $\times 3$ separation.

The (3×1) reconstruction found on the GaInAsN(100) in less As-rich conditions is also rather exceptional; to our knowledge it appears only on the AlAs(100) in the less As-rich concentrations, as shown in the diagram of Fig. 1.7. LEED and XPS experiments showed that the heating of the GaInAsN in UHV yielded a (3×1) surface structure and concomitantly led to indium segregation toward the surface.

1.4.5 In-Situ Calibrations of Growth Rate and Composition of Multinary Compounds

RHEED has been instrumental for the development of MBE technology and is nowadays a standard part of MBE growth chambers. Besides its use for obtaining real-time information on the growth modes and surface morphology, it provides

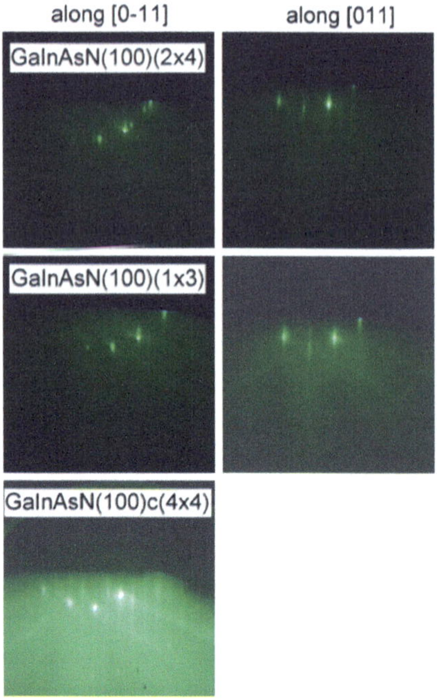

Fig. 1.9 RHEED patterns of GaInAsN films (N content around 3%) lattice matched to the GaAs(100) substrate during growth stops under the As flux in two perpendicular directions

also an easy way to calibrate the growth rates of binary alloys with atomic layer precision [84]. The growth rate can be determined by measuring the time required to complete the deposition of an atomic layer. This is done by monitoring the intensity variation of the specular RHEED spot, which depends on the surface roughness: the maximum intensity is obtained when an atomic layer is completed, while a minimum is obtained for an incomplete layer with a maximum disorder [a growth front is still reconstructed during the growth, e.g., the (2×4)-reconstructed GaAs]. The intensity oscillations are analyzed with an image processing software, which plots the oscillation as a function of time. An example of such intensity oscillations is shown in Fig. 1.10. In order to avoid errors due to the fluctuations of the growth conditions, the growth rate is determined by measuring the time required to complete several oscillations. The procedure is repeated several times by closing and opening the shutter controlling the flux of the material that determines the growth rate (e.g., Ga in the case of GaAs).

The variation of RHEED intensity oscillations and their amplitude damping behavior could reveal other important aspects regarding the growth kinetics [84, 85] and can be used for the composition calibration of ternary (or multinary) alloys, as discussed next. Usually, the MBE growth is performed in excess flux of one of the components (e.g., As in the case of arsenides and Ga in the case of GaN). When growing a ternary compound, i.e., adding a third element, the growth rate of the layer increases proportionally to the additional flux. Knowing the sticking coefficient for the given flux (usually equal to one for the III elements in the MBE growth), one

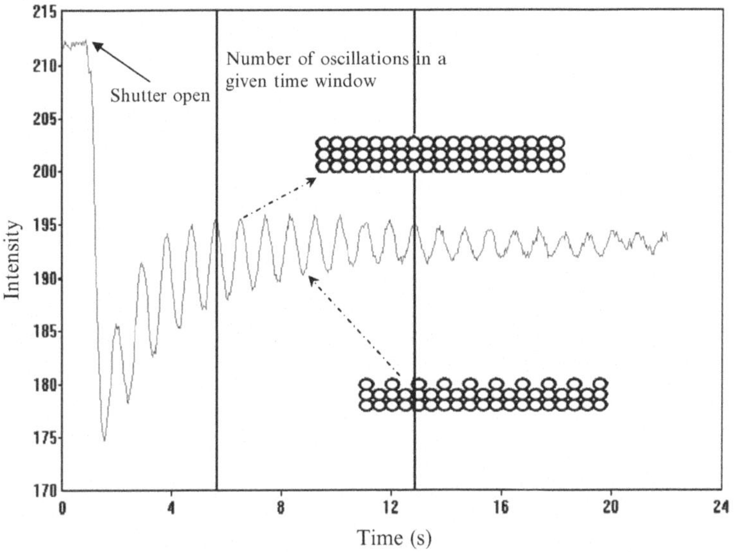

Fig. 1.10 An example of possible intensity oscillation of the specular RHEED beam from a GaAs(100) surface. The period corresponds to the growth rate of a GaAs monolayer

may calibrate the composition of the ternary compound in a simple way. Next we show how this method can be applied for calibrating the growth rate of InGaAs and GaSbAs ternary alloys.

In the case of the $In_xGa_{1-x}As$, by measuring the growth rates of GaAs, r_{GaAs}, and InGaAs, r_{InGaAs}, we can determine the In content (x), as being equal to $x = (r_{InGaAs} - r_{GaAs})/r_{InGaAs}$. For a concrete example, Fig. 1.11 shows the oscillation of the RHEED specular spot during deposition of GaAs and subsequent deposition of InGaAs. The growth was performed on the GaAs(100) using the substrate temperature of 480°C. The oscillations have been recorded with the RHEED beam parallel to the [010] azimuth. The estimated composition of In was ~14%.

We should note that applicability of this method for composition evaluation is dependent on the lattice mismatch between the compound grown for the calibration purposes (InGaAs in this case) and the substrate/buffer layer. For example, it is not possible to grow several ML-thick InAs binary alloy on GaAs(100) because the critical thickness for growing two-dimensional layer-by-layer InAs is below two MLs. For an InGaAs ternary alloy, the critical thickness for growth on the GaAs(100) substrate depends on the In content; for the In content exceeding 30%, it drops below 7 nm rendering difficult the use of RHEED for composition calibration. In this case, higher Ga flux is used to test the InGaAs growth rate and In composition while for the final target structure, the Ga flux is decreased. In the case of quaternary alloy, such as $Al_xIn_yGa_{1-x-y}As$, the Al contents (x) and In contents (y) can be estimated by measuring the growth rates of AlAs, GaAs, and InGaAs and r_{GaAs}, r_{InGaAs}, and r_{AlAs}, respectively. Then the composition of

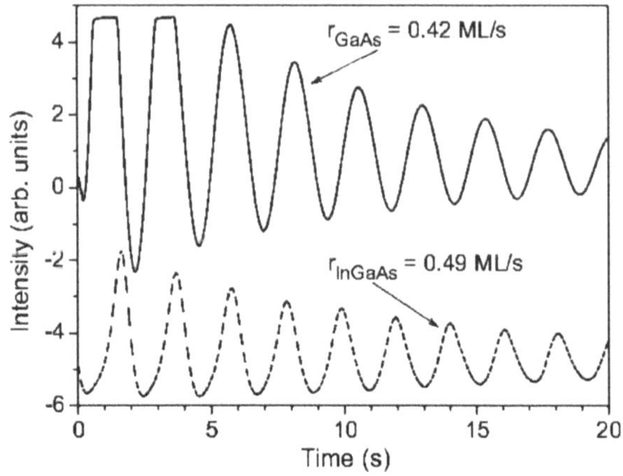

Fig. 1.11 RHEED intensity oscillations (specular spot in the [010] azimuth) recorded for GaAs (*upper panel*) and subsequently grown InGaAs (*lower panel*). The In contents calculated from the increase of the InGaAs growth rate with respect to GaAs is equal to 14%

the quaternary alloy can be estimated as being $x = r_{AlAs}/(r_{InGaAs} + r_{AlAs})$ and $y = (r_{InGaAs} - r_{GaAs})/(r_{InGaAs} + r_{AlAs})$.

The applicability of this method is limited to the multinary alloys with the intermixed III elements (e.g., InGaAs and AlGaAs). For alloys with the mixed V elements, like GaAsSb, GaAsP, and GaAsN, the use of this method is much more difficult, or sometimes impossible, because the growth rates are limited by the elements of the group III (i.e., they do not depend on the group V fluxes). In some special cases, this problem can be overcome. For example, reducing the group-V flux and depositing 2–3 MLs of the III element (Ga in the case of GaAs), one gets a group III-rich surface (e.g., Ga). Exposing this Ga-rich surface to the flux of V element will result in 2–3 RHEED intensity oscillations. Then the growth rate determined from the period of these oscillations is proportional to the impinging V element flux.

1.5 Concluding Remarks

Crystalline interfaces and thin films are the building blocks of many advanced semiconductor devices as well as devices not existing currently. Any improvement in the ordering at the junction will decrease the defect density, having clear implications on the device lifetime and energy consumption. These well-defined materials also make it possible to elucidate their fundamental atomic and electronic structures with detailed measurements and calculations. The engineering of the well-defined growth fronts has a key role in the manufacturing of crystalline interfaces and films. This chapter introduced the LEED and RHEED methods which are routinely used by crystal growers to monitor the surfaces of crystalline films. The LEED and RHEED patterns visualize the reciprocal lattice of the surface layer studied, from which the surface lattice can be readily constructed. The examples discussed dealt with the III–V semiconductor surfaces, which exhibit various reconstructions as a function of the substrate temperature and III/V atom ratio. The literature abounds in many results showing how the starting surface reconstructions are linked to the quality of the III–V interfaces and grown films. Here we have presented a series of diffraction patterns for a few technologically important III–V(100) substrate surfaces. Knowing the type of reconstruction required one could identify adequate growth conditions for the fabrication of novel crystalline films and related device structures.

Acknowledgements Dr. Antti Tukiainen, Arto Aho, Ville-Markus Korpijärvi, and Janne Puustinen from the Optoelectronics Research Centre, Tampere University of Technology are gratefully acknowleged for fabricating the InGaAsN structures and for their help in performing some of the RHEED experiments presented in this chapter. We would like to thank Emeritus Prof. Markus Pessa for useful discussions. We thank Prof. Changsi Peng (at present with the Soochow University, China) and Dr. Janne Pakarinen (at present with the VTT Technical Research Centre of Finland) for the ideas and experiments concerning the GaAs/AlAs studies. Also we would like to thank Dr. Mikhail Kuzmin (at present with the Ioffe Institute, Russia) for helpful discussion regarding LEED.

References

1. J. Mannhart et al., Science **327**, 1607 (2010)
2. A. Zangwill, *Physics at Surfaces* (Cambridge University Press, Cambridge, 1988)
3. W.N. Unertl (ed.), in *Physical Structure. Handbook of Surface Science*, vol. 1, (Elsevier Science B.V., The Netherlands, 1996)
4. H. Lüth, *Solid Surfaces, Interfaces and Thin Films* (Springer, New York, 2001)
5. F. Bechstedt, *Principles of Surface Physics* (Springer, New York, 2003)
6. P.K. Larsen, P.J. Dobson (eds), *Reflection High-Energy Electron Diffraction and Reflection Electron Imaging of Surfaces* (Plenum, New York, 1988)
7. K. Heinz, Progr. Surf. Sci. **27**, 239 (1988)
8. J.B. Pendry, Surf. Sci. Rep. **19**, 87 (1993)
9. Q.-K. Xue, T. Hashizume, T. Sakurai, Progr. Surf. Sci. **56**, 1 (1997)
10. A. Ohtake, Surf. Sci. Rep. **63**, 295 (2008)
11. C. Davisson, L.H. Germer, Nature **119**, 558 (1927)
12. C. Davisson, L.H. Germer, Phys. Rev. **30**, 705 (1927)
13. T. Uda, H. Shigekawa, Y. Sugawara, S. Mizuno, H. Tochihara, Y. Yamashita, J. Yoshinobu, K. Nakatsuji, H. Kawai, F. Komori, Progr. Surf. Sci. **76**, 147 (2004)
14. S.-H. Lee, W. Moritz, M. Scheffler, Phys. Rev. Lett. **85**, 3890 (2000)
15. C. Kumpf, L.D. Marks, D. Ellis, D. Smilgies, E. Landemark, M. Nielsen, R. Feidenhans'l, J. Zegenhagen, O. Bunk, J.H. Zeysing, Y. Su, R.L. Johnson, Phys. Rev. Lett. **86**, 3586 (2001)
16. C. Kumpf, D. Smilgies, E. Landemark, M. Nielsen, R. Feidenhans'l, O. Bunk, J.H. Zeysing, Y. Su, R.L. Johnson, L. Cao, J. Zegenhagen, B.O. Fimland, L.D. Marks, D. Ellis, Phys. Rev. B **64**, 075307 (2001)
17. J.J.K. Lång, M.P.J. Punkkinen, P. Laukkanen, M. Kuzmin, V. Tuominen, M. Pessa, M. Guina, I.J. Väyrynen, K. Kokko, B. Johansson, L. Vitos, Phys. Rev. B **81**, 245305 (2010)
18. T. Anan, S. Sugou, K. Nishi, Appl. Phys. Lett. **63**, 1047 (1993)
19. H.-S. Kim, I. Ok, M. Zhang, F. Zhu, S. Park, J. Yum, H. Zhao, J.C. Lee, P. Majhi, N. Goel, W. Tsai, C.K. Gaspe, M.B. Santos, Appl. Phys. Lett. **93**, 062111 (2008)
20. B. Shin, D. Choi, J.S. Harris, P.C. McIntyre, Appl. Phys. Lett. **93**, 052911 (2008)

21. M.D. Pashley, K.W. Haberern, W. Friday, J.M. Woodall, P.D. Kirchner, Phys. Rev. Lett. **60**, 2176 (1988)
22. D.K. Biegelsen, R.D. Bringans, J.E. Northrup, L.-E. Swartz, Phys. Rev. B **41**, 5701 (1990)
23. S.L. Skala, J.S. Hubacek, J.R. Tucker, J.W. Lyding, S.T. Chou, K.-Y. Cheng, Phys. Rev. B **48**, 9138 (1993)
24. C.W. Snyder, J. Sudijono, C.-H. Lam, M.D. Johnson, B.G. Orr, Phys. Rev. B **50**, 18194 (1994)
25. J.E. Northrup, S. Froyen, Phys. Rev. B **50**, R2015 (1994)
26. T. Hashizume, Q.-K. Xue, J. Zhou, A. Ichimiya, T. Sakurai, Phys. Rev. Lett. **73**, 2208 (1995)
27. T. Hashizume, Q.-K. Xue, A. Ichimiya, T. Sakurai, Phys. Rev. B **51**, 4200 (1995)
28. H. Yamaguchi, Y. Horikoshi, Phys. Rev. B **51**, 9836 (1995)
29. W.G. Schmidt, F. Bechstedt, Phys. Rev. B **54**, 16742 (1996)
30. Y. Garreau, M. Sauvage-Simkin, N. Jedrecy, R. Pinchaux, M.B. Veron, Phys. Rev. B **54**, 17638 (1996)
31. Q.-K. Xue, T. Hashizume, T. Sakurai, Prog. Surf. Sci. **56**, 1 (1997)
32. V.P. LaBella, H. Yang, D.W. Bullock, P.M. Thibado, P. Kratzer, M. Scheffler, Phys. Rev. Lett. **83**, 2989 (1999)
33. G.R. Bell, J.G. Belk, C.F. McConville, T.S. Jones, Phys. Rev. B **59**, 2947 (1999)
34. W.G. Schmidt, S. Mirbt, F. Bechstedt, Phys. Rev. B **62**, 8087 (2000)
35. A. Ohtake, J. Nakamura, S. Tsukamoto, N. Koguchi, A. Natori, Phys. Rev. Lett. **89**, 206102 (2002)
36. A. Ohtake, N. Koguchi, Appl. Phys. Lett. **83**, 5193 (2003)
37. A. Ohtake, P. Kocán, K. Seino, W.G. Schmidt, N. Koguchi, Phys. Rev. Lett. **93**, 266101 (2004)
38. P. Laukkanen, M. Kuzmin, R.E. Perälä, M. Ahola, S. Mattila, I.J. Väyrynen, J. Sadowski, J. Konttinen, T. Jouhti, C.S. Peng, M. Saarinen, M. Pessa, Phys. Rev. B **72**, 045321 (2005)
39. A. Ohtake, Phys. Rev. B **74**, 165322 (2006)
40. K. Seino, W.G. Schmidt, A. Ohtake, Phys. Rev. B **73**, 035317 (2006)
41. C. Kendrick, G. LeLay, A. Kahn, Phys. Rev. B **54**, 17877 (1996)
42. M. Göthelid, Y. Garreau, M. Sauvage-Simkin, R. Pinchaux, A. Cricenti, G. Le Lay, Phys. Rev. B **59**, 15285 (1999)
43. C. Ratsch, W. Barvosa-Carter, F. Grosse, J.H.G. Owen, J.J. Zinck, Phys. Rev. B **62**, R7719 (2000)
44. R.H. Miwa, G.P. Srivastava, Phys. Rev. B **62**, 15778 (2000)
45. F. Grosse, W. Barvosa-Carter, J.J. Zinck, M.F. Gyure, Phys. Rev. B **66**, 75321 (2002)
46. P. Laukkanen, M.P.J. Punkkinen, M. Ahola-Tuomi, J. Lång, M. Kuzmin, J. Sadowski, J. Adell, R.E. Perälä, M. Ropo, K. Kokko, L. Vitos, B. Johansson, M. Pessa, I.J. Väyrynen, J. Electron. Spectros. Relat. Phenom. **177**, 52 (2010)
47. A.G. de Oliveira, S.D. Parker, R. Droopad, B.A. Joyce, Surf. Sci. **227**, 150 (1990)
48. C.F. McConville, T.S. Jones, F.M. Leibsle, S.M. Driver, T.C.Q. Noakes, M.O. Schweitzer, N.V. Richardson, Phys. Rev. B **50**, 14965 (1994)
49. P.R. Varekamp, M.C. Håkansson, J. Kanski, D.K. Shuh, M. Björkqvist, M. Gothelid, W.C. Simpson, U.O. Karlsson, J.A. Yarmoff, Phys. Rev. B **54**, 2101 (1996)
50. P. Laukkanen, M.P.J. Punkkinen, N. Räsänen, M. Ahola-Tuomi, J. Sadowski, J. Adell, M. Kuzmin, J. Lång, R.E. Perälä, M. Ropo, K. Kokko, L. Vitos, B. Johansson, M. Pessa, I.J. Väyrynen, Phys. Rev. B **81**, 035310 (2010)
51. G.E. Franklin, D.H. Rich, A. Samsavar, E.S. Hirschorn, F.M. Leibsle, T. Miller, T.-C. Chiang, Phys. Rev. B **41**, 12619 (1990)
52. M.T. Sieger, T. Miller, T.-C. Chiang, Phys. Rev. B **52**, 8256 (1995)
53. J. Olde, K.-M. Behrens, H.-P. Barnscheidt, R. Manzke, M. Skibowski, J. Henk, W. Schattke, Phys. Rev. B **44**, 6312 (1991)
54. U. Resch-Esser, N. Esser, B. Brar, H. Kroemer, Phys. Rev. B **55**, 15401 (1997)
55. W. Barvosa-Carter, A.S. Bracker, J.C. Culbertson, B.Z. Nosho, B.V. Shanabrook, L.J. Whitman, H. Kim, N.A. Modine, E. Kaxiras, Phys. Rev. Lett. **84**, 4649 (2000)
56. V.P. LaBella, Z. Ding, D.W. Bullock, C. Emery, P.M. Thibado, J. Vac. Sci. Tech. A **18**, 1492 (2000)

57. C.D. MacPherson, R.A. Wolkow, C.E.J. Mitchell, A.B. McLean, Phys. Rev. Lett. **77**, 691 (1996)
58. W.G. Schmidt, F. Bechstedt, N. Esser, M. Pristovsek, Ch. Schultz, W. Richter, Phys. Rev. B **57**, 14596 (1998)
59. W.G. Schmidt, F. Bechstedt, Surf. Sci. **409**, 474 (1998)
60. S. Mirbt, N. Moll, K. Cho, J.D. Joannopoulos, Phys. Rev. B **60**, 13283 (1999)
61. N. Esser, W.G. Schmidt, J. Bernholc, A.M. Frisch, P. Vogt, M. Zorn, M. Pristovsek, W. Richter, F. Bechstedt, Th. Hannappel, S. Visbeck, J. Vac. Sci. Tech. B **17**, 1691 (1999)
62. L. Li, Q. Fu, C.H. Li, B.-K. Han, R.F. Hicks, Phys. Rev. B **61**, 10223 (2000)
63. K. Lüdge, P. Vogt, O. Pulci, N. Esser, F. Bechstedt, W. Richter, Phys. Rev. B **62**, R11046 (2000)
64. P. Vogt, K. Ludge, M. Zorn, M. Pristovsek, W. Braun, W. Richter, N. Esser, Phys. Rev. B **62**, R12601 (2000)
65. M.R. Pillai, S.-S. Kim, S.T. Ho, S.A. Barnett, J. Vac. Sci. Tech. B **18**, 1232 (2000)
66. S. Tixier, M. Adamcyk, E.C. Young, J.H. Schmid, T. Tiedje, J. Cryst. Growth **251**, 449 (2003)
67. G. Feng, K. Oe, M. Yoshimoto, J. Cryst. Growth **301–302**, 121 (2007)
68. T. Liu S. Chandril, A.J. Ptak, D. Korakakis, T.H. Myers, J. Cryst. Growth **304**, 402 (2007)
69. A.J. Ptak, R. France, C.-S. Jiang, R.C. Reedy, J. Vac. Sci. Tech. B **26**, 1053 (2008)
70. E.C. Young, S. Tixier, T. Tiedje, J. Cryst. Growth **279**, 316 (2005)
71. S.W. Jun, C.M. Fetzer, R.T. Lee, J.K. Shurtleff, G.B. Stringfellow, Appl. Phys. Lett. **76**, 2716 (2000)
72. S.W. Jun, G.B. Stringfellow, J.K. Shurtleff, R.-T. Lee, J. Cryst. Growth **235**, 15 (2002)
73. M.P.J. Punkkinen, P. Laukkanen, H.-P. Komsa, M. Ahola-Tuomi, N. Räsänen, K. Kokko, M. Kuzmin, J. Adell, J. Sadowski, R.E. Perälä, M. Ropo, A. Tukiainen, T.T. Rantala, I.J. Väyrynen, M. Pessa, L. Vitos, J. Kollár, S. Mirbt, B. Johansson, Phys. Rev. B **78**, 195304 (2008)
74. P. Laukkanen, M.P.J. Punkkinen, H.-P. Komsa, M. Ahola-Tuomi, K. Kokko, M. Kuzmin, J. Adell, J. Sadowski, R.E. Perälä, M. Ropo, T.T. Rantala, I.J. Väyrynen, M. Pessa, L. Vitos, J. Kollár, S. Mirbt, B. Johansson, Phys. Rev. Lett. **100**, 086101 (2008)
75. M. Ahola-Tuomi, P. Laukkanen, M.P.J. Punkkinen, R.E. Perälä, I.J. Väyrynen, M. Kuzmin, K. Schulte, M. Pessa, Appl. Phys. Lett. **92**, 011926 (2008)
76. J. Pakarinen, V. Polojärvi, P. Laukkanen, A. Tukiainen, A. Laakso, C.S. Peng, P. Tuomisto, V.-M. Korpijärvi, J. Puustinen, M. Pessa, Appl. Surf. Sci. **255**, 2985 (2008)
77. E. Kaxiras, K.C. Pandey, Y. Bar-Yam, J.D. Joannopoulos, Phys. Rev. Lett. **57**, 923 (1986)
78. K.W. Haberern, M.D. Pashley, Phys. Rev. B **41**, 3226 (1990)
79. J.M.C. Thornton, P. Weightman, D.A. Woolf, C.J. Dunscombe, Phys. Rev. B **51**, 14459 (1995)
80. A. Taguchi, K. Shiraishi, T. Ito, Phys. Rev. B **60**, 11509 (1999)
81. A. Ohtake, J. Nakamura, T. Komura, T. Hanada, T. Yao, H. Kuramochi, M. Ozeki, Phys. Rev. B **64**, 045318 (2001)
82. J. Neugebauer, T.K. Zywietz, M. Scheffler, J.E. Northrup, H. Chen, R.M. Feenstra, Phys. Rev. Lett. **90**, 056101 (2003)
83. J.C. Thomas, N.A. Modine, J. Mirecki Millunchick, A. Van der Ven, Phys. Rev. B **82**, 165434 (2010)
84. J.H. Neave, B.A. Joyce, P.J. Dobson, N. Norton, Appl. Phys. A **81**, 1 (1983)
85. B.A. Joyce, J. Zhang, T. Shitara, J.H. Neave, A. Taylor, S. Armstrong, M.E. Pemble, C.T. Foxon, J. Cryst. Growth **115**, 338 (1991)

Chapter 2
High-Resolution Electron Microscopy of Semiconductor Heterostructures and Nanostructures

David L. Sales, Ana M. Beltrán, Juan G. Lozano, José M. Mánuel, M. Paz Guerrero-Lebrero, Teresa Ben, Miriam Herrera, Francisco M. Morales, Joaquín Pizarro, Ana M. Sánchez, Pedro L. Galindo, David González, Rafael García, and Sergio I. Molina

Abstract This chapter briefly describes the fundamentals of high-resolution electron microscopy techniques. In particular, the Peak Pairs approach for strain mapping with atomic column resolution, and a quantitative procedure to extract atomic column compositional information from Z-contrast high-resolution images are presented. It also reviews the structural, compositional, and strain results obtained by conventional and advanced transmission electron microscopy methods on a number of III–V semiconductor nanostructures and heterostructures.

2.1 Introduction

Transmission electron microscopy (TEM) and associated techniques based on the analysis of a transmitted electron beam through a specimen are basic tools in the field of Materials Science and Engineering. These techniques give information on the structure, composition, strain, and even the bonding in materials from the mesoscale to nanoscale, presently reaching even atomic column resolution as routine task. Excellent books and reviews exist in the literature on this field [1–3]. Here, we focus on some methodologies mainly based on high-resolution electron microscopy techniques developed by the authors, as well as the application to semiconductor materials of these and other methodologies based on the analysis of an electron beam transmitted through an electron-transparent sample. The studied semiconductor materials consist of III–V semiconductor nanostructures and heterostructures grown by molecular beam epitaxy (MBE).

D.L. Sales (✉) · S.I. Molina (✉)
Departamento de Ciencia de los Materiales e I.M. y Q.I., Facultad de Ciencias, Universidad de Cádiz, Campus Río San Pedro, s/n, 11510 Puerto Real, Cádiz, Spain
e-mail: david.sales@uca.es; sergio.molina@uca.es

A. Patanè and N. Balkan (eds.), *Semiconductor Research*, Springer Series in Materials Science 150, DOI 10.1007/978-3-642-23351-7_2,

Section 2.2 explains a method developed by the authors to map the composition in III–V materials with atomic column spatial resolution. This and other related methods have been applied by the authors to a number of semiconductor epitaxial layers and nanostructures; some examples showing the potential of this approach are also presented in this section. High-resolution compositional measurements combined with strain measurements at nanoscale are very helpful to improve the design of the semiconductor nanostructures and, more generally, of semiconductor materials. Section 2.3 presents the Peaks Pairs approach, which has been developed by the authors, to measure with high spatial resolution the strain in materials from the analysis of high-resolution electron microscopy images. Sections 2.4–2.8 summarize several selected examples where the above-mentioned high-resolution methodologies and other more conventional electron microscopy methods are applied to improve the understanding of a number of layers and nanostructures composed by III–V (antimonides, arsenides, and nitrides) semiconductor materials.

2.1.1 *Transmission Electron Microscopy, a Powerful Tool for Semiconductor Research*

TEM is a material characterization technique that is based on irradiating a very thin sample—thin enough to be transparent to electrons—with a parallel and uniform electron beam, which passes through the sample generating some sort of contrast that allows observation of its internal structure. Among the different signals originated from the electron–matter interaction (shown in Fig. 2.1), the transmitted direct beam and the elastic and inelastically scattered electrons can be collected and analyzed when using this technique. Electron diffraction can obtain accurate information about the materials under study, mainly because of two reasons (1) electrons, accelerated by several hundred kV potential, have a much smaller wavelength than the typical interatomic spacing, and (2) the availability of adequate coherent beams allows parallel and monochromatic irradiation of the sample.

2.1.1.1 High-Resolution Transmission Electron Microscopy

The phase-contrast electron microscopy is the origin of the technique known as high-resolution transmission electron microscopy (HRTEM). In this technique, the image is formed by the interference of multiple beams so that both the phase and amplitude of the diffracted beams contribute to the contrast in the image. For example, a phase-contrast image formed with any two diffracted beams, separated by $g = 1/d$ in reciprocal (or diffraction) space, reveals the periodicity of the crystal, d, due to their interference. This image is the most directly related to the structure and is linearly related when one of the beams is the directly (not scattered) transmitted beam and the sample is thin enough. The lattice image formed by the two beams can be extended to an image that seems to show a complete projection of the crystal structure when collecting multiple beams in all directions of higher

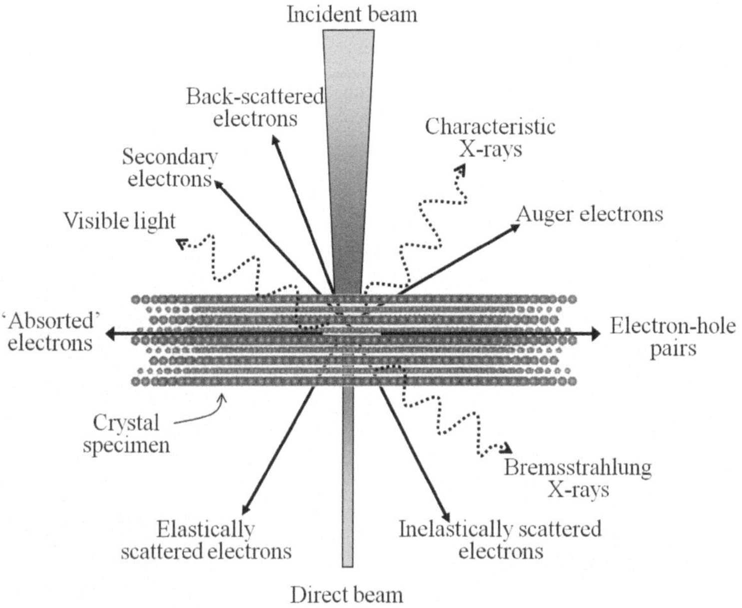

Fig. 2.1 Signals generated from the interaction between a high energy electron beam and a electron transparent (*very thin*) sample. Most of them are able to be detected by TEM

symmetry, assuming that the imaging and emission systems are adequate. Then, HRTEM can be exploited to image the atomic structure of thin specimens.

2.1.1.2 Scanning Transmission Electron Microscopy and *Z*-Contrast Imaging

This technique also collects information from the transmitted beam to characterize the micro- and nanostructure of the material, although in this case the incident beam is concentrated (focused) in a tiny probe which is scanned over the sample. Thus, each point of the image is obtained sequentially (pixel to pixel), and several detectors record multiple signals. Therefore, this technique is optimal for extracting maximum information from a single sample point and detecting multiple signals simultaneously [4], such as bright field, dark field, or energy loss spectra as shown in the example configuration in Fig. 2.2.

This basic difference in image acquisition makes scanning transmission electron microscopy (STEM) settings different with respect to conventional TEM. The most important differences are as follows:

1. There is no need of post-specimen lenses to form the image, which makes this technique less sensitive to chromatic aberrations.
2. Scanning coils are needed to move the beam across the sample.

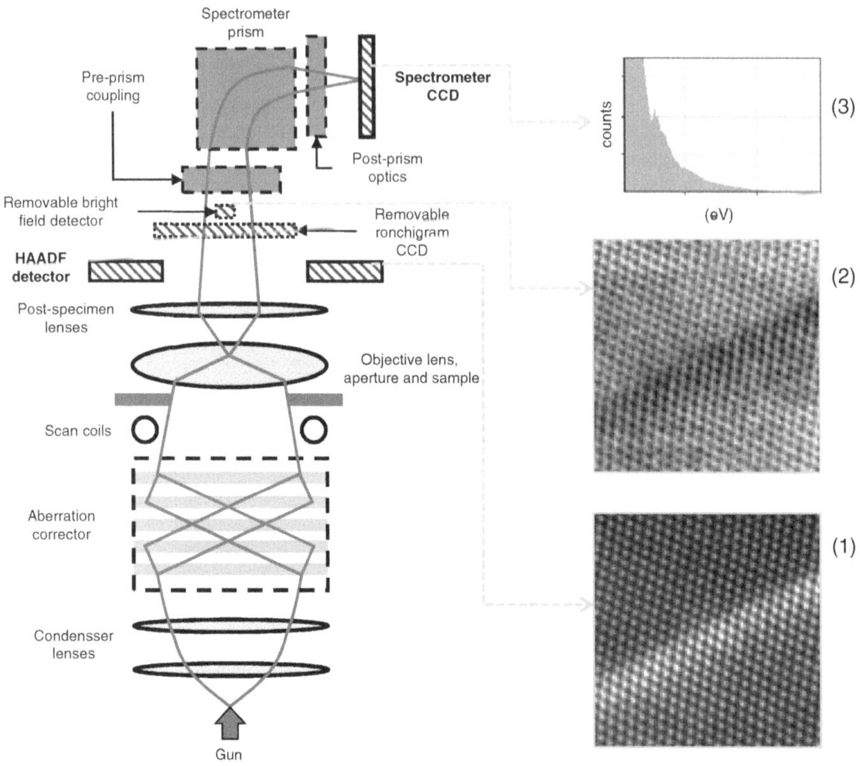

Fig. 2.2 Schematic of an aberration-corrected STEM with electron energy loss spectrometer. (1–3) Examples of HAADF and BF images, and EELS spectrum, respectively (adapted from [5])

STEM images can also be bright field or dark field, depending on whether the signal comes from the direct beam or one or more diffracted beams, respectively. Normally, these microscopes have installed a central detector to capture the direct beam, and annular dark field (ADF) detectors, with the hole centered on the optical axis, to capture diffracted beams.

Probably, the main advantage of the STEM configuration, in addition to the variety of available signals, is the possibility of Z-contrast imaging, where Z refers to the atomic number. A Z-contrast image is formed by collecting high-angle scattered electrons using an annular dark field detector (HAADF) [6, 7]. Images acquired in this way very well approximate the convolution of the intensity probe with the object function representing the sample [8], being considered as *incoherent* images—i.e., formed without interaction of the scattered waves from surrounding atoms or atomic columns. Then, high-resolution images acquired with this detector are connected in a more direct and simple way to the material structure than a phase-contrast image, being the last subjected to the complexity of coherent interference interactions [8, 9]. The underlying physical principle is that atoms with higher atomic number scatter more prominently at higher angles than lighter ones. When the scattering is Rutherford scattering, this is proportional to Z^2, the image contrast being modified in this ratio. Although the precise contrast depends on experimental setup, and is affected by several factors such as the *channeling* effect or the detector size, it is demonstrated in the next section that this approach is adequate to perform a quantitative compositional analysis of III–V semiconductors.

2.1.1.3 Aberration-Corrected Electron Microscopes

Magnetic lenses of electron microscopes are far from being perfect. They are still affected by aberrations, mainly spherical and chromatic ones. The incorporation of aberration correctors in new generation microscopes has made a breakthrough in the resolution limit, which seemed to be stacked. These correctors were applied to conventional TEM microscopes (CEOS) and STEM (Nion and CEOS) for correction of geometric aberrations. One of the most obvious improvement of the incorporation of aberration correctors in STEM, additionally to the increased resolution [4, 10], is the efficiency of images acquisition with enhanced signal-to-noise ratio.

2.2 Compositional Quantification Column-To-Column in III–V Semiconductors

Chemical composition is a crucial attribute of functional materials. Specifically in the case of nanostructured heteroepitaxial lasers, chemical composition of the nano-objects constituting their active layers strongly influences the emission energy of the final device. Normally, the real composition differs from the nominal one (i.e., from the desired one, for which the growth process is set up) [11]. Segregation and inter-diffusion occurring during upper confining layer growth are two important processes in this scenario. Not only real or effective composition is required, but also the compositional distribution. Shumway et al. show that for InGaAs/GaAs quantum dots (QDs), a difference of 10% of In content (from 50 to 60%) leads to a wavelength redshift of around 20–40 meV [12]. This strong effect justifies the effort on developing new methodologies for compositional analysis at the nanoscale.

Energy dispersive X-ray spectroscopy (EDX) and electron energy loss spectroscopy (EELS) are common analytical techniques based on electron–matter interaction. Although these techniques have greatly improved in recent years [13, 14], they still have some limitations, namely (1) the intrinsic electron probe broadening effect, (2) a poor energy resolution, and (3) the need for high acquisition times, and hence a high stability (i.e., radiation-resistant materials, extreme stability of the microscope setup, hydrocarbons free sample surfaces, etc.). It is therefore necessary to look for another compositional sensitive signal, requiring less acquisition time with high resolution.

HAADF images provide significant information on the position of atomic columns in a crystal with sub-angstrom resolution (when the microscope is equipped with aberration correctors), requiring a low acquisition time (high-quality images in less than 16 s). Furthermore, contrast in HAADF images is approximately proportional to the square of the average atomic number of the atomic column. Therefore, this technique is also called *Z-contrast imaging*.

Several works in the literature show qualitative compositional mapping with high resolution from Z-contrast images of various material systems [15, 16].

In this section, a method to determine quantitatively with atomic column spatial resolution the composition of a semiconducting ternary alloy is presented. The method is based on the analysis of local integrated intensities from aberration-corrected Z-contrast images, and was first presented in [17]. It is necessary to use a set of reference samples of known composition to quantify the relationship between the Z-contrast intensity with the thickness and composition of the material analyzed. This method is applied as an example to an $InAs_xP_{1-x}$ alloy. It uses a series of $InAs_xP_{1-x}$ epilayers grown on InP(001) by Molecular Beam Epitaxy (MBE) as reference samples.

2.2.1 *Reference Samples Study*

Three MBE samples consisting of $InAs_xP_{1-x}$ epilayers grown on InP(001) were considered. The As molar fraction was precisely measured by high-resolution X-ray diffraction (XRD), giving $x = 0.27$, 0.59, and 0.87, respectively, for each sample. Specimens for Z-contrast imaging were produced for cross-sectional view by mechanical thinning and Ar^+ ion milling using a precision ion polishing system (PIPS). The Ar^+ PIPS was used in such a way that the beam always hits the sample from the same direction, which has been found to produce cleaner surfaces in the region of interest. A beam energy less than 3.5 kV has been selected to reduce amorphization of InP. As a final step, the sample was introduced in a *Fischione* ion mill at 15 and 0.5 kV to reduce surface damage. The thickness of each analyzed region was determined from the analysis of the corresponding spatially resolved low-loss EELS signal, by using the mean free path determined for each x composition of the three $InAs_xP_{1-x}$/InP reference samples, according to the method described in [18].

A characterization procedure was defined to ensure (1) the systematicity and reproducibility of the study, (2) the minimization of radiation damage, and (3) the minimization of errors by deposition of hydrocarbons during the observation. The procedure is summarized below:

- Previous checking:
  - Although, in principle, the manufacturer asserts that the intensity ratio is linear, the detector settings (gain and black level) were not changed between sessions.

  - The objective lens current remains constant through the analysis of the three samples, which means that the distance from the virtual objective aperture does not vary.

- Acquisition of images without sample to evaluate the detector black level.
- Astigmatism correction on an amorphous close to the area of interest.
- Acquisition of spectrum image centered at the zero loss peak, for the thickness determination of the area of interest.
- Acquisition of HAADF images from various points of the sample with different thicknesses. Images are to be acquired in the interface, and both InP and the $InAs_xP_{1-x}$ layers must always appear.
- Localization of the HAADF images within the thickness map, identifying the mean specimen thickness for every image.

2.2.2 *Image Analysis and Comparative Index*

This paragraph describes how to analyze the acquired HAADF images in order to extract chemical information. The starting point is a HAADF image from the $InAs_xP_{1-x}$/InP interface with atomic column resolution and known composition (x) and thickness (t), as the one shown in Fig. 2.3a.

HAADF detectors normally register a background signal even if the sample is not being imaged. Then, the first step is to subtract the mean value of the detector black level, which depends on the detector gain and might differ from one microscope session to another. Next, pixels with the local intensity maxima associated to In atomic columns (higher Z) are detected on the image. For this, the Peak Finding tool of the *Strain Determination Software* is used [19]. Once these maxima intensity

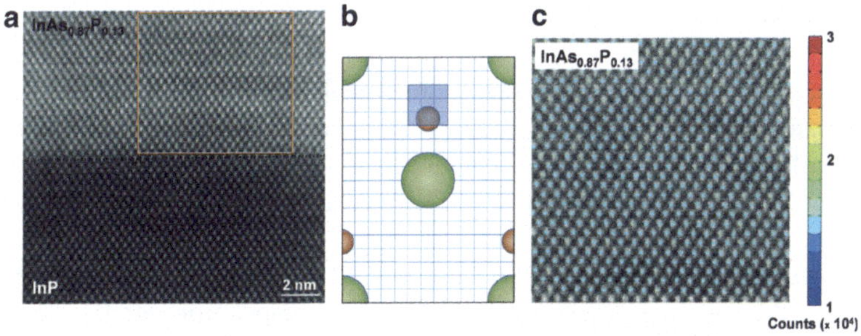

Fig. 2.3 (**a**) Raw data high-resolution aberration-corrected Z-contrast image taken along [110] from the $InAs_{0.87}P_{0.13}$/InP interface. (**b**) Area selected (*blue square* of nine pixels) for each group V column to integrate intensities and calculate R ratios, drawn on the projected unit cell. Projected positions of In and group V atomic columns are represented with *red* and *green color circles*. (**c**) Integrated intensities in the area selected in (**b**) superimposed on the high-resolution Z-contrast image from which the intensities were determined for each analyzed group V column. Note that the intensity distribution is very homogeneous throughout the whole analyzed image

pixels are located, it is straightforward, with the help of image processing software, to measure the intensities integrated within a selected area of the projected unit cell, as shown in Fig. 2.3b. In this case, the number of pixels that an $InAs_xP_{1-x}$ unit cell occupies in the 500,000 × image is 18 × 13 pixels. Taking into account the crystal orientation, and dimensions of the unit cell, pixels that correspond to the position of the anion are localized within the image. These pixels will be, in principle, more sensitive to the change in x (arsenic molar fraction) in an $InAs_xP_{1-x}$ alloy with variable composition.

The resulting integrated intensities of an $InAs_{0.87}P_{0.13}$ region are shown superimposed on the experimental image in Fig. 2.3c. The integrated intensity for each pair of In-AsP columns is shown in this figure by a colored circle located in the brightest pixel, near the center of the projected position of the In atomic column. The integration area used is the one marked with a blue box in Fig. 2.3b. It is noted that the dispersion of the value is very low. This process is repeated for an InP area of the same image. Finally, two integrated intensity mean values, one for the $InAs_{0.87}P_{0.13}$ ($I_{S0.87}$) and one for the InP(I_0), are determined. Defining the normalized integrated intensity parameter as given in (2.1), the ratio $R_e(0.87)$ is obtained, which defines the image in Fig. 2.3a.

$$R_e(x) = \frac{I_{sx}}{I_0}. \tag{2.1}$$

The error is minimized in this way, as the specimen thickness variations and other local fluctuations—that may contribute to the intensity values—are small within the area covered by a single HAADF image.

The selected integration area for the analysis has to be (1) linearly dependent on the As content, (2) independent of specimen thickness, and (3) independent on the surrounding atomic columns so that the signal is essentially due to the contribution of As and P from the selected atomic column. These three aspects were evaluated in several integration areas, the marked one in Fig. 2.3b resulting the most satisfying.

The same procedure for measuring integrated intensities of hundreds of anion–cation column pairs was performed in regions of different specimen thickness in the three reference samples (different x). Then, the values of R_e for certain x are obtained averaging the normalized integrated intensity values from different thickness areas of the same reference sample. The maximum observed standard deviation of $R_e(x)$ due to the specimen thickness was 1.5%. The $R_e - x$ plot shows a clear linear dependency (see Fig. 2 of [17]). A fitting equation with a correlation coefficient of 0.9974 is obtained.

Thus, measuring the values of R_e according to the procedure in a heterogeneous alloy $InAs_xP_{1-x}$, it is possible to determine the composition, atomic column to atomic column, simply from the linear regression equation relating $R_e - x$ in the calibration samples.

2.2.3 Simulation of Integrated Intensities

Image simulation would help to corroborate the conclusions of experimental analysis when a good correlation between calculated and experimental images exists. However, quantitative matching between experimental and simulated aberration-corrected Z-contrast images has not been reached up to now. The inclusion of spatial incoherence has been shown to give quantitative agreement between non-aberration-corrected HAADF-STEM images and theoretical simulations [20, 21]. It is shown here that, using the same approach, a significant improvement in the correlation between calculated and experimental normalized integrated intensities is obtained in the $InAs_xP_{1-x}$ ternary semiconductor alloy, but residual discrepancies still remain [22].

An important difference between aberration-corrected and non-corrected Z-contrast images is the dependence of normalized intensities on the thickness of the electron-transparent specimen. This dependence is significant for non-corrected images, whereas it is very low for a range of specimen thicknesses in aberration-corrected images. As shown in Fig. 2.4, the R values calculated without including the effect of spatial incoherence clearly overestimate the experimental R values. On the contrary, experimental and calculated R values are much closer when a Gaussian is used to take into account the effect of spatial incoherence. In order to determine

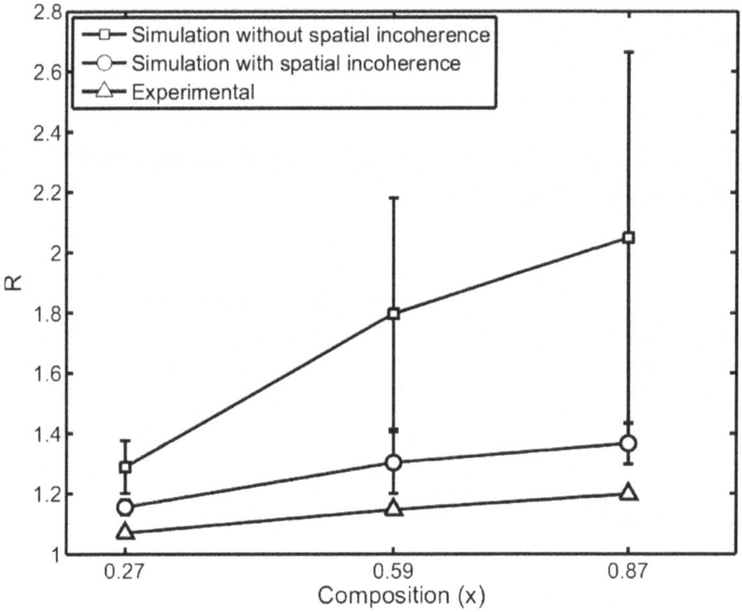

Fig. 2.4 Experimental and simulated R ratios vs. As atomic content (x), with and without introducing the effect of spatial incoherence, for three compositions of the $InAs_xP_{1-x}$ alloy (reproduced from [22])

the optimum value of the width (standard deviation) of the Gaussian function that gives the best fit between experimental and calculated images, an error function defined as the sum of squared differences between simulated and experimental Z-contrast images is considered. The minimum value of the error function occurs when the optimum width of the Gaussian function is 0.06 nm, which corresponds to a source size with a full width at half maximum (FWHM) of 0.14 nm.

Integrated intensities in aberration-corrected HAADF-STEM images have been simulated by including the effect of spatial incoherence. In this way, intensities in both simulated and experimental images acquired in $InAs_xP_{1-x}$ thin layers with calibrated compositions show improved correlation. A residual error of $\sim$10%, however, cannot be accounted for, and the most obvious origin of this error would be the effect of non-round aberrations that were not recorded when the experimental data were obtained. These non-round aberrations are expected to be quite dependent on the exact tuning of the aberration corrector, and further work is required to determine whether they could quantitatively account for the residual discrepancy. The calculated normalized intensities are found to be almost independent of the specimen thickness in a practical range of useful thickness values, in good agreement with the experimental results, and in contrast with the observed behavior in non-aberration-corrected images. Based on dynamical simulations of the channeling process that electron beams suffer after propagating along an atomic column, our findings are explained and predicted to be valid for a wide range of materials studied by aberration-corrected Z-contrast imaging. Our results pave the way for a direct interpretation of aberration-corrected high-resolution Z-contrast images, in terms of elemental compositions with atomic column spatial resolution.

2.3 Strain Measurements from High-Resolution Electron Microscopy Images

Recent advances in digital imaging have offered the possibility of locally determining the elastic strain of materials from HRTEM images very precisely. Strain mapping can be used to analyze materials at the atomic column level, measuring local displacements and so revealing lattice translations, dislocations, and/or rotations.

2.3.1 Techniques

Algorithms to map elastic strains from HRTEM images may be classified into two broad categories: real space methods (also called peak-finding techniques) based on the detection of peaks of intensity in real space, and Fourier methods based on Takeda's technique for phase retrieval [23].

Peak-finding techniques were first described by Bierwolf et al. [24] and further developed by Jouneau et al. [25], Seitz et al. [26], Kilaas et al. [27], Robertson et al. [28], and Rosenauer et al. [29]. All these techniques were based on superimposing a two-dimensional reference lattice extrapolated from a non-distorted region of the material to the experimental one, built up from the set of intensity maxima in the HRTEM image, and calculating the local discrete displacement field at each node [30]. Subsequently, the strain field is calculated as the derivative of the displacement field. Peak pairs [19] can be considered an improved version of previous peak-finding techniques in the sense that the reference lattice is not extrapolated, but calculated at each pair of neighboring peaks. The neighborhood function is calculated in the affine-transformed space described by the reference basis vectors. This transformation greatly reduces potential errors in the determination of pairs and allows the proper calculation of the displacement and strain fields in the presence of defects.

Geometric phase [31] makes use of the Fourier transform fringe analysis method. The method was first introduced by Takeda et al. [23] and adapted by Bone et al. [32], who provided a full 2D version. This technique was applied a decade later to lattice-distortion measurement in crystallography [31, 33], achieving excellent results [34]. It consists of applying an asymmetric filter at the chosen peak (Bragg spot) of the Fourier transform of a HRTEM lattice image and performing an inverse Fourier transform. The resulting complex image holds the phase information which can be calculated by the logarithm of an inverse tangent calculation. The spatial resolution of the phase map in the Fourier transform fringe analysis method is determined by the size of the filter's window. Let us note that a larger window provides a better frequency resolution but a poorer spatial resolution, and vice versa. Hence a suitable window size must be selected for a compromise between the two resolutions. When noisy strain maps are obtained, researchers should resist the temptation to produce smoother maps by decreasing window size, because spatial resolution (which is directly linked to window size) could be, in this case, too small, and accuracy of strain results would be seriously affected.

2.3.2 Methodology

In order to determine strain from a HRTEM image, the first step is image filtering, a critical issue in peak-finding methods. All techniques work by first determining a base vector in an unstrained region and calculating the corresponding displacement map with respect to this reference. It is important to note that the reference area should be taken on the same image, but outside deformed regions, because strain is determined from the distortions in the image compared to the reference (unstrained) vectors. By calculating the displacement map across two non-collinear directions, it is possible to determine the displacement field to derive the strain field easily.

2.3.2.1 Basis Vectors Determination

In real space, image filtering is usually needed not only to reduce noise in the image, but also to smooth the image around the peaks. Special attention should be paid to the filtering stage in order to avoid loss of significant information. Currently, different techniques are frequently used, such as low-pass, Wiener, and/or Bragg filtering. Once the image has been filtered, the direction of a basis vector is obtained by selecting two neighboring peaks in the desired direction within the reference area. The coordinates of the vector connecting both peaks define the basis vector. Two basis vectors are needed in order to be able to calculate the displacement map in any direction. In order to improve the accuracy, the average base vector in a desired region of the reference area is calculated.

In Fourier space, the direction is obtained by selecting a peak of intensity in the power spectra of the image. Once the peak location has been chosen, its position determines the coordinates of the base vector. In order to improve the accuracy, Takeda's technique is used to calculate the 2D phase. A slight error in the location of the peak will produce a flat but non-constant phase in the reference area, assuming it is unstrained. Discrete character of images, where the maximum resolution is limited by pixel size, usually generates a non-constant phase because of the limited resolution of the power spectra. However, it is possible to improve the accuracy by fitting a plane to the phase and using its coefficients accordingly to modify the coordinates of the base vector so that the refined phase is constant in the chosen region (reference area). In doing this, subpixel resolution is achieved for the coordinates of the base vector.

2.3.2.2 Displacement Field Calculation

Once the basis vectors have been obtained with subpixel resolution, the displacement field can be calculated using different approaches.

In the peak pairs approach, all the intensity peaks are automatically located using eight-neighborhood maxima detection at the pixel level. In order to get sub-pixel resolution, a 2D quadratic function is fitted by using gray levels in the neighborhood of each peak and solving the resulting equations for the maxima. Using all pairs of peaks, the displacement field is calculated at the positions of atomic columns using the difference between local peak pairs and basis vectors.

In the geometric phase approach, the phase is calculated using Takeda's approach to determine the displacement field directly.

2.3.2.3 Strain Field Calculation

The strain field components can be easily calculated by solving the following set of linear equations:

$$\left.\begin{aligned} u_x &= a_x e_{xx} + a_y e_{xy} \\ u_y &= a_y e_{yy} + a_x e_{yx} \\ \upsilon_x &= b_x e_{xx} + b_y e_{xy} \\ \upsilon_y &= b_y e_{yy} + b_x e_{yx} \end{aligned}\right\} \begin{aligned} \begin{bmatrix} e_{yx} \\ e_{yy} \end{bmatrix} &= \begin{bmatrix} a_x & a_y \\ b_x & b_y \end{bmatrix}^{-1} \begin{bmatrix} u_y \\ \upsilon_y \end{bmatrix} \\ \begin{bmatrix} e_{xx} \\ e_{xy} \end{bmatrix} &= \begin{bmatrix} a_x & a_y \\ b_x & b_y \end{bmatrix}^{-1} \begin{bmatrix} u_x \\ \upsilon_x \end{bmatrix} \end{aligned}, \quad (2.2)$$

where (a_x, a_y) and (b_x, b_y) are the coordinates of the basis vectors; (u_x, u_y) and (υ_x, υ_y) are the coordinates of the displacement field in the direction defined by each basis vector, respectively; and e_{xx}, e_{yy}, e_{xy}, and e_{yx} are defined as follows:

$$e_{xx} = \frac{\partial u}{\partial x}, \quad e_{xy} = \frac{\partial u}{\partial y}, \quad e_{yy} = \frac{\partial \upsilon}{\partial y}, \quad e_{yx} = \frac{\partial \upsilon}{\partial x}. \quad (2.3)$$

2.3.2.4 Alternative Applications

Strain mapping is often used not only to determine strain and identify defects and/or distortions in the image, but also to rectify distorted images. It is well known that CTEM projector lens systems, STEM scanning systems, and CCD cameras or scanners often have distortions that affect the images they form. When these distortions are systematic, it is possible to eliminate them by measuring the displacement map of an unstrained specimen and using it to define a warp function to restore any distorted image taken in the same conditions [35].

2.3.2.5 Problems

Strain determination from HRTEM images depends upon the assumption that the intensity maxima in the image and the position of atomic columns in the sample are directly related. There are many situations where this assumption is violated, and image conditions should be carefully optimized. The most important rule to determine optimal conditions is choosing, when possible, thickness and defocus values where its variation does not modify the location of the intensity peaks in the image. These values should be determined by image simulation. Other practical rules are choosing thickness and defocus conditions where fringe contrast is high, and avoiding regions where the fringe contrast changes rapidly. Let us note that frequently, Scherzer defocus might not be the optimal choice. Error quantification in strain mapping methods has shown reduced accuracy near abrupt interfaces, independently of the strain mapping technique [36]. Therefore, depending on the chosen imaging conditions, strain profiles may contain severe artifacts. Nevertheless, average strain can be determined with sufficient accuracy on experimental micrographs of thicker layers [37]. The reliability of strain mapping from HRTEM images is also affected by thin foil relaxation and lattice fringes shifting due to variations in composition in the sample. In order to take into account all these effects, accurate simulation using Molecular Dynamics or Finite Element

Method is often a requirement to corroborate the measured strain from HRTEM images. Only when these factors are taken into account, the method offers a reliable way of extracting quantitative strain information at the atomic scale [38, 39].

2.3.3 Applications

Strain mapping has been used in different fields during the last few decades. Some recent applications of peak pairs include the analysis of confinement potential of III–V quantum dots [40], the determination of surface-induced disorder in magnetic FeCoPd nanoparticles [41], the determination of precise atomic-resolution compositional distribution in HAADF [42], the control of growth directions of InAs nanowires on Si substrate [43], the determination of composition of $In_xGa_{1-x}N$ from strain measurements [44], etc. Geometric phase has also been applied to a number of different problems, including the measurement of strain in silicon devices [45], the mapping of strain in MOSFETs at the nanometer scale [46], the measurement of strain in strained-silicon MOSFETs [47], and the measurement of strain in aberration-corrected HAADF-STEM images of InAs/GaAs dot-in-well heterostructures [35].

Different plug-ins for DigitalMicrograph^{TM} implementing strain mapping algorithms have been developed by HREM Research [48], including peak pairs analysis (PPA) and geometric phase analysis (GPA). Its features include high-resolution peak detection (only available in PPA), image filtering, displacement and strain map calculation, atomic column (only available in PPA) and/or continuous strain maps, histogram and lattice point's analysis (only available in PPA), and geometric distortion determination and correction.

2.4 Results on III-Sb Hetero- and Nanostructures

Fiber-optic systems require preferably uncooled high-speed lasers and detectors which are matched to the 1.55-μm fiber transmission window. So far, data rates in telecommunication applications at 1.55 μm are limited by the use of costly InP-based high-speed devices [49]. Huge research efforts have been made to develop improved and lower cost optoelectronic devices on inexpensive large-area GaAs substrates. Thus, GaSb/InAs/GaAs heterostructures—with or without the growth of an intermediate GaAs barrier layer—are emerging technologies to lower costs and increase broadband and network speeds. Currently, there is an increasing interest in combining the electronic properties of more than two semiconductor compounds. More specifically, heterostructures which combine InAs with antimonides offer outstanding band-gap engineering possibilities which are not available in the more traditional InAs/GaAs system. Of particular interest are

GaSb/InAs/GaAs and GaSb/GaAs quantum dot heterostructures due to the type-II band alignment expected between arsenides and antimonides [50].

The semiconductor-based optoelectronic devices face significant challenges in material quality and fabrication routes which must be solved to realize commercially viable technologies. One of the main barriers is the effect of the surface segregation, which can impede the achievement of abrupt heterointerfaces. It has been demonstrated that a competitive incorporation between As and Sb atoms occurs during the growth of GaSb on GaAs [51]. Furthermore, in GaSb/InAs/GaAs heterostructures, both cation and anion segregation can occur. A clear blocking of indium incorporation by antimony has recently been reported in the literature [52]. These effects can degrade interface sharpness considerably, and therefore alter band offsets.

Significant difficulties could arise in systems where quantitative compositional measurements at the nanometer scale or below are required. The use of a very thin specimen (tens of nm or less) in TEM means that large-scale interface roughness does not affect compositional analysis, but the interpretation of images and analytical spectra is not straightforward. An exhaustive compositional analysis requires the combination of different techniques, such as EELS, HAADF images collected by STEM, and conventional dark-field imaging. This section focuses on the compositional determination of GaSb/InAs/GaAs and GaSb/GaAs systems, mainly using spatially resolved EELS and aberration-corrected HAADF-STEM.

The combination of low-loss EELS and high-resolution HAADF-STEM can clearly demonstrates Sb incorporation inside GaSb-capped InAs QDs grown on GaAs substrates by MBE. These techniques show the presence of the four elements (Ga, In, Sb, and As) in the QDs, presumably as a result of an intermixing process during the growth of the QDs and the GaSb capping layer. In this sense, the more intense contrast in the HAADF-STEM image of Fig. 2.5a is associated with high Z-number, depicting the presence of In-rich cationic columns and/or Sb anionic columns inside the observed quantum dot. For a statistical analysis of the integrated intensities associated with anionic and cationic columns, the process described in Sect. 2.2 was followed. The intensity profile along the growth direction shows a well-defined maximum which supports the data obtained by EELS (see [52]). The intensity maxima of anionic and cationic columns are located in the same areas where EELS measurements showed maximum In and Sb concentration. The analysis also demonstrated a tendency of the Sb to extend toward the apex of the QDs. Although no clear GaSb capping layer was observed in the heterostructure, the analytical results showed a $GaAs_ySb_{1-y}$ on top of the $Ga_xIn_{1-x}As_ySb_{1-y}$ core.

A possible route to optimize the luminescence properties of GaSb-capped InAs/GaAs quantum dot heterostructures is through the incorporation of a GaAs barrier layer between the InAs QDs and the GaSb capping layer. Composition profiles were obtained in heterostructures containing a GaAs barrier using EELS, HAADF-STEM, and dark-field 002 diffraction contrast TEM. Figure 2.6b shows EELS maps of elemental distribution for Ga, In, and Sb. The measured scenario—two In-containing layers separated by a thin Sb-containing layer—does not correspond to the desired (nominal) structure. This striking difference gives insight into the

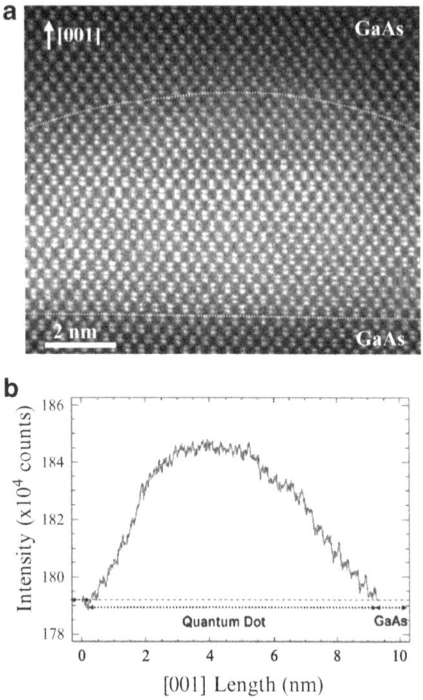

Fig. 2.5 (**a**) High-resolution HAADF-STEM image of an InAs QD capped by a GaSb layer. The approximate location of interface with the GaAs capping layer is indicated on the image with *dotted lines*. (**b**) Intensity profile taken along [001] across the central part of the HAADF-STEM image of (**a**) (reproduced from [52])

complex interplay between segregation and desorption. Two conclusions are evident (1) surface-segregated Sb competes strongly with In, and effectively blocks In incorporation into the lattice, and (2) when the Sb concentration falls below a certain threshold, In incorporation into the lattice resumes [53]. This is the first clear evidence of a strong influence of Sb on the incorporation of In—which occurs despite the fact that In is incorporated only on group III sites, while Sb is incorporated on group V sites.

Finally, GaSb QDs grown directly on GaAs substrate were analyzed by high-resolution HAADF-STEM images, resolving the atomic columns and obtaining a Sb distribution map. Figure 2.7 corresponds to integrated HAADF-STEM intensities for each projected unit cell superimposed on the original HAADF image, where a higher intensity corresponds to unit cells with higher Sb content. This type of analysis is very helpful to determine the lateral size and height of a nano-object. In this case, it demonstrates the formation of a discontinuous layer composed of $GaAs_xSb_{1-x}$ nanostructures with a lateral size in the region of 1–10 nm and heights of around 1 nm [54]. These measured dimensions mean that there has been a volume reduction of the nanostructure by three orders of magnitude during the GaAs capping process. This proves the strong Sb segregation occurring during the growth of the GaAs capping layer.

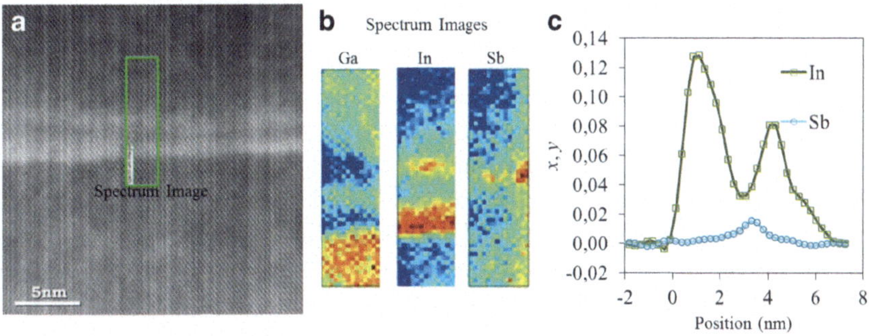

Fig. 2.6 (**a**) HAADF-STEM image of a nominal GaSb/6 ML GaAs/InAs/GaAs heterostructure viewed along [110]. The *rectangle* indicates the region analyzed by EELS. (**b**) EELS-Independent Component Analysis images of Ga, In, and Sb distribution of the marked area. (**c**) Composition profiles taken from (**b**). Note that two In-containing layers are registered in (**b**) and (**c**), showing the reduction of In incorporation by Sb (reproduced from [53])

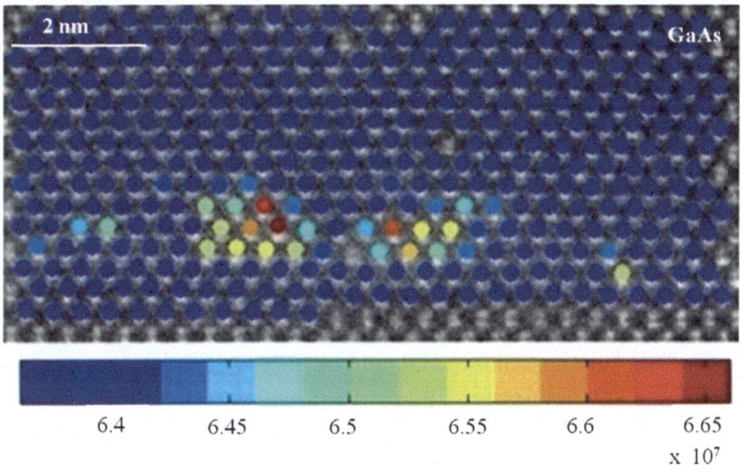

Fig. 2.7 Integrated intensities for each projected unit cell, superimposed on a high-resolution HAADF-STEM image of the GaSb/GaAs layer (adapted from [54])

In summary, these studies show the formation of $Ga_xIn_{1-x}As_ySb_{1-y}$ in the core of QDs, likely to be a strain-driven process. A clear influence of Sb on the In incorporation in heterostructures containing a GaAs barrier layer is revealed, as well as Sb segregation during the capping process of GaSb QDs grown directly on a GaAs substrate. In all cases, the difference between the nominal structure and the one observed experimentally is very clear, thus demonstrating the need for high-resolution analysis in the development of these semiconductor systems.

2.5 Results on InAs Quantum Wires

This section summarizes our contribution to the understanding of the growth of InAs self-assembled stacked quantum wires on InP(001) substrate by MBE using different techniques. Morphological and compositional changes from the initial stages have been characterized. Experimental and simulated results of the inhomogeneous strain field in the structures have proved it as the driving force for the heterogeneous morphology and As distribution, and for their nucleation on preferential sites on each stacked layer and with them, their vertical distribution along the stacked heteroepitaxial layers.

To manufacture improved three-dimensional (3D) semiconductor nanostructures for telecommunication devices which has to respond to increasing requirements of data transmission speed and stability, the self-assembly, through the Stranski–Krastanov (SK) growth mode, is recognized as one of the most remarkable approach to be followed. For the heteroepitaxial system InAs/InP grown by MBE, the strong stress anisotropy at the InAs/InP interface along <110> directions [55,56] can lead, under particular growth conditions, to the formation of horizontal quantum wires (QWRs) instead of QDs. Due to its potential applications for telecommunication technologies, an intensive research has been carried out on this type of nanostructures during the last years. In fact, it is well demonstrated that QWR emission properties can be tailored from 1.2 to 1.9 μm [57] by controlling several parameters such as the substrate temperature, the distribution of the different layers during the vertical arrangement, or their density [58–60].

In this context, the control of the shape, size, position, compositional information, and strain of the semiconductor nano-objects constituting these nanostructures is crucial to improve their design and functionality [61]. The development and implementation of methods to extract this information at the nanoscale, and in some cases at the atomic scale, should help us to be able to know and relate their characteristics with their functional properties.

Results obtained by high-resolution phase-contrast and Z-contrast electron microscopy techniques, in coordination with peak-finding methods, finite elements analysis (FEA), and EELS, are reviewed next. They are applied to measure the above-mentioned structural properties at the nano- and atomic scale, to predict the preferential nucleation sites of single and stacked layers, and to contribute to understand the growth of single and stacked layers of QWRs.

2.5.1 Nucleation and Initial Growth Stages of InAs/InP(001) QWRs

With the aim of evaluating the wires' growth at their initial stages, samples with 2.3, 2.5, and 2.7 InAs monolayers (ML) grown on InP(001) were analyzed. The first sample corresponds to the onset of the elastic relaxation process due to the

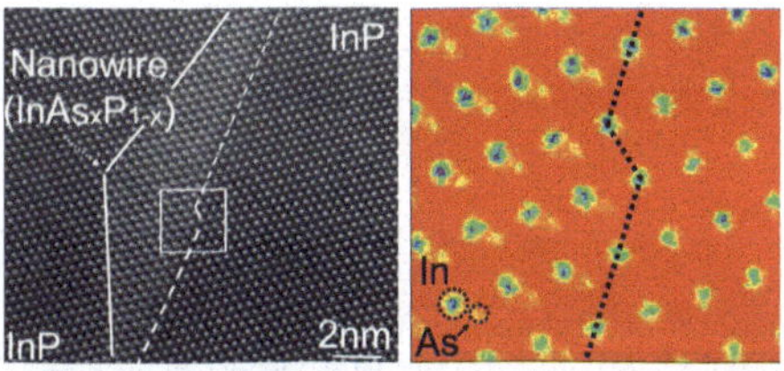

Fig. 2.8 Color low pass-filtered high-resolution Z-contrast image obtained from a magnified region of the central part of the InAs(P)/InP wire interface. The existence of a diatomic step is clearly visible on the *dotted line* (reproduced from [52])

spontaneous self-assembling of QWRs, and the last sample to the well-developed QWRs. The growth conditions are reported in [58, 62].

The analysis of atomic force microscopy images [52] and high-resolution Z-contrast micrographs of capped and uncapped nanowires reveal that atomic steps located at the surfaces of substrates or buffer layers play a central role in controlling the growth mechanism. In fact, they show that most of the nanowires are positioned on the upper terraces of the growth surface steps close to the step edge. In particular, Fig. 2.8 shows a high-resolution aberration-corrected Z-contrast image of a capped nanowire located just on InAs(P)/InP(001) interfacial steps. In this case, InAs-rich regions are demonstrated to be situated on the region described below. The strain at the growth surface on the upper and lower terraces of diatomic steps during the very initial stage (2D) of InAs deposition is calculated by FEA, using anisotropic elasticity theory. Larger positive strains are found at the upper terrace surface, which explains the lower chemical potential for further growth on the upper terrace of the diatomic step, and therefore, the preferential nucleation for the nanowires.

Regarding the morphological evolution of the initial QWRs, maps of the integrated intensity around As–P columns (Fig. 2.9b, d) determined from the high-resolution Z-contrast images in Fig. 2.9a, c—as described in Sect. 2.2—display that (001) is the main facet of the nano-objects at their very initial stage of formation; however, it is gradually reduced in favor of $\{114\}$ or $\{118\}$. It is clearly observable from this analysis the compositional asymmetry of the nanowires since their initial steps of growth, which has a strong effect on the final arrangement, composition, and strain of the nanowires.

Three-dimensional models built by FEA explain the facets' evolution in terms of variation of the elastic energy. These calculations locate energy minima on (001) facets and maxima around the QWR perimeter where it joins the wetting layer (WL), favoring diffusion of arsenic atoms from these edges toward the QWR top (to increase its height) and outside (for the nucleation of new wires).

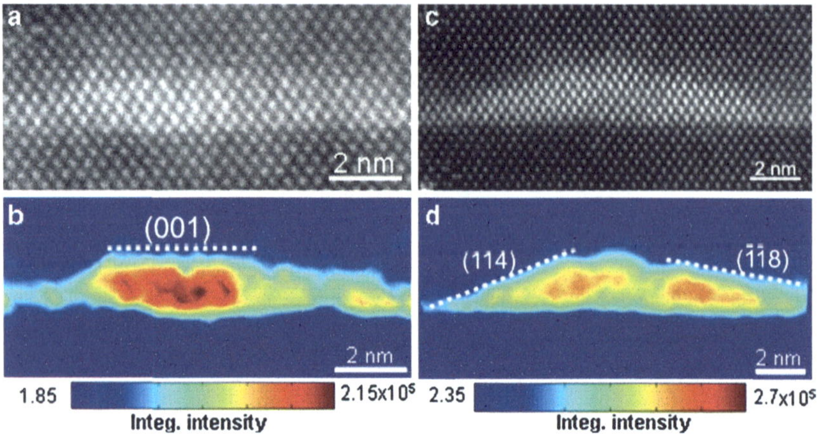

Fig. 2.9 Z-contrast images and their maps of integrated intensities around As–P columns of samples with 2.5 ML (**a**–**c**) and 2.7 ML (**b**–**d**) of InAs. Higher values of integrated intensity denote higher As content (reproduced from [62])

2.5.2 Simulated and Experimental Determination of Strain Map and Prediction of Nucleation Sites for the Growth of the Stacked Structures

For self-assembled nanostructures, the size, homogeneity, density, and spatial organization degree can be improved by stacking several layers. The InP-capped $InAs_xP_{1-x}$ nanowires of each layer underneath each new stacked layer of nanowires produce inhomogeneous strain fields that propagate toward the capping layer surface where the next nanostructures will be formed, controlling the distribution of the strain at the InP growth surface.

By combining in situ stress measurements and TEM, the effect of this strain field has been reported and the wire growth explained by considering stress-driven mass transport and P/As exchanges [60]. The thickness and the growth temperature for layers capping nanowires are critical parameters that permit the control of the heterogeneous strain field associated with the buried nanowires and the nucleation sites on each stacked layer at the growth surface.

The understanding of the wires growth can be completed following two approaches: (1) Measuring the strain field by applying peak-finding methods [19, 30] (see Sect. 2.3) to HRTEM images of stacked nano-objects. (2) Calculating the strain field by solving the anisotropic elastic theory equations by FEA using a compositional model of the studied nano-objects. This method has been proven to be one of the most efficient procedures to model the strain associated with these buried nanostructures [63–66]. Inputs for this compositional model are given by both spatially resolved EELS and aberration-corrected high-resolution Z-contrast images. Figure 2.10 depicts the process followed to find the preferential nucleation sites of stacked epitaxial nanowires.

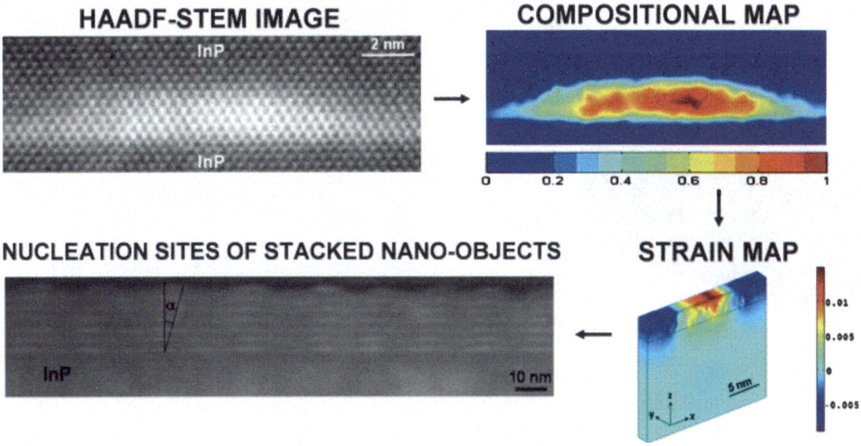

Fig. 2.10 Scheme of the sequential process to predict preferential nucleation sites of stacked InAs/InP nanowires by the analysis of high-resolution HAADF-STEM images (adapted from [67])

Both approaches have been applied to InAs/InP single and stacked nanowires for two different growth temperatures (380 and 515°C) and thicknesses (between 3 and 20 nm) of the capping layer (see growth details in [59] and [60]).

Compositionally sensitive TEM images of low capping temperature (380°C) samples show nanowires randomly distributed for 20-nm-thick spacer layers. On the contrary, for 5-nm-thick layers, they are displayed well ordered, and arranged along a direction that forms 9° with the [001] vertical direction. For high capping temperatures (515°C), a similar arrangement is only found in the sample with 3-nm-thick capping layer, resulting in a stacking angle of about 16°.

The location of the minimum chemical potential at the growth surface (preferential nucleation sites) corresponds approximately to the positions of minimum strain energy calculated by FEA. Since experimental and simulated values are in good agreement, one can conclude that this estimate allows the successful prediction of the stacking angles of the studied nanowires [67, 68].

Strain from HRTEM images has also been mapped in the same stacked layers of nanowires. The InAs(P) strained area related to the wire decreases as the growth temperature of the capping layer is risen. This is due to As/P exchange process during wire capping, which is enhanced at higher temperatures. In general, the maximum strain is not in the wire center, which is associated with the existence of an inhomogeneous composition distribution in the wires. This effect confers an asymmetric profile to the strain spreading on the InP surface that explains the shifted nucleation sites found by FEA. Moreover, quantitative agreement between the strain values based on simulated maps and experimental ones is found (see Fig. 2.11) [69].

In summary, high-resolution phase-contrast and Z-contrast electron microscopy techniques, complemented by peak-finding methods, FEA, and EELS, have led to a better understanding of the horizontal InAs/InP quantum wires' growth process

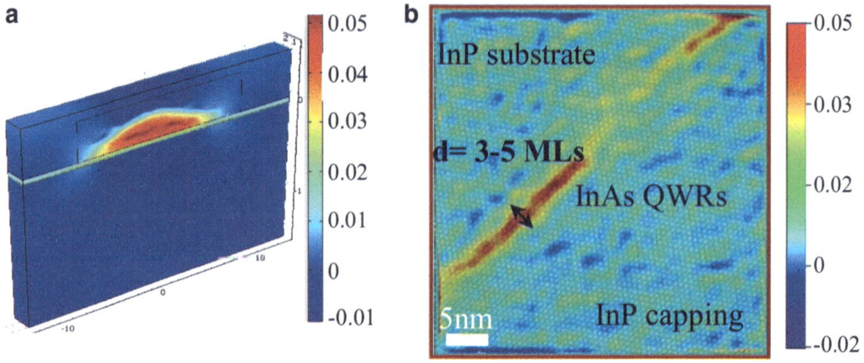

Fig. 2.11 Simulated [001] strain component of nanowires buried under 5 nm of spacer layer grown at 380°C (**a**), and the same strain component determined from the analysis of HRTEM images of a single layer of nanowires (**b**). Experimental contour plots of the [001] strain component measured for both samples with different growth temperatures show similar strain maxima (5.7%)

since its initial stages. They can be successfully applied to measure the structural characteristics of the wires at the atomic scale and to predict the preferential nucleation sites of single and stacked layers.

2.6 Analysis of the N Distribution in GaAsN

Recently, there has been great interest in extending the possibilities for band-structure engineering for GaAs-based materials by alloying with N. The addition of a small amount of N (lower than 5%) to Ga(In)As produces a substantial extension of the emission wavelength [70], allowing operation at the transmission windows of optical fibers. However, point defects such as the N–N split interstitials [71] and the AsGa antisites [72] have been identified as electron traps in Ga(In)NAs, and the local clustering of N atoms has been suggested to create a range of defect states within the energy gap [73]. Therefore, the characterization at the atomic scale of the behavior of N at the few percent level in GaAs is critical to understand and optimize the performance of this system.

In order to shed light on the behavior of N in GaAs, Ga(AsN) quantum wells with different N compositions (0.1, 0.4, 1, and 2.5%) grown between GaAs barriers by MBE were analyzed. The N composition was controlled by monitoring the optical intensity of the atomic N plasma emission. The high-resolution HAADF-STEM study was performed using an aberration-corrected VGHB501 dedicated STEM operated at 100 kV; the low magnification analysis was carried out in a JEOL 2500 (S)TEM, operated at 200 kV.

Initial HAADF-STEM analysis of the Ga(AsN) QWs at low magnification led to an interesting result. As can be observed in the inset in Fig. 2.12a, the Ga(AsN)

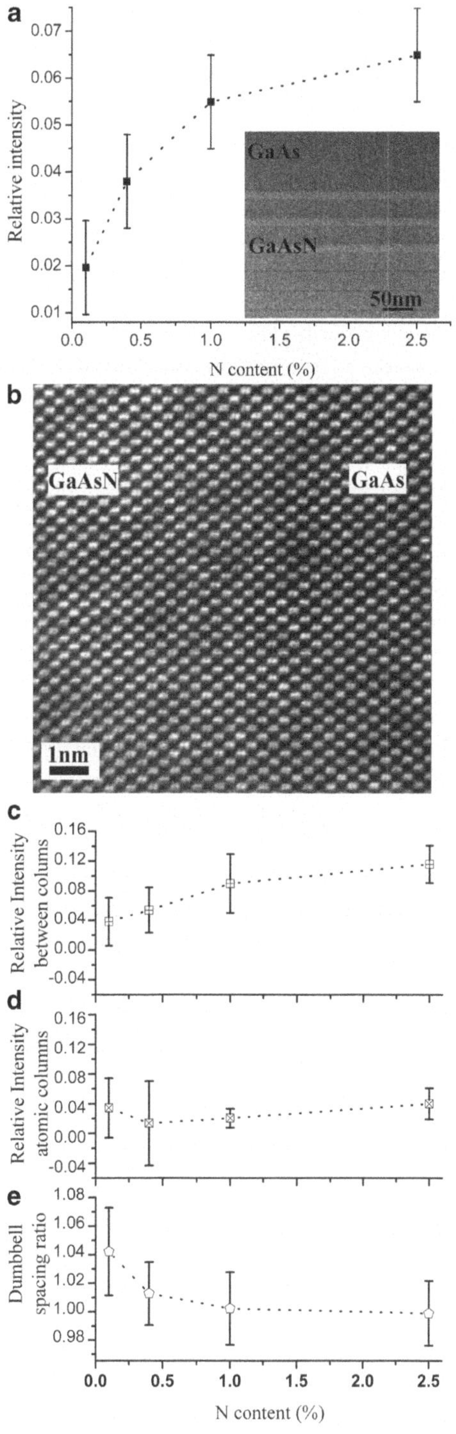

Fig. 2.12 (**a**) Plot of the evolution of the contrast $(I_{GaAsN} - I_{GaAs})/I_{GaAs}$ vs. N content from the low magnification images of Ga(AsN) QWs. The inset shows a micrograph of a QW with N = 2.5%; (**b**) high-resolution HAADF-STEM image of a QW with N = 2.5%; plot of the contrast measured from the high-resolution images separately from the valley between the atomic columns (**b**) and from the atomic columns (**c**) vs. N content; (**d**) plot of the experimental average dumbbell spacing ratio vs. N content (reproduced from [74])

QWs with N $= 2.5\%$ appear brighter than the GaAs barriers when imaged in HAADF-STEM despite the reduced average atomic number. As pointed by early studies by Perovic et al. [75] and Treacy et al. [76] and more recently by Grillo et al. [77], the local distortion of the lattice plays a major role in HAADF-STEM imaging of non-perfect crystals. Thus, the origin of the observed contrast may rely on the local distortion of the lattice due to the introduction of nitrogen, as pointed by Wu et al. [78]. Figure 2.12a shows a plot of the contrast GaAsN–GaAs [$C = (I_{\mathrm{GaAsN}} - I_{\mathrm{GaAs}})/I_{\mathrm{GaAs}}$] vs. the N content. It is observed that the evolution of contrast with the N content does not follow a linear behavior: the slope of the curve for low N concentration is relatively large and for the higher N content, the curve levels off. The origin of this dependence is not immediately clear for what is supposed to be a random substitutional alloy.

To investigate the origin of the observed evolution of contrast with increasing N content, the QWs were analyzed by high-resolution HAADF-STEM with aberration corrector, as can be observed in Fig. 2.12b for N $= 2.5\%$. The contrast from the valley between atomic columns and that from the atomic columns were separately measured and plotted vs. the N content in Fig. 2.12c, d, respectively. The intensity valley/peak was measured individually for each dumbbell in each image for up to six images per concentration; the error bars correspond to the standard error over all these measurements. It is clear that the contrast from the valleys between columns shows a strong increase and then flattens off for higher N concentration, similar to the behavior obtained from the low magnification images, whereas the increase in contrast from the atomic columns is negligible. This reveals that the increased intensity found in Ga(AsN) originates specifically from the valleys between atomic columns rather than from the columns themselves. Additionally, the average dumbbell spacing ratio was also measured from individual dumbbells, finding a nonlinear reduction when increasing the N content (Fig. 2.12e). This behavior is likely caused by the introduction of an additional strain component due to the incorporation of N in the lattice.

In order to explain the origin of these results, simulations of the HAADF-STEM images were performed for different likely N configurations in diluted nitride Ga(AsN). Thus, the existence of N–N split interstitials, N–As split interstitials, or $(\mathrm{AsGa{-}NAs})_{nn}$ pairs in Ga(AsN) has been suggested theoretically [79] and verified experimentally [80,81]. Also, the low solubility of N in GaAs [82] has been reported to induce the clustering of N atoms [83] in the alloy. Both interstitial N and N clustering could produce a change in the thermal diffuse scattering (TDS) from the Ga(AsN) region. According to this, the simulation of Ga(AsN) structures was performed, N atoms being incorporated in the group-V lattice sites (substitutional N) both randomly and in incipient clusters (2N cluster), and also in the form of the above interstitial N complexes (a schematic of these complexes is shown in Fig. 2.13).

The equilibrium lattice positions of the atoms in fully relaxed Ga(AsN) complexes used as input in the HAADF-STEM simulations were calculated using density functional theory within the local density approximation, as implemented in

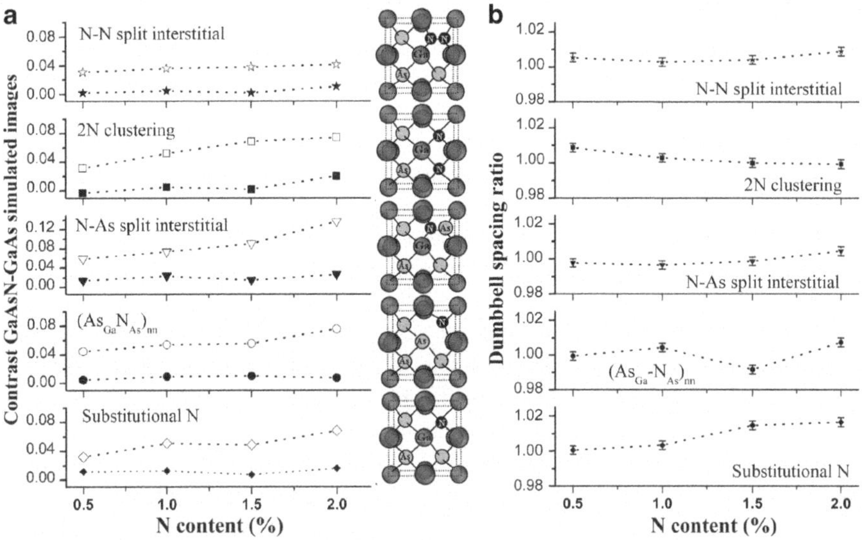

Fig. 2.13 (**a**) Plot of the contrast vs. nitrogen content from the simulated images of the different N-complexes. Open symbols correspond to the valley between the columns and solid symbols to the atomic columns; a schematic picture of each complex is included where the dark atoms represent N, the *dark gray* represent Ga and the *light gray* represent As; (**b**) averaged dumbbell spacing ratio measured from the simulated images for the different complexes vs. N content (reproduced from [74])

VASP [84]. All the calculations were performed using 64-atom cubic cells. All the atoms were relaxed to minimize the Feynman–Hellmann forces to below 0.02 eV/Å.

HAADF-STEM images were computed using the simulation software SICSTEM [85]. Elastic scattering was implemented following Kirkland's approach, while TDS was included using a local TDS absorptive potential approach. Images corresponding to 70-nm-thick structures (which is the thickness of the sample foil where the experimental images were acquired) for the different complexes were simulated, where the N complexes were located in random positions in three dimensions. The dumbbells' spacing and intensity valley/peak have been measured individually for each dumbbell in each image, and then averaged. It is worth noting that a mismatch of 2–4 times between STEM images and simulations [86] has been reported previously, associated with an overestimation of the degree of coherence in the electron source [87]. Because of this, the comparison of the experimental and simulated images will be based on the evolution of contrast with composition, not aiming to compare absolute values of intensity between theory and experiment.

The simulated images for substitutional N have shown an increased intensity coming from the N containing area, presumably due to the local distortion of the lattice. In order to separate the effect of the Z number from that due to the deformation of the structure, the simulation of a GaAs cell mimicking the local distortion of substitutional N was performed, where the N atom of a relaxed

Ga(AsN) structure was substituted by As without changing the position of the atoms. The simulation of such a structure exhibited exactly the same intensity as the simulated substitutional N in Ga(AsN). This result shows the strong effect of the static atomic displacements in HAADF-STEM images. This distortion of the lattice has an effect similar to freezing in lattice vibrations, enhancing the effective Debye–Waller factor of the host lattice [88] and, in some cases as shown in this work, it can dominate over the Rutherford-scattering component.

Figure 2.13 shows the plots of the contrast Ga(AsN)–GaAs (a) and the average dumbbell spacing ratio (b) vs. N content derived from the simulated images. As can be observed, the contrast for substitutional N shows an increasing slope for higher N contents, whereas the experimental study showed saturation for increasing N. The dumbbell spacing ratio also follows the opposite tendency with respect to the experimental study, increasing with N content. This different behavior reveals the existence of alternative N configurations in the alloy besides random substitutional N.

According to Fig. 2.13a, both the N–As split interstitial and the $(As_{Ga}–N_{As})_{nn}$ pair complexes show a pronounced increase in contrast for the higher N compositions, whereas the experimental results show the opposite. Moreover, the evolution of dumbbell spacing ratio with the N content does not agree with the experimental observations for any of these complexes. For the N–N split interstitial, both the Ga(AsN)–GaAs contrast and the dumbbell spacing ratio stay very stable with increasing N content because of the small local distortion when substituting As for two N. Consequently, the existence of this complex in the alloy cannot explain the experimental results.

The 2N substitutional complex is the only one for which the simulations agree with the experimental results. Thus, as can be observed, the evolution of contrast with composition increases with N content, with a reduction in the slope for higher N contents. The dumbbell spacing ratio, on the contrary, decreases for higher N contents with the same tendency as in the experimental curve in Fig. 2.12e. These results clearly point to the absence of interstitials in the alloy and to the onset of N clustering. Substitutional N pairing up to four neighbor position in $GaAs_{1-x}N_x$ alloys with $x < 0.025$ has been reported previously [89], in good agreement with our experimental observations.

2.7 Review on InN Nanostructures

Despite being traditionally a material discriminated against the rest of the family of III-nitrides, the interest in InN started growing after the re-establishment in 2002 of its band gap to a value close to 0.7 eV [90–93] from the 1.9 eV that was assumed before [94]. This observation opened a new field of possibilities, since the combination of InN with GaN and/or AlN to form ternary or quaternary alloys could lead to high-efficiency optoelectronic devices covering a wide range of wavelengths, from the infrared to the ultraviolet including the whole visible spectral range [95]. In the same way, the gradual improvement of the different growth techniques led to the fabrication of high-quality crystalline InN layers that presented largely improved electrical properties compared with the past [96,97]. The attempt to join the promising properties of InN with those derived from quantum confinement effects made the fabrication of InN-based nanostructures the next challenge. Unlike some other systems such as InAs, (InGa)As, or GaN, results about the development of InN nanostructures have emerged only during the last few years. It was not until 2003 when Briot et al. [98] first achieved the growth of InN QDs grown on GaN by metalorganic vapor phase epitaxy, and later in the same group, also on AlN and Si substrates [99]. In the same year, Cao et al. reported the growth of InN QDs and their shape and size dependence on the MBE deposition conditions [100]. After this, important improvements were achieved regarding smaller sizes and higher densities [101–107], and recently the emission, even very poor, of InN QDs [108, 109] and growth and optical properties of cubic InN dots [110] have been reported. Other types of InN nanostructures fabricated so far include single [111] and multiple [112–115] quantum wells, nanocolumns [116, 117], and nanowires [118–120]. However, even if the number of publications related to these nanomaterials has increased considerably during the last years, the number of them including characterization by HRTEM still remains scarce. Some of the results that will be described later in this section correspond to the first ones reported in the literature about structural and chemical characterization by (HR)TEM of InN QDs. This technique was also used to analyze the crystalline structure and quality [117], or the Si/InN and Si/AlN/InN interfaces in InN nanocolumns [116]; to confirm the cubic structure of the QDs [110] and the existence of oxide precipitates in InN nanostructures [121]; or to investigate the relaxation mechanisms of InN nanowires on GaN [120]. Our main contribution to the field is related to the characterization by means of electron beam-related techniques of uncapped and GaN-capped InN QD heterostructures grown on GaN/Sapphire substrates by metalorganic vapor phase epitaxy [98]. Even though a wider number of techniques were employed and an extensive work has been carried out, here we will mainly concentrate on the results derived from the HRTEM analysis.

The investigated InN/GaN quantum dots present a well-defined flat hexagonal shape with an average height of 12 nm and diameters of 72 and 120 nm for uncapped and capped QDs, respectively [122]. When observed in plan-view TEM (PVTEM) geometry, the presence of three-directional moire fringes patterns is revealed in all the uncapped QDs. These patterns were used to estimate the average degree of plastic relaxation of InN in the heterostructure, showing that it is close to a full relaxation, which was confirmed later by HRTEM [123]. As in other cases of highly mismatched heterostructures, this high degree of relaxation is related to the presence of a dense misfit dislocation (MD) network at the interface between the two materials. The application of conventional diffraction contrast TEM in plain view orientation to study the MD network is not feasible due to the large number of dislocations and their associated strain fields. However, these problems were overcome after the application of the GPA method [31] to high-resolution PVTEM micrographs, which allowed us to obtain strain maps at the atomic scale of the heterostructure and successfully perform a complete characterization of this MD network [124]. Figure 2.14a displays a HR-PVTEM micrograph of an uncapped

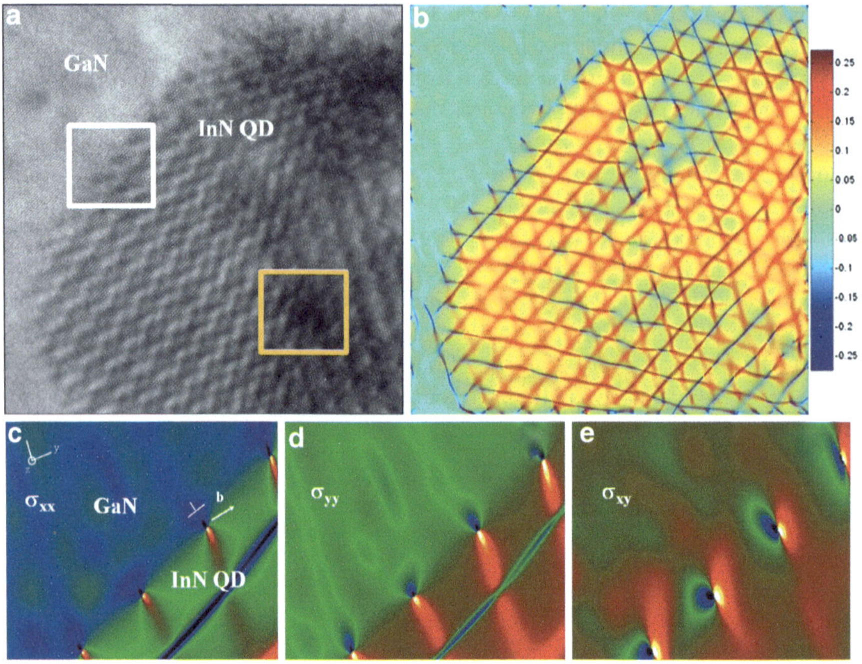

Fig. 2.14 (**a**) Planar view HRTEM micrograph of an uncapped InN QD and (**b**) the corresponding strain map calculated using the GPA method; (**c**–**e**) calculated stress components from the strain map in the area marked with a *white square*

InN QD, and Fig. 2.14b the associated strain map. In the latter, three families of MDs (red lines) are distinguished, without interaction between them or node generation, but forming a sort of "stars of David" network. Moreover, these maps also revealed the behavior of the MDs when they are close to the edge of the QD. From the strain maps and the corresponding elastic constants, stress maps were built in areas containing the boundary of the QD, as the one marked with a white square in Fig. 2.14a. The results, displayed in Fig. 2.14c–e, turned out to be in perfect agreement with the stress distribution around an edge threading dislocation. This means that not only a dense network of MDs exists at the interface InN/GaN, but there is also another network of threading dislocations surrounding the system, as a consequence of the bending of the MDs when they are close to the edge of the QDs due to the free surface forces.

Additionally, the above-mentioned Moiré fringes revealed extra information about the system: the interruption of the Moiré lines in the region close to the center of the QD (see arrow in Fig. 2.15a) is a consequence of the presence of a threading dislocation in that area. To assess if that TD is located in the GaN, the QD, or both, inverse fast Fourier transform of the area marked with an orange square in Fig. 2.14a was built considering the reflections that correspond to GaN and InN separately, and drew for both cases a Burgers circuit around the area of interest.

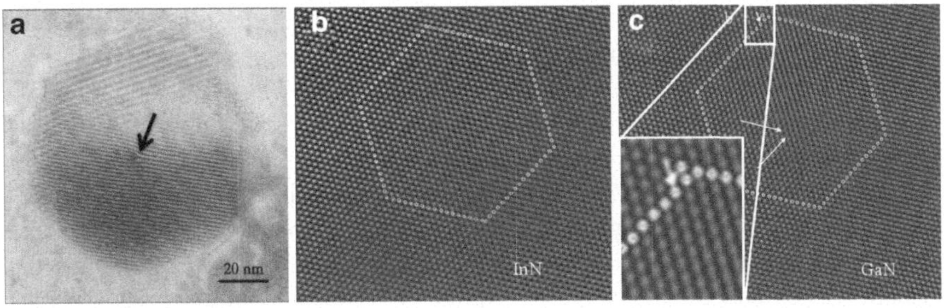

Fig. 2.15 (**a**) PVTEM micrograph of an uncapped InN QD showing one family of Moiré fringes. The *arrow* indicates a discontinuity of one of these lines. (**b** and **d**) Inverse fast Fourier transforms of the area containing the discontinuity marked with an *orange square* in Fig. 2.14a for InN and GaN, respectively

Figure 2.15b, c shows the results: for InN the circuit closes perfectly, whereas for GaN the same closure fails. This means that the TD is located in the GaN, does not propagate into the InN QD, and has at least an edge component of the Burgers vector. Finally, applying the invisibility criterion to conventional diffraction contrast micrographs confirms that the dislocation is pure edge type. This was not an isolated observation, but a general rule for every dot observed in large areas of the sample, so these pure edge-type TDs in GaN are assumed to be related to the nucleation mechanism of the InN QDs [125]. None of the classical nucleation models found in the bibliography (Volmer–Weber, Stranski–Krastanow, and Burton–Cabrera–Frank) fulfilled our observations, so even though the basic interactions that lead to the nucleation of the InN QDs on top of pure edge-type dislocations in GaN are still not clear, our observations suggest a new mechanism never reported before.

Finally, HRTEM was also used to investigate the aging process of the uncapped InN QDs after 36 months of exposure to environmental conditions [126, 127]. While the GaN capping layer prevented the capped QDs from interacting with air and then no structural changes are observed, the uncapped dots suffered a transformation from the *hcp* structure to cubic phases in the more external layers, while the core of the dot remained as wurtzite InN. These cubic phases turned out to be composed of indium oxide, and highly oxygen-doped cubic InN was proposed as an intermediate phase between the original wurtzite InN and the final product, In_2O_3.

2.8 Crystalline, Compositional, and Strain TEM Assessments of High-Quality Epilayers of Ternary and Quaternary III-N Alloys

Semiconductor stoichiometric solid solutions composed of half of N atoms and half of a single element or a random mix of specific elements of the group III of the periodic table (i.e., Al, Ga, and/or In) energetically tend to stabilize in the hexagonal wurtzite-type phase (2H-polytype) and constitute the base of numerous recent and foreseen advantageous technological applications. These new materials have mainly been developed through the last 15 years and were commonly fabricated as thick deposits on different substrates or pseudosubstrates such as bare or modified sapphire, silicon carbide, gallium arsenide, or silicon wafer surfaces since no technique for obtaining single crystals existed. Due to many deleterious effects for their growth, such as lattice mismatch-induced stress and other thermodynamical restrictions, these materials develop and suffer from the formation of extended defects and crystalline or compositional phase separations. Nevertheless, the use of MBE and other techniques is starting to allow epitaxial depositions of III-N binary compounds (AlN, GaN, and InN) and their ternary and quaternary alloys (AlGaN, InAlN, InGaN, and InAlGaN) in the form of single crystals. The current trend on semiconductors downscaling to get device miniaturization—and hence improvements by quantum confinement and/or electron accumulation effects—compels to obtain nano-sized motifs with high quality and controlled size and shape of layers, wires, or dots depending on the desired performance. In the described context, any heteroepitaxial volume of III-N under a critical thickness, even unfaulted, is subjected to a tetragonal-like distortion. The amount of this deformation can be calculated through the elastic theory, and its determination depends on the fine fourfold knowledge of material features, namely, lattice mismatch, surface orientation, chemical composition, and directional anisotropic stiffness constants. TEM-related techniques permit to collect important structural and chemical information to solve these problems. This section focuses on the importance of these aspects and describes some examples of how TEM can contribute to characterize heterosystems consisting of mismatched and nearly lattice-matched thin films of hexagonal InGaN, InAlN, AlGaN, and InAlGaN grown on the basal plane of unstrained GaN buffer thick layers grown on sapphire.

2.8.1 Previous Considerations

Films based on a hexagonal III-N compound or alloy, $In_xAl_yGa_zN$ $(x + y + z = 1)$, with a definite composition $(x,\ y,\ z)$, are often subjected to biaxial stresses. On polar surfaces, these stresses promote a strain parallel to the basal plane (ε_a) which is opposite and related to the perpendicular strain to this (0001) plane (ε_c). Provided that the quality of the layer is high enough (single phased with crystalline and compositional homogeneity and stoichiometry), the anisotropic elastic theory assumes that the experimental lateral strain coefficient $R^{\mathrm{S}}(x, y, z) = \varepsilon_c/\varepsilon_a$ should be equal to the biaxial strain relaxation coefficient $R^{\mathrm{B}}(x, y, z)$, which is dependent on the ratio of the crystal elastic constants, $C_{ij}(x, y, z)$. Lattice and elastic constants of ternaries and quaternaries can be calculated for any composition within contour conditions which correspond to the different binary III-N, using Vegard's law $(d_0\mathrm{In}_x\mathrm{Al}_y\mathrm{Ga}_z = x\ d_0\mathrm{InN} + y\ d_0\mathrm{AlN} + z\ d_0\mathrm{GaN})$. Table 2.1 lists reliable elastic constants statistically calculated as far as recently claimed accurate lattice constants

Table 2.1 Reference material parameters used in this work

	InN	AlN	GaN
2H- a_0 (Å)	3.53774	3.11197	3.18840
2H- c_0 (Å)	5.70374	4.98089	5.18500
C_{11} (GPa)	237 ± 7	395 ± 5	374 ± 4
C_{12} (GPa)	106 ± 4	137 ± 3	138 ± 4
C_{13} (GPa)	85 ± 3	107 ± 3	101 ± 4
C_{33} (GPa)	236 ± 6	404 ± 6	395 ± 5
C_{44} (GPa)	53 ± 3	117 ± 2	98 ± 3
Reference	[44]	[128]	[44]

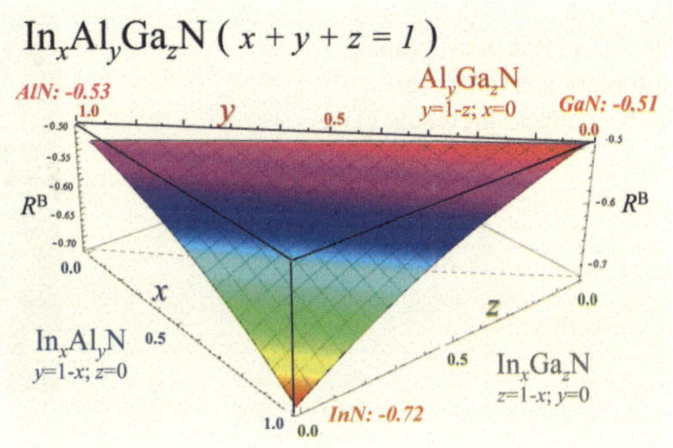

Fig. 2.16 Values of $R^B(x, y, z)$ for the whole $In_xAl_yGa_zN$ compositional range. These values should theoretically be coincident with the experimental ratio of lateral strains, $R^S(x, y, z) = \varepsilon_c/\varepsilon_a$

for hexagonal binaries: InN ($y = 0$ and $z = 0$), AlN ($x = 0$ and $z = 0$), and GaN ($x = 0$ and $y = 0$).

Figure 2.16 shows the expected values of R^B for any perpendicularly strained polar wurtzite $In_xAl_yGa_zN$. This plot and further calculations have been obtained by using the data in Table 2.1 and that $R^B(x, y, z) = -2C_{13}(x, y, z)/C_{33}(x, y, z)$. It is very important to note that the assumption of a linear trend for $C_{ij}(x, y, z)$ among binaries implies a bowing behavior for $R^B(x, y, z)$. From the assumption that $R^B(x, y, z) = R^S(x, y, z)$, compositions can be indirectly calculated by the use of the lattice parameters of the films. However, the erroneous applications of Vegard's rule have led to big underestimations and overestimations of sets of x, y (InAlN); x, z (InGaN); y, z (AlGaN); or x, y, z (InAlGaN) values in the past.

2.8.2 *Briefly, a Complete (S)TEM Study*

In a first stage, careful analyses of conventional diffraction contrast micrographs (DCTEM) and selected area electron diffraction (SAED) patterns might be enough to know if the analyzed III-N materials are of a sufficient crystalline quality and worth to deserve a deeper insight [44, 128–130]. In this case, being the nitride layers of good quality, sharp diffraction spots associated with the substrate materials together with a unique set coming from a single crystalline ternary or quaternary

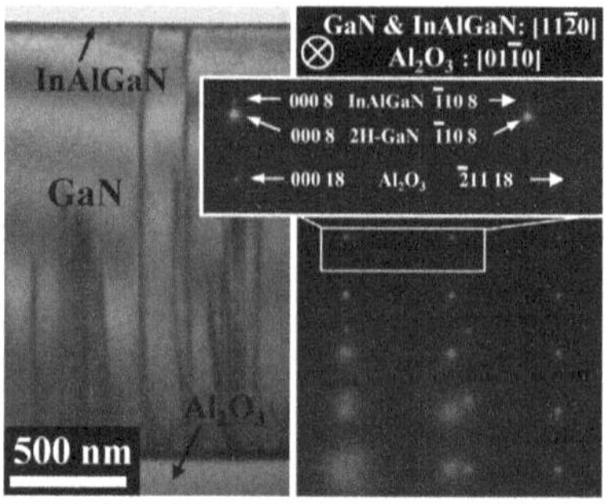

Fig. 2.17 XTEM micrograph of a good quality InAlGaN/GaN/Al_2O_3 heterostructure (*left*) and corresponding SAED pattern with indexed reflections (*right*)

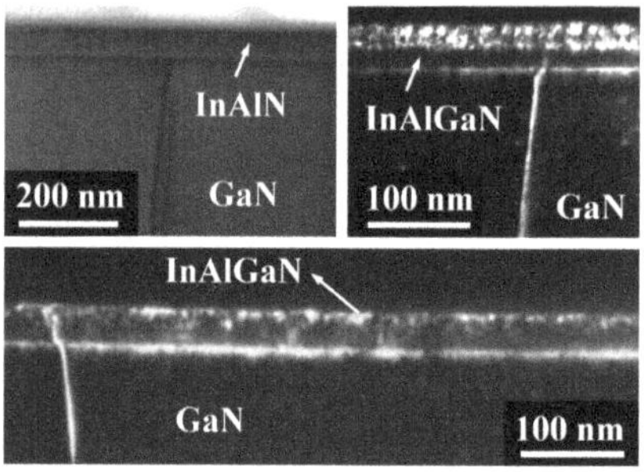

Fig. 2.18 BF (*top left*) and DF DCXTEM micrographs of ternary and quaternary III-N epilayers grown on GaN

III-N layer are only expected to be present in an electron diffraction pattern. Figure 2.17 presents a cross-section (XTEM) panoramic image representative of an InAlGaN/GaN/sapphire heterostructure, and its respective SAED pattern with reflections indexed into the inset.

It is well known that the low miscibility of binary compounds, especially when using InN, often leads to phase separation effects. Contrasts associated with these effects have been reported by means of DCTEM and SAED for ternary and quaternary alloys, for example, in extensive studies made by Prof. Mahajan's group in the last decade [131–137].

The only defects that are still acceptable in III-Ns for their adequate performance are the threading dislocations (TDs). It is not easy to get rid of them in film fabrications mainly due to the involved misfit strains among layers. Although some scientists found that they are electronically beneficial, the main goal is to have the lower possible densities; and a study of their null, occasional or common presence is feasible by DCTEM. If the semiconductor material is properly oriented in the electron microscope, the reflections 000N or $01\underline{1}N$ could be used on bright-field (BF), dark-field (DF), or weak-beam (WB) modes, and imaging is acquired for screw component (their Burger vectors are "c" or "$a + c$") or for edge plus mixed ("a" or "$a+c$") TDs, respectively. Figure 2.18 shows DC-XTEM images ($g = 0002$ near the [$11\underline{2}0$] zone axis) representative of non-defective alloys in InAlN/GaN and InAlGaN/GaN heterostructures, and of a horizontally phase-separated quaternary alloy (top right image). Screw component TDs presenting different behaviors, when stopping at or crossing the interfaces, are also visible.

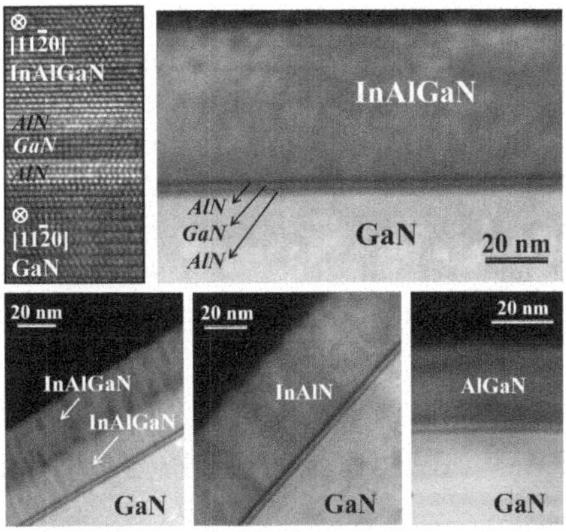

Fig. 2.19 HRTEM (*top left*) and HAADF images of III-N/GaN interfaces

HRTEM and STEM techniques are applied to demonstrate the quality of the material. While HRTEM images permit a structural comprehension of layers' sequence and their interfaces at an atomic level, to analyze the chemical homogeneity in layers, a combination of nano-probe energy dispersive X-ray (EDX) spectra and HAADF images, both in STEM mode, is useful and needed. EELS or energy-filtered TEM can also be helpful for this purpose. The 2D arranged atomic columns in the HRTEM detail of Fig. 2.19 illustrate an example of sharp interfaces among single crystals. The nano-interlayers labeled as "spacer" served as a buffer between the epilayer and the GaN in these cases. See in the same figure HAADF micrographs of various alloys with constant compositions [deduced from smooth and invariable atomic mass (Z)-contrasts], except for the one horizontally phase-separated InAlGaN bilayer. Note that for any of the epilayers claimed as homogeneous, for 10–20 EDX-STEM spectra taken for nano-sized regions of every ternary and quaternary alloy in different positions, the measured compositions spread less that 2% of the average for each film. This spread is on the level of the absolute error of the technique, giving this fact another hint on the uniformity of the composition.

Nevertheless, these examples of high homogeneity are still unusual in thin III-N epilayers and similar good qualities as those presented here have been rarely claimed, as, for example, in InAlN [138] or in InGaN [139], but for a very rough surface in the latter case. Some representative examples in the recent literature can be found for the state-of-the-art, but defective, ternary and quaternary III-Ns studied by some combination of HRTEM, HAADF, EDX, and EELS analyses. In this way, for InAlN nearly lattice matched to GaN, there are reports on horizontal composition separation [128,129], or on In segregation to vertical domain boundaries [140, 141] or to surface V-defects [142, 143]. For AlGaN, composition modulation [136], phase-separated domains [144] sometimes having tunnel defects and associated dislocations with erratic behaviors [145] have also been measured. On the other hand, surface-grown InGaN thin films showed bilayered structures of In- and Ga-rich InGaN parallel to the growth plane [146]. For InGaN/GaN and

AlGaN/GaN, most of these kinds of studies focused on the presence of V-defects, atomic fluctuations or clustering, and electron beam damage in multi-quantum wells [147, 148]. Only a few advanced TEM studies begin to consider InAlGaN alloys, a system where the control of composition is less well understood during the growth. Nano-scale In clustering in InAlGaN was observed by HRTEM [149], and a high-quality InAlGaN layer was elsewhere claimed [150]. However, the contents of both In and Al in the latter cases are in practice very low, with a lower probability of defects promoted by the lack of affinities among In, Ga, and Al atoms.

2.8.3 Analyses of Lateral Strains

Once the high compositional homogeneity is confirmed, the combination of chemical and lattice measurements can facilitate a complete analysis of the perpendicular deformations in the layers, assuming they behaved biaxially strained as expected by the elastic theory. Accurate measurements of lattice parameters are often supplied by high-resolution XRD analyses, but they can also be estimated by TEM with precisions and accuracies below 1% [151]. In this way, the comparison of the positions of sharp spots corresponding to nitrides with respect to those of a present unstrained internal standard in the same SAED pattern is needed for claiming such accuracies. In the example shown in Fig. 2.20 for an InGaN/GaN/AlN/sapphire heterostructure, having one pixel resolution for InGaN- and Al_2O_3-related far reflections, lattice measurements of the ternary alloy with an error of ±0.001 Å are possible [44]. In the case of phase-separated films or for those having small solid solution inhomogeneities, the variation in the composition can create a distribution of d-spacing for a crystallographic plane and this approach becomes complex, although multi-peak-finding retrieval methods can be applied as it is well accepted in XRD theory.

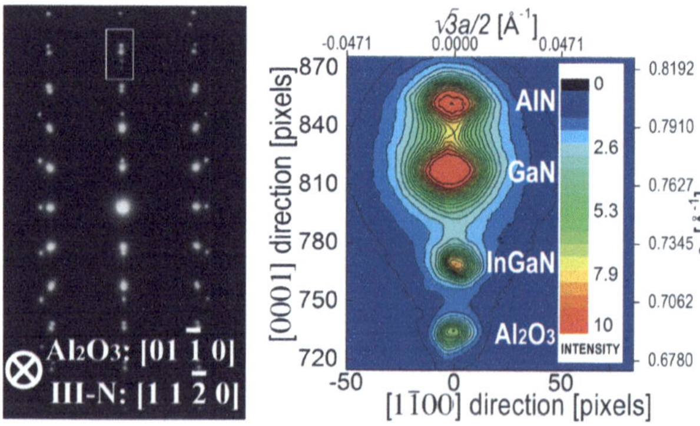

Fig. 2.20 Brightness profile for specific diffraction spots from a SAED pattern taken for a InGaN/GaN/AlN/Al_2O_3 heterostructure (adapted from [44])

To determine the compositions x, y, z, the general (2.4) can be built from the considerations given in Sect. 2.8.1. The only required parameters to solve (2.4) are those listed in Table 2.1 together with measured c and a lattice parameters. Moreover, a second input is needed for solving a system of two equations, which is straightforward for the ternary alloys being $x = 0$ if AlGaN; $y = 0$ if InGaN; or $z = 0$ if InAlN. In the case of a quaternary alloy, the value of any x/y, x/z, or y/z experimental ratio can be used

$$\frac{c-(xc_0^{\mathrm{InN}}+yc_0^{\mathrm{AlN}}+zc_0^{\mathrm{GaN}})}{xc_0^{\mathrm{InN}}+yc_0^{\mathrm{AlN}}+zc_0^{\mathrm{GaN}}}\times\frac{xa_0^{\mathrm{InN}}+ya_0^{\mathrm{AlN}}+za_0^{\mathrm{GaN}}}{a-(xa_0^{\mathrm{InN}}+ya_0^{\mathrm{AlN}}+za_0^{\mathrm{GaN}})}$$
$$=-2\frac{xC_{13}^{\mathrm{InN}}+yC_{13}^{\mathrm{AlN}}+zC_{13}^{\mathrm{GaN}}}{xC_{33}^{\mathrm{InN}}+yC_{33}^{\mathrm{AlN}}+zC_{33}^{\mathrm{GaN}}}. \quad (2.4)$$

When an agreement is found among calculated and experimental compositions, the high quality of the layers are confirmed since the expected elastic response is fulfilled. Then, a derivation of lateral strains is further comprehensible. Note that the lack of agreement between experimental and calculated compositions, even for single crystals, may indicate an absence of perfect stoichiometry of the layers [128]. A collection of the EDX-measured compositions of ternary and quaternary alloys, together with their thicknesses and strains are listed in Table 2.2. The utilized measured lattice parameters to make the composition calculations come from the best quality samples of our own research, and except for the micrographs related to the bilayers (inhomogeneous layers), the images shown in this section come from any of these listed samples [44, 128–130]. Note the agreement inside the EDX error as a proof of their good quality (i.e., expected elastic response).

Let us notice to finish that the solutions of compositions calculated with other well-accepted but simpler approaches for quaternaries [152] lead to unrealistic values, sometimes very far away from those demonstrated in this work. In conclusion, the application of Vegard's rule among binaries without any modification

Table 2.2 Composition and strain of the best quality InGaN and InAlGaN alloys from own studies with their reference source

Composition	Thick (nm)	ε_c (%)	ε_a (%)	Reference
$In_{0.42}Ga_{0.58}N$	25	0.81	−1.75	[44]
]$In_{0.51}Ga_{0.49}N$	25	0.21	−0.65	[44]
$In_{0.62}Ga_{0.38}N$	25	0.75	−1.07	[44]
$In_{0.70}Ga_{0.30}N$	25	0.33	−0.55	[44]
$In_{0.81}Ga_{0.19}N$	25	0.08	−0.14	[44]
$In_{0.18}Al_{0.82}N$	48	−0.01	0.01	[128]
$Al_{0.42}Ga_{0.58}N$	20	−0.43	0.81	[129]
$In_{0.10}Al_{0.67}Ga_{0.23}N$	50	−0.01	0.36	[129]
$In_{0.01}Al_{0.42}Ga_{0.57}N$	45	−0.36	0.67	[129]
$In_{0.04}Al_{0.39}Ga_{0.57}N$	45	−0.22	0.57	[129]
$In_{0.08}Al_{0.37}Ga_{0.55}N$	20	0.006	−0.04	[129]

should be the only method for calculating ternary and quaternary lattice and elastic constants.

Acknowledgements This work was supported by the European Commission (SANDiE European Network of Excellence NMP4-CT-2004-500101 and COST MP0805), the Spanish MICINN (TEC2005-05781-C03-02/MIC, TEC2008-06756-C03-02/TEC, and CONSOLIDER CSD2009-00013), and the Junta de Andalucía (Ref. P08-TEP-03516 and P09-TEP-5403) with EU-FEDER cofinancing. The authors would like to thank their collaborators and other coauthors of the articles reviewed here (A.G. Taboada, A.J. Papworth, A. Yáñez, B. Alén, D. Fuster, D.G. Morgan, F. de la Pena, J.M. Ripalda, L. González, L. Kirste, M.F. Chisholm, M.H. Du, M.H. Gass, M. Hopkinson, M. Varela, M. Walls, N.D. Browning, O. Ambacher, P.J. Goodhew, Q.M. Ramasse, R. Aidam, S.B. Zhang, S. Hauguth-Frank, S.J. Pennycook, T. Lim, V. Cimalla, V. Lebedev, and Y. González).

References

1. M. De Graef, *Introduction to Conventional Transmission Electron Microscopy* (Cambridge University Press, Cambridge, 2003)
2. D.B. Williams, C.B. Carter, *Transmission Electron Microscopy: A Textbook for Materials Science* (Springer, New York, 2009)
3. S.J. Pennycook et al., in *The Oxford Handbook of Nanoscience and Technology: Frontiers and Advances: in Three Volumes*, ed. by A.V. Narlikar, Y.Y. Fu (Oxford University Press, New York, 2010), Volume II: Materials
4. S.J. Pennycook, M. Varela, C.J.D. Hetherington, A.I. Kirkland, MRS Bull. **31**, 36 (2006)
5. M. Varela, A.R. Lupini, K.V. Benthem, A.Y. Borisevich, M.F. Chisholm, N. Shibata, E. Abe, S.J. Pennycook, Annu. Rev. Mater. Res. **35**, 539 (2005)
6. S.J. Pennycook, S.D. Berger, R.J. Culbertson, J. Microsc. Oxford **144**, 229 (1986)
7. A. Howie, J. Microsc. Oxford **117**, 11 (1979)
8. P.D. Nellist, S.J. Pennycook, J. Microsc. **190**, 159 (1998)
9. S.J. Pennycook, D.E. Jesson, Phys. Rev. Lett. **64**, 938 (1990)
10. P.D. Nellist, et al., Science **305**, 1741 (2004)
11. M. Galluppi, A. Frova, M. Capizzi, F. Boscherini, P. Frigeri, S. Franchi, A. Passaseo, Appl. Phys. Lett. **78**, 3121 (2001)
12. J. Shumway, A.J. Williamson, A. Zunger, A. Passaseo, M. DeGiorgi, R. Cingolani, M. Catalano, P. Crozier, Phys. Rev. B **64**, 125302 (2001)
13. H. E, P.D. Nellist, S. Lozano-Perez, D. Ozkaya, J. Phys. Conf. Ser. **241**, 012067 (2010)
14. L.J. Allen, A.J. D'Alfonso, S.D. Findlay, J.M. LeBeau, N.R. Lugg, S. Stemmer, J. Phys. Conf. Ser. **241**, 012061 (2010)
15. E. Carlino, S. Modesti, D. Furlanetto, M. Piccin, S. Rubini, A. Franciosi, Appl. Phys. Rev. **83**, 662 (2003) doi:10.1063/1.1592314
16. D. Zhi, P.A. Midgley, R.E. Dunin-Borkowski, B.A. Joyce, D.W. Pashley, A.L. Bleloch, P.J. Goodhew, in *Group-IV Semiconductor Nanostructures*, vol. 832, ed. by L. Tsybeskov, D.J. Lockwood, C. Delerue, M. Ichikawa (2005), p. 105
17. S.I. Molina, D.L. Sales, P.L. Galindo, D. Fuster, Y. Gonzalez, B. Alen, L. Gonzalez, M. Varela, S.J. Pennycook, Ultramicroscopy **109**, 1315 (2009)
18. R.F. Egerton, *Electron Energy-Loss Spectroscopy in the Electron Microscope* (Plenum, New York, 1986)
19. P.L. Galindo, S. Kret, A.M. Sanchez, J.Y. Laval, A. Yanez, J. Pizarro, E. Guerrero, T. Ben, S.I. Molina, Ultramicroscopy **107**, 1186 (2007)
20. J.M. LeBeau, S.D. Findlay, L.J. Allen, S. Stemmer, Phys. Rev. Lett. **100**, 206101 (2008)
21. J.M. LeBeau, S. Stemmer, Ultramicroscopy **108**, 1653 (2008)

22. S.I. Molina, M.P. Guerrero, P.L. Galindo, D.L. Sales, M. Varela, S.J. Pennycook, J. Electron Microsc. **60**, 29 (2011)
23. M. Takeda, H. Ina, S. Kobayashi, J. Opt. Soc. Am. **72**, 156 (1982)
24. R. Bierwolf, M. Hohenstein, F. Phillipp, O. Brandt, G.E. Crook, K. Ploog, Ultramicroscopy **49**, 273 (1993)
25. P.H. Jouneau, A. Tardot, G. Feuillet, H. Mariette, J. Cibert, J. Appl. Phys. **75**, 7310 (1994)
26. H. Seitz, M. Seibt, F.H. Baumann, K. Ahlborn, W. Schroter, Phys. Stat. Sol. A Appl. Res. **150**, 625 (1995)
27. R. Kilaas, S. Paciornik, A.J. Schwartz, L.E. Tanner, J. Comput. Assist. Microsc. **6**, 129 (1994)
28. M.D. Robertson, J.M. Corbett, J.B. Webb, J. Jagger, J.E. Currie, Micron **26**, 521 (1995)
29. A. Rosenauer, S. Kaiser, T. Reisinger, J. Zweck, W. Gebhardt, D. Gerthsen, Optik **102**, 63 (1996)
30. S. Kret, P. Ruterana, A. Rosenauer, D. Gerthsen, Phys. Stat. Sol. B Basic Res. **227**, 247 (2001)
31. M.J. Hÿtch, E. Snoeck, R. Kilaas, Ultramicroscopy **74**, 131 (1998)
32. D.J. Bone, H.A. Bachor, R.J. Sandeman, Appl. Opt. **25**, 1653 (1986)
33. M. Takeda, J. Suzuki, J. Opt. Soc. Am. A Opt. Image Sci. Vis. **13**, 1495 (1996)
34. M.J. Hytch, J.L. Putaux, J.M. Penisson, Nature **423**, 270 (2003)
35. A.M. Sanchez, P.L. Galindo, S. Kret, M. Falke, R. Beanland, P.J. Goodhew, Microsc. Microanal. **12**, 285 (2006)
36. E. Guerrero, P. Galindo, A. Yanez, T. Ben, S.I. Molina, Microsc. Microanal. **13**, 320 (2007)
37. K. Tillmann, M. Lentzen, R. Rosenfeld, Ultramicroscopy **83**, 111 (2000)
38. M.J. Hytch, T. Plamann, Ultramicroscopy **87**, 199 (2001)
39. M.M.J. Treacy, J.M. Gibson, J. Vac. Sci. Tech. B **4**, 1458 (1986)
40. S. Shusterman, A. Raizman, A. Sher, A. Schwarzman, O. Azriel, A. Boag, Y. Rosenwaks, P.L. Galindo, Y. Paltiel, EPL **88** (2009)
41. A. Kovacs, K. Sato, V.K. Lazarov, P.L. Galindo, T.J. Konno, Y. Hirotsu, Phys. Rev. Lett. **103** (2009)
42. P.D. Robb, A.J. Craven, Ultramicroscopy **109**, 61 (2008)
43. K. Tomioka, J. Motohisa, S. Hara, T. Fukui, Nano Lett. **8**, 3475 (2008)
44. F.M. Morales, D. Gonzalez, J.G. Lozano, R. Garcia, S. Hauguth-Frank, V. Lebedev, V. Cimalla, O. Ambacher, Acta Mater. **57**, 5681 (2009)
45. J.H. Chung, G.D. Lian, L. Rabenberg, IEEE Electron Dev. Lett. **31**, 854 (2010)
46. F. Hue, M. Hytch, F. Houdellier, E. Snoeck, A. Claverie, Mater. Sci. Eng. B Adv. Funct. Solid State Mater. **154**, 221 (2008)
47. F. Hue, M. Hytch, H. Bender, F. Houdellier, A. Claverie, Phys. Rev. Lett. **100**, 156602 (2008)
48. HREM Research, http://www.hremresearch.com/Eng/plugin/PPAEng.html
49. M. Henini, M. Bugajski, Microelectron. J. **36**, 950 (2005)
50. H.Y. Liu, M.J. Steer, T.J. Badcock, D.J. Mowbray, M.S. Skolnick, P. Navaretti, K.M. Groom, M. Hopkinson, R.A. Hogg, Appl. Phys. Lett. **86** (2005)
51. M. Yano, M. Ashida, Y. Iwai, M. Inoue, Appl. Surf. Sci. **41–42**, 457 (1989)
52. S.I. Molina et al., Appl. Phys. Lett. **91** (2007)
53. A.M. Sanchez et al., Nanotechnology **21** (2010)
54. S.I. Molina, A.M. Beltran, T. Ben, P.L. Galindo, E. Guerrero, A.G. Taboada, J.M. Ripalda, M.F. Chisholm, Appl. Phys. Lett. **94** (2009)
55. J.M. Garcia, L. Gonzalez, M.U. Gonzalez, J.P. Silveira, Y. Gonzalez, F. Briones, J. Cryst. Growth **227**, 975 (2001)
56. L. Gonzalez, J.M. Garcia, R. Garcia, F. Briones, J. Martinez-Pastor, C. Ballesteros, Appl. Phys. Lett. **76**, 1104 (2000)
57. D. Fuster, L. Gonzalez, Y. Gonzalez, M.U. Gonzalez, J. Martinez-Pastor, J. Appl. Phys. **98** (2005)
58. D. Fuster, B. Alen, L. Gonzalez, Y. Gonzalez, J. Martinez-Pastor, M.U. Gonzalez, J.M. Garcia, Nanotechnology **18** (2007)
59. D. Fuster, L. Gonzalez, Y. Gonzalez, J. Martinez-Pastor, T. Ben, A. Ponce, S. Molina, Eur. Phys. J. B **40**, 433 (2004)

60. D. Fuster, M.U. Gonzalez, L. Gonzalez, Y. Gonzalez, T. Ben, A. Ponce, S.I. Molina, Appl. Phys. Lett. **84**, 4723 (2004)
61. M.S. Skolnick, D.J. Mowbray, Annu. Rev. Mater. Res. **34**, 181 (2004)
62. D.L. Sales, M. Varela, S.J. Pennycook, P.L. Galindo, L. Gonzalez, Y. Gonzalez, D. Fuster, S.I. Molina, Nanotechnology **1**, 325706 (2010)
63. G.R. Liu, S.S.Q. Jerry, Semicond. Sci. Tech. **17**, 630 (2002)
64. P. Liu, Y.W. Zhang, C. Lu, Phys. Rev. B **68**, 195314 (2003)
65. Q.X. Pei, C. Lu, Y.Y. Wang, J. Appl. Phys. **93**, 1487 (2003)
66. Y.W. Zhang, S.J. Xu, C.H. Chiu, Appl. Phys. Lett. **74**, 1809 (1999)
67. S.I. Molina et al., J. Nanosci. Nanotechnol. **8**, 3422 (2008)
68. S.I. Molina et al., Nanotechnology **17**, 5652 (2006)
69. T. Ben et al., Microsc. Microanal. **14**, 344 (2008)
70. M. Kondow, K. Uomi, A. Niwa, T. Kitatani, S. Watahiki, Y. Yazawa, Jpn. J. Appl. Phys. 1 Regular Papers Short Notes Rev. Papers **35**, 1273 (1996)
71. S. Dhar, N. Halder, A. Mondal, Thin Solid Films **515**, 4427 (2007)
72. W.M. Chen, I.A. Buyanova, C.W. Tu, H. Yonezu, Acta Phys. Pol. A **108**, 571 (2005)
73. R. Intartaglia et al., IEE Proc. Optoelectron. **151**, 365 (2004)
74. M. Herrera et al., Phys. Rev. B **80**, 125211 (2009)
75. D.D. Perovic, C.J. Rossouw, A. Howie, Ultramicroscopy **52**, 353 (1993)
76. M.M.J. Treacy, J.M. Gibson, K.T. Short, S.B. Rice, Ultramicroscopy **26**, 133 (1988)
77. V. Grillo, E. Carlino, F. Glas, Phys. Rev. B **77** 054103 (2008)
78. X. Wu, M.D. Robertson, J.A. Gupta, J.M. Baribeau, J. Phys. Condens. Matter **20**, 075215 (2008)
79. S.B. Zhang, A. Zunger, Appl. Phys. Lett. **71**, 677 (1997)
80. P. Krispin, V. Gambin, J.S. Harris, K.H. Ploog, J. Appl. Phys. **93**, 6095 (2003)
81. M. Reason, H.A. McKay, W. Ye, S. Hanson, R.S. Goldman, V. Rotberg, Appl. Phys. Lett. **85**, 1692 (2004)
82. I.H. Ho, G.B. Stringfellow, J. Cryst. Growth **178**, 1 (1997)
83. P.R. Chalker, T.J. Bullough, M. Gass, S. Thomas, T.B. Joyce, J. Phys. Condens. Matter **16**, S3161 (2004)
84. G. Kresse, J. Furthmuller, Phys. Rev. B **54**, 11169 (1996)
85. J. Pizarro, P.L. Galindo, E. Guerrero, A. Yanez, M.P. Guerrero, A. Rosenauer, D.L. Sales, S.I. Molina, Appl. Phys. Lett. **93**, 153107 (2008)
86. D.O. Klenov, S. Stemmer, Ultramicroscopy **106**, 889 (2006)
87. R.A. Herring, Ultramicroscopy **106**, 960 (2006)
88. C.R. Hall, P.B. Hirsch, G.R. Booker, Phil. Mag. **14**, 979 (1966)
89. A. Carvalho et al., Phys. B Condens. Matter **401**, 339 (2007)
90. T. Matsuoka, H. Okamoto, M. Nakao, H. Harima, E. Kurimoto, Appl. Phys. Lett. **81**, 1246 (2002)
91. J. Wu, W. Walukiewicz, K.M. Yu, J.W. Ager, E.E. Haller, H. Lu, W.J. Schaff, Y. Saito, Y. Nanishi, Appl. Phys. Lett. **80**, 3967 (2002)
92. M. Hori, K. Kano, T. Yamaguchi, Y. Saito, T. Araki, Y. Nanishi, N. Teraguchi, A. Suzuki, Phys. Stat. Sol. (B) **234**, 750 (2002)
93. V.Y. Davydov et al., Phys. Stat. Sol. (B) **230**, R4 (2002)
94. T.L. Tansley, C.P. Foley, J. Appl. Phys. **59**, 3241 (1986)
95. A.G. Bhuiyan, A. Hashimoto, A. Yamamoto, J. Appl. Phys. **94**, 2779 (2003)
96. N. Khan, A. Sedhain, J. Li, J.Y. Lin, H.X. Jiang, Appl. Phys. Lett. **92**, 172101 (2008)
97. T.B. Fehlberg, G.A. Umana-Membreno, C.S. Gallinat, G. Koblmüller, S. Bernardis, B.D. Nener, G. Parish, J.S. Speck, Phys. Stat. Sol. (C) **4**, 2423 (2007)
98. O. Briot, B. Maleyre, S. Ruffenach, Appl. Phys. Lett. **83**, 2919 (2003)
99. S. Ruffenach, B. Maleyre, O. Briot, B. Gil, Phys. Stat. Sol. (C) **2**, 826 (2005)
100. Y.G. Cao, M.H. Xie, Y. Liu, Y.F. Ng, H.S. Wu, S.Y. Tong, Appl. Phys. Lett. **83**, 5157 (2003)
101. E. Dimakis et al., Phys. Stat. Sol. (C) **3**, 3983 (2006)
102. S. Ruffenach, O. Briot, M. Moret, B. Gil, Appl. Phys. Lett. **90**, 153102 (2007)

103. C. Shen, H. Lin, H. Lee, C. Wu, J. Hsu, S. Gwo, Thin Solid Films **494**, 79 (2006)
104. M. Kumar, B. Roul, T.N. Bhat, M.K. Rajpalke, N. Sinha, A.T. Kalghatgi, S.B. Krupanidhi, Adv. Sci. Lett. **3**, 379 (2010)
105. Y. Romanyuk, R. Dengel, L. Stebounova, S. Leone, J. Cryst. Growth **304**, 346 (2007)
106. J.Y. Chen et al., Appl. Phys. Lett. **92**, 162103 (2008)
107. A. Yoshikawa, N. Hashimoto, N. Kikukawa, S.B. Che, Y. Ishitani, Appl. Phys. Lett. **86**, 153115 (2005)
108. C.Y. Chen, L. Lee, S.K. Tai, S.F. Fu, W.C. Ke, W.C. Chou, W.H. Chang, M.C. Lee, W.K. Chen, Jpn. J. Appl. Phys. **48**, 031001 (2009)
109. S. Yuanping, W. Hui, W. Lili, Z. Shuming, Y. Hui, C. Yong-Hoon, J. Kor. Phys. Soc. **57**, 128 (2010)
110. Y.K. Huang, C.P. Liu, Y.L. Lai, C.Y. Wang, Y.F. Lai, H.C. Chung, Appl. Phys. Lett. **91**, 091921 (2007)
111. S.B. Che, W. Terashima, T. Ohkubo, M. Yoshitani, N. Hashimoto, K. Akasaka, Y. Ishitani, A. Yoshikawa, Phys. Stat. Sol. (C) **2**, 2258 (2005)
112. W. Lin, S. Li, J. Kang, Appl. Phys. Lett. **96**, 101115 (2010)
113. A. Yoshikawa, S.B. Che, W. Yamaguchi, H. Saito, X.Q. Wang, Y. Ishitani, E.S. Hwang, Appl. Phys. Lett. **90**, 073101 (2007)
114. H. Naoi, M. Kurouchi, D. Muto, S. Takado, T. Araki, T. Miyajima, H. Na, A.Y. Nanishi, Phys. Stat. Sol. (a) **203**, 93 (2006)
115. S. Che, A. Yuki, H. Watanabe, Y. Ishitani, A. Yoshikawa, Appl. Phys. Expr. **2**, 021001 (2009)
116. E. Calleja, J. Ristić, S. Fernández-Garrido, L. Cerutti, M.A. Sánchez-García, J. Grandal, A. Trampert, U. Jahn, G. Sánchez, A. Griol, B. Sánchez, Growth, morphology, and structural properties of group-III-nitride nanocolumns and nanodisks. Phys. Status Solidi B **244**, 2816–2837 (2007)
117. C. Denker, J. Malindretos, F. Werner, F. Limbach, H. Schuhmann, T. Niermann, M. Seibt, A. Rizzi, Phys. Stat. Sol. C – Curr. Top. Solid State Phys. **5**(6), 1706 (2008)
118. Y.L. Chang, F. Li, Z. Mi, J. Vacuum Science & Technology B **28**(3) DOI: 10.1116/1.3292560 (2010)
119. J. Zhang, B. Xu, F. Jiang, Y. Yang, J. Li, Phys. Lett. A **337**, 121 (2005)
120. Y.H. Kim, H.J. Park, K. Kim, C.S. Kim, W.S. Yun, J.W. Lee, M.D. Kim, Appl. Phys. Lett. **95**, 033112 (2009)
121. S.Y. Kwon et al., Appl. Phys. Lett. **91**, 234102 (2007)
122. J.G. Lozano, D. González, A.M. Sánchez, D. Araújo, S. Ruffenach, O. Briot, R. García, Phys. Stat. Sol. (C) **3**, 1687 (2006)
123. J.G. Lozano, A.M. Sanchez, R. Garcia, D. Gonzalez, O. Briot, S. Ruffenach, Appl. Phys. Lett. **88**, 151913 (2006)
124. A.M. Sánchez, J.G. Lozano, R. García, M. Herrera, S. Ruffenach, O. Briot, D. González, Adv. Funct. Mater. **17**, 2588 (2007)
125. J.G. Lozano, A.M. Sanchez, R. García, D. González, D. Araujo, S. Ruffenach, O. Briot, Appl. Phys. Lett. **87**, 263104 (2005)
126. D. González, J.G. Lozano, M. Herrera, N.D. Browning, S. Ruffenach, O. Briot, R. Garíca, J. Appl. Phys. **105**, 013527 (2009)
127. D. González, J.G. Lozano, M. Herrera, F.M. Morales, S. Ruffenach, O. Briot, R. García, Nanotechnology **21**, 185706 (2010)
128. J.M. Mánuel, F.M. Morales, J.G. Lozano, D. González, R. García, T. Lim, L. Kirste, R. Aidam, O. Ambacher, Acta Mater. **58**, 4120 (2010)
129. J.M. Mánuel, F.M. Morales, R. García, R. Lim, L. Kirste, R. Aidam, O. Ambacher, Cryst. Growth and Design **11**, 2588 (2011)
130. J.M. Mánuel, F.M. Morales, J.G. Lozano R. García, T. Lim L. Kirste, R. Aidam, O. Ambacher, Phys. Stat. Sol. (C) **8**(7–8), 2500–2502 (2011)
131. F.Y. Meng, M. Rao, N. Newman, R. Carpenter, S. Mahajan, Acta Mater. **56**, 4036 (2008)
132. F.Y. Meng, M. Rao, N. Newman, S. Mahajan, Acta Mater. **56**, 5552 (2008)
133. M. Rao, D. Kim, S. Mahajan, Appl. Phys. Lett. **85**, 1961 (2004)

134. M. Rao, N. Newman, S. Mahajan, Scripta Mater. **56**, 33 (2007)
135. A.N. Westmeyer, S. Mahajan, Appl. Phys. Lett. **79**, 2710 (2001)
136. A. Wise, R. Nandivada, B. Strawbridge, R. Carpenter, N. Newman, S. Mahajan, Appl. Phys. Lett. **92**, 261914 (2008)
137. A.T. Wise, D.W. Kim, N. Newman, S. Mahajan, Scripta Mater. **54**, 153 (2006)
138. L. Zhou, D.A. Cullen, D.J. Smith, M.R. McCartney, A. Mouti, M. Gonschorek, E. Feltin, J.F. Carlin, N. Grandjean, Appl. Phys. Lett. **94**, 121909 (2009)
139. N. Li, S.J. Wang, E.H. Park, Z.C. Feng, H.L. Tsai, J.R. Yang, I. Ferguson, J. Cryst. Growth **311**, 4628 (2009)
140. S.L. Sahonta et al., Appl. Phys. Lett. **95**, 021913 (2009)
141. L. Zhou, D.J. Smith, M.R. McCartney, D.S. Katzer, D.F. Storm, Appl. Phys. Lett. **90**, 081917 (2007)
142. T. Kehagias et al., Appl. Phys. Lett. **95**, 071905 (2009)
143. Z.L. Miao et al., Appl. Phys. Lett. **95**, 231909 (2009)
144. M.E. Hawkridge, Z. Liliental-Weber, H.J. Kim, S. Choi, D. Yoo, J.H. Ryou, R.D. Dupuis, Appl. Phys. Lett. **94**, 071905 (2009)
145. M.E. Hawkridge, Z. Liliental-Weber, H.J. Kim, S. Choi, D. Yoo, J.H. Ryou, R.D. Dupuis, Appl. Phys. Lett. **94**, 171912 (2009)
146. J. Park, S.I. Baik, D.S. Ko, S.H. Park, E. Yoon, Y.W. Kim, J. Electron. Mater. **38**, 518 (2009)
147. Z.C. Feng, J.R. Yang, A.G. Li, I.T. Ferguson, in *III-Nitride Device and Nano-Engineering*, ed. by Z.C. Feng (Imperial College Press, London, 2008)
148. M. Shiojiri, Chiang Mai J. Sci. **35**, 25 (2008)
149. J. Wu et al., Nanotechnology **17**, 1251 (2006)
150. H. Okuno, M. Takeguchi, K. Mitsuishi, Y. Irokawa, Y. Sakuma, K. Furuya, Surf. Interface Anal. **40**, 1660 (2008)
151. E. Mugnaioli, G. Capitani, F. Nieto, M. Mellini, Am. Mineral. **94**, 793 (2009)
152. C. Williams, T. Glisson, J. Hauser, M. Littlejohn, J. Electron. Mater. **7**, 639 (1978)

Chapter 3
Hot Electron Energy and Momentum Relaxation

Naci Balkan

Abstract This chapter reviews common techniques used to determine the energy and momentum relaxation rates of hot carriers in semiconductors where the carrier heating is achieved by either the application of an electrical field or optical excitation. Steady-state spectral techniques aim at determining the carrier temperature as a function of the optical or electrical power. Transient measurements provide the relaxation rates via probing the time evolution of the hot carrier population.

3.1 Introduction

The physics of hot carriers deals with nonlinear effects associated with the energy and momentum exchange between electrons and holes, between carriers and acoustic and optical phonons, impurities, imperfections, and photons. As the sizes of functional devices shrink to dimensions comparable with the de-Broglie wavelength, even the application of a small voltage produces very large electric fields in the device. Hence Ohm's law breaks down and nonlinear effects become the norm.

Following Ridley [1,2], the average energy of electron $\langle E \rangle$ increases with applied electric field, F, as

$$\langle E \rangle = e\mu F^2, \tag{3.1}$$

and the momentum $\langle p \rangle$, as

$$\langle p \rangle = eFt, \tag{3.2}$$

where μ is the electron mobility. However, the gain in the energy and momentum of the electrons from the applied field is balanced by the losses via the elastic and

N. Balkan (✉)
School of Computer Science and Electronic Engineering, University of Essex, Essex, UK
e-mail: balkan@essex.ac.uk

A. Patanè and N. Balkan (eds.), *Semiconductor Research*, Springer Series in Materials Science 150, DOI 10.1007/978-3-642-23351-7_3,

inelastic scattering. Therefore, the simple rate equations describing the conservation of momentum and energy are

$$\frac{\mathrm{d}p}{\mathrm{d}t} = \frac{\mathrm{d}(m^*v)}{\mathrm{d}t} = eF - \frac{m^*v}{\tau_\mathrm{m}}, \tag{3.3}$$

$$\frac{\mathrm{d}E}{\mathrm{d}t} = eFv - \frac{\Delta E}{\tau_\mathrm{E}}, \tag{3.4}$$

where m^* is the effective mass, $\Delta E = E - E_0$ is the excess energy with respect to that at zero field (E_0), and τ_m and τ_E are the momentum and energy relation times.

In the steady state

$$v = \frac{e\tau_\mathrm{m}}{m^*}F = \mu F, \tag{3.5}$$

$$E = E_0 + \frac{(eF)^2\tau_\mathrm{m}\tau_\mathrm{E}}{m^*}. \tag{3.6}$$

If the conduction band is parabolic, the effective mass does not depend on the energy of the electron. The energy dependences of the momentum and energy relaxation times can be represented by [2]

$$\tau_\mathrm{m} = AE^s \quad \tau_\mathrm{E} = BE^{sr}, \tag{3.7}$$

$$E = \left(\frac{(eF)^2AB}{m^*}\right)^{s/1-s-r}, \tag{3.8}$$

$$v = \frac{eFA}{m^*}\left(\frac{(eF)^2AB}{m^*}\right)^{s/1-s-r}. \tag{3.9}$$

Hence the field dependences of energy and the drift velocity are determined by the energy dependences of the scattering time constants. Furthermore, steady-state solutions are possible only for $1 - s - r > 0$. For polar interactions, both s and r are positive quantities and this condition is not satisfied. However, for nonpolar optical phonons s is negative, $r = 1 + s$, and for equipartitioned acoustic modes $s = r$ and the stability condition is met.

There are a number of spectroscopic and nonspectroscopic techniques employed in the measurements of hot electron energy relaxation rates. The techniques that are based on the heating of carriers utilize the following experimental methods: optical heating and photoluminescence (PL) in the steady state, electrical heating and PL in the steady state [3–5], time-resolved photoluminescence with femtosecond and picosecond resolutions using direct detection (streak camera), photoluminescence correlation spectroscopy, photoluminescence up conversion spectroscopy [3, 4], time-resolved Raman spectroscopy [6–8], hot electron noise spectroscopy measurements [9, 10], heat pulse measurements [11], mobility mapping [12], and Shubnikov de Haas (SdH) oscillations for the acoustic phonon-assisted energy

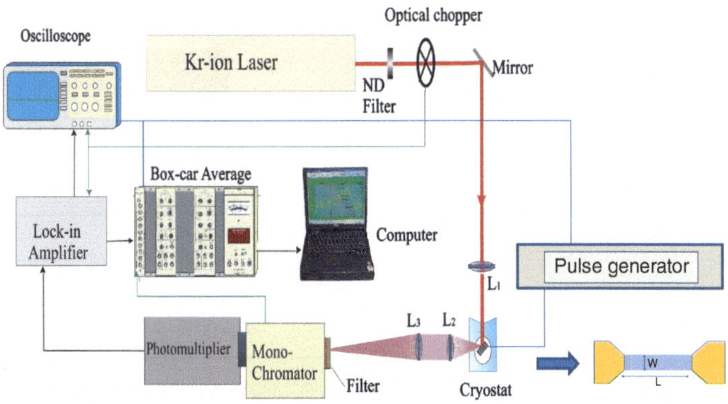

Fig. 3.1 Experimental setup for Hot Electron Photoluminescence experiments

relaxation [13]. We start with the review of the steady-state hot-electron energy relaxation measurements where the carriers are heated up by the application of high electric fields.

3.2 Hot Electron Photoluminescence in the Steady State

A typical experimental setup for the hot electron photoluminescence (HEPL) measurements is shown in Fig. 3.1. In the HEPL experiments, samples are fabricated in the form of two terminal simple bars (GaAs in the experiment shown in Fig. 3.1 [5]) with the ohmic contacts formed by alloying Indium or Au–Ge–Ni. In order to minimize the Joule heating, short electric field pulses (<1-μs duration) with a small duty cycle ($<10^{-5}$) are applied along the layers. In the experiment shown in Fig. 3.1, a low intensity 676-nm line of a Krypton laser is used for excitation. The disturbance of the Fermi level by photoexcitation is minimal. The PL is collected and dispersed by simple gating techniques only when the electric field is on, as illustrated in Fig. 3.2. The photoluminescence is therefore due to the recombination of the majority hot electrons with the photoexcited minority holes.

The high energy tail of a typical PL spectrum is shown in Fig. 3.3. In the figure, the logarithm of the PL intensity is plotted against the emitted photon energy at a lattice temperature of $T_L = 13.6\,\text{K}$ [5]. At finite electric fields and for photon energies $\hbar\omega > E_G + E_F$, the logarithm of the PL intensity drops linearly with increasing energy while it drops sharply at around $E_G + E_F$ when no electric field is applied. Most of the luminescence arises from an electron recombining with a heavy hole whose distribution in **k**-space extends much further than that for the electrons as a consequence of the disparity in effective masses.

Figure 3.4 shows the energy–momentum diagram together with the electron and hole distributions at $F > 0$, and the photoexcitation and the recombination

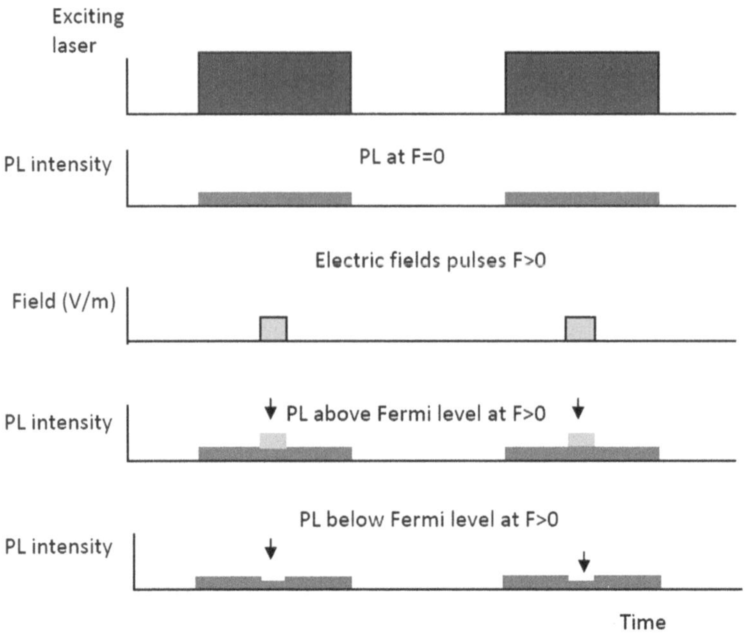

Fig. 3.2 Schematics of photoluminescence ($F = 0$) and the gating technique for the measurement of the HEPL (when the finite electric field is applied to the sample). The *arrows* indicate the points where the HEPL pulses are detected

processes. Also illustrated in the figure is the Fermi-Dirac distribution function for the electrons when $F = 0$ and $F > 0$. At energies a few $k_B T_e$ above the Fermi Energy, E_F, the distribution function can be represented by a Maxwellian. Since the radiative transitions are vertical in the momentum space, (because of the low momentum of the photon) electrons up to high energies will be radiatively coupled to holes at low energies. The shape of the spectral dependences will depend on the product of the electron and hole distribution functions. Thus for a given energy E and a Maxwell–Boltzmann distribution, the most rapidly varying part of the spectral function $f(E)$ is

$$f(E) = \exp\left[-E\left(\frac{1}{k_B T_e} + \frac{m_e^*}{m_{hh}^*}\frac{1}{k_B T_{hh}}\right)\right], \tag{3.10}$$

and the effective temperature is

$$\frac{1}{T_{eff}} = \frac{1}{T_e} + \frac{m_e^*}{m_{hh}^*}\frac{1}{T_{hh}}, \tag{3.11}$$

where T_e and T_{hh} are the electron and heavy hole temperatures, m_e^* and m_{hh}^* are the electron and heavy hole effective masses and spherical parabolic bands are assumed.

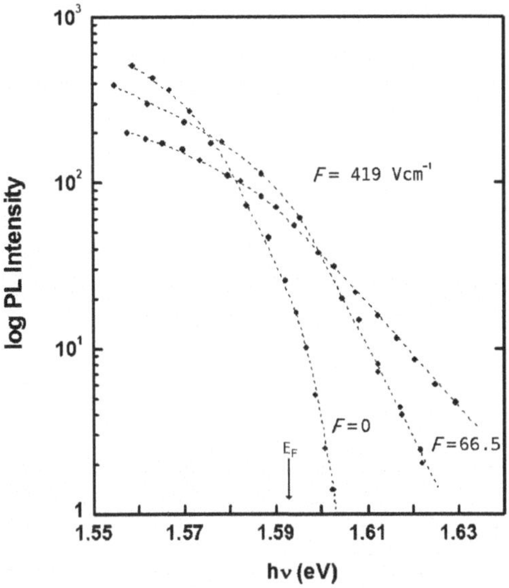

Fig. 3.3 High-energy tail of the PL spectrum of a GaAs sample at a lattice temperature, $T_L = 6\,\text{K}$ [5]

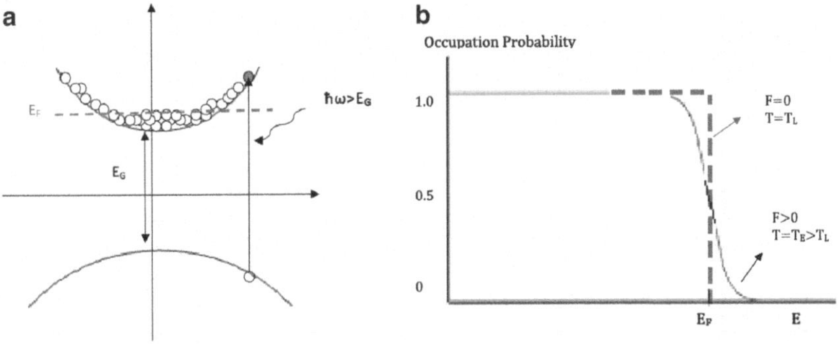

Fig. 3.4 **(a)** Energy momentum diagram showing the electron and hole distributions at $F > 0$ and the photoexcitation and the recombination processes and **(b)** the Fermi-Dirac distribution function for the electrons when $F = 0$ and $F > 0$

If the electron and hole scattering equalize the temperatures in the steady state, then

$$T_{\text{eff}} = \frac{m^*_{\text{hh}}}{m^*_{\text{hh}} + m^*_{\text{e}}} T_{\text{e}}. \tag{3.12}$$

For GaAs $m^*_{\text{hh}} \cong 7.3\, m^*_{\text{e}}$, thus taking $T_{\text{e}} = T_{\text{eff}}$ corresponds to an error of about 12%. Therefore, the linear behavior is consistent with a Maxwellian distribution characterized by a carrier temperature $T_{\text{E}} > T_{\text{L}}$, i.e.,

$$I_{\text{PL}} \propto \exp\left(-\frac{\hbar\omega}{k_{\text{B}} T_{\text{e}}}\right), \tag{3.13}$$

where I_{PL}, $\hbar\omega$, T_{e}, and k_{B} are the PL intensity, emitted photon energy, electron temperature, and Boltzmann constant, respectively. The slopes of the linear regions can therefore be used to extract the temperature of the electrons which are in thermal equilibrium with each other through electron–electron scattering [14]. Power input per electron at these carrier temperatures can also be obtained from the simultaneously measured I–V characteristics. Since, in the steady state, power input per carrier is equal to that lost to the lattice through scattering processes, we can obtain the energy loss rate per electron, provided we assume that the carrier density remains constant as a function of the electron temperature. This is shown in Fig. 3.5. Also plotted in the same figure is the calculated energy loss rate in the form [5]

$$P = \frac{\hbar\omega_{\mathrm{LO}}}{\langle\tau\rangle}\exp\left(-\frac{\hbar\omega_{\mathrm{LO}}}{k_{\mathrm{B}}T_{\mathrm{e}}}\right), \tag{3.14}$$

where P, $\hbar\omega_{\mathrm{LO}}$, and $\langle\tau\rangle$ are the power input per electron, the longitudinal optical (LO) phonon energy (36.5 meV), and the average time for an electron to emit an LO phonon (108 fs), respectively.

It is evident from the figure that the observed loss rates are about an order of magnitude slower over most of the temperature range. One immediate reason for the slow loss rates may be the overestimation of the carrier density in the calculation

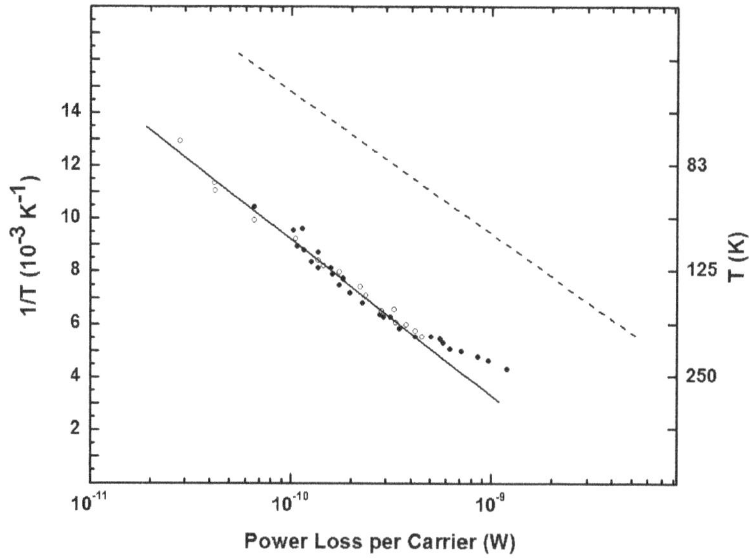

Fig. 3.5 Inverse electron temperature vs. power loss per carrier for a modulation doped GaAs/GaAlAs QW sample at $T = 13.6\,\mathrm{K}$ [5]. The *full line* drawn over the experimental points has a slope of ~36 meV. *Open circles* represent the electron temperatures obtained directly from the HEPL experiments. *Full circles* (dark) are the electron temperatures obtained from the high-field mobility measurements as described in the text. The *broken line* is the loss rate calculated by using (3.14)

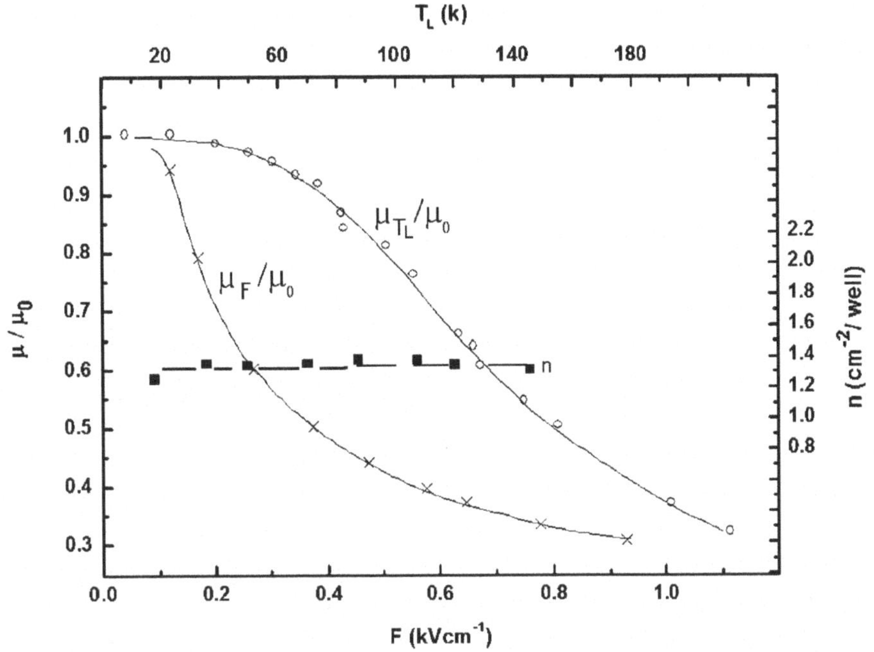

Fig. 3.6 Upper horizontal axis: normalized Hall mobility as a function of the lattice temperature at low electric fields (*open circles*). Lower horizontal axis: normalized Hall mobility (*crosses*) and sheet carrier concentration (*ellipses*) as a function of electric field at $T = 6\,\text{K}$ [5]

of the loss per carrier from the experimental data by assuming a constant electron density over the whole field range. Indeed it was pointed out by Van Welzenis et al. [15] and by Schubert et al. [16] that, as a result of the increased probability of the capture of hot carriers at interface states, the sheet carrier density at high electric fields can be much lower than that at low fields. Therefore, the carrier concentration has to be measured as a function of the applied electric field. The results are illustrated in Fig. 3.6. It is clear from the figure that for this particular sample, the carrier concentration in the field range investigated remains constant within the experimental accuracy [5].

3.3 Mobility Mapping

Another technique for the determination of the electron temperatures as a function of power input is to compare the measured Hall mobility vs. the electric field curve with the measured Hall mobility vs. the lattice temperature curve, as shown also in Fig. 3.6. Here μ_{T_L}/μ_0 and μ_F/μ_0 are the lattice temperature and field-dependent mobilities normalized with respect to the low-field mobility (μ_0) at

$T_L = 6\,\mathrm{K}$. It is clear from the figure that at electric fields $F > 100\,\mathrm{V\,cm^{-1}}$ and at lattice temperatures $T_L > 40\,\mathrm{K}$, where the dominant energy loss is by LO phonon scattering, both curves show similar behavior. This suggests that the mobility change, which is caused by the applied electric field, can be described in terms of electric-field-induced electron heating [17].

Electron temperatures can therefore be determined accurately by comparing directly the μ_{T_L}/μ_0 and μ_F/μ_0 curves. The results obtained from this comparison, both when the sample is in darkness and when illuminated by the 676-nm line of a Krypton laser, are also plotted in Fig. 3.5. The agreement between the power-loss curves obtained from the two experiments is excellent. The slight bending of the experimental curve indicates that the effective scattering time depends somewhat on the carrier temperature. At intermediate carrier temperatures, at $T_e < 150\,\mathrm{K}$, however, where the slope of the loss curve is close to the LO phonon energy (36 meV), an effective scattering time can be obtained by replacing $\langle\tau\rangle$ with τ_{eff} in (3.14). It is then found that $\tau_{\mathrm{eff}} = 1.17\,\mathrm{ps}$, which should be compared with the scattering time $\langle\tau\rangle = 108\,\mathrm{fs}$.

3.4 Nonequilibrium Phonons (Hot Phonons)

The difference between the experimental results and the theory can be explained in terms of the nonequilibrium phonon model [18–21]. Figure 3.7 shows schematically the energy relaxation in the presence and absence of hot phonons. The model uses the concept of phonon temperature to describe the phonon number per active mode. The rate of production can be written as

$$\frac{\mathrm{d}n(T_p)}{\mathrm{d}t} = G_e(n(T_p), T_e) - \frac{n(T_p)}{\tau_p}, \tag{3.15}$$

where $n(T)$ is the phonon number per mode at temperature T, i.e., the Bose–Einstein factor, $G_e(n(T_p), T_e)$ is the production rate from the electron gas, and τ_p is the phonon lifetime.

Equation (3.15) can be solved for the steady state to give $n(T_p)$ as a function of the electron temperature. The power input per electron is eFv_d, where F is the electric field and v_d is the drift velocity, and in the steady state is equal to the energy-loss rate per electron to optical phonons via the polar interaction, namely

$$eFv_d = \frac{\mathrm{d}E}{\mathrm{d}t} \quad \text{with} \quad v_d = e\tau_m\frac{F}{m^*}, \tag{3.16}$$

where $\tau_m(n(T_p), T_e)$ is the momentum relaxation time averaged over the distribution. Hence, we can obtain a relation between the electron temperature and field. For $\hbar\omega/k_B T_e > 1$, we have

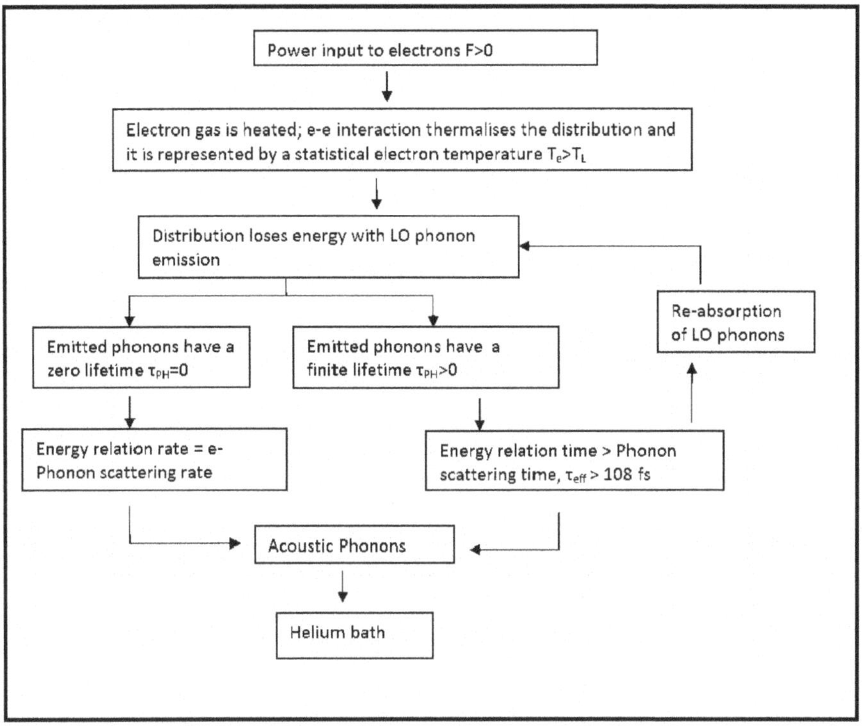

Fig. 3.7 Schematic representation of hot electron energy relaxation process

$$\frac{\mathrm{d}E}{\mathrm{d}t} = -\left(\frac{\hbar\omega}{\tau_{\mathrm{eff}}}\right)\left[\exp\left(\frac{\hbar\omega}{k_{\mathrm{B}}T_{\mathrm{e}}}\right) - \exp\left(\frac{\hbar\omega}{k_{\mathrm{B}}T_{\mathrm{L}}}\right)\right], \tag{3.17}$$

where τ_{eff} is an effective energy relaxation time, which takes into account all the hot-phonon effects. For intra-subband scattering in a deep quantum well via the interaction with the LO interface mode in the nondegenerate case [22]:

$$\tau_{\mathrm{eff}} = \tau_{02\mathrm{D}}(1+\gamma), \tag{3.18}$$

where $\tau_{02\mathrm{D}} = \tau_0(\hbar\omega/E_{\mathrm{L}})^{1/2}$, $E_{\mathrm{L}} = h^2\pi^2/2m^*L^2$, and $\gamma = (\tau_{\mathrm{p}}/2\tau_0)(1/q_0L)(\pi k_{\mathrm{B}}\ T_{\mathrm{e}}/\hbar\omega)^{1/2}(n/N_{\mathrm{C}})[1 - \exp(-(\hbar\omega)/(k_{\mathrm{B}}T_{\mathrm{e}}))]$. Here $\tau_{02\mathrm{D}}$ is the characteristic scattering time in the quantum well, q_0 is the phase matching wave vector $((h^2q_0^2)/\ (2m^*) = \hbar\omega)$, L is the well width, n is the electron density, N_{C} is the effective density of states $(m^*k_{\mathrm{B}}T_{\mathrm{e}}/\pi h^2L)$ and τ_0 is the time constant for polar interaction. Note that in general τ_{eff} is dependent on the electron temperature and therefore dependent on the field. In the highly degenerate case [22], $\tau_{\mathrm{eff}} = \tau_{02\mathrm{D}}(E_{\mathrm{F}}/\hbar\omega(1+\gamma'))$, E_{F} is the Fermi energy and $E_{\mathrm{F}}/\hbar\omega > 1$, and

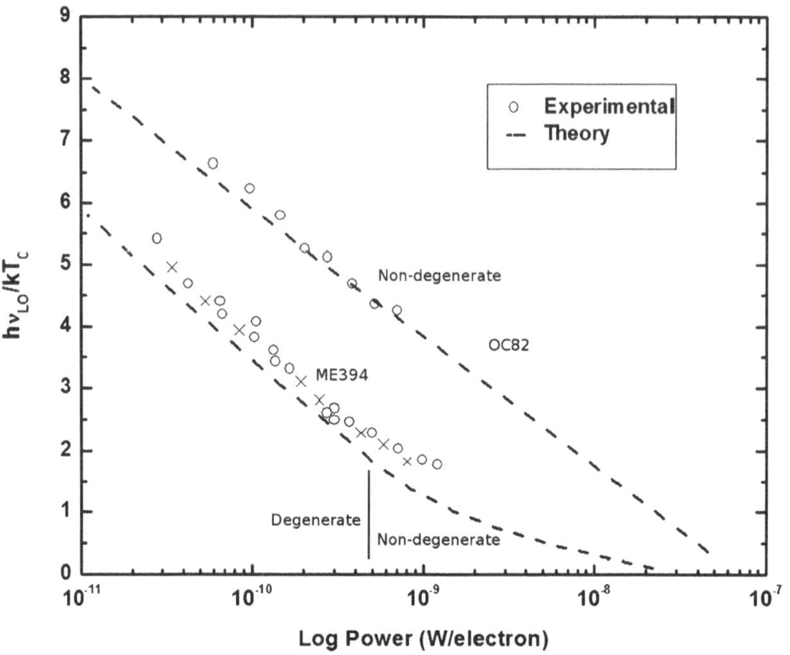

Fig. 3.8 Energy loss rates obtained from (3.15) to (3.19) by using the sample parameters investigated. Experimental results are also illustrated [5, 18, 19]

$$\gamma' = \frac{\tau_p}{2\tau_0} \frac{1}{q_0 L} \frac{1}{((E_F/\hbar\omega) - (1/2))^{1/2}}. \tag{3.19}$$

Note that τ_{eff} is independent of electron temperature.

The effect of the production of hot phonons and the phonon lifetime on energy relaxation is shown schematically in Fig. 3.7. In Fig. 3.8, the calculated electron energy relaxation rates are plotted together with the experimental results. The agreement between the theory and the experiment is excellent for the sample (OC82). The data for the degenerate sample (ME394), although fitting well to the calculations at low carrier temperatures, tend to deviate from the calculated relaxation rates at higher carrier temperatures, $T > 180$ K. The bending of energy relaxation rates (ERR) was reported by various groups [4]. One possible reason for the behavior appears to be the reduction of the lifetime of the LO phonons at high temperatures. Kash et al. [7] reported $\tau_p = 3.5$ ps at $T = 300$ K, compared with $\tau_p = 7$ ps at $T = 77$ K, in bulk GaAs. Evidently, the energy relaxation rates calculation is also very sensitive to the phonon lifetime. If a phonon lifetime of $\tau_p = 3.5$ ps is used in the calculations at $T > 180$ K where the experimental curve bends upwards, we see that the calculations agree well with the experimental results at high temperatures too.

In the late 1980s, Ridley [22] suggested that at high electric fields when the nonequilibrium phonons (hot phonons) are produced, the randomisation of the hot phonon distribution may take place via the elastic scattering of phonons interacting for instance with interface imperfections and well-width fluctuations. Hence the drift of hot phonons can be neglected. Conventional treatments of hot phonon effects consider a forward displaced distribution of the nonequilibrium phonons in momentum space that arises from the drift of hot electrons. The emission and reabsorption of drifting nonequilibrium phonons are shown to reduce the overall energy relaxation rates for electrons [23].

When the momentum of the phonons is parallel to the electron motion, then the emission and the subsequent reabsorption of the phonon by the electron will leave the electron momentum unaltered. The implicit assumption is that the momentum relaxation rate for phonons is the same as the lifetime as determined by the phonon decay process. On the other hand, if the phonon relaxes its momentum faster than the decay rate, the change in the electron momentum between the emission and reabsorption can be quite large. It was shown by Gupta et al. [24] that in 2D structures, the almost inevitable presence of interface roughness as well as alloy fluctuations provide an additional source for the elastic scattering and hence the momentum relaxation of phonons. Therefore, the drift of the nonequilibrium phonon population may be negligible especially in the region where hot phonon effects are important [23, 24]. Consequently, the production of hot phonons not only reduces the energy relaxation rate, as predicted by conventional theories, but also enhances the momentum relaxation rate. Experimental work on the hot electron energy and momentum relaxation of electrons in 2D GaAs suggests strongly that this is indeed the case at least in these structures [20, 21, 23], where the electron drift velocity at high fields is reduced drastically (see Sect. 3.5).

3.5 Cooling of Hot Electron Hole Plasma by LO Phonon Emission Using the CW and Transient Photoluminescence Spectroscopy

In this section, we address the cooling of optically heated hot carriers via the emission of LO phonons with reference to dilute nitrides (GaInNAs/GaAs) quantum wells (QWs) and nitrogen-free InGaAs/GaAs samples using the steady-state and time-resolved photoluminescence techniques [25]. The excitation intensity dependence of carrier temperature, determined via the CW-PL spectroscopy, can be used to estimate the LO-phonon energies that are emitted by the cooling hot electron–hole plasma. The carrier cooling rates in the quantum wells can be probed directly by using time-resolved photoluminescence (TRPL). The TRPL reveals that the carrier dynamics in the GaInNAs material are dominated by nonradiative processes [25].

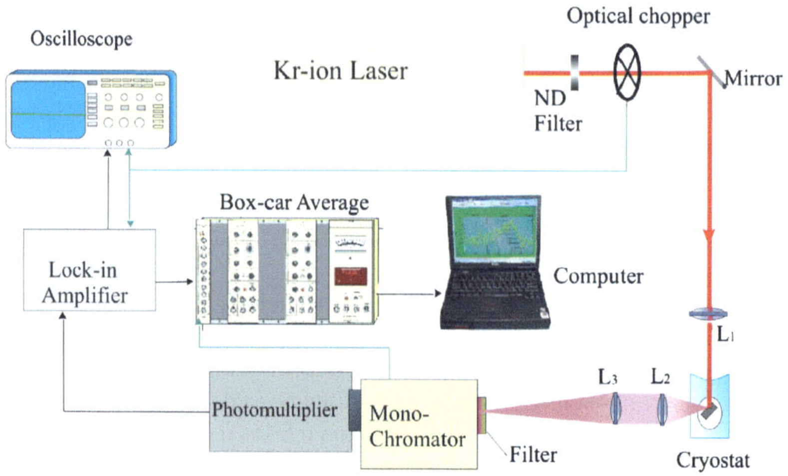

Fig. 3.9 Experiment setup for CW-PL measurement, where L_1 is the lens to focus the excitation laser on the sample; L_2 and L_3 are the lenses to collect and focus the emission of sample into the monochromator [25]

3.5.1 Optical Heating in the Steady State Using CW Photoluminescence

Samples studied in [25] are #752, which consists of sequentially grown 9-nm-thick $In_{0.2}Ga_{0.8}As$ and $Ga_{0.8}In_{0.2}N_{0.015}As_{0.985}$ QWs, and #1087, which consists of sequentially grown 6-nm-thick $In_{0.3}Ga_{0.7}As$ and $Ga_{0.7}In_{0.3}N_{0.015}As_{0.985}$ multiple quantum wells (MQW). For CW-PL measurements, the samples are excited using, for example, the 647-nm line of a Krypton-ion laser. The beam is mechanically chopped for the purpose of collecting the signal using lock-in techniques for a better signal-to-noise ratio. It is focused onto the sample defining a spot size of diameter of about 0.25 mm. The CW-PL is spectrally dispersed by a high resolution monochromator and detected using a cooled InP/InGaAs near infrared photomultiplier in conjunction with a lock-in amplifier or a boxcar averager. Figure 3.9 shows the block diagram of the setup for the CW-PL experiments. The effects of optical pump power upon the steady-state PL emission investigated at $T_L = 4.4$ K for the 9-nm sample indicate that in low excitation conditions, the "hot" carriers tend to thermalise rapidly, reaching thermal equilibrium with the lattice before they radiatively recombine. However, as the excitation rate increases, populations of hot carriers are generated, which are manifested in the emission spectra as high-energy exponential tails. The carrier temperature T_e of the hot electrons can be obtained from the exponential tail by assuming a Maxwellian distribution given by (3.13). In Fig. 3.10, the electron temperatures are extracted from emission spectra by plotting the logarithm of intensity against photon energy, and taking the gradient of the slope in the high-energy tail region.

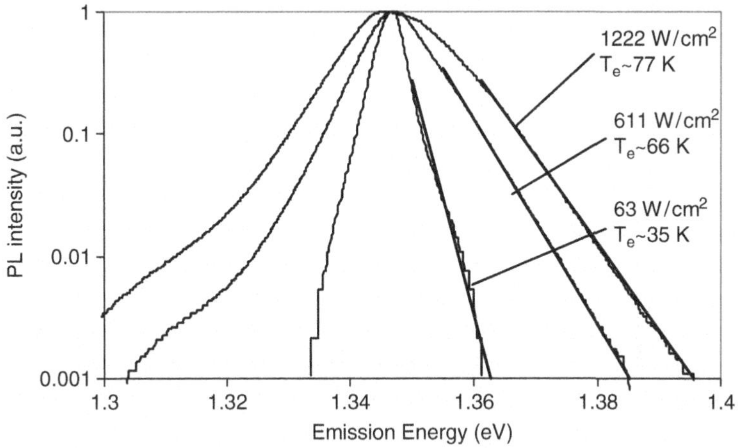

Fig. 3.10 Normalized InGaAs PL spectra taken at different pump powers, plotted on a semilog scale, illustrating how the carrier temperature is estimated [25]

Under steady-state excitation, the power given to the carriers by the optical field must be balanced by the power lost to the lattice. If the electron temperature is much greater than the lattice temperature, then polar optical phonon scattering tends to dominate the energy loss. Under such conditions and when the hot phonon effects are neglected then $T_e > T_L$ and the average rate of energy lost to the lattice in (3.17) becomes

$$\frac{\mathrm{d}E}{\mathrm{d}t} = \left(\frac{\hbar\omega_{\mathrm{LO}}}{\tau_{\mathrm{LO}}}\right)\exp\left(\frac{-\hbar\omega_{\mathrm{LO}}}{k_{\mathrm{B}}T_{\mathrm{e}}}\right), \tag{3.20}$$

where $\hbar\omega_{\mathrm{LO}}$ is the phonon energy and τ_{LO} is the phonon scattering time constant. Since the power lost to the lattice equals the power added to the carrier system, it is clear that the power lost can be equated to the pump intensity. Hence, by plotting the inverse electron temperature as a function of the logarithm of excitation power, as illustrated in Fig. 3.11, and taking the slope of such plots, it is possible to estimate the LO phonon energy. Based upon best fits, this approach gives LO phonon energies of $\hbar\omega \sim 30.8$ meV in InGaAs and $\hbar\omega \sim 29.7$ meV in GaInNAs.

While it is not possible to extract quantitative estimates of the electron-LO-phonon scattering times by using the CW-PL of optically heated carriers, the relative scatting times in the samples can be investigated.

3.5.2 *Time-Resolved PL Measurements*

The rate of carrier cooling can be obtained using low temperature ($T \sim 10$ K) time-resolved PL. The experimental setup for studying the time-resolved PL is shown in Fig. 3.12. Pulses from a mode-locked Ti–Sapphire laser pumped by an Argon

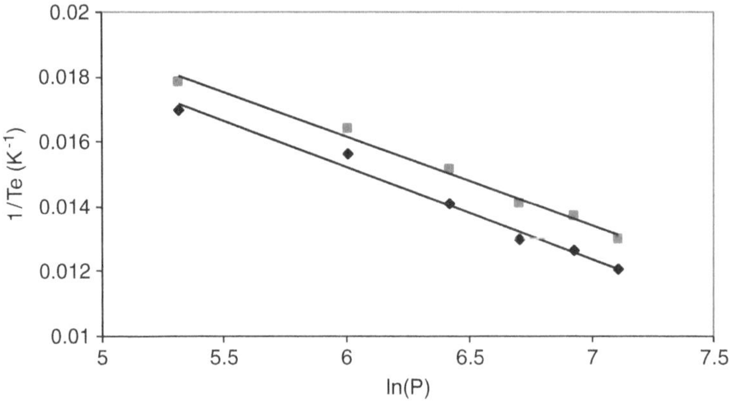

Fig. 3.11 Logarithm of pump power vs. inverse electron temperature the 9 nm sample at T_v = 4.4 K. For (*diamond*) GaInNAs and (*square*) InGaAs [25]

ion laser are used to excite the sample with a repetition rate of 82 MHz and pulse duration of ~1.4 ps. The Ti–sapphire laser is tuned to 800-nm (1.55 eV) exciting carriers within the GaAs barriers of the sample. The pump beam was focused to define a 200-μm diameter spot on the sample and pump power was controlled using a variable density neutral density (ND) filter. In the absence of an appropriate long pass filter, polarisers (P1 and P2) are used to reduce reflected pump light from entering the monochromator. The luminescence is spectrally dispersed by a 1/3 m monochromator and temporally resolved by a 2D streak camera with an S1 photocathode. The time-resolved PL results are calibrated to account for the photocathode spectral response prior to analysis.

The luminescence is collected and focused onto the monochromator slits using a two-lens system. The luminescence is spectrally dispersed about the desired central wavelength using a monochromator and the output temporally resolved by a 2D streak camera. It should be noted that the S1 photocathode response is highly dependent upon wavelength; hence spectral data requires normalization before analysis.

Figures 3.13 and 3.14 show the carrier cooling curves for the InGaAs QW (9 nm sample) and the GaInNAs QW (6-nm sample), respectively. In both cases, carrier cooling proceeds in a near linear fashion until the carrier temperature reaches around 50 K. In the case of the InGaAs QW, this initial cooling takes around 1,000 ps, while for the GaInNAs QW it takes less than 200 ps. The slow cooling of carriers in the InGaAs QW may be due to the large population of photo-excited carriers in the well. This may inhibit effective cooling of the hot carrier population via hot-phonon effects and screening of the long-range polar carrier-LO-phonon coupling by the high density free carriers. These effects are thought to have much less influence on carrier cooling in the GaInNAs QWs due to strong nonradiative recombination, which rapidly reduces the population of excited carriers. After the initial cooling, further thermalisation of the carriers appears to be relatively

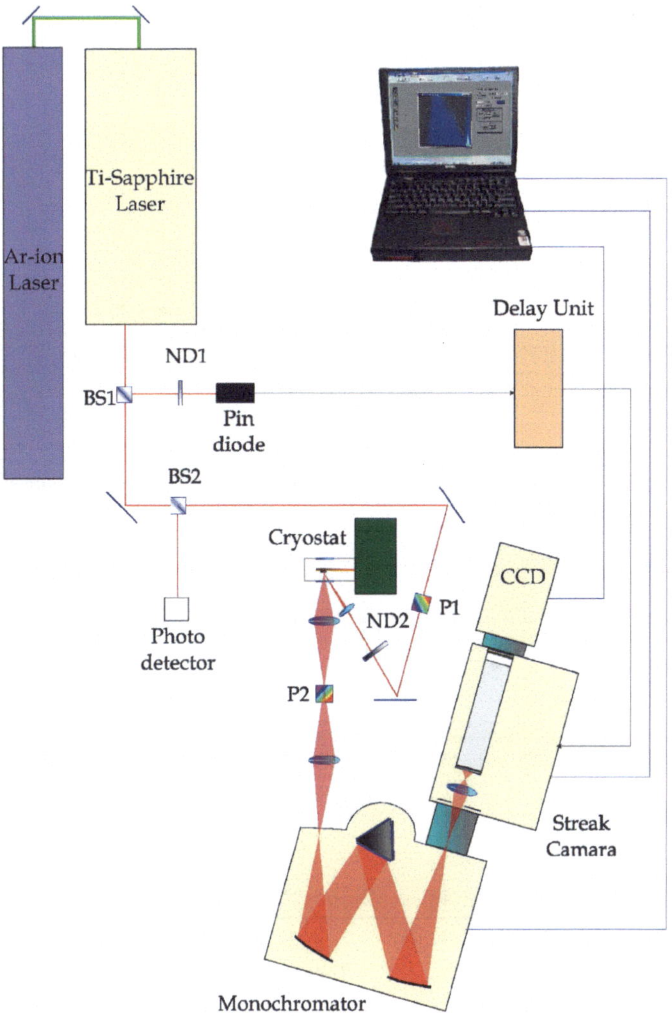

Fig. 3.12 Experimental setup for time-resolved PL, where: BS1 and BS2 are beam splitters, ND1 is a fixed neutral density filter, ND2 is a variable neutral density filter used to adjust pump power, P1 and P2 are polarisers [25]

slow. It is believed that the initial relatively rapid cooling process is dominated by LO-phonon scattering. Once the carriers have cooled sufficiently, it is assumed that acoustic phonon scattering, which is relatively slow, becomes the dominant relaxation mechanism. It is likely that spectral broadening induces some errors in the measured carrier temperatures. This explains why the recorded carrier temperatures stabilize at around 50 K rather than falling to the lattice temperature.

It is clear from these experiments that the incorporation of small fractions of nitrogen into InGaAs quantum wells does not appear to have much effect upon the LO-phonon energy. However, the addition of nitrogen does have a considerable

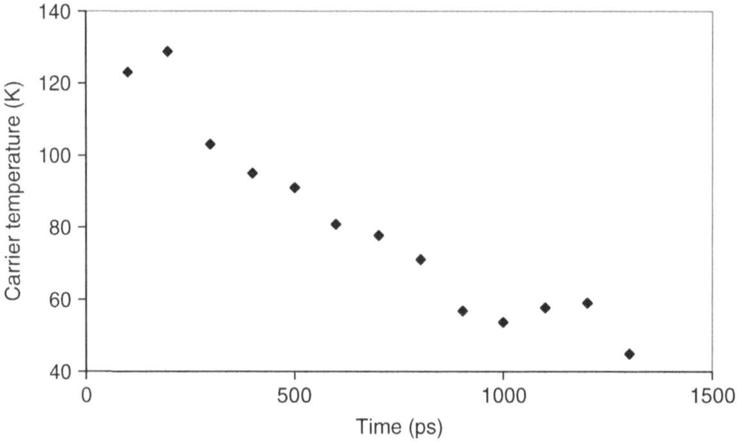

Fig. 3.13 Carrier temperature vs. time for emission from the InGaAs QW of sample #752 at 10 K with 5 mW pumping [25]

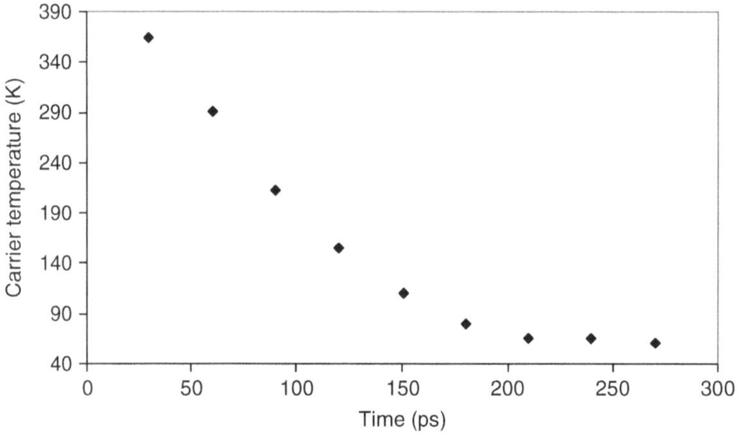

Fig. 3.14 Carrier temperature vs. time for emission from the GaInNAs QWs at 10 K with 5 mW pumping [25]

effect upon carrier cooling rates, leading to a significant increase in the apparent cooling rate. This rapid carrier cooling is believed to result from strong nonradiative recombination in the GaInNAs QW.

3.6 Hot Electron Momentum Relaxation

In order to analyze and improve the design of functional devices, and further predict the ultra-fast novel devices based on III–V compounds, a full understanding of electron momentum relaxation at high electric fields is necessary. As we have presented in Sects. 3.1–3.5, the hot-electron energy relaxation in 2D GaAs and other III-N structures are well documented, where the production of nonequilibrium LO phonons (hot phonons) are well known to slow down the hot electron energy relaxation. It is also shown that at high electric fields, the randomisation of the hot phonon distribution may take place via the elastic scattering of phonons with, for instance, interface roughness and alloy fluctuations; hence the drift of hot phonons can be neglected. If the momentum relaxation of phonons is faster than their decay time, the change in the electron momentum between the emission and reabsorption of hot phonons can be quite large. Hence the production of hot phonons with a finite lifetime may also enhance the momentum relaxation rate. The electron drift velocity at high fields is therefore reduced and, consequently, inter-valley and real-space transfer negative differential resistance may be quenched [18–24, 26].

In this section, we present the results of hot-electron momentum relaxation studies for longitudinal transport with reference to modulation-doped dilute nitride ($Ga_xIn_{1-x}N_yAs_{1-y}/GaAs$) QWs. The aim is to explore whether the electron drift velocity at high fields can be explained by invoking a mechanism involving the production of non-drifting hot phonons. In dilute nitrides, the almost inevitable presence of nitrogen clusters, impurities, interface imperfections, and dislocations provides ample sources for phonon momentum scattering. The population of nonequilibrium phonons is therefore expected to be strongly non-drifting, particularly in the region where hot phonon effects are important. The experimental technique used is the high speed, pulsed I–V measurements on two terminal devices. We also address how the experimental results are compared with a simple theoretical model for high-field transport taking into account the effect of nonequilibrium phonon production.

3.6.1 Experimental Techniques

The experimental setup for the hot electron momentum relaxation for longitudinal transport experiments is illustrated in Fig. 3.15. Here the optical part of the experiment (monochromator, photomultiplier, and lenses) are included in order to investigate light emission associated with impact ionization [27] and not directly relevant to I–V measurements. The samples are n-modulation-doped quantum wells of GaInNAs/GaAs. Samples are mounted on gold-coated holders to ensure good thermal contact to the heat sink. Ohmic contacts to the conducting quantum well channels are made by alloying Au–Ge–Ni. High speed I–V measurements are carried out using simple bars as shown in Fig. 3.15. In order to achieve high electric fields at small operating voltages, the samples are fabricated in lengths of between 1 and 100 μm. The aspect ratio (length/width) of the bars is usually >4 to ensure uniform electric field distribution. In order to reduce the Joule heating, electric field pulses of between 10 and 100 ns duration with a duty cycle less than 10^{-5} s are applied along the QW layers. The drift velocity (V_d) as a function of the applied field is then obtained directly from the measured I–V characteristics. In the

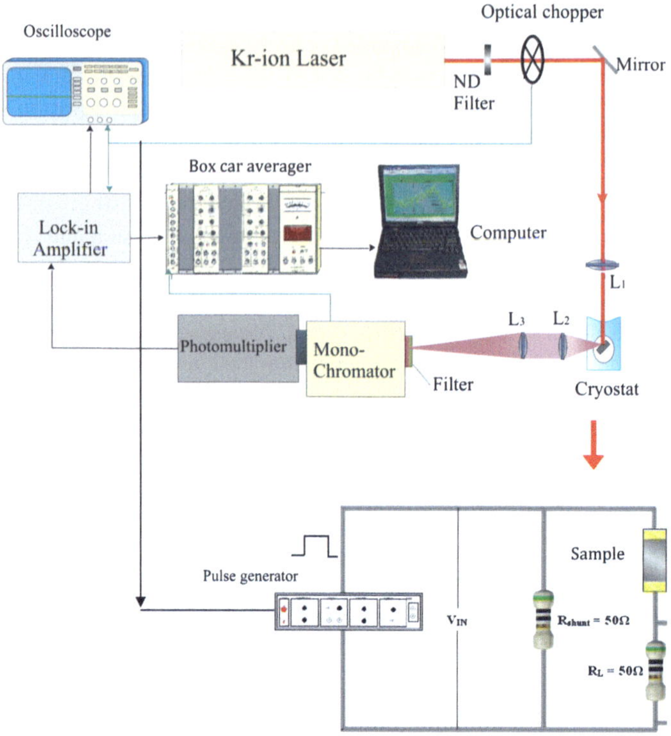

Fig. 3.15 Experimental set up for transport at high electric fields [28]

Table 3.1 Saturation drift velocities at different lattice temperatures [29]

$Ga_xIn_{1-x}N_yAs_{1-y}$	Temperature			
	12.4 K		77 K	
	V_{ds} (cm/s)	F (kV/cm)	V_{ds} (cm/s)	F (kV/cm)
$x = 30\%$; $y = 1.0\%$	1.28×10^7	14.9	1.17×10^7	14.6
$x = 30\%$; $y = 1.5\%$	0.62×10^7	16.9	0.71×10^7	16.7

measurements, it is assumed that the electron concentration remains constant in the measurement range of electric fields and that the I–V characteristics reflect solely the change of electron mobility with electric field.

Figure 3.16 shows typical curves of the measured drift velocity plotted against the applied field for two samples with different nitrogen composition [29] at lattice temperatures of 12.4 and 77 K, respectively. It is evident from the Fig. 3.16a, b that the drift velocity first increases linearly with electric field F at low fields (Ohmic region), it then deviates from linearity as expected from increased momentum scattering of hot electrons with LO phonons and finally reaches saturation. The approximate saturation drift velocities (V_{ds}) obtained in the two structures are given in Table 3.1.

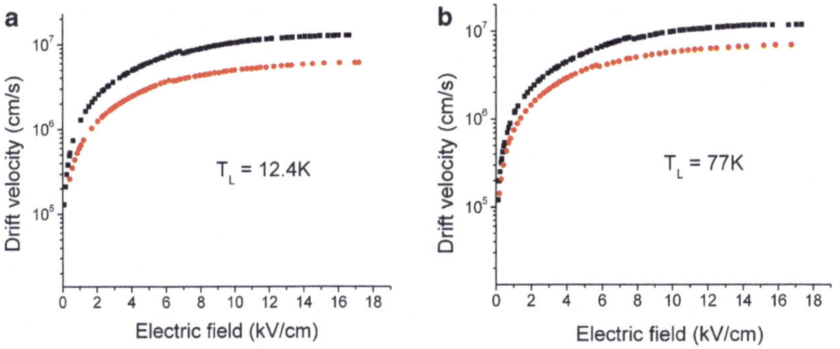

Fig. 3.16 Measured drift velocity vs. applied field at $T_L = 12.4$ K and $T_L = 77$ K for two samples with different nitrogen composition. Sample B (*solid squares*), and A (*solid circles*) [29]

3.6.2 *Theoretical Modeling of Experimental Results*

The results can be compared with a simple theoretical model based on the energy and momentum balance method, in which the energy and momentum rates are averaged over a drifted Maxwellian distribution [30]. Assuming the average displacement of the electron distribution is in the direction of the applied field F, the energy and momentum balance equations are

$$\left\langle \frac{\mathrm{d}E}{\mathrm{d}t} \right\rangle = ev_\mathrm{d}F = R_\mathrm{E} = f(T_\mathrm{e}) \tag{3.21}$$

$$\left\langle \frac{\mathrm{d}\hbar k}{\mathrm{d}t} \right\rangle = eF = R_\mathrm{p},$$

where v_d, E, and $\hbar k$ are the drift velocity, electron energy, and momentum, respectively. For simplicity, we may restrict attention to electrons in a single parabolic band interacting solely with polar optical phonon scattering for energy relaxation. For momentum relaxation, a further elastic scattering rate is incorporated to correct the low field mobility. It is found that for a parabolic band, after averaging over the drifted electron distribution, the momentum relaxation rate for a given process i is a function of electron temperature (T_e):

$$R_\mathrm{p} = \sum_i R_\mathrm{p}^\mathrm{i} = v_\mathrm{d} \sum_i h_\mathrm{i}(T_\mathrm{e}) = v_\mathrm{d} h(T_\mathrm{e}). \tag{3.22}$$

For simplicity, we assume that electrons are confined to the lowest subband in a deep square QW and the interaction with phonons is confined to the same QW. The energy relaxation rate for polar optical scattering then can be written as [22]

$$eFv_d = f_{PO}(T_e) = \frac{\hbar\omega}{\tau_0}\left(\frac{E_L}{\hbar\omega}\right)^{1/2}\left[(n(\omega)+1)\exp\left(-\frac{\hbar\omega}{k_B T_e}\right) - n(\omega)\right], \tag{3.23}$$

where $\hbar\omega$ is the LO phonon energy, $n(\omega)$ is the phonon occupancy (determined by electron temperature), $E_L = \hbar^2\pi^2/2m^*L^2$ is the energy shift of the lowest subband in an infinitely deep well of width L and k_B is the Boltzmann's constant. Here τ_0 is a characteristic time constant for the polar interaction given by

$$\tau_0 = \frac{e^2\omega}{2\pi\hbar}\left(\frac{m^*}{2\hbar\omega}\right)^{1/2}\left(\frac{1}{\varepsilon_\infty} - \frac{1}{\varepsilon_s}\right), \tag{3.24}$$

where ε_∞ and ε_s are the high-frequency and static permittivities, respectively and m^* is the effective electron mass. The effect of a population of non-drifted hot phonons can be incorporated by substituting for $n(\omega)$ using the expression for hot phonon occupation derived by Ridley [22]

$$n(\omega) = \frac{n_0(\omega) + (\tau_p/2\tau_0 q_0 L)(\pi k_B T_e/\hbar\omega)^{1/2}(n/N_c)\exp(-\hbar\omega/k_B T_e)}{1 + (\tau_p/2\tau_0 q_0 L)(\pi k_B T_e/\hbar\omega)^{1/2}(n/N_c)(1 - \exp(-\hbar\omega/k_B T_e)}, \tag{3.25}$$

where τ_p is the model phonon lifetime, n is the bulk electron density, $q_0 = (2m^*\hbar\omega)^{1/2}/\hbar$ is the phase-matched phonon wave vector and $N_c = m^* k_B T_e/\pi\hbar^2 L$ is an effective density of states. The equilibrium phonon occupation number $n_0(\omega)$ is given as usual by the Bose–Einstein factor $n_0(\omega) = 1/(\exp(\hbar\omega/k_B T_0) - 1)$. Note that it is the electron density that essentially determines the hot phonon effects. Substituting for $n(\omega)$ from (3.25) into (3.23) gives

$$eFV_d = f_{PO}(T_e) = \frac{\hbar\omega}{\tau^*}\left(\frac{E_L}{\hbar\omega}\right)^{1/2}\left[(n_0(\omega) + 1)\exp\left(-\frac{\hbar\omega}{k_B T_e}\right) - n_0(\omega)\right], \tag{3.26}$$

where the effective energy relaxation time τ^* which takes into account all the hot phonons effects is given by

$$\tau^* = \tau_0 + \left(\frac{\tau_p n}{2q_0 L N_c}\right)\left(\frac{\pi k_B T_e}{\hbar\omega}\right)^{1/2}\left\{1 - \exp\left(-\frac{\hbar\omega}{k_B T_e}\right)\right\}. \tag{3.27}$$

The momentum relaxation rate via the interaction with optical phonons can be found using [22]

$$h_{PO}(T_e) = \frac{m^*(\hbar\omega E_L)^{1/2}}{\tau_0 k_B T_e}[n(\omega) + 1]\exp(-\hbar\omega/k_B T_e). \tag{3.28}$$

Substituting for $n(\omega)$ from (3.5) gives

$$h_{PO}(T_e) = \frac{m^*(\hbar\omega E_L)^{1/2}}{\tau^* k_B T_e}\left[n_0(\omega) + 1 + \left(\frac{\tau_p}{2q_0\tau_0 L}\right)\left(\frac{\pi k_B T_e}{\hbar\omega}\right)^{1/2}\frac{n}{N_c}\right]\exp\left(\frac{-\hbar\omega}{k_B T_e}\right). \tag{3.29}$$

In addition to the electron-LO phonon scattering, a further elastic scattering rate with an arbitrary temperature dependence can be included by defining $h_e(T_e) = h_e(T_0)(T_e/T_0)^r$, where T_0 is the equilibrium lattice temperature and r can be determined from the best fit to the experimental data. We then have $h(T_e) = h_{PO}(T_e) + h_e(T_e)$. The value of $h_e(T_0)$ can be determined in accordance with the low-field mobility μ_0 by

$$h_e(T_0) = \left(\frac{e}{\mu_0} - h_{PO}(T_0)\right), \tag{3.30}$$

where $h_{PO}(T_0)$ can be obtained from (3.9) in the limit $T_e \to T_0$

$$h_{PO}(T_0) = \frac{m^*(\hbar\omega E_L)^{1/2}}{\tau_0 k_B T_0} n_0(\omega). \tag{3.31}$$

The drift velocity can then be found as a function of electron temperature

$$v_d(T_e) = \frac{eF(T_e)}{h_{PO}(T_e) + ((e/u_0) - h_{PO}(T_0))(T_e/T_0)^r}. \tag{3.32}$$

Combining (3.26) and (3.32) gives an expression relating the electron temperature to electric field, hence drift velocity vs. electric field curves can be deduced. The model is applied using $\tau_P = 7$ ps. The phonon energy and dielectric constant are taken to be linear interpolations of the GaAs and InAs values. The effective mass is calculated from the band-anticrossing model [31] using parameters from Fahy and O'Reilly [32] together with the effective mass for InGaAs interpolated using bowing parameters taken from the work of Vurgaftman et al. [33]. In Fig. 3.17a, b, the calculated electron drift velocity vs. electric field curves are shown for sample B. Also shown in the figures are the experimental results, represented by solid squares. It is clear from the figures that at low fields, (low electron temperatures) the experimental data agree very well with the calculated drift velocity curves with $r = 0.17$. At high electric fields, as the electron temperature rises, the experimental drift velocities are suppressed and move closer to the calculated curves with $r = 0$. The model without hot phonons only agrees well with experimental data at low lattice temperature (12.4 K) and predicts a higher drift velocity than measured values at 77 K.

3.7 Hot Electron Energy Relaxation via Acoustic Phonon Emission

At temperatures below about 35 K, the population of optical phonons is negligibly small; therefore, acoustic-phonon scattering provides the only inelastic scattering mechanism. At such low temperatures, the electron distribution is often degenerate and Pauli exclusion is important in limiting the allowed scattering [1, 2, 14, 34–38].

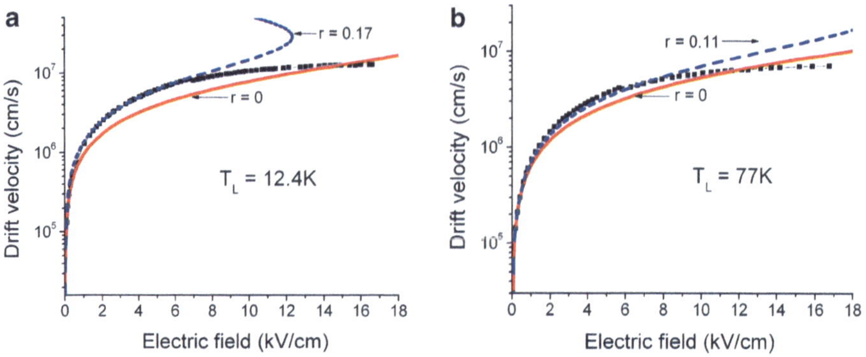

Fig. 3.17 Drift velocity vs. applied electric field at lattice temperatures of 12.4 and 77 K, for sample B. The *solid* and *broken curves* show the calculated drift velocity with $r = 0.17$ (*broken line*) and $r = 0$ (*solid line*), respectively. The *solid squares* represent the experimental data [29]

Experimental work concerning the energy-loss rates of 2D electrons in the acoustic-phonon scattering regime has been reported by a large number of groups [34,39–52]. Most of the work reported in the literature was carried out on high electron mobility transistor (HEMT) structures.

There are two main experimental techniques employed in the investigations of electron energy relaxation in the acoustic phonon regime:

1. In modulation-doped structures, where a highly degenerate electron gas exists, the amplitude variation of quantum oscillations like the SdH. Effect with the applied electric field and the lattice temperature can be used in the determination of the electron temperature power loss characteristics [43, 53].
2. In nondegenerate material where the momentum relaxation is dominated by ionized impurity remote impurity or interface roughness scattering, electron temperatures can be obtained as a function of the applied electric field by a simple comparison of the electric field dependent and lattice temperature dependent mobility curves [53, 54].

In this chapter, we aim to address the first of the two techniques above, i.e., the measurement of acoustic phonon-assisted power loss vs. temperature curves using the SdH technique. In the measurements, extra caution should be taken to avoid the temperature regime where extra loss mechanisms such as plasma effects or optical phonon scattering may interfere. The experimental results are published in detail elsewhere [34] therefore, unless otherwise stated, they will represent acoustic phonon assisted cooling of 2D electron gas in GaAs/GaAlAs MQW structures [1, 34, 43]. We will also show how the experimental results are interpreted within the framework of the current theoretical models concerning electron energy-loss rates in 2D and 3D semiconductors. The results provide concrete information about the relative magnitude of the deformation potential and piezoelectric contributions to power loss.

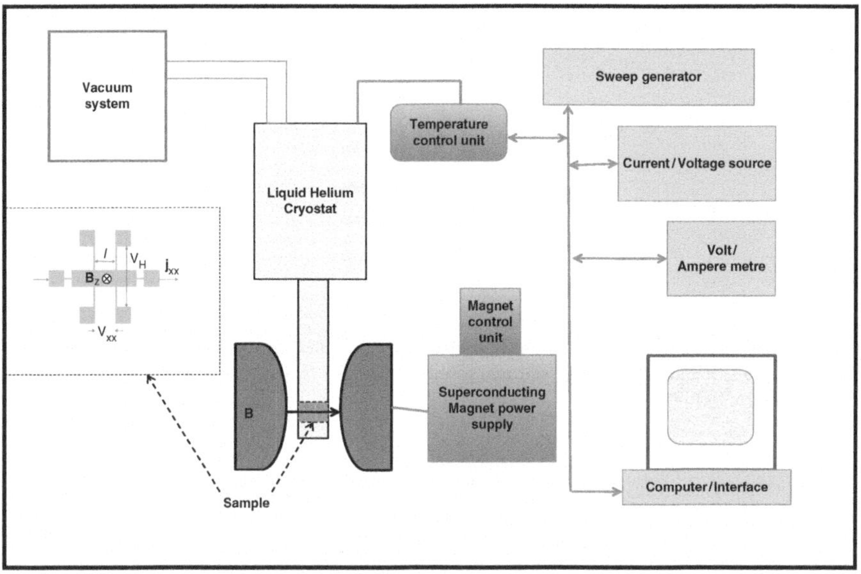

Fig. 3.18 The experimental setup and the Hall bar sample used in the investigations of electron energy relaxation via the emission of acoustic phonons

3.7.1 Experimental Procedures

The experimental setup for the orthodox DC-SdH measurements is illustrated schematically in Fig. 3.18. The technique for the power loss measurements is based on the assumption that scattering mechanisms such as interface-roughness scattering and ionized impurity scattering, which determine the low-temperature transport mobility of 2D electrons in GaAs/$Ga_{1-x}Al_xAs$ MQWs, are all elastic in nature. Consequently, the energy which is gained by electrons in an applied electric field is dissipated via emission of acoustic phonons. There are a few important points to be considered in the experiments:

1. The base lattice temperature. This is required to be much lower than that when loss mechanisms other than the acoustic phonon assisted one may interfere and in the range when the piezoelectric interactions are not screened out. In the experiments referred to here the base temperature is $T_0 = 1.55$ K.
2. The range of the applied magnetic fields. The magnetic field, B, has to be high enough for the Landau level separation to be greater than the thermal energy $((\hbar eB/m^*) > kT)$.
3. The samples used in the SdH measurements are highly degenerate so that the reduced Fermi energy-first subband separation is much greater than the thermal energy $[(E_F - E_1)/k_B T] \gg 1$. The measurements are carried out as a function of (1) the applied electric field F at a fixed lattice temperature T_0 and (2) the lattice

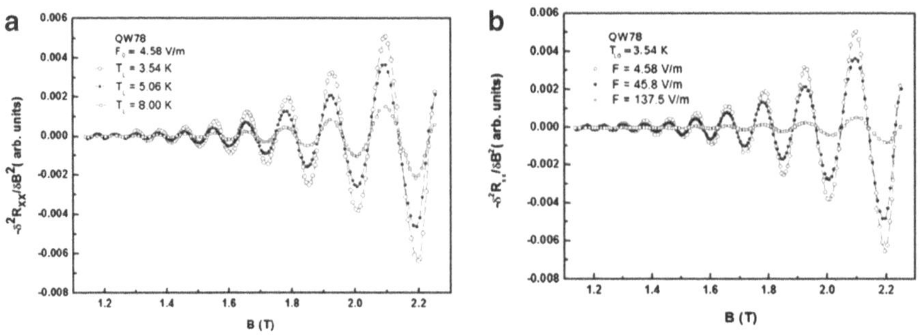

Fig. 3.19 Second derivatives of the SdH oscillations measured at the lowest electric field as a function of lattice temperature (**a**) and measured at the base lattice temperature as a function of applied electric field (**b**) [34, 55]

temperature T_L ($T_L > T_0$) at a fixed electric field F_0 which is low enough to ensure ohmic conditions and hence to avoid carrier heating. The current flows in the plane of the 2D electron gas. Steady magnetic fields (B) are applied perpendicular to the plane of the samples and hence to the plane of the 2D electron gas. The data are taken at equal intervals of $1/B$.

Figure 3.19 shows the experimental recordings of the negative second derivative of the oscillations at different lattice temperatures (1) and applied electric fields (2). Here in order to exclude the effects of the background magnetoresistance and separate the oscillatory components, the second-derivative technique, $-\partial^2 R_{xx}/\partial B^2$, and digital filtering are employed. The second-derivative technique removes the effects of the background magneto-resistance, which varies slowly with magnetic field, suppresses the amplitude of the long-period oscillations, and amplifies that of the short-period oscillations [34, 52–57]. The amplitude of SdH oscillations decreases with increasing T_L and applied electric field and it can be represented by the function

$$\frac{\Delta\rho_{xx}}{\rho_0} \propto \exp\left(\frac{-\pi}{\omega_c \tau_q}\right) \cos\left[\frac{2\pi(E_F - E_1)}{\hbar\omega_c} - \pi\right] D(\chi), \tag{3.33}$$

with

$$\mu_q = \frac{e\tau_q}{m^*} \quad \text{and} \quad D(\chi) = \frac{\chi}{\sinh\chi}, \tag{3.34}$$

where $\Delta\rho_{xx}$, ρ_0, E_F, E_1, $\hbar, k_B$, $\omega_c (= eB/m^*)$, m^*, and τ_q are the oscillatory magneto-resistivity, zero magnetic-field resistivity, Fermi energy, first subband energy, Planck's constant, Boltzmann constant, cyclotron frequency, effective mass, and quantum lifetime of 2D electrons, respectively. Equation (3.33) describes well the experimental SdH oscillations data for all $\omega_c\tau_q$ values as long as $\Delta\rho_{xx} < \rho_0$ [4, 34, 55]. The temperature dependence of the envelope function of the oscillations is totally contained in the function $D(\chi)$ with $\chi = 2\pi^2 k_B T/\hbar\omega_c$.

The temperature dependence of the amplitude of the oscillations is, therefore, determined by temperature, magnetic field and the effective mass via:

$$\frac{A(T, B_n)}{A(T_0, B_n)} = \frac{T \sinh\left(2\pi^2 k_B T_0 m^*/\hbar e B_n\right)}{T_0 \sinh\left(2\pi^2 k_B T m^*/\hbar e B_n\right)}, \tag{3.35}$$

where $A(T, B_n)$ and $A(T_0, B_n)$ are the amplitudes of the oscillation peak observed at a magnetic field B_n and at temperatures T and T_0. The derivation of this equation assumes that the quantum lifetime or the quantum mobility is independent of lattice temperature, and applied electric and magnetic field. The accuracy of this assumption can be verified experimentally by ascertaining the constancy of the Dingle temperature T_D as a function of temperature [34, 54, 55] where

$$T_D = \frac{\hbar}{2k_B \pi \tau_q}. \tag{3.36}$$

The value of T_D is determined by plotting

$$\ln\left[A(T)B^{-1/2}\frac{\sinh\chi}{\chi}\right], \tag{3.37}$$

vs. $1/B$ from (3.33).

The slopes of the lines give the Dingle temperature T_D, and, according to (3.33) and (3.35), the quantum lifetime τ_q as a function of temperature. The change in the amplitude of SdH oscillations with applied electric field, in Fig. 3.20 can be described in terms of electric field-induced electron heating. The temperature T in (3.33)–(3.35) can be replaced by the electron temperature T_e. Therefore T_e can be determined by comparing the relative amplitudes of the SdH oscillations measured as functions of the lattice temperature ($T = T_L$) and the applied electric field (F) using (3.35):

$$\left[\frac{A(T_L, B_n)}{A(T_{L0}, B_n)}\right]_{F=F_0} = \left[\frac{A(F, B_n)}{A(F_0, B_n)}\right]_{T_L=T_{L0}}. \tag{3.38}$$

The in-plane effective mass (m^*) and the quantum lifetime (τ_q) of 2D electrons in the MQW samples, can be determined, respectively, from the temperature and magnetic field dependences of the amplitude of the SdH oscillations using (3.32)–(3.36). The 2D electron density (N_{2D}) and the Fermi energy with respect to the subband energy ($E_F - E_1$) may also be obtained from the period of the SdH oscillations. The analysis of the SdH oscillations measured as a function of both the lattice temperature and the applied electric field assumes that N_{2D} and τ_q for all the samples are essentially independent of both lattice temperature and applied electric field in the range of measurements. Recent experimental work shows in detail how these parameters are obtained from the experimental data for the GaInNAs/GaAs QWs and GaN HEMT structures [58, 59]. Figure 3.20a, b shows the normalized amplitudes of the SdH oscillations at a fixed magnetic field as a function of lattice

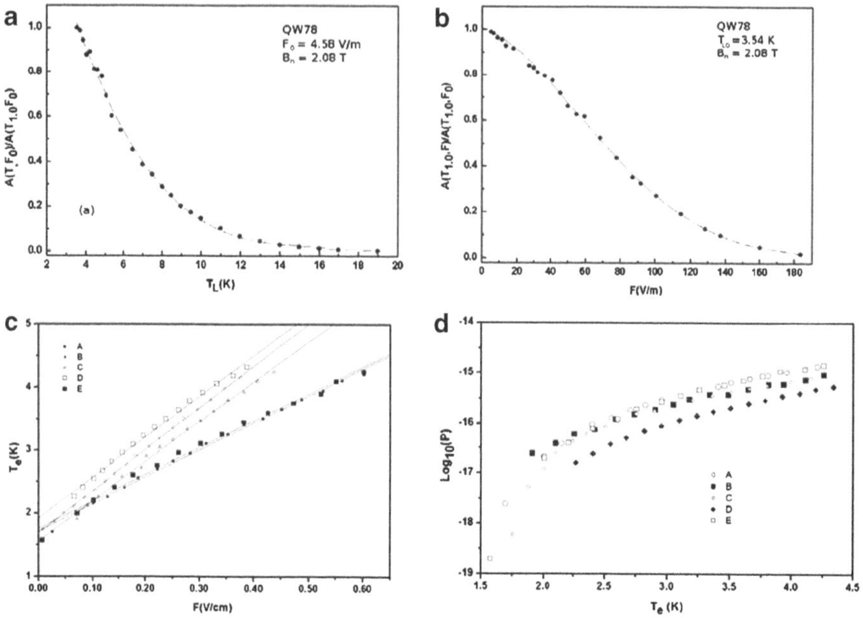

Fig. 3.20 Normalized amplitudes of the SdH oscillations at a fixed magnetic field as a function of lattice temperature at low electric field (**a**) and as a function of applied electric field at the base lattice temperature (**b**). Electron temperatures as a function of applied field for the five samples with different quantum well widths studied (**c**) and energy loss rate per electron vs. electron temperature for all the samples (**d**) [34, 55]

temperature at low electric field and as a function of applied electric field at the base lattice temperature, respectively.

The 2D electron gas represents a system in thermal equilibrium characterized by an electron temperature T_e greater than the lattice temperature T_L [1, 14]. The electron temperature T_e, can be obtained by comparing directly the curves similar to those in Fig. 3.20a, b. T_e is plotted as a function of the applied electric field in Fig. 3.20c for a range of samples studied [34]. In the analysis of the data, it is important to make sure that T_e for each sample is independent of magnetic field and that the magnetic field used in the experiments does not affect significantly the energy relaxation processes.

The power loss per electron, P, supplied by the applied electric field can be calculated using

$$P = e\mu_t F^2, \tag{3.39}$$

where μ_t is the transport mobility.

In the steady state, the power supply is equal to the power loss via the emission of acoustic phonons. In the calculations of power loss, we used the transport mobility determined from resistivity and low magnetic-field Hall Effect measurements. It is vital to establish that μ_t is independent of both lattice temperature and the applied electric field in the ranges of measurements. The power loss vs. electron temperature is plotted in Fig. 3.20d for all the samples studied. The magnitude of power loss varies significantly from one sample to another. In the QW samples studied, the barrier parameters of all the samples were identical and the 2D electron densities were similar, hence the observed variation in power loss is indicative of an association with the only variable parameter in the samples, i.e., the quantum well width L_z [34, 52].

3.7.2 Theoretical Modeling of Experimental Results

The power loss from a degenerate electron gas due to scattering by acoustic phonons has been calculated by many authors (see for instance [1, 40, 48, 60–62]) in two distinct temperature regimes:

1. The low-temperature (Bloch–Grüneisen) regime, where the phonon population diminishes, such that $n(\omega_q) \ll 1$, and the Pauli exclusion increasingly restricts the allowed scattering processes.
2. The high-temperature (equipartition) regime, where $\hbar\omega_q/k_B T \ll 1$, and hence the phonon distribution can be approximated by

$$n(\omega_q) = \frac{1}{\exp(\hbar\omega_q/k_B T) - 1} \approx \frac{k_B T}{\hbar\omega_q}, \tag{3.40}$$

where $\hbar\omega_q$ is the acoustic-phonon energy at wave vector q.

In the 2D calculations, where the scattering by absorption of acoustic phonons was neglected and only spontaneous emission was considered to be important, the infinite-well approximation was used in the extreme quantum limit, and the phonons were assumed to be bulk phonons. Approximate analytical expressions derived for the power loss in both the low-temperature and high-temperature regimes are given in [1,34,62] for the unscreened interactions in 2D and 3D semiconductors. When an electron undergoes intra-subband transition by the emission of acoustic phonons, the electron changes its momentum and at the same time loses energy to the phonon. At low temperatures the Fermi gas has a sharp boundary curve and consequently any momentum changes which might involve the emission of an acoustic phonon of energy much greater than $k_B T$ are hindered greatly by the Pauli Exclusion Principle. Therefore, only small-angle scattering is allowed at very low temperatures [63, 64]. In this regime, the power loss is characterized by the dependences [65]: $P_{np} \propto (T_e^5 - T_L^5)$ for the unscreened deformation-potential scattering and $P_p \propto (T_e^3 - T_L^3)$ for the unscreened piezoelectric scattering.

For a 2D electron gas, $P = P_{np} + P_p$, in the low-temperature regime can be represented by [1, 34]

$$P = C_{np}[(k_B T_e)^5 - (k_B T_L)^5] + C_p[(k_B T_e)^3 - (k_B T_L)^3], \tag{3.41}$$

where

$$C_{\rm np} = \frac{6\Xi^2 m^{*^2} L_z}{\pi^3 \rho \hbar^7 V_S^4 N_{\rm 2D}}, \tag{3.42}$$

and

$$C_{\rm p} = \frac{e^2 K_{\rm av}^2 m^{*^2}}{2\pi^2 \varepsilon \hbar^5 k_{\rm F} N_{\rm 2D}}, \tag{3.43}$$

are the magnitudes of the deformation potential (nonpolar acoustic) and piezoelectric (polar acoustic) interactions, respectively. Here Ξ is the acoustic deformation potential, ρ is the density and V_S is the speed of sound in the material, ε is the static dielectric constant, and $k_{\rm F}$ $(= (2\pi N_{\rm 2D})^{1/2})$ is the Fermi wave vector of 2D electrons. The average electromechanical coupling constant $K_{\rm av}^2$ is given by [4]

$$K_{\rm av}^2 = \frac{e_{14}^2}{\varepsilon}\left(\frac{12}{35C_{\rm L}} + \frac{16}{35C_{\rm T}}\right), \tag{3.44}$$

where e_{14} is the piezoelectric stress constant, and $C_{\rm L}$ and $C_{\rm T}$ are the spherical-average elastic stiffness constants of a single-crystal with cubic symmetry. The screening of the electron–phonon interaction in the 2D case, which is not included in the above calculations, is predicted to increase the exponent of the $k_{\rm B}T_{\rm e}$ and $k_{\rm B}T_{\rm L}$ terms in (3.41) by 2 [1, 66]. In the high-temperature regime, when Maxwell–Boltzmann statistics are applicable and equipartition is assumed, the electron-temperature dependence of the energy-loss rate can be

$$P = (C_{\rm np} + C_{\rm p})(k_{\rm B}T_{\rm e} - k_{\rm B}T_{\rm L}), \tag{3.45}$$

where

$$C_{\rm np} = \frac{3\Xi^2 m^{*^2}}{2\rho\hbar^3 L_z}, \tag{3.46}$$

and

$$C_{\rm p} = \frac{3e^2 K_{\rm av}^2 m^{*^2} V_S^2}{4\pi\varepsilon\hbar^3 L_z N_{\rm 2D}}, \tag{3.47}$$

are the magnitudes of the deformation potential and piezoelectric interactions, respectively.

Inspection of (3.41)–(3.47) shows that the experimental data for the power loss vs. electron temperature can be represented in the form

$$P = A\left(T_{\rm e}^{\gamma} - T_{\rm L}^{\gamma}\right), \tag{3.48}$$

where the proportionality constant A depends on the coupling constants, 2D carrier density, and quantum well width. Equation (3.48) fits well to the experimental $P(T_{\rm e})$ with a constant value for γ over the whole electron temperature range up to $T = 15\,{\rm K}$. For the full analysis of data please see [34, 52, 54].

The agreement between (3.48) and the experimental $P(T_e)$ data may not be good and the fitting procedure may yield relatively higher values of γ. This indicates that the temperature range of the experiments is intermediate between the low-temperature and high-temperature limits (see also [1, 40, 48]). In this intermediate-temperature regime, therefore, an approximation, which is often used [1, 34, 45] for the energy-loss rate in the intermediate-temperature regime, is represented by

$$P = f(T_e, T_L)(C_{np} + C_p)(k_B T_e - k_B T_L), \tag{3.49}$$

where C_{np} and C_p are the magnitudes of the deformation potential and piezoelectric interactions in the high-temperature regime, as given in (3.44) and (3.45), and

$$f(T_e, T_L) = \frac{\sinh(x_L - x_e)}{\sinh x_L \sinh x_e}\left[\frac{x_L x_e}{x_L - x_e}\right], \tag{3.50}$$

with $x_e = \hbar\omega/2k_B T_e$, $x_L = \hbar\omega/2k_B T_L$, and $\hbar\omega = \sqrt{2}\hbar V_S k_F$.

The function $f(T_e, T_L)$ approaches unity at high temperatures and falls down exponentially at low temperatures. Equation (3.50) provides an approximation extending the high-temperature regime toward low temperatures.

3.8 Conclusions

The physics of hot carriers deals with nonlinear effects associated with the energy and momentum exchange between electrons and holes, between carriers and acoustic and optical phonons, impurities imperfections and photons. As the sizes of functional devices shrink to nanoscale dimensions even the application of a small voltage results in the establishment of very large electric fields in the device where Ohm's Law breaks down and hot electron transport takes over. At higher electric fields, other effects such as negative differential resistance, impact ionization, and dielectric breakdown occur. Hot electrons, coupled with the high carrier densities, result in the production of nonequilibrium phonons and the nonlinear effects involving coupled hot electron, and phonon gases become the norm. Hence the operation of such modern devices may become very complex and the quasiclassical treatment of the operation of both electrical and optoelectronic devices gives way to quantum transport of dense hot electron and hole gases. Therefore, the study of hot electron energy and momentum is important not only from a fundamental physics point of view but also is vital for understanding the operation of and predicting novel devices operating in highly nonlinear transport regimes. In this chapter, we presented a review of the experimental techniques commonly used for the determination of hot carrier energy and momentum relaxation with reference to the GaAs/GaAlAs and GaInNAs/GaAs QW structures. Theoretical models used in the interpretation of the experimental data were also presented.

Acknowledgements I am grateful to my students, Y. Sun, R. J. Potter, D. Zanato, and M. Vaughan, and to my colleagues, B. K. Ridley, A. Erol, M. C. Arikan, H. Celik, M. Cankurtaran, and E. Tiras, for many useful discussions and for the fruitful collaboration in hot electron research over the last 25 years.

References

1. B.K. Ridley, Rep. Prog. Phys. **54**, 169 (1991)
2. B.K. Ridley, in *Hot Electrons in Semiconductors*, ed. by N. Balkan (Oxford University Press, New York, 1998), pp. 3–13
3. G. Weber, A.M. Depaula, J.F. Ryan, Semicond. Sci.Technol. **6**, pp 397–400 (1991)
4. For a review see Hot Carriers in Semiconductor Nanostructures, ed. by J. Shah (Academic, New York, 1992)
5. N. Balkan, B.K. Ridley, M. Emeny, I. Goodridge, Semicond. Sci. Technol. **4**, 852–857 (1989)
6. J.A. Kash, J.C. Tsang, J.M. Hvam, Phys. Rev. Lett. **54**, 2151–2154 (1985)
7. J.A. Kash, S.S. Shah, J.C. Tsang, Phys. Rev. Lett. **58**, 1869–1872 (1987)
8. M.C. Tatham, J.F. Ryan, C.T. Foxon, Surf. Sci. **228**, 127–130 (1990)
9. M. Ramonas, A. Matulionis, J. Liberis, L. Eastman, X. Chen, Y.-J. Sun, Phys. Rev. B **71**, 075324 (2005)
10. A. Matulionis, J. Liberis, I. Matulionien, M. Ramonas, L.F. Eastman, J.R. Shealy, V. Tilak, A. Vertiatchikh, Phys. Rev. B **68**, 035338 (2003)
11. A.J. Kent, in *Hot Electrons in Semiconductors: Physis and Devices*, ed. by N. Balkan (Oxford University Press, Oxford, 1997)
12. H. Celik, M. Cankurtaran, N. Balkan, A. Bayrakli, Semicond. Sci. Technol. **17**, 18–29 (2002)
13. D. Zanato, N. Balkan, B.K. Ridley, G. Hill, W.J. Schaff, Semicond. Sci. Technol. **19**, 1024–1028 (2004)
14. S. Lyon, J. Lumin. **35**, 121 (1986)
15. R.G. van Welzenis, H. Wifshoff, K. Ploog, Physica B **134**, 347 (1985)
16. E.F. Schubert, K. Ploog, H. Dambkes, K. Heime, Appl. Phys. A **33**, 63 (1984)
17. J. Shah, A. Pinczuk, H.L. Stormer, A.C. Gossard, W. Wiegmann, Appl. Phys. Lett. **42**, 55 (1983)
18. R. Gupta, B.K. Ridley, Solid-State Electron. **32**, 1241 (1989)
19. R. Gupta, N. Balkan, B.K. Ridley, Semicond. Sci. Technol. **7**, 8274–6278 (1992)
20. Y. Sun, M. Vaughan, A. Agarwal, M. Yilmaz, B. Ulug, A. Ulug, N. Balkan, M. Sopanen, O. Reentilä, M. Mattila, C. Fontaine, A. Arnoult, Phys. Rev. B **75**, 205316 (2007)
21. Y. Sun, N. Balkan, J. Appl. Phys. **106**, 073704 (2009)
22. B.K. Ridley, Semicond. Sci. Technol. **4**, 142 (1989)
23. N. Balkan, R. Gupta, M.E. Daniels, B.K. Ridley, M. Emeny, Semicond. Sci. Technol. **5**, 986 (1990)
24. R. Gupta, B.K. Ridley, Proc. SPIE **1362**, 790 (1990)
25. R.J. Potter, N. Balkan, J. Phys.: Condens. Matter **16**, 3387–3412 (2004)
26. B.K. Ridley, W.J. Schaff, L.F. Eastman, J. Appl. Phys. **96(3)**, 1499 (2004)
27. S. Chung, N. Balkan, J. Appl. Phys. **104**, 073101 (2008)
28. R.J. Potter, N. Balkan, J. Phys.: Condens. Matter **16**, 3387 (2004)
29. Y. Sun, M. Vaughan, A. Agarwal, M. Yilmaz, B. Ulug, A. Ulug, N. Balkan, M. Sopanen, O. Reentilä, M. Mattila,C. Fontaine, A. Arnoult, Phys. Rev. B **75**, 205316 (2007)
30. B.K. Ridley, *Quantum Process in Semiconductors*, 4th edn. (Clarendon, Oxford, 1999)
31. W. Shan, W. Walukiewicz, J.W. Ager III, E.E. Haller, J.F. Geisz, D.J. Friedman, J.M. Olson, S.R. Kurtz, Phys. Rev. Lett. **82**, 1221 (1999)
32. S. Fahy, E.P. O'Reilly, Appl. Phys. Lett. **83**, 3732 (2003)
33. I. Vurgaftman, J.R. Meyer, L.R. Ram-Mohan, J. Appl. Phys. **89**, 5815 (2001)

34. N. Balkan, H. Çelik, A.J. Vickers, M. Cankurtaran, Phys.Rev. B **52**, 17210 (1995)
35. J. Shah, A. Pinczuk, H.L. Störmer, A.C. Gossard, W. Wiegmann, Appl. Phys. Lett. **44**, 322 (1984)
36. C.H. Yang, J.M. Carlson-Swindle, S.A. Lyon, J.M. Worlock, Phys. Rev. Lett. **55**, 2359 (1985)
37. P. Hawker, A.J. Kent, O.H. Hughes, L.J. Challis, Semicond. Sci. Technol. **7B**, 29 (1992)
38. B.N. Murdin, W. Heiss, C.J.G.M. Langerak, S.-C. Lee, I. Galbraith, G. Strasser, E. Gornik, M. Helm, C.R. Pidgeon, Phys. Rev. B **55**, 5171 (1997)
39. K.M. Wennberg, S.N. Ytterboe, C.M. Gould, H.M. Bozler, J. Klem, H. Morkoç, Phys. Rev. B **34**, 4409 (1986)
40. K. Hirakawa, H. Sakaki, Appl. Phys. Lett. **49**, 889 (1986)
41. M.G. Blyumina, A.G. Denisov, T.A. Polyanskaya, I.G. Savel'ev, A.P. Senichkin, Yu V. Shmartsev, JETP Lett. **44**, 331 (1986)
42. S.J. Manion, M. Artaki, M.A. Emanuel, J.J. Coleman, K. Hess, Phys. Rev. B **35**, 9203 (1987)
43. M.J. Barlow, B.K. Ridley, M.J. Kane, S.J. Bass, Solid State Electron. **31**, 501 (1988)
44. A.M. Kreshchuk, M.Yu Martisov, T.A. Polyanskaya, I.G. Savel'ev, I.I. Saidashev, A.Ya Shik, Yu V. Shmartsev, Sov. Phys. Semicond. **22**, 377 (1988)
45. M.E. Daniels, B.K. Ridley, M. Emeny, Solid State Electron. **32**, 1207 (1989)
46. A. Straw, A.J. Vickers, J.S. Roberts, Solid State Electron. **32**, 1539 (1989)
47. D.R. Leadley, R.J. Nicholas, J.J. Harris, C.T. Foxon, Semicond. Sci. Technol. **4**, 879 (1989)
48. Y. Ma, R. Fletcher, E. Zaremba, M. D'Iorio, C.T. Foxon, J.J. Harris, Phys. Rev. B **43**, 9033 (1991)
49. R. Fletcher, J.J. Harris, C.T. Foxon, R. Stoner, Phys. Rev. B **45**, 6659 (1992)
50. K. Hirakawa, M. Grayson, D.C. Tsui, Ç. Kurdak, Phys. Rev. B **47**, 16651 (1993)
51. A.M. Kreshchuk, S.V. Novikov, I.G. Savel'ev, T.A. Polyanskaya, B. Pödör, G. Remenyi, Gy Kovacs, Act. Phys. Pol. A **94**, 415 (1998)
52. E. Tiras, M. Cankurtaran, H. Çelik, N. Balkan, Phys. Rev. B **64**, 085301 (2001)
53. A.J. Vickers, Phys. Rev. B **46**, 3315 (1992)
54. A. Straw, A.J. Vickers, N. Balkan, J.S. Robert, Superlatt. Microstruct. **10**, 203 (1991)
55. E. Tiras, M. Cankurtaran, H. Çelik, A. Boland Thoms, N. Balkan, Superlatt. Microstruct. **29**, 147 (2001)
56. M. Cankurtaran, H. Çelik, E. Tiras, A. Bayrakli, N. Balkan, Phys. Status Solidi B **207**, 139 (1998)
57. H. Çelik, M. Cankurtaran, A. Bayrakli, E. Tiras, N. Balkan, Semicond. Sci. Technol. **12**, 389 (1997)
58. E. Tiras, N. Balkan, S. Ardali, M. Gunes, C. Fontaine, A. Arnoult, Philos. Mag. **91(4)**, 628 (2011)
59. S.B. Lisesivdin, N. Balkan, O. Makarovsky, A. Patane, A. Yildiz, M.D. Caliskan, M. Kasap, S. Ozcelik, E. Ozbay, J. Appl. Phys. **105(9)**, Article Number 093701 (2009)
60. S. Das Sarma, J.K. Jain, R. Jalabert, Phys. Rev. B **37**, 6290 (1988)
61. M. Kogan Sh, Sov. Phys. Solid State **4**, 1813 (1963)
62. B.K. Ridley, *Electrons and Phonons in Semiconductor Multilayers* (Cambridge University Press, New York, 1997)
63. V. Karpus, Sov. Phys. Semicond. **22**, 268 (1988)
64. Y. Okuyama, N. Tokuda, Phys. Rev. B **40**, 9744 (1989)
65. P.K. Milsom, P.N. Butcher, Semicond. Sci. Technol. **1**, 58 (1986)
66. P.J. Price, J. Appl. Phys. **53**, 6863 (1982)

Chapter 4
Optical Modulation Spectroscopy

Robert Kudrawiec and Jan Misiewicz

Abstract This chapter describes the principles of optical modulation spectroscopy. Special attention is focused on photo- and contactless electro-reflectance techniques, which are nondestructive for samples, and are widely applied to study the band structure of various semiconductor materials and low-dimensional heterostructures. For these methods, experimental setups are described and theoretical approaches to analyze the experimental data are discussed. In addition, examples of the application of photo- and contactless electro-reflectance spectroscopies to study optical transitions in III–V(-N) bulk-like epilayers, quantum wells, quantum dots, and device structures are presented.

4.1 Principles of Optical Modulation Spectroscopy

Optical modulation spectroscopy utilizes a general principle of experimental physics, in which a periodically applied perturbation to the sample leads to derivate-like features in the optical response of the sample. Therefore, the objective of optical modulation spectroscopy is to change a parameter inside the sample to produce a change in the optical reflectance (or transmittance) spectrum of the sample. The derivate nature of this technique emphasizes features localized in the photon energy region of interband transitions of semiconductor structures and suppresses uninteresting background effects [1–4]. It means that weak features that may not have been detected in the absolute spectra (reflectance or transmittance spectra) are often enhanced and a large number of sharp spectral features can be observed even at room temperature. An example of the application of optical modulation

R. Kudrawiec (✉)
Institute of Physics, Wrocław University of Technology, Wybrzeże Wyspiańskiego 27, 50-370 Wrocław, Poland
e-mail: robert.kudrawiec@pwr.wroc.pl

A. Patanè and N. Balkan (eds.), *Semiconductor Research*, Springer Series in Materials Science 150, DOI 10.1007/978-3-642-23351-7_4,

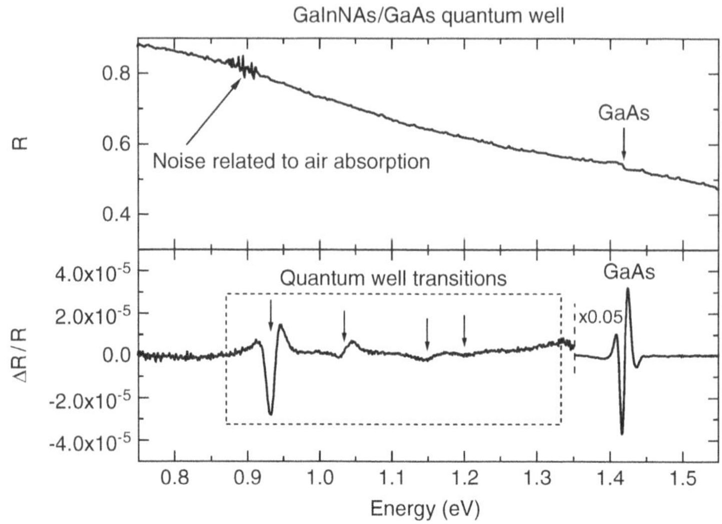

Fig. 4.1 Reflectance and modulated reflectance spectra of a GaInNAs/GaAs quantum well structure measured at room temperature

spectroscopy to study a GaInNAs/GaAs quantum well is shown in Fig. 4.1. In addition, a reflectance (R) spectrum measured for this sample at the same conditions is shown in the figure. In this spectrum, spectral features, which could be associated with optical transitions into the GaInNAs quantum well, are not visible. Only a weak feature, which is related to the GaAs energy gap transition, can be identified in this spectrum. Moreover, it is worth noting that a strong background signal is observed for this structure in the transparency region ($E < 1.4$ eV) since the sample was glued to a metal holder which is mirror-like. Such a background signal is not observed in the modulated reflectance spectrum (photoreflectance spectrum in this case) due to its differential-like character. In this case, only changes in the reflectance spectrum, which are associated with the modulation of the built-in electric field inside the sample, can be detected. As seen in Fig. 4.1, it is possible to detect these changes with a sensitivity better than $10^{-5}\Delta R/R$. Therefore, spectral features related to optical transitions in the GaInNAs quantum well can be observed in this spectrum besides the strong GaAs-related signal, which is visible at the energy of $\sim$1.42 eV. The quantum well-related transitions are not observed in reflectance spectrum due to their weak intensity in comparison to the GaAs-related and background signal. These signals can be eliminated in modulated reflectance spectra since they do not vary with the built-in electric field, which is modulated inside the sample in photoreflectance measurements.

The shape of the spectral features observed in modulation spectra depends on the type of optical transitions as well as on the parameter which is modulated inside the sample. Depending on the modulated parameter, the optical modulation spectroscopy can be classified as (1) electro-, (2) piezo-, and (3) thermo-modulation spectroscopy. In the case of electromodulation spectroscopy, the built-in electric field is modulated inside the sample, whereas in the case of piezo- and thermo-modulation spectroscopy, the modulated parameter is the internal strain and the sample temperature. It is also worth noting that it is also possible to modulate an outside parameter (e.g., the polarization of incident light or its wavelength). This kind of modulation is an *external modulation* in contrast to the *internal modulation* (modulation of a parameter inside the investigated sample), which takes place in electro-, piezo-, and thermo-modulation spectroscopy. In this chapter, we focus on electromodulation (EM) spectroscopy since this method is widely applied to study various semiconductor heterostructures.

4.1.1 Built-in Electric Field in Semiconductor Structures and Its Modulation

The fundamental condition to obtain electro-modulation inside a semiconductor structure and measure EM spectra is the presence of a built-in electric field inside the sample under investigation. This condition is usually fulfilled in majority semiconductor structures since such samples posses a surface electric field or a built-in electric field at internal interfaces. This field results from different level of intentional (or unintentional) doping in individual layers as well as different concentration of native point defects in these layers. It is worth noting that it is usually more difficult to obtain such conditions in narrow gap semiconductors at room temperature and, therefore, the modulation of built-in electric field in such structures can be inefficient at room temperature.

The electric field inside the investigated sample can be modulated by using electric contacts made on the sample or without such contacts in a contactless manner. The last approach is nondestructive for samples and, therefore, experimental techniques with noncontact electromodulation are more desirable. Such techniques are contactless electroreflectance (CER) and photoreflectance (PR) spectroscopies.

In the case of PR spectroscopy, the modulation of the internal electric field in the sample is caused by photoexcited electron–hole pairs created by the pump source (usually a laser) which is chopped with a given frequency. The mechanism of the photo-induced modulation of the built-in electric field (F_{DC}) is illustrated in Fig. 4.2 for the case of an n-type semiconductor. Because of the pinning of the Fermi energy (E_F) at the surface, there exists a space-charge layer. The occupied surface states contain electrons from the bulk (Fig. 4.2a). Photoexcited electron–hole pairs are separated by the built-in electric field, with the minority carriers (holes in this case) being swept toward the surface. At the surface, the holes neutralize the trapped charge, reducing the built-in field from F_{DC} to $F_{DC}-F_{AC}$, where F_{AC} is a change in the built-in electric field (Fig. 4.2b). As it is seen, a built-in electric field in the sample is necessary for photomodulation of the band bending.

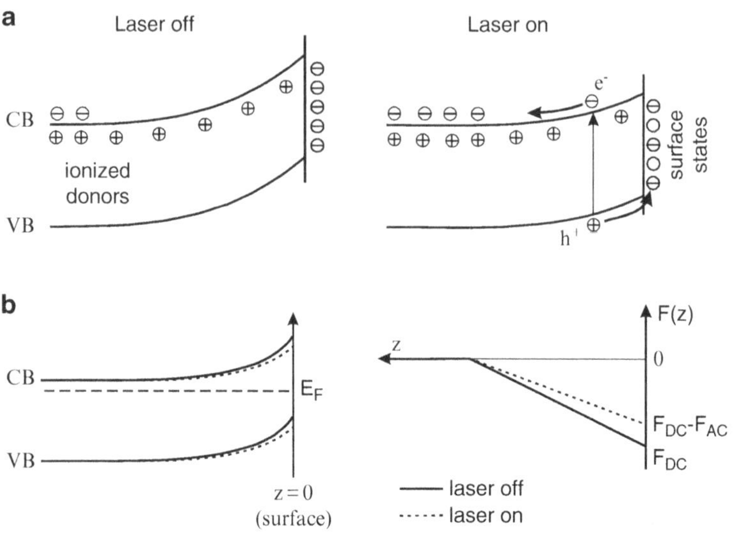

Fig. 4.2 (**a**) Sketch of band bending modulation in photoreflectance for an n-type semiconductor (**b**) shows the position of the Fermi level and the electric field along the z-direction

The built-in electric field can be also modulated in a contactless manner in a capacitor system, see Fig. 4.3. It is expected that the mechanism of band bending modulation in this approach is very similar to that taking place in electroreflectance with electric contacts, i.e., the applied voltage leads to a carrier redistribution which influences the built-in electric field inside the sample. In electroreflectance measurements with electric contacts the maximal amplitude of modulation voltage should be smaller than the Schottky barrier for a given structure. Usually this amplitude is much smaller than the Schottky barrier since only a small modulation of the built-in electric field is needed for electroreflectance measurements. Too large band bending modulation can lead to signals which cannot be analyzed within standard approaches used in EM spectroscopy. In the case of CER measurements, a high voltage is used to modulate the band bending inside the sample. However, most of the voltage drop appears in the air gap between the semitransparent electrode and the sample surface and, therefore, inside the sample only a small band bending modulation takes place at the sample surface.

4.1.2 Experimental Setup for Photo- and Contactless Electro-Reflectance

Both the bright and dark experimental configurations can be applied to measure PR and CER spectra [5]. The classical approach to measure these spectra is the dark configuration [1–4] where the sample is illuminated with monochromatic light.

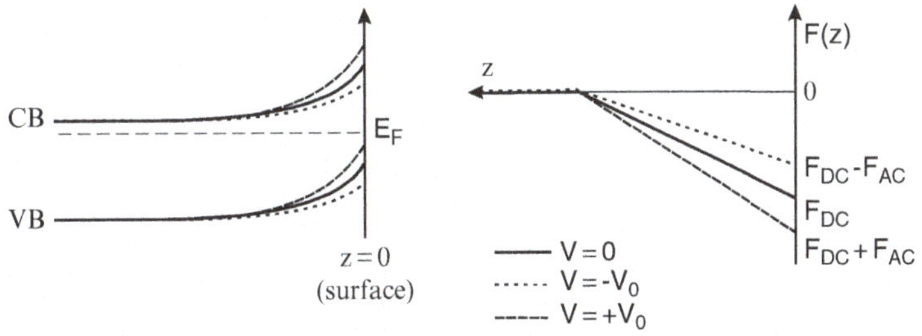

Fig. 4.3 Sketch of band bending modulation in contactless electroreflectance for an n-type semiconductor

Fig. 4.4 Experimental setup for photoreflectance and contactless electroreflectance measurements in the bright configuration

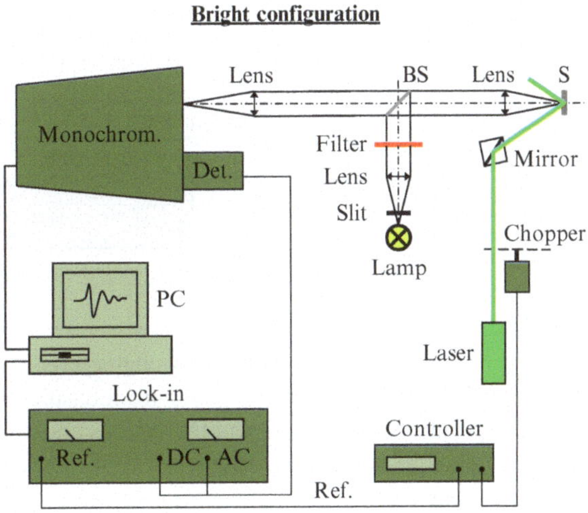

Recently, the bright configuration has been widely applied to measure PR and CER spectra [6–13]. This approach, where the sample is illuminated with white light, possesses several advantages and is very promising in fast measurements with a multichannel detection of PR (or CER) signal [14]. Setups for PR and CER measurements in the two experimental configurations are schematically shown in Figs. 4.4 and 4.5. In addition, the method of PR measurements with Fourier-transform spectrometer is presented in this section.

4.1.2.1 Bright Configuration

The sample is illuminated with white light from a halogen lamp at the near normal incidence [5] as shown in Fig. 4.4. The reflected light is dispersed through the monochromator and detected by a detector. The signal measured by the detector has two components: (1) the DC component which is proportional to $I_0 R$ and (2) AC component which is proportional to $I_0 \Delta R$ (I_0 is the spectral characteristic of

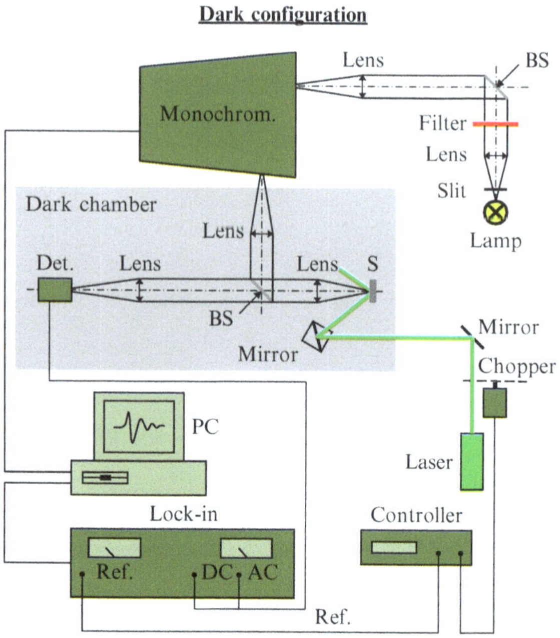

Fig. 4.5 Experimental setup for photoreflectance and contactless electroreflectance measurements in the dark configuration

experimental setup, which results from characteristics of all optical elements in this setup). Both DC and AC components are measured with a lock-in amplifier. The computer program records AC and DC signals and divides the two components thus giving the PR spectrum, $\Delta R/R(E)$, where E is the photon energy of the incident beam. In order to control the spectrum of the probing beam, various edge filters can be used. In addition, the intensity of the probing beam can be controlled by the voltage on the halogen lamp. It is worth noting that the total power of the white spectrum can be below a few mW, i.e., smaller than the power of the pump beam (laser beam) in PR experiment. In this way, the probing beam-induced photovoltaic effect can be significantly reduced. It is worth noting that the photovoltaic effect related to the pump beam is eliminated in CER spectroscopy since the band bending modulation in this technique is induced by an external electric field generated in a capacitor system, i.e., the sample is not illuminated by the pump beam.

4.1.2.2 Dark Configuration

In the classical approach for the measurement of the PR and CER spectra (i.e., the dark configuration), the sample is illuminated with monochromatic light [1–4] as shown in Fig. 4.5. The rest of the measurement concept is the same as in the bright configuration. It is worth noting that the illumination of the sample with the monochromatic probing beam instead of the beam of white spectrum leads to a weaker band flattering, which is related to the photovoltaic effect.

The issue of photovoltage effect can be important when the PR or CER spectroscopy is applied to study the surface electric field since this field is reduced by the photovoltaic effect. In the case of applications of PR and CER spectroscopy to study energies of QW and/or QD transitions, the white light-induced photovoltaic effect does not play any significant role. The other difference between the dark and bright configurations of PR measurements is the contribution of the unwanted photoluminescence signal. In the dark configuration, this signal leads to a constant background which can be electronically compensated. If the photoluminescence signal is too strong, the electronic compensation does not help and special optics have to be used to eliminate this signal. In the bright configuration, the problem of unwanted photoluminescence does not exist in spectral ranges where the photoluminescence is not observed. In the spectral range of photoluminescence signal, this signal has to be eliminated by using special optics or other methods.

It is also possible to measure PR (and CER) spectra with two monochromators. One monochromator for making monochromatic probing beam like in the dark configuration and the second monochromator for selective spectral detection of PR (and CER) signal like in the bright configuration. Such an approach includes advantages of the dark configuration (i.e., the weak photovoltaic effect) as well as features of the bright configuration (easier elimination of photoluminescence signal in PR measurements).

4.1.2.3 Photoreflectance

In the case of photoreflectance, the modulation of the built-in electric field inside the sample is caused by photoexcited electron–hole pairs created by the pump source (usually a laser). Photon energy of the pump beam is generally larger than the band gap of the semiconductor being under study. However, there is a possibility to use a below band gap modulation through the excitation of impurity or surface states [15–17]. In general, the efficiency of the band bending modulation with the phonon energy smaller than the energy gap of the investigated sample is much smaller than the efficiency of photomodulation with the phonon energy higher than the energy gap, see, e.g., [16]. It is worth noting that the below band gap modulation does not generate unwanted photoluminescence, which complicates PR measurements especially at low temperatures when the photoluminescence signal is very strong. The pump beam is defocused on the sample to the diameter of the probing beam (2–3 mm) and chopped by a mechanical chopper with a frequency in the range of 100–400 Hz.

4.1.2.4 Contactless Electroreflectance

The CER method utilizes a capacitor-like system with one top semitransparent electrode and one bottom metal-block electrode. A typical capacitor for CER measurements is shown in Fig. 4.6. The semitransparent electrode is made from

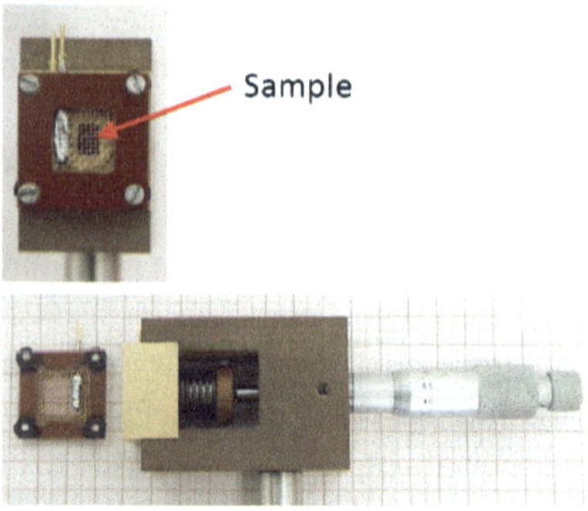

Fig. 4.6 Capacitor used for contactless electroreflectance measurements

a metal wire mesh. The sample is glued to the bottom electrode by using a silver paste. The front electrode is separated from the sample surface by a spacer (e.g., 0.1–1.0 mm). Thus, there is nothing in direct contact with the sample. It means that the sample does not conduct any current and the external electric field is able to change only the carrier distribution inside the sample. The limit for the applied voltage is the electric breakdown in this air gap. It means that the maximal amplitude of electromodulation in the CER technique is usually more limited than the maximal electromodulation amplitude in PR or electroreflectance with electric contacts. It is expected that the amplitude of electromodulation in CER is very weak (few or few tens of kV/cm) since most of the applied voltage drops in the air gap between the semitransparent front electrode and the sample. Typical amplitudes and frequencies of AC square-like voltage used to measure CER spectra are usually in the range of 500–3,500 V and 100–500 Hz, respectively.

4.1.2.5 Fast Measurements with the Multichannel Detection

The multichannel detection of ΔR signal can be realized in the bright experimental configuration. Such a detection allows measurements of PR (or CER) spectrum within a few seconds instead of a few dozen minutes or a few hours typically needed for "step-by-step" measurements with the phase sensitive lock-in detection of ΔR signal. The experimental setup for fast measurements with multichannel detection is analogous to the bright experimental setup with the one-channel detection system, which is shown in Fig. 4.4. The main difference between the two measurements approached lies in the detection of ΔR and R signal. In order to detect the change in the reflectance (i.e., ΔR) by using a multichannel detector (i.e., a CCD camera), two reflectance spectra are recorded: one spectrum with laser off (R^{off}) and the second spectrum with laser on (R^{on}). This procedure takes less than 1 s since the time of R^{off} and R^{on} collections is the order of a few dozen of milliseconds (this integration time is controlled electronically directly on CCD detector; the control of the integration time by a mechanical chopper is rather inaccurate in this regime of times). The difference between the two reflectance spectra (R^{off} and R^{on}) is proportional to changes in the reflectance according to (4.1)

$$I_0 \Delta R = I_0 R^{\mathrm{off}} - I_0 R^{\mathrm{on}}, \tag{4.1}$$

where I_0 is the spectral characteristic of the experimental setup. The PR spectrum can be obtained from R^{off} and R^{on} measurements according to (4.2)

$$\frac{\Delta R}{R} = \frac{I_0 \Delta R}{0.5\,(I_0 R^{\mathrm{off}} + I_0 R^{\mathrm{on}})}. \tag{4.2}$$

In order to improve the signal-to-noise ratio for $\Delta R/R$ spectra, the measurements of R^{off} and R^{on} can be repeated a few (or a few hundreds) times.

Unfortunately, the sensitivity of PR measurements within this approach is two or three orders worse than the sensitivity obtained with the phase sensitive lock-in detection. However, it is expected that the development of multichannel detectors and systems for multichannel signal processing will be able to improve this sensitivity.

4.1.2.6 Fourier Transformed Photoreflectance

PR measurements in the mid-infrared spectral range is very difficult due to weaker sensitivity of detectors in this spectral range, a strong absorption of gases (CO_2 and others) at characteristic wavelengths as well as weaker PR signals for narrow gap semiconductors, which is usually related to a weaker band bending modulation in these semiconductors. In addition, the classical approach with a grating spectrometer is inconvenient in the application to PR measurements in a broad spectral range (e.g., 2–7 μm) due to higher orders of diffraction. Because of this inconvenience, PR measurements with a Fourier transformer [18–20] are recommended for this spectral range. A typical experimental setup for PR measurements with the Fourier-transformer spectrometer is shown in Fig. 4.7. The collimated beam of light from a halogen (or glowbar) lamp is sent to the Michelson interferometer, where the position of the moving mirror can be controlled in the "step-scan" mode. The outgoing beam from the Michelson interferometer is the probing beam which is used to illuminate the investigated sample. This beam is reflected from the sample and detected by a detector in the lock-in technique with the frequency corresponding to the modulation of the pump beam, which also illuminates the sample and is responsible for changes in built-in electric field inside the sample (i.e., changes in the reflectance spectrum). In the "step scan" mode, it is possible to measure the change in the interferogram for a given position of the moving mirror. Such measurements performed for various positions of the moving mirror leads to a "differential" interferogram. The Fourier transformation of such an interferogram leads to a spectrum of elctromodulated changes in reflectance (i.e., $I_0 \Delta R$ spectrum, where I_0 is the spectral characteristic of the experimental setup). In order to obtain $\Delta R/R$ spectrum, the reflectance ($I_0 R$) spectrum has to be measured

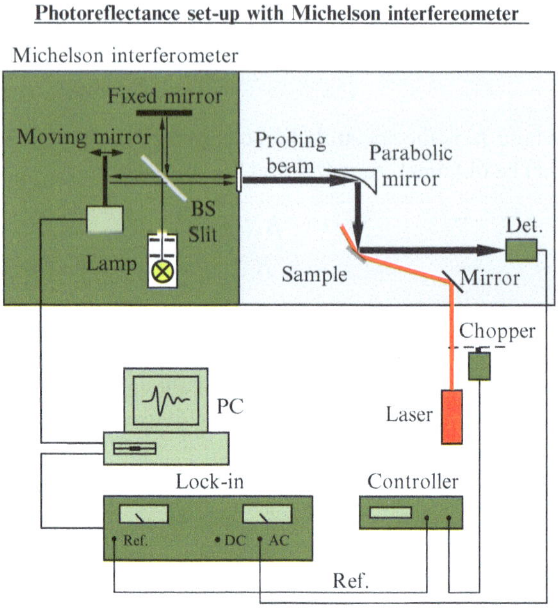

Fig. 4.7 Experimental setup for photoreflectance measurements with Fourier-transformer spectrometer

separately at the same conditions. It is worth noting that the problem of unwanted photoluminescence signal does not appear in PR measurements with the Fourier transformer, which is also very important advantage of this approach to measure PR spectra.

4.1.3 Analysis of Photo- and Electro-Reflectance Spectra

In order to extract maximum information from experimental data, an appropriate analysis of spectral features has to be performed. In PR and CER spectroscopy, relative changes in the reflectivity coefficient are measured. These changes are related to the perturbation of the dielectric function ($\varepsilon = \varepsilon_1 + i\varepsilon_2$) and they are described by the following expression [1–3]

$$\frac{\Delta R}{R} = \alpha(\varepsilon_1, \varepsilon_2)\Delta\varepsilon_1 + \beta(\varepsilon_1, \varepsilon_2)\Delta\varepsilon_2, \tag{4.3}$$

where α and β are the Seraphin coefficients, related to the dielectric function, and $\Delta\varepsilon_1$ and $\Delta\varepsilon_2$ are related by Kramers–Kronig relations.

The detailed lineshape of the PR and CER features related to optical transitions can be discussed in terms of electromodulation mechanisms. Electromodulation can be classified into three categories depending on the relative strengths of characteristic energies [21]. In the low-field regime $|\hbar\Omega| \leq \Gamma$, where Γ is the broadening parameter and $\hbar\Omega$ is the electro-optic energy given by (4.4)

$$(\hbar\Omega)^3 = \frac{q^2\hbar^2 F^2}{2\mu}. \tag{4.4}$$

In the above equation, F is the electric field and μ is the reduced interband mass in the direction of the field. In the intermediate-field case, when $|\hbar\Omega| \geq \Gamma$ and $qFa_0 \ll E_g$ (a_0 is the lattice constant), the Franz–Keldysh oscillations (FKO) appear in the spectrum. In the high-field regime, the electro-optic energy is much greater than the broadening Γ but $qFa_0 \approx E_g$ so that Stark shifts are produced. A theoretical description of the spectral features observed in the EM spectra are discussed in detail in [21–28]. A summary of this discussion is presented below.

4.1.3.1 Low-Field Limit: Third-Derivative Spectroscopy

Electro- and photoreflectance spectra of simple, lightly doped systems, measured under low-field conditions, can often be modeled using Aspnes' third derivate functional form [21], so-called Lorentzian line shape

$$\frac{\Delta R}{R} = \mathrm{Re}\left[A\mathrm{e}^{\mathrm{i}\theta}(E - E_0 + \mathrm{i}\Gamma)^{-m}\right], \tag{4.5}$$

where E_0 is the critical point (CP) energy, Γ is the broadening parameter ($\Gamma \sim \hbar/\tau$), and A and θ are the amplitude and phase factor, respectively. The term m refers to the type of CPs, i.e., the nature of optical transitions, namely: $m = 2, 2.5$, and 3 for an excitonic transition, a three-dimensional CP one-electron transition, and a two-dimensional CP one-electron transition, respectively. This formula is appropriate at low temperatures for high quality structures. At room temperature, the Lorentzian dielectric function is inappropriate and (4.5) must be replaced by more general formula given by (4.6)

$$\frac{\Delta R}{R} = \mathrm{Re}\left[\frac{C\mathrm{e}^{\mathrm{i}\theta}\left[G(z) + \mathrm{i}\cdot F(z)\right]}{E^2}\right], \tag{4.6}$$

where C is a constant, $G(z)$ and $F(z)$ are the electro-optic functions, and the broadening is included via the normalized energy $z = (E_0 - E + i\cdot\Gamma)/\hbar\theta$. Equation (4.6) becomes complex for the Gaussian-like form of the dielectric function. Such form of (4.6) is called as first derivate Gaussian. Note that for some temperatures, the line shape, and this form is the most appropriate at room temperature and/or for a sample with significant inhomogeneities. Note that in the same range of temperatures, the line shape is an intermediate form between Lorentzian and Gaussian [2, 3]. Also it is worth noting that the shape of a Lorentzian line with $m = 3$ is very close to the shape of a Gaussian line.

4.1.3.2 Low-Field Limit: First-Derivative Spectroscopy

The perturbation due to the changes of the electric field does not accelerate carriers in their bound states, such as excitons, carriers in quantum wells, or impurities. These types of particles are confined in space which does not have a translational symmetry. For bound states, the EM line shape has first-derivative character [25], and the changes in the dielectric function may be expressed as

$$\Delta\varepsilon = \left[\frac{\partial\varepsilon}{\partial E_g}\frac{\partial E_g}{\partial F_{AC}} + \frac{\partial\varepsilon}{\partial\Gamma}\frac{\partial\Gamma}{\partial F_{AC}} + \frac{\partial\varepsilon}{\partial I}\frac{\partial I}{\partial F_{AC}}\right] F_{AC}, \tag{4.7}$$

where F_{AC} is the change in the built-in electric field and I is the intensity of the optical transition. Equation (4.7) can be rewritten as [28]

$$\Delta\varepsilon_i = \left[A_E f_E^i + A_\Gamma f_\Gamma^i + A_I f_I^i\right]\frac{I}{\Gamma}F_{AC}, \quad i = 1, 2, \tag{4.8}$$

with

$$\begin{aligned} A_E &= \frac{1}{E_g}\frac{\partial E_g}{\partial F_{AC}}, & f_E^i &= \frac{\partial\varepsilon_i}{\partial E_g}, \\ A_\Gamma &= \frac{1}{\Gamma}\frac{\partial\Gamma}{\partial F_{AC}}, \text{ and } & f_\Gamma^i &= \frac{\partial\varepsilon_i}{\partial\Gamma}, \\ A_I &= \frac{1}{I}\frac{\partial I}{\partial F_{AC}}, & f_I^i &= \frac{\partial\varepsilon_i}{\partial I}. \end{aligned} \tag{4.9}$$

Depending on the broadening mechanism, the unperturbated dielectric function can be either Lorenzian or Gaussian. For quantum wells, the dielectric function is excitonic even at elevated temperatures. Thus, for the quantum microstructures, the Lorenzian or Gaussian profiles of dielectric function are appropriate. The Lorenzian dielectric function can be written as [24]

$$\varepsilon = 1 + \frac{I}{E - E_g + \mathrm{i}\Gamma}. \tag{4.10}$$

The modulation terms of (4.8) are given by

$$\begin{aligned} f_E^1 &= \frac{y^2-1}{(y^2+1)^2}, \; f_\Gamma^1 = f_E^2, \quad f_I^1 = \frac{y}{y^2+1}, \\ f_E^2 &= \frac{-2y}{y^2+1}, \quad f_\Gamma^2 = -f_E^1, \; f_I^2 = \frac{-1}{y^2+1}, \end{aligned} \tag{4.11}$$

where $y = (E - E_g)/\Gamma$. If the intensity modulation terms are ignored, only two independent line-shape factors [see (4.11)] do not vanish. The combined spectral dependence can then be expressed by (4.5) with $m = 2$ [24].

The unperturbated dielectric function of a Gaussian profile is given by [28]

$$\varepsilon = 1 + I(L_1 + \mathrm{i}L_2), \tag{4.12}$$

where

$$L_1 = \frac{y}{\Gamma}\Phi(1, 3/2, -y^2/2),$$
$$L_2 = \sqrt{\frac{\pi}{2}}\frac{1}{\Gamma}\exp(-y^2/2). \tag{4.13}$$

Here Φ is the confluent hypergeometric function. In this case, the modulation terms of (4.8) can be written as

$$f_E^1 = -\Phi(1, 1/2, -y^2/2),$$
$$f_E^2 = -\sqrt{\frac{\pi}{2}}y\exp(-y^2/2),$$
$$f_\Gamma^1 = -2y\Phi(2, 3/2, -y^2/2),$$
$$f_\Gamma^2 = -\sqrt{\frac{\pi}{2}}(y^2 - 1)\exp(-y^2/2), \tag{4.14}$$
$$f_I^1 = y\Phi(1, 3/2, -y^2/2),$$
$$f_I^2 = -\sqrt{\frac{\pi}{2}}\exp(-y^2/2).$$

Hence, for the dielectric function of Gaussian type one can derive

$$\frac{\Delta R}{R} = \left[Af_E^1 + Bf_E^2\right]. \tag{4.15}$$

4.1.3.3 Intermediate Field Limit: Franz–Keldysh Oscillations

When the low-field criteria are not satisfied, but $eFa_0 \ll E_g$, the dielectric function can exhibit FKO. Although the exact form of $\Delta R/R$ for the intermediate-field case with the broadening is quite complicated, Aspnes and Studna [22] have derived a relatively simple expression

$$\frac{\Delta R}{R}(E) \propto \frac{1}{E^2(E - E_g)}\exp\left[-2(E - E_g)^{1/2}\frac{\Gamma}{(\hbar\Omega)^{3/2}}\right] \times \cos\left[\frac{4}{3}\frac{(E - E_g)^{3/2}}{(\hbar\Omega)^{3/2}} + \chi\right], \tag{4.16}$$

where E is the energy, E_g is the energy gap for the considered optical transition, $\hbar\Omega$ is the electro-optic energy defined by (4.4), Γ is the linewidth, and χ is an arbitrary phase factor. From the above equation, the position of a nth extreme in the FKO is given by

$$n\pi = \frac{4}{3}\left[\frac{E_n - E_g}{\hbar\Omega}\right]^{3/2} + \chi, \tag{4.17}$$

where E_n is the photon energy of the nth extreme [23]. A plot $4/3\pi(E_n - E_g)^{3/2}$ vs. the index number n will yield a straight line with slope $(\hbar\Omega)^{3/2}$. Therefore, the electric field F can be directly obtained from the period of FKO if μ [i.e. the electron–hole reduced mass, see (4.4)] is known. Conversely, μ can be measured if the electric field is known.

The dominant field in the structure determines the period of the FKO. In the above expressions, the nature of that field was not specified. There are two limiting cases to be considered. If modulation is from a flat band, i.e., no presence of a DC field, then the field is clearly the modulating field F_{AC}. A more interesting situation occurs when there exists a large DC electric field in the material and a small modulating field is applied, i.e., $F_{AC} \ll F_{DC}$. In this case, the period of the FKO is given by F_{DC} and not by F_{AC} [27]. Shen and Pollak [27] have even considered the case when F_{AC} is not small compared to F_{DC}. They have shown that even for F_{AC}/F_{DC} as large as 0.15, the first few FKO are still determined by F_{DC}.

4.1.3.4 Kramers–Kronig Analysis in the Low-Field Limit

An alternative method for estimating E_g, Γ, and A (but not θ) related to the lineshape described by (4.5) or (4.6), in the so called "low field limit", has been developed based on a Kramers–Kronig transformation of the PR (or CER) spectrum [29, 30]. In this case, the complex PR function is defined as below:

$$\Delta\tilde{\rho}(E) = \Delta\rho_R(E) + i\Delta\rho_I(E) = \Delta\rho(E)e^{i\Theta(E)}, \tag{4.18}$$

where the measured value of the PR (or CER) signal is equal to

$$\frac{\Delta R}{R} = \Delta\rho_R = \Delta\rho \cdot \cos\Theta. \tag{4.19}$$

After mathematical considerations similar to that carried out for other optical constants functions [30], the Kramers–Kronig relation for the complex PR function, (4.18), can be written:

$$\Delta\rho_I(E_0) = \frac{2E_0}{\pi} P \int_{E_a}^{E_b} \frac{\Delta R}{R} \frac{1}{E_0^2 + E^2} \mathrm{d}E, \tag{4.20}$$

where $P\int$ means the principal value of the integral and (E_a, E_b) is the energy range in which $\Delta R/R$ is measured. The values of E_a and E_b should be chosen in a way that $\Delta R/R(E_a) = \Delta R/R(E_b) = 0$ having all the oscillations interesting for us inside this range. Knowing the values of $\Delta\rho_I$, the modulus $\Delta\rho$ can be determined by means of the simple formula:

$$\Delta\rho = \sqrt{\left(\frac{\Delta R}{R}\right)^2 + (\Delta\rho_I)^2}, \tag{4.21}$$

which can be treated as the modulus of the PR resonance.

In the framework of fitting procedure with (4.5), the modulus of PR resonance is defined by (4.22)

$$|\Delta\rho(E)| = \frac{|A|}{[(E - E_0)^2 + \Gamma^2]^{\frac{n}{2}}}. \tag{4.22}$$

In order to plot the modulus of the PR (or CER) resonance, A, E_0, θ, and Γ parameters have to be determined by using the fitting procedure. The advantage of Kramers Kronig analysis (KKA) is to avoid the fitting procedure. It is very useful if the line shape of EM data changes between Lorentzian- and Gaussian-like, i.e., between (4.5) and (4.6). Notice that the integrated modulus of the EM resonance is interpreted as the oscillator strength of the optical transition while E_0 and Γ are the transition energy and the transition broadening, respectively. The broadening is related to the sample quality and temperature.

4.1.3.5 Fitting Procedure Vs. Kramers–Kronig Analysis

A comparison of the analysis of EM data by using standard fitting procedure and KKA is shown in Fig. 4.8. Figure 4.8a shows experimental data (open points) which are fitted by (4.5) with $m = 3$ (dashed line), together with the modulus of the resonance (4.22) (solid line), whereas Fig. 4.8b shows the same experimental data (open points) with the modulus of PR resonance obtained by using KKA (4.21) (solid line). The dashed line in Fig. 4.8b shows $\Delta\rho_I$ obtained from (4.20). It is clearly visible that the modulus of the PR resonance obtained by the two ways are almost the same.

4.2 Applications of Photo- and Contactless Electro-Reflectance

PR and CER spectroscopies can be applied to study various issues in semiconductors materials and their low dimensional structures [6–10, 13, 17, 31–46]. In this section, we present examples of the application of PR and CER spectroscopy to study bulk-like epilayers and low dimensional heterostructures including quantum wells and quantum dots. In addition, some examples of the application PR or CER spectroscopy to study semiconductor devices are presented in this section.

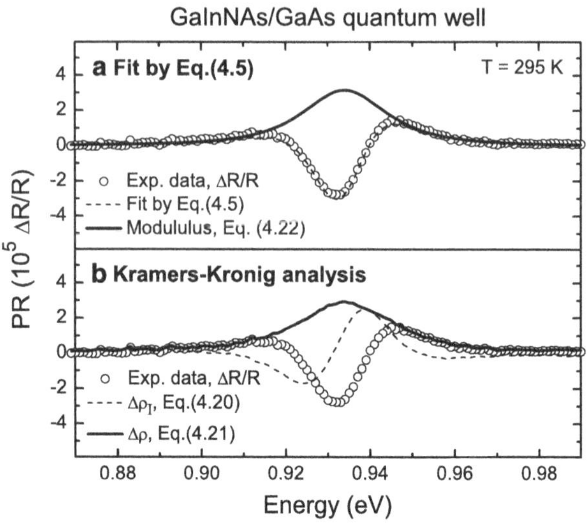

Fig. 4.8 Analysis of the ground state transition for a GaInNAs/GaAs quantum well

4.2.1 Bulk-Like Epilayers

PR and CER spectroscopies probe optical transitions between extended states (i.e., optical transitions at CPs of the band structure) in contrast to photoluminescence which is very sensitive to localized states. For bulk-like materials, these techniques are excellent to study the energy gap, its homogeneity as well as energies of optical transitions at other CPs of the band structure. Because of these features PR and CER spectroscopies are often used to determine the spin–orbit splitting and the energy separation between light holes and heavy holes in strained epitaxial layers. For high quality materials, free exciton transitions can be also studied by PR and CER techniques.

Figure 4.9 shows PR spectra measured at 10 K for 2-μm-thick GaN epilayers grown by metalorganic chemical vapor deposition on sapphire and truly bulk GaN substrate obtained by the ammonothermal method [17]. In order to avoid the problem with unwanted photoluminescence signal, the band bending was modulated by a green laser with the photon energy of 2.3 eV. In this case, the modulation mechanism can be related to an ionization of donor- and/or acceptor-like states at the surface and in the whole volume of GaN crystal. This mechanism is very efficient for this sample since the PR signal is quite strong ($\sim 10^{-2}\ \Delta R/R$). It is worth noting that the efficiency of band bending modulation with the photon energy smaller than the energy gap of the investigated sample varies very significantly between samples and, therefore for many samples, such a modulation is very difficult (or even impossible) to obtain. However, it is rather recommended to try this possibility of band bending modulation for low temperature PR measurements since such a modulation very easily solves problems with the unwanted photoluminescence signal.

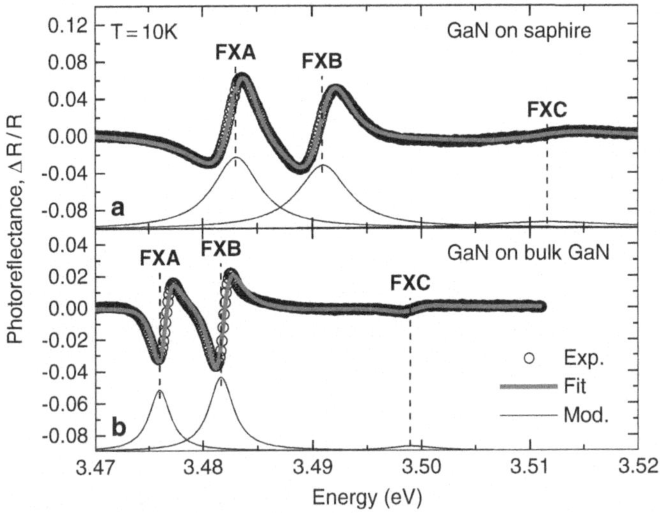

Fig. 4.9 Photoreflectance spectra of heteroepitaxial (**a**) and homoepitaxial (**b**) GaN layers measured at 10 K [17]

In Fig. 4.9, three excitonic transitions (FXA, FXB, and FXC related to the $\Gamma_9^V - \Gamma_7^C, \Gamma_7^V$, (upper band)–$\Gamma_7^C$ and Γ_7^V (lower band)–Γ_7^C interband transitions, respectively) are clearly observed for GaN layers. For the homoepitaxial layer (GaN layer deposited on truly bulk GaN substrate), the exciton transitions are observed at lower energies, see Fig. 4.9b. In addition, it is clearly visible that the linewidth of exciton resonances is much smaller for the homoepitaxial layer. In order to extract the energy and linewidth of excitonic transitions from PR spectra, the standard fitting procedure assuming Lorentzian lineshape (4.5) with $m = 2$ can be applied. The fitting curves are shown by thick gray lines in Fig. 4.9 together with the moduli of individual resonances (thin solid lines) obtained according to (4.22). The energy position for the FXA, FXB, and FXC excitonic transitions in the heteroepitaxial layer have been determined to be 3.483, 3.491, and 3.512 eV, respectively. For the homoepitaxial layer, these transitions are observed at lower energies: $E_{\mathrm{FXA}} = 3.4760$, $E_{\mathrm{FXB}} = 3.4817$, and $E_{\mathrm{FXC}} = 3.4991$ eV, because of no residual strain in this layer [17]. In addition, it is worth noting that the exciton linewidth for the homoepitaxial layer is about twice smaller than the linewidth for the heteroepitaxial layer, see moduli of PR resonances in Fig. 4.9. Such narrow PR resonances correspond to a homogeneous broadening of excitonic transitions and indicate no residual strain (or negligible small residual strain) in the epilayer grown on truly bulk GaN substrate. In the case of heteroepitaxial layers, it is difficult to eliminate the residual strain and therefore the low temperature broadening of PR resonances corresponds to the strain distribution in these layers. The presented results clearly show that PR spectroscopy is an excellent tool to study energies of excitonic transitions as well as their broadenings, i.e., the sample

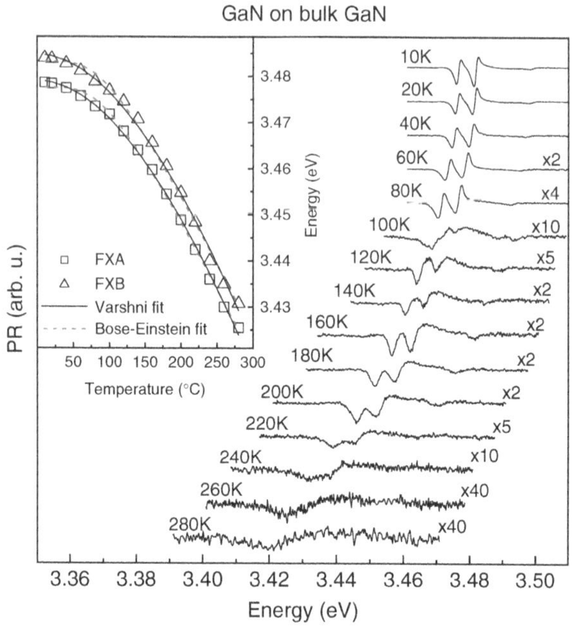

Fig. 4.10 Temperature dependence of photoreflectance spectrum measured for the homoepitaxial GaN layer together with energies of excitonic transitions (*inset*) extracted from photoreflectance spectra and fitted by Varshni and Bose–Einstein formulas [17]

quality/homogeneity. Because of these advantages, PR (and CER) spectroscopy is often used to study the temperature dependence of excitonic transitions and their broadenings. An example of such studies for the homoepitaxial GaN layer is shown in Fig. 4.10. A detailed discussion of this issue including the fitting of experimental data by Varshni and Bose–Einstein formulas can be found in [17].

For semiconductor alloys, the carrier localization phenomenon is very often observed and its magnitude can be studied comparing the energy of photoluminescence peak with the energy of PR (or CER) transition (i.e., evaluating the Stokes shift, which is defined as the energy difference between absorption and emission). Figure 4.11 shows such studies for InGaN layers with various indium concentrations. For all samples, the photoluminescence peak is observed at lower energy than the CER transition because of indium content fluctuations typical of this alloy. In addition, the strong alloy inhomogeneities are manifested by very large broadening of CER resonances. In this case, CER resonance was fitted by (4.5) with $m = 3$ (Lorenzian-like line), which can be applied to very inhomogeneous systems instead of the Gaussian-like line given by (4.6). For this alloy, the Stokes shift has been found to increase with the increase in indium concentration, see Fig. 4.11 and [31]. In this case, CER spectroscopy is also an excellent tool to study the bowing parameter for the InGaN alloy since the energy of the CER resonance corresponds to the average energy gap for a given sample, whereas photoluminescence probes mainly In-rich parts of InGaN layers and, hence, it leads to an overestimated bowing parameter, see details in [31].

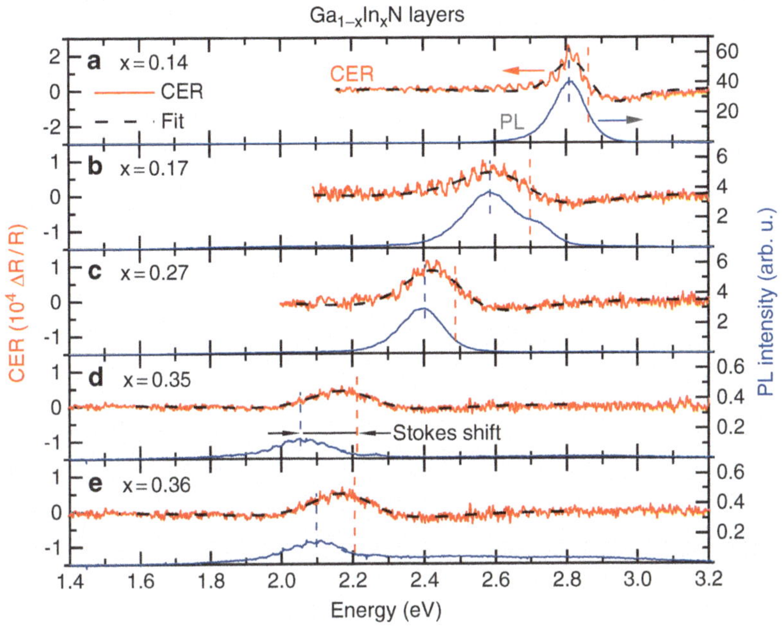

Fig. 4.11 Contactless electroreflectance and photoluminescence spectra measured at room temperature for $Ga_{1-x}In_xN$ layers of various indium concentrations [31]

Besides the fundamental transition, the optical transitions between the spin–orbit splitted band and other bands can be studied by PR and CER spectroscopies. Figure 4.12 shows PR spectrum measured at 10 K for a GaNAs layer grown on GaAs substrate. In addition to the sharp PR resonance at ~1.5 eV, which is associated with the photon absorption in GaAs cap/buffer layer, three quite broad optical transitions are clearly visible in this spectrum. Beside the fundamental transition (E_-) observed at the energy of 1.25 eV, the optical transition between the spin–orbit splitted band and the conduction band is clearly visible at the energy of 1.57 eV. In addition, the optical transition between the valence band and the E_+ band is observed at the energy of 1.95 eV. The E_- and E_+ transitions are typical of GaNAs layers and they can be interpreted in the framework of the band anticrossing model [47] introduced for dilute nitrides (i.e., III–V alloys with a few percent of nitrogen atoms). For this layer, the light- and heavy-hole related bands should be splitted since this layer is tensely strained; however, this splitting is smaller than the broadening of CER resonance and, therefore, it is rather difficult to resolve two individual resonances for the E_- and E_+ transition. In order to observe the strain-related splitting between the light and heavy hole bands, the alloy homogeneity of the investigated sample should be very high. An example of the application of PR spectroscopy to study the

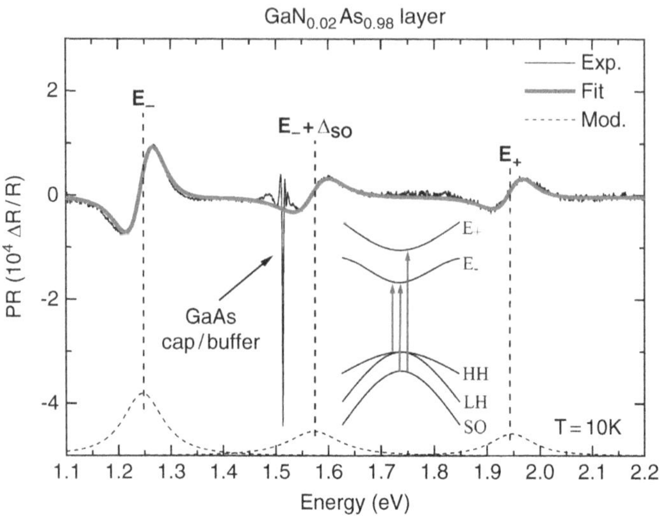

Fig. 4.12 Photoreflectance spectrum of the $Ga_{0.98}N_{0.02}As$ layer measured at 10 K in broad spectral range

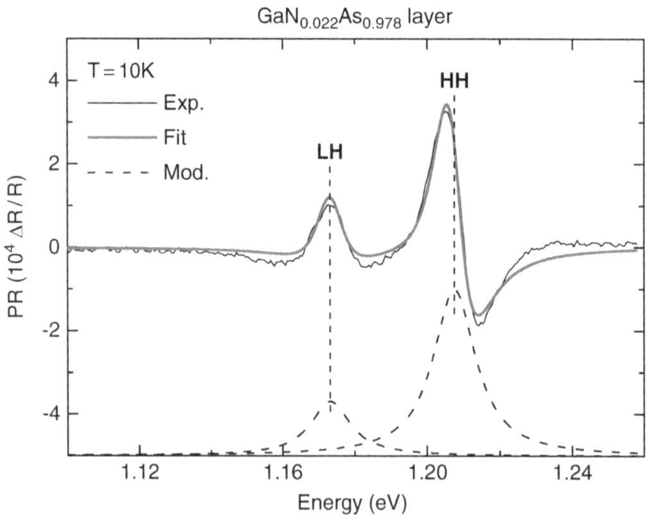

Fig. 4.13 Photoreflectance spectrum of high quality $Ga_{0.978}N_{0.022}As$ layer measured at 10 K in the range of the fundamental transition

strain-related splitting between light and heavy holes in high quality GaNAs layer is shown in Fig. 4.13. For this layer, the broadening of PR resonances is smaller than the splitting between the light and heavy holes and therefore two resonances are visible in PR spectrum. The two resonances have been fitted by (4.5) with $m = 2$.

4.2.2 *Quantum Wells*

The application of PR (and CER) spectroscopy to quantum well (QW) structures allows to determine the type of QW (type I, II, or III) as well as the band gap discontinuities at QW interfaces. It is possible since PR (and CER) probes energies of optical transitions between both the ground and excited states. Energies of QW transitions between excited states strongly depend on the band gap alignment between the well and barrier materials and hence PR (and CER) spectroscopy can be used to determine the conduction band offset Q_C (or valence band offset Q_V), which is defined by (4.23a) as

$$Q_C = \frac{\Delta E_C}{(\Delta E_C + \Delta E_V)} \times 100\% \tag{4.23a}$$

$$Q_V = 100\% - Q_C \tag{4.23b}$$

where ΔE_C and ΔE_V are the conduction- and valence-band discontinuities at the heterojunction for unstrained materials, as illustrated in Fig. 4.14. They are designated *unstrained* because they are the "natural" band offset corresponding to the ab initio calculations. From a laser device perspective, the most interesting values are the band gap discontinuities with the strain corrections, shown in Fig. 4.14 as ΔE_C^* and ΔE_V^{HH}. Generally, the most appropriate approach to heterojunction band offsets is to determine the Q_C (or Q_V) since a QW can be grown on different substrates. For identical QWs grown on different substrates, the Q_C does not vary whereas the ΔE_C^* and ΔE_V^{HH} discontinuities vary due to different lattice strains. Moreover, the value of ΔE_C^* and ΔE_V^{HH} discontinuities can be calculated if the *unstrained* "natural" offset, Q_C (or Q_V), is known. Thus, the Q_C value is more universal.

In order to find the band structure for the investigated QW, energies of QW transitions have to be measured by PR (or CER) spectroscopy and next the measured

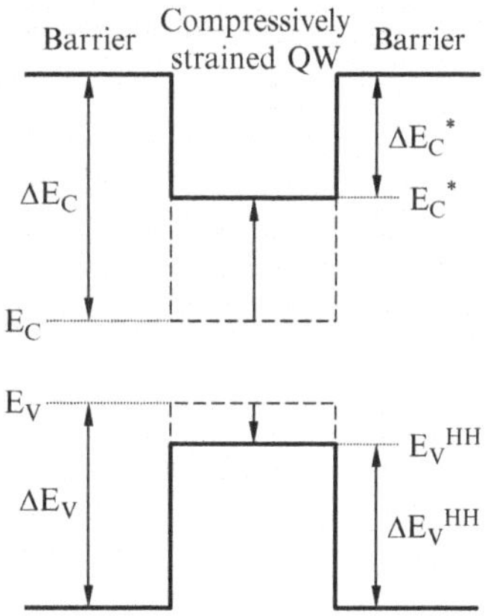

Fig. 4.14 Band gap alignment in a strained quantum well before (*dashed lines*) and after (*solid lines*) strain corrections

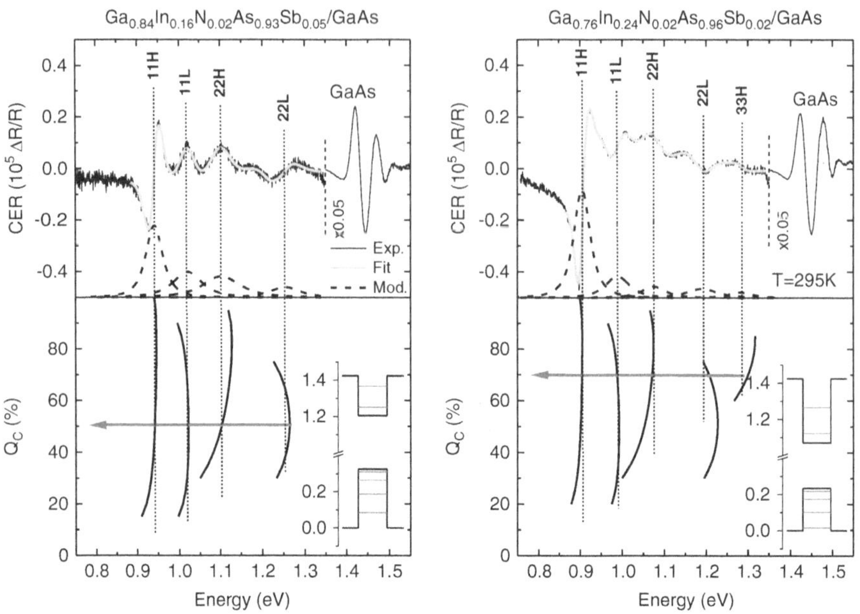

Fig. 4.15 Contactless electroreflectance spectra measured at room temperature for $Ga_{0.84}In_{0.16}N_{0.02}As_{0.93}Sb_{0.05}/GaAs$ (*left panel*) and $Ga_{0.76}In_{0.24}N_{0.02}As_{0.96}Sb_{0.02}/GaAs$ (*right panel*) quantum well together with theoretical calculations of energies of optical transitions for these quantum wells [32]. The *insets* show energy levels for electrons and heavy holes in these quantum wells

energies have to be compared with theoretical calculations performed for various conduction (valence) band offsets. Usually, the agreement between experimental data and theoretical calculations is observed in a narrow range of the conduction band offset. This range can be better determined if a set of QWs of various widths and the same content is studied by PR (or CER) spectroscopy. Such an approach to study the band structure of QWs was applied to different material systems [7, 9, 32–35]. An example of such studies for two GaInNAsSb/GaAs QW structures with various GaInNAsSb contents [32] are shown in Fig. 4.15 and discussed below.

In general, the number of CER resonances necessary to simulate experimental data is established arbitrary analyzing individual CER spectra by nagged eye (note that preliminary calculations help to make a reasonable decision). It is obvious that larger number of CER resonances better simulate experimental data but each resonance has to be justified since imaginary transitions can be generated if too many resonances were used to simulate CER spectrum. Therefore, the moduli of individual CER resonances are plotted and evaluated. The evaluation means a comparison of transition intensities and the energy difference between individual resonances. Usually, the allowed transitions are much stronger than partially allowed ones. Moreover, the broadening of each CER resonances should be comparable. Also it is acceptable that the broadening is higher for excited state transitions, e.g.

the partially allowed transitions. Next energies of CER resonances are compared with theoretical calculations for various Q_C. This procedure allows us to identify CER resonances and determine the conduction band offset.

For the investigated GaInNAsSb/GaAs QW structures the strongest CER signal is observed at an energy of ~1.42 eV originating from the band-to-band absorption in the GaAs barrier/cap layers. Below the GaAs signal, CER features associated with optical transitions in GaInNAsSb/GaAs SQW are clearly observed. These features are fitted by using the (4.5). The fitting curves are shown in Fig. 4.15 as thick solid grey lines. In addition, the moduli of individual resonances (dashed lines) obtained according to (4.22) are shown in this figure. As was mentioned the identification of CER resonances is possible due to a series of calculations with various Q_C. The notation *kl*H(L) denotes the transition between the *k*th heavy-hole (light-hole) valence subband and the *l*th conduction subband. The resonance at the lowest energy originates from the 11H transition, which is a fundamental transition for the two samples. In addition to the 11H transition, the CER spectra show an 11L transition (the lowest energy transition for light-holes) and transitions between the excited QW states such as 22H, 33H, and 22L transition. The partially forbidden transitions, such as the 31H transition, were also considered. However, it was concluded that they are weak in comparison with the allowed transitions and, therefore, they can be neglected in the fitting procedure. In this manner, the number of CER resonances was reduced to the number of allowed transitions. Finally, for the first sample (GaInNAsSb/GaAs QW with 16% In content) four resonances have been resolved in the CER spectrum. With the indium content increase, all QW transitions shift to the red. For the second sample (GaInNAsSb/GaAs QW with 24% In content) an additional CER resonance related to the 33H transition appears between 22L and GaAs transitions.

The best agreement between experimental data and theoretical calculations for the $Ga_{0.84}In_{0.16}N_{0.02}As_{0.93}Sb_{0.05}/GaAs$ and $Ga_{0.76}In_{0.24}N_{0.02}As_{0.96}Sb_{0.02}/GaAs$ QW is observed when the Q_C equals 50 and 70%, respectively. The energy level structure for the two samples for the determined Q_C is shown in insets in Fig. 4.15. It is clearly visible that two electron and four heavy-hole states are confined for these QWs. Moreover it is clearly visible that the change of GaInNAsSb content significantly changes quantum confinement in the conduction and valence band.

4.2.3 *Quantum Dots*

Quantum dots (QDs) usually contain more than one confined state for electrons and holes. Optical transitions between these states can be probed by photoluminescence spectroscopy despite the emission-like character of this technique. It is possible because of small volume of QDs and their low density in comparison to photon density used in standard photoluminescence measurements. It is worth noting that the same photon density applied to photoluminescence measurements of a QW allows to detect the fundamental transition only because of larger volume of QW in

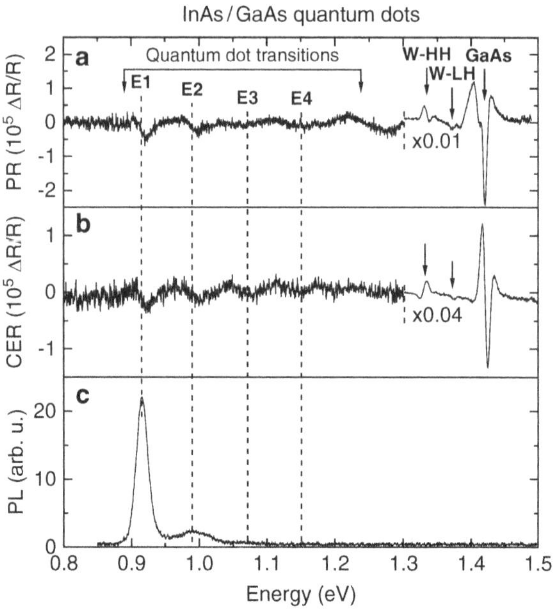

Fig. 4.16 Photoreflectance (**a**), contactless electroreflectance (**b**), and photoluminescence (**c**) spectra measured at room temperature for the InAs/GaAs quantum dot sample [16]

comparison to the QD volume and faster carrier thermalization and recombination. PR and CER spectroscopy due to their high sensitivity and absorption-like character are able to probe each allowed optical transition in QDs as well as optical transitions in the wetting layer [16, 36, 37]. For QDs embedded in a QW it is also possible to probe the band structure of the surrounded QW [38, 39]. Because of high photoluminescence signal from zero dimensional objects CER spectroscopy is more recommended to study QD structures than the PR spectroscopy since any problems with the unwanted photoluminescence signal do not exist in CER measurements. A few examples of the application of PR and CER spectroscopy to study QD structures is presented below.

Figure 4.16 shows PR, CER and PL spectra measured for the structure with InAs/GaAs QDs. The strongest PR and CER signal is observed for the GaAs barrier. Below the GaAs-related resonance optical transitions associated with the photon absorption in QDs are clearly visible. These transitions correlate very well with peaks observed in the PL spectrum. Between the QD- and GaAs-related transitions two resonances are clearly observed at the energy of 1.33 and 1.37 eV. These resonances are associated with the photon absorption in the InAs wetting layer (W-HH transition is attributed to the optical transition between the heavy-hole and electron level whereas the W-LH transition is attributed to the optical transition between the light-hole and electron level which are confined in the wetting layer). Such transitions are not observed in PL spectrum since the photogenerated carriers relax to QDs before recombination. A similar situation takes place for QDs embedded in a QW. For such structures PR or CER can be used to investigate the band structure of the surrounding QW.

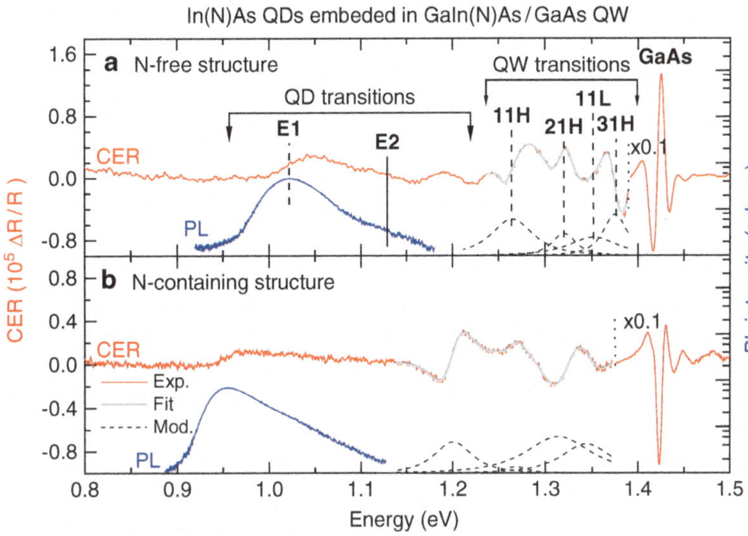

Fig. 4.17 Contactless electroreflectance (*red curves*) and photoluminescence (*blue curves*) spectra measured at room temperature for the InAs/GaInAs/GaAs (**a**) and InNAs/GaInNAs/GaAs (**b**) quantum well structure [39]

Figure 4.17 shows CER and PL spectra measured for In(N)As QDs embedded in GaIn(N)As QW [39]. First it is visible that the QD emission redshifts due to the incorporation of N atoms into the InAs/InGaAs system. A redshift of CER resonance, which is related to QD absorption, is also observed in these spectra. The strongest CER signal, which is observed at ~1.42 eV, is associated with bulk-like absorption in GaAs barriers. In addition to the GaAs transition and QD transitions, CER resonances, which are associated with optical transitions in the In(N)As/GaIn(N)As/GaAs QW, are clearly visible in CER spectra (see CER resonances between ~1.15 and 1.4 eV). It is worth noting that the optical transitions in the surrounded QW are not observed in the PL spectra whereas QD transitions (both the ground and excited state ones) are usually visible in PL spectra, see for example the PL spectrum for the N-free sample. Therefore, the most interesting part of the CER spectrum, in this case, is associated with the optical transitions in the In(N)As/GaIn(N)As/GaAs QW. A knowledge of the number of confined states in the surrounded QW and their energies is important from the point view of such processes as (1) the thermal quenching of QD emission and (2) the carrier relaxation to the ground state in QDs.

In order to extract energies of the QW transitions, CER spectra in the range of QW transitions have been fitted by (4.5) with $m = 3$. Fitting curves and modulus of the CER resonances are plotted in Fig. 4.17. The identification of the resonances was possible on the basis of the calculations performed in the framework of the effective mass approximation, see details in [39]. As in the previous case, the notation klH(L) in Fig. 4.17 denotes the transition between the kth heavy-hole (*light-hole*) valence subband and the lth conduction subband. In addition to the fundamental QW transition (11H) the CER spectra show an 11L transition (i.e., the fundamental transition for light-holes) and transitions between excited QW states (21H and 31H). Note that the selection rules for a square-like QW do not work for this QW due to strong asymmetry in this system.

The calculated band structure and wavefunctions for In(N)As/GaIn(N)As/GaAs QWs can be found in Ref.[39]. It has been found that the QWs confine one electron, one light-hole and three heavy-hole states. The energy separation between the QD and QW ground state transitions in this system has been found to be $\sim$250 meV. Taking into account that QD excited state transitions are observed in CER spectra, it can be concluded that in this system the energy separation between the electron (heavy-hole) QD and QW ground states is much higher than the thermal energy at room temperature. It means that this design of gain medium is very perspective for laser applications.

4.2.4 Device Structures

PR spectroscopy is often used to characterize the full active part in device structures. For laser structures it is possible to probe the band structure of a given QW (or QD) inside the structure as well as the energy gap of cladding layers and other layers in the structure [40,41]. In addition, it is possible to determine the built-in electric field in p–i–n junction from measurements of the period of Franz–Keldysh oscillations [41]. However it should be stressed that the deepness of sample probing is much smaller in CER spectroscopy and, therefore, this technique cannot be applied to measure the built-in electric field in p–i–n junctions. The internal electric field is also very important issue for field effect transistor heterostructures. In this case, PR and CER spectroscopy can be used to measure the distribution of the built-in electric field in transistor heterostructures [42]. A few examples of the application of PR and CER spectroscopy to study transistor heterostructures and laser structures are presented below.

Figure 4.18a, b show CER and PR spectra, respectively, measured at room temperature for an AlGaN/GaN transistor heterostructure grown by metalorganic vapor phase epitaxy on the [111] oriented Si substrate. The strong CER and PR signal below 3.4 eV is uninteresting since it is associated with the Fabry–Perot (F–P) oscillation, which is very strong for this heterostructure due to the high contrast of refractive index between the Si substrate and epitaxial layers. The most interesting part of PR and CER spectra for this heterostructure is associated with photon absorption in the AlGaN layer. The AlGaN-related resonance is almost the same in CER and PR spectra and exhibits a strong Franz–Keldysh oscillation which is typical of photon absorption in a layer with the built-in electric field. The period of FKO is directly connected with the value of the electric field. A conventional method to determine the built-in electric field from FKO is to use an asymptotic expression for electro-reflectance which is given by (4.16). In this case

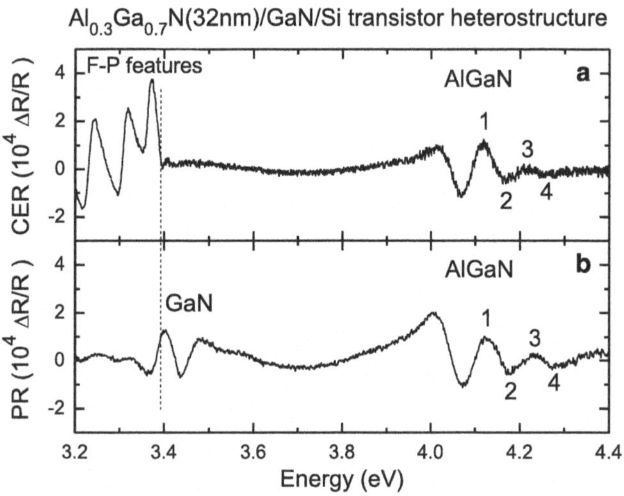

Fig. 4.18 Contactless electroreflectance (**a**) and photoreflectance (**b**) spectra measured at room temperature for the AlGaN/GaN transistor heterostructure [42]

Fig. 4.19 Analysis of the period of Franz–Keldysh oscillations for the AlGaN-transition

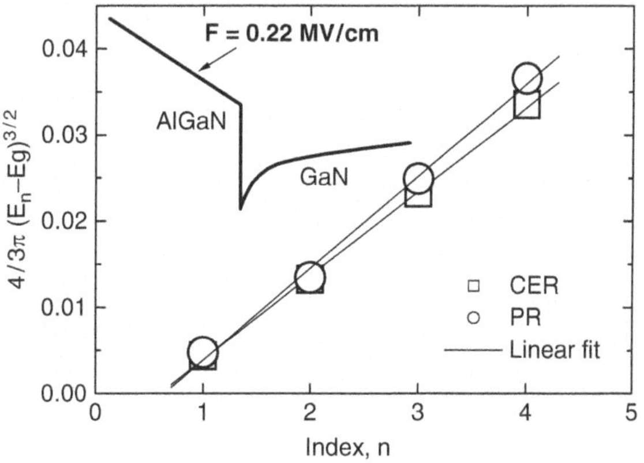

the extrema of FKO are given by (4.17) and a plot of $(E_n - E_g)^{3/2}$ vs. n yields a straight line with a slope proportional to F. An analysis of the AlGaN-related FKO period for the AlGaN/GaN heterostructure is shown in Fig. 4.19. Assuming that the internal electric field in AlGaN is homogeneous (as shown in the sketch in Fig. 4.19) it has been determined that the built-in electric field in the AlGaN layer equals ~0.22 MV/cm.

It is also worth noting that any GaN-related CER signal (i.e., a well-defined CER resonance at ~3.4 eV) is not observed in the CER spectrum whereas such a signal is observed in the PR spectrum. This behavior of CER spectra is typical of AlGaN/GaN structures with the two dimensional electron gas (2DEG) at the interface. This phenomenon is associated with a screening of the band bending

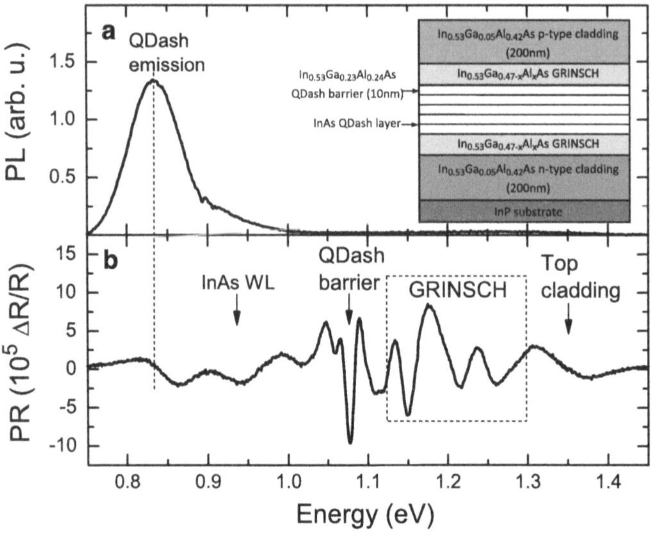

Fig. 4.20 Photoluminescence (**a**) and photoreflectance (**b**) spectra measured at room temperature for the quantum dash laser structure emitting at 1.5 μm [40]

modulation in the GaN layer by the 2DEG existing at the AlGaN/GaN interface from the external electromodulation generated by the capacitor-like system [6].

Figure 4.20 shows photoluminescence and PR spectra measured at room temperature for an InAs quantum dash (QDash) laser structure designed for emission at 1.5 μm [40]. In the case of photoluminescence spectrum, only one broad peak is observed at 0.833 eV, whereas a lot of spectral features are visible in the PR spectrum. The photoluminescence peak is associated with the carrier recombination inside QDashes. The PR signal observed at the lowest energy is quite well correlated with the emission peak. Thus this spectral feature is attributed to the photon absorption inside InAs QDashes. The next PR feature, observed between 0.9–1.0 eV, is also associated with the photon absorption in InAs. This resonance is connected with the InAs wetting layer, i.e., a very thin QW which is formed before the appearance of QDash islands [48]. The PR resonance observed at ~1.07 eV is associated with the photon absorption inside the 10 nm thick QDash barriers, i.e., $In_{0.53}Ga_{0.23}Al_{0.24}As$ layers. The PR signal observed between 1.1 and 1.3 eV is attributed to the $In_{0.53}Ga_{0.47-x}Al_xAs$-GRINSCH (GRaded INdex Separate Confinement Heterostructure) layer. The broad PR resonance at the energy of 1.3–1.4 eV is related to the photon absorption in the top cladding layer. This example clearly shows that PR spectroscopy is an excellent tool to study optical transitions in various layers in complex laser structures.

Acknowledgements The authors acknowledge many colleagues (Alfred Forchel from Wurzburg University, James Harris from Stanford University, Klaus Ploog from Paul-Drude-Institut für Festkörperelektronik, Czeslaw Skierbiszewski from UNIPRESS, Robert Dwilinski from

AMMONO company, Mariusz Rudzinski from ITME, and their coworkers) for delivering samples for PR and CER studies as well as the COST Action MP0805 and the MNiSW grant related to this action for financial support.

References

1. M. Cardona, *Modulation Spectroscopy*, (Academic, New York, 1969)
2. O.J. Glembocki, B.V. Shanabrook, in *Photoreflectance Spectroscopy of Microstructures*, ed. by D.G. Seiler, C.L. Littler Semiconductors and Semimetals, vol 36, p. 221 (Academic, New York, 1992)
3. F.H. Pollak, in *Modulation Spectroscopy of Semiconductors and Semiconductor Microstructures*, ed. by M. Balkanski, Handbook on Semiconductors, vol 2, p. 527 (Elsevier Science B.V., Amsterdam, 1994)
4. J. Misiewicz, P. Sitarek, G. Sek, R. Kudrawiec, Mater. Sci. **21**, 264 (2003)
5. R. Kudrawiec, J. Misiewicz, Rev. Sci. Instrum. **80**, 096103 (2009)
6. R. Kudrawiec, M. Syperek, M. Motyka, J. Misiewicz, R. Paszkiewicz, B. Paszkiewicz, M. Tlaczala, J. Appl. Phys. **100**, 013501 (2006)
7. R. Kudrawiec, M. Gladysiewicz, J. Misiewicz, H.B. Yuen, S.R. Bank, M.A. Wistey, H.P. Bae, J.S. Harris Jr, Phys. Rev. B **73**, 245413 (2006)
8. R. Kudrawiec, M. Motyka, J. Misiewicz, A. Somers, R. Schwertberger, J. P. Reithmaier, A. Forchel, A. Sauerwald, T. Kümmell, G. Bacher, J. Appl. Phys. **101**, 013507 (2007)
9. R. Kudrawiec, M. Motyka, J. Misiewicz, M. Hummer, K. Rossner, T. Lehnhardt, M. Muller, A. Forchel, Appl. Phys. Lett. **92**, 041910 (2008)
10. R. Kudrawiec, J. Misiewicz, M. Rudzinski, M. Zajac, Appl. Phys. Lett. **93**, 061910 (2008)
11. R. Kudrawiec, J. Misiewicz, J. Phys.: Confer. Ser. **146**, 012029 (2009)
12. R. Kudrawiec, R. Kucharski, M. Rudzinski, M. Zajac, J. Misiewicz, W. Strupinski, R. Doradzinski, R. Dwilinski, J. Vac. Sci. Technol. A **28**, L18 (2010)
13. R. Kudrawiec, Phys. Stat. Sol. (b) **247**, 1616 (2010)
14. R. Kudrawiec, A. Khachapuridze, G. Cywinski, T. Suski, J. Misiewicz, Phys. Stat. Sol. (a) **206**, 847 (2009)
15. P.J. Klar, C.M. Townsley, D. Wolverson, J.J. Davies, D.E. Ashenford, B. Lunn, Semicond. Sci. Technol. **10**, 1568 (1995)
16. M. Motyka, G. Sek, R. Kudrawiec, J. Misiewicz, L.H. Li, A. Fiore, J. Appl. Phys. **100**, 073502 (2006)
17. R. Kudrawiec, M. Rudzinski, J. Serafinczuk, M. Zajac, J. Misiewicz, J. Appl. Phys. **105**, 093541 (2009)
18. T.J.C. Hosea, M. Merrick, B.N. Murdin, Phys. Stat. Sol. A **202**, 1233 (2005)
19. J. Shao, W. Lu, F. Yue, X. Lu, W. Huang, Z. Li, S. Guo, Rev. Sci. Instrum. **78**, 013111 (2007)
20. M. Motyka, G. Sek, J. Misiewicz, A. Bauer, M. Dallner, S. Hofling, A. Forchel, Appl. Phys. Express **2**, 126505 (2009)
21. D.E. Aspnes, Surf. Sci. **37**, 418 (1973)
22. D.E. Aspnes, A.A. Studna, Phys. Rev. B **7**, 4605 (1973)
23. D.E. Aspnes, Phys. Rev. **B10**, 4228 (1974)
24. D.E. Aspnes, in *Handbook on Semiconductors*, vol. 2, ed. by M. Balkanski (North-Holland, Amsterdam, 1980), p. 109
25. B.V. Shanabrook, O.J. Glembocki, W.T. Beard, Phys. Rev. **B35**, 2540 (1987)
26. O.J. Glembocki, Proc. SPIE Proc. **1286**, 2 (1990)
27. H. Shen, F.H. Pollak, Phys. Rev. **B42**, 7097 (1990)
28. Y.S. Huang, H. Qiang, F.H. Pollak, J. Lee, B. Elman, J. Appl. Phys. **70**, 3808 (1991)
29. T.J.C. Hosea, Phys. Stat. Sol. (b) **182**, K43 (1994)

30. K. Jezierski, P. Markiewicz, J. Misiewicz, M. Panek, B. Sciana, T. Korbutowicz, M. Tlaczala, J. Appl. Phys. **77**, 4139 (1995)
31. R. Kudrawiec, M. Siekacz, M. Krysko, G. Cywinski, J. Misiewicz, C. Skierbiszewski, J. Appl. Phys. **106**, 113517 (2009)
32. R. Kudrawiec, H.B. Yuen, M. Motyka, M. Gladysiewicz, J. Misiewicz, S.R. Bank, H.P. Bae, M.A. Wistey, J.S. Harris, J. Appl. Phys. **101**, 013504 (2007)
33. R. Kudrawiec, M. Gladysiewicz, J. Misiewicz, F. Ishikawa, K.H. Ploog, Appl. Phys. Lett. **90**, 041916 (2007)
34. R. Kudrawiec, M. Motyka, M. Gladysiewicz, J. Misiewicz, J.A. Gupta, G.C. Aers, Solid State Commun. **138**, 365 (2006)
35. R. Kudrawiec, P. Podemski, M. Motyka, J. Misiewicz, J. Serafinczuk, A. Somers, J.P. Reithmaier, A. Forchel, Superlatt. Microstruct. **46**, 425 (2009)
36. R. Kudrawiec, M. Motyka, J. Misiewicz, A. Somers, R. Schwertberger, J. P. Reithmaier, A. Forchel, A. Sauerwald, T. Kümmell, G. Bacher, J. Appl. Phys. **101**, 013507 (2007)
37. G. Sek, R. Kudrawiec, M. Motyka, P. Poloczek, W. Rudno-Rudzinski, P. Podemski, J. Misiewicz, Phys. Stat. Sol. (a) **204**, 400 (2007)
38. M. Motyka, R. Kudrawiec, G. Sek, J. Misiewicz, I.L. Krestnikov, S. Mikhrin, A. Kovsh, Semicond. Sci. Technol. **21**, 1402 (2006)
39. M. Motyka, R. Kudrawiec, G. Sek, J. Misiewicz, D. Bisping, B. Marquardt, A. Forchel, M. Fischer, Appl. Phys. Lett. **90**, 221112 (2007)
40. M. Motyka, W. Rudno-Rudzinski, G. Sek, R. Kudrawiec, J. Misiewicz, A. Komers, R. Schwertberger, J.P. Reithmaier, A. Forchel, Phys. Stat. Sol. (c) **4**, 350 (2007)
41. A. Podhorodecki, J. Andrzejewski, R. Kudrawiec, J. Misiewicz, J. Wojcik, B.J. Robinson, T. Roschuk, D.A. Thompson, P. Mascher, J. Appl. Phys. **100**, 013111 (2006)
42. R. Kudrawiec, B. Paszkiewicz, M. Motyka, J. Misiewicz, J. Derluyn, A. Lorenz, K. Cheng, J. Das, M. Germain, J. Appl. Phys. **104**, 096108 (2008)
43. M. Munoz, S. Guo, X. Zhou, M.C. Tamargo, Y.S. Huang, C. Trallero-Giner, A.H. Rodriguez, Appl. Phys. Lett. **83**, 4399 (2003)
44. R. Kudrawiec, J. Misiewicz, Q. Zhuang, A.M.R. Godenir, A. Krier, Appl. Phys. Lett. **94**, 151902 (2009)
45. J.D. Wu, Y.S. Huang, B.S. Li, A. Shen, M.C. Tamargo, K.K. Tiong, J. Appl. Phys. **108**, 123105 (2010)
46. E. Cánovas, D. Fuertes Marrón, A. Martí, A. Luque, A.W. Bett, F. Dimroth, S.P. Philipps, Appl. Phys. Lett. **97**, 203504 (2010)
47. W. Shan, W. Walukiewicz, J.W. Ager III, E.E. Haller, J.F. Geisz, D.J. Friedman, J.M. Olson, S.R. Kurtz, Phys. Rev. Lett. **82**, 1221 (1999)
48. W. Rudno-Rudzinski, G. Sek, K. Ryczko, R. Kudrawiec, J. Misiewicz, A. Somers, R. Schwertberger, J.P. Reithmaier, A. Forchel, Appl. Phys. Lett. **86**, 101904 (2005)

Chapter 5
Photoluminescence: A Tool for Investigating Optical, Electronic, and Structural Properties of Semiconductors

G. Pettinari, A. Polimeni, and M. Capizzi

Abstract Theoretical basis and typical experimental setups of photoluminescence, PL, are briefly described. The investigation by PL of some of the fundamental properties of compound semiconductors and alloys—e.g., optical gap, type and density of shallow impurities, effects of structural disorder in alloys and at heterostructure interfaces, and carrier effective masses—is illustrated. The effects on PL spectra of magnetic fields are discussed, together with the validity limits of perturbation and numerical models for different ratios of magnetic and excitonic energies.

5.1 Introduction

In a luminescence process, a system decays from an excited to a lower electronic state, usually the ground state, by emitting a photon. In photoluminescence, PL, the excitation energy is carried by a photon, in cathodoluminescence by accelerated electrons, in electroluminescence by carriers electrically injected into the material. Therefore, luminescence spectroscopy is a sensitive, non destructing tool—provided exceedingly high excitation power densities are not used—for characterizing the optical and electronic properties of either "ideal" or "real" semiconductors and insulators. Hereafter, we will refer only to photoluminescence (a rich and detailed review of photoluminescence processes can be found in [1–3]).

Photoluminescence is the counterpart of absorption; however, while absorption is determined by high density-of-states, DOS, and energy levels, photoluminescence is ruled mainly by fast relaxation processes: At low temperature, transitions from

M. Capizzi (✉)
CNISM-Dipartimento di Fisica, Sapienza Università di Roma, P.le A. Moro 2, 00185 Roma, Italy
e-mail: mario.capizzi@roma1.infn.it

A. Patanè and N. Balkan (eds.), *Semiconductor Research*, Springer Series in Materials Science 150, DOI 10.1007/978-3-642-23351-7_5,

the lowest excited state or from states with the shortest *radiative* lifetime are usually observed. Photoluminescence is then most suited to investigate transitions at, or just below, the energy gap in intrinsic materials or deeper states in doped or defected materials. PL linewidth and peak energy provide information about impurity concentration and type [4] and alloy inhomogeneity in bulk materials or size/composition fluctuations in low-dimensional systems. Moreover, PL measurements as a function of temperature allow to investigate nonradiative centers and their thermal activation energy. Second, heterostructures based on thin layers, which cannot be probed directly by optical absorption, can be investigated by PL or PL excitation, PLE, spectroscopy, a technique that mimics optical absorption.

Photoluminescence has been first used [5][1] to investigate the optical properties of a semiconductor, Ge, in the early 1950s. However, it became a widespread technique only in the 1960s when different laser sources became available. Lasers provide, indeed, stable and intense optical sources easily collimated to micron-sized spots as to permit the investigation of compositional homogeneity in alloys [6]. Moreover, lasers provide optical sources that can be modulated in time down to the picosecond domain.

In the following, we will summarize the main features of a PL apparatus and the different relaxation and recombination mechanisms. Then, we will provide some examples of how this technique can be used to investigate the material electronic properties and their dependence on external perturbations, such as magnetic fields.

5.2 Generalities

5.2.1 Experimental Apparatus

A typical PL setup consists of a light source, at fixed or variable energy, a variable temperature cryostat, a single or double monochromator, detectors, and related electronics. A few optical components (polarizing and interferential filters, variable retarders, lenses, and beam-splitter cubes) are also required.

As exemplified in Fig. 5.1a, the light sources are usually gas- or solid-state lasers, whose emitted light is filtered by a narrow-band filter or an interferential high-energy pass filter, F1, in order to avoid the collection of spurious laser lines at the sample emission energies. The laser light is usually directed onto the sample by a couple of mirrors, M1 and M2, and focused onto the sample by a lens L1 of focal length f_1, thus resulting in a small laser spot on the sample. A microscope objective replaces L1 when the laser spot has to be reduced to micrometer size for micro-photoluminescence (μ-PL) measurements. A cube beam-splitter is inserted

[1] It was the subject of an abstract presented at the 1951 Autumn Meeting of the American Physical Society (Trinity College, Hartford, Connecticut, November 3, 1951)

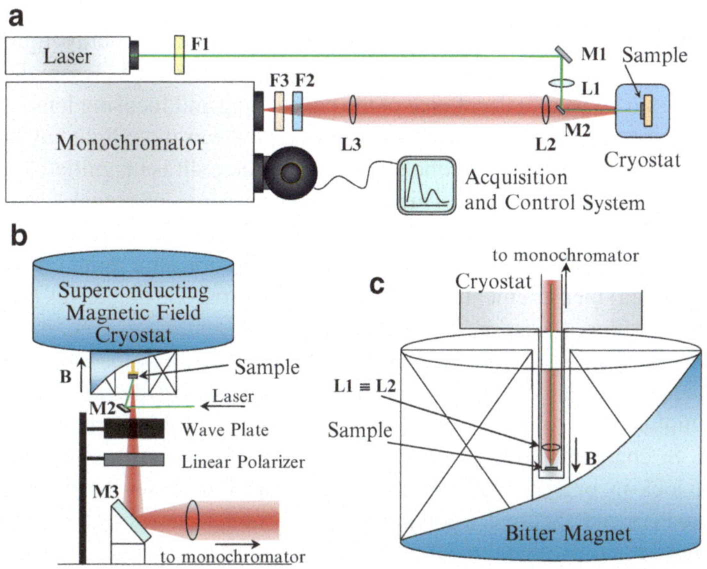

Fig. 5.1 (**a**) Sketch of the experimental setup for PL measurements. M*i* label mirrors, L*i* indicate lenses and F*i* filters. (**b**) Sketch of the cryostat for measurements under low-to-medium (up to ∼12 T) magnetic fields as obtained by superconducting coils. A linear polarizer and a wave plate are used to resolve the polarization state of the emitted photons. (**c**) Sketch of the cryostat for measurements under low-to-high (up to ∼30 T) magnetic fields as obtained in a Bitter magnet by conventional copper disks

before the sample whenever a continuous measurement of the exciting power or wavelength is required.

The sample is cooled down in a pumped liquid-He bath cryostat when the temperature should be ramped between ∼1 and 300 K, or in a closed-cycle variable-temperature cryostat if temperatures below 10 K are not required. A continuous-flux disk-shaped cryostat is commonly used in the case of μ-PL measurements.

The light emitted from the sample is collected by a collimating lens, L2, placed at distance from the sample equal to its focal length, f_2. A second lens, L3, of focal length f_3, focuses the parallelized light onto the entrance slit of the monochromator. The F-number of the focusing lens, F_3, should match that of the monochromator, F_M, in order to fill up exactly the monochromator grating: This condition maximizes monochromator energy resolution and avoids loss of signal and unwanted additional stray light in the monochromator. These two different lenses, a collimating and a focusing one, are used to get flexibility in the geometrical disposition of cryostat and monochromator. Moreover, polarizing filters and variable retarders may be easily inserted between those two lenses to analyze the PL signal.

A cold filter F2 and/or a low energy-pass filter, F3, are placed just before the monochromator entrance slit. This cleans the spectra from unwanted signals due, e.g., to laser light scattered and/or reflected from the sample and falling onto the filters, as well as from monochromator second-order transmissions. This is particularly important when the PL signal is very weak and stray light in the monochromator is comparable to, or even higher than, the sample signal. Particular attention should be paid to the choice of the collimating and focusing lenses. A high numerical aperture, short focus collimating lens L2 collects more light but the image of the emitting spot size on the monochromator entrance slit is magnified by a factor f_3/f_2 by the focusing lens L3. This may result in a loss of signal for a narrow entrance slit, or a loss of energy resolution if the slit is widened to accept the full magnified image of the emitting spot. Such considerations are quite critical in μ-PL (or μ-Raman) measurements, where high NA microscope objectives (with focal lengths of the order of 1 cm or less) are used.

The energy-dispersed PL signal is collected at the exit slit of the monochromator and converted into an electric signal by a single channel detector, either as a photomultiplier or a solid-state photodetector, or by a multiple channel detector, such as a Si charged coupled device or an InGaAs linear array. A single photon counter, a lock-in, or a dedicated electronic equipment increases the signal-to-noise ratio of the PL signal. Second, a computer controls sample and detector temperature, exciting power and PL signal reading, monochromator-wavelength change, etc.

A similar setup is used to perform PL measurements under an applied magnetic field, either up to 14–18 T, as obtained by a superconducting coil, see Fig. 5.1b, or up to 33 T, as typically obtained in a Bitter magnet by conventional copper disks, see Fig. 5.1c. In the former case, a wave-plate and a linear polarizer are inserted before any optical component (mirrors or lenses) in order to perform PL measurements as a function of light polarization. In the latter case, an insert is placed in the magnet bore (usually $3 \div 5$ cm in diameter). In this insert, a same short-focus lens (L1 $\equiv$ L2, $f \sim 1$ cm) focuses the laser light onto the sample and collects/collimates the light emitted from the sample.

A similar setup is used for PLE measurements. In this case, however, a variable energy source is used, namely, either a high brilliance tungsten lamp monochromatized by a short-focal-length monochromator with a high through-output, or a tunable laser. In an automatized system, a Fourier interferometer reads and records the exciting wavelength.

5.2.2 *Absorption Spectroscopy*

The absorption coefficient $\alpha(\hbar\omega)$ is intimately connected with the electronic DOS and band structure of a material. According to the Fermi Golden Rule, the probability per unit time and volume, $R(\hbar\omega)$, that a given system will absorb an incoming photon of energy $\hbar\omega$ with the transition of an electron from the valence band, VB, to the conduction band, CB, is given by

$$R(\hbar\omega) = \frac{2\pi}{\hbar} \sum_{\boldsymbol{k}_{\mathrm{CB}},\boldsymbol{k}_{\mathrm{VB}}} |\langle \mathrm{CB}|H_{\mathrm{eR}}|\mathrm{VB}\rangle|^2 \cdot \delta[E(\boldsymbol{k}_{\mathrm{CB}}) - E(\boldsymbol{k}_{\mathrm{VB}}) - \hbar\omega]. \tag{5.1}$$

$E(\boldsymbol{k}_{\mathrm{CB}})$ $[E(\boldsymbol{k}_{\mathrm{VB}})]$ gives the energy dispersion of the conduction (valence) band, the Hamiltonian H_{eR} accounts for the electron–radiation interaction, and $|\langle \mathrm{CB}|H_{\mathrm{eR}}|\mathrm{VB}\rangle|^2$ is the probability for an electron in a valence band state $|\mathrm{VB}\rangle$ to be excited into a conduction band state $|\mathrm{CB}\rangle$ by absorbing a photon. The crystal translational symmetry requires that $\boldsymbol{k}_{\mathrm{CB}} = \boldsymbol{k}_{\mathrm{VB}} + \boldsymbol{G}$, where $\boldsymbol{G}$ is a reciprocal lattice vector. If the energy dispersion curves are restricted to the first Brillouin zone, $\boldsymbol{G} = 0$ and $|\langle \mathrm{CB}|\boldsymbol{H}_{\mathrm{eR}}|\mathrm{VB}\rangle|^2$ $(= |\langle \mathrm{CB}|\hat{\varepsilon}\cdot\boldsymbol{p}|\mathrm{VB}\rangle|^2$ in the *dipole approximation*) is nonzero only for transitions between states with a same crystal momentum $\boldsymbol{k}$ (the photon momentum can be usually disregarded). Here $\hat{\varepsilon}$ is a unit vector parallel to the potential vector $\boldsymbol{A}$ and $\boldsymbol{p}$ is the electron momentum. Moreover,

$$|\langle \mathrm{CB}|\boldsymbol{H}_{\mathrm{eR}}|\mathrm{VB}\rangle|^2 = \left(\frac{e}{mc}\right)^2 |A(\hbar\omega)|^2 \cdot |P_{\mathrm{VB,CB}}|^2, \tag{5.2}$$

where $A(\hbar\omega)$ is the amplitude of the potential vector $\boldsymbol{A}$. If the momentum matrix element is assumed to weakly depend on $\boldsymbol{k}$ and $|P_{\mathrm{VB,CB}}|^2$ is set equal to a constant, $R(\hbar\omega)$ can be rewritten as

$$R(\hbar\omega) = \frac{2\pi}{\hbar}\left(\frac{e}{m\omega}\right)^2 \left|\frac{\mathcal{E}(\hbar\omega)}{2}\right|^2 |P_{\mathrm{VB,CB}}|^2 \sum_{\boldsymbol{k}_{\mathrm{CB}},\boldsymbol{k}_{\mathrm{VB}}} \delta[E(\boldsymbol{k}_{\mathrm{CB}}) - E(\boldsymbol{k}_{\mathrm{VB}}) - \hbar\omega], \tag{5.3}$$

where $\mathcal{E}(\hbar\omega)$ is the amplitude of the incident electric field.

The *absorption coefficient* of the material, $\alpha(\hbar\omega)$, is defined through the *Lambert relation* $\mathrm{d}I/\mathrm{d}x = -\alpha(\hbar\omega)I$, which yields $I(x,\hbar\omega) = I(0,\hbar\omega)\mathrm{e}^{-\alpha(\hbar\omega)x}$, where $I(x,\hbar\omega)$ is the intensity of the electromagnetic radiation as it has propagated through a distance x in the material. $\alpha(\hbar\omega)$ is connected to $R(\hbar\omega)$ through the relation

$$\alpha(\hbar\omega) = \frac{8\pi}{c\cdot n}\frac{\hbar\omega}{|\mathcal{E}(\hbar\omega)|^2}\cdot R(\hbar\omega), \tag{5.4}$$

where n is the refractive index of the material. We can write then

$$\alpha(\hbar\omega) = \frac{\hbar}{c\cdot n}\left(\frac{2\pi e}{m}\right)^2 \cdot \frac{1}{\hbar\omega}\cdot |P_{\mathrm{VB,CB}}|^2 \sum_{\boldsymbol{k}_{\mathrm{CB}},\boldsymbol{k}_{\mathrm{VB}}} \delta[E(\boldsymbol{k}_{\mathrm{CB}}) - E(\boldsymbol{k}_{\mathrm{VB}}) - \hbar\omega]. \tag{5.5}$$

If the summation over $\boldsymbol{k}$ is replaced by an integral, we obtain

$$\alpha(\hbar\omega) = \frac{\hbar}{c\cdot n}\left(\frac{2\pi e}{m}\right)^2 \cdot \frac{1}{\hbar\omega}\cdot |P_{\mathrm{VB,CB}}|^2 \int_{\mathrm{BZ}} \frac{\mathrm{d}^3k}{4\pi^3}\delta[E(\boldsymbol{k}_{\mathrm{CB}}) - E(\boldsymbol{k}_{\mathrm{VB}}) - \hbar\omega]. \tag{5.6}$$

The integral in (5.6) resembles an electronic DOS and can be defined directly in terms of the gradient of the difference between the energy dispersion of the conduction and valence bands, $E_{\mathrm{CV}}(\boldsymbol{k}) = E(\boldsymbol{k}_{\mathrm{CB}}) - E(\boldsymbol{k}_{\mathrm{VB}})$. Thus, one obtains

$$\alpha(\hbar\omega) = \frac{\hbar}{c\cdot n}\left(\frac{2\pi e}{m}\right)^2 \cdot \frac{1}{\hbar\omega}\cdot |P_{\mathrm{VB,CB}}|^2 \int_{S_{\mathrm{CV}}(\hbar\omega)} \frac{\mathrm{d}S}{4\pi^3}\frac{1}{|\nabla E_{\mathrm{CV}}(\boldsymbol{k})|}, \tag{5.7}$$

where $S_{CV}(\hbar\omega)$ is the portion of the energy isosurface $E_{CV}(\boldsymbol{k}) = \hbar\omega$ that lies within the first Brillouin zone, *BZ*. This integral is a measure of the number of states of the conduction and valence bands with the same wave vector $\boldsymbol{k}$ that are separated by an energy difference $\hbar\omega$. This quantity is usually referred to as the *joint density of states* $[D_J(\hbar\omega)]$ of the crystal.

Measurements of the absorption coefficient provide a direct determination of $D_J(\hbar\omega)$ and, therefore, information on the band structure of a material: Singularities in $D_J(\hbar\omega)$ appear as structures in $\alpha(\hbar\omega)$ and indicate the energy position of critical points of $E_{CV}(\boldsymbol{k})$. In particular, the *minimum* energy difference between CB and VB defines the energy gap E_g of the material. There $\alpha(\hbar\omega)$ assumes the form

$$\alpha(\hbar\omega) = \begin{cases} 0 & \hbar\omega \leq E_g \\ \frac{\hbar}{cn}\left(\frac{2\pi e}{m}\right)^2 \cdot \frac{1}{\hbar\omega} \cdot |P_{VB,CB}|^2 \frac{1}{2\pi^2 \cdot \sqrt{\alpha_1\alpha_2\alpha_3}} \sqrt{\hbar\omega - E_g} & \hbar\omega > E_g \end{cases}, \quad (5.8)$$

where α_1, α_2, and α_3 are the coefficients of a quadratic expansion of $E_{CV}(\boldsymbol{k})$ near the minimum $E_{CV}(\boldsymbol{k}) = E(0) + \alpha_1 k_1^2 + \alpha_2 k_2^2 + \alpha_3 k_3^2$. The onset of optical absorption is known as the *absorption edge* of the material. Equation (5.8) has been obtained in the framework of the dipole approximation for transitions between CB and VB states with a same crystal momentum and therefore applies only to the case of *direct gap* semiconductors. The contribution to the absorption coefficient of phonon-assisted transitions between CB and VB states with different crystal momentum, namely, *indirect* transitions, is very small with respect to that of direct transitions [7]. The former contribution, however, cannot be neglected when the energy of the indirect transition is smaller than the energy of *all* direct transitions, as it is the case in the vicinity of the absorption edge of an *indirect gap* semiconductor, such as GaP or Si. In this case, it can be shown [8] that $\alpha(\hbar\omega)$ is well described by the following equation:

$$\alpha(\hbar\omega) = \begin{cases} 0 & \hbar\omega \leq E_g \pm E_{ph} \\ C \cdot \hbar\omega \cdot (\hbar\omega \pm E_{ph} - E_g)^2 & \hbar\omega > E_g \pm E_{ph} \end{cases}. \quad (5.9)$$

Here C is a constant and E_{ph} is the energy of the *phonon* involved in the transition in order to ensure the conservation of $\boldsymbol{k}$. Of course, the absorption coefficient of an indirect material will also present a direct absorption edge at higher energy, in correspondence of the energy separation between the valence band maximum, VBM, which is found at the Γ point of all zincblende semiconductors, and the lowest energy minimum at Γ of the CB.

5.2.3 Photoluminescence Spectroscopy

Considering an atomic two-level system in thermodynamic equilibrium with an electromagnetic radiation, Albert Einstein demonstrated first that the spontaneous emission rate, R_{sp}, the stimulated emission rate, and the absorption rate are

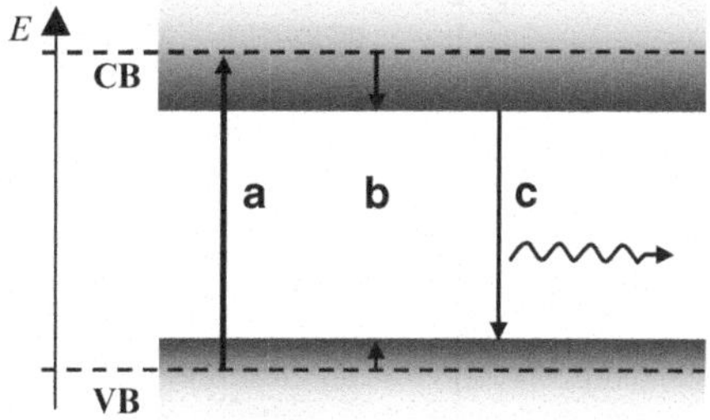

Fig. 5.2 Scheme of the three-step photoluminescence process: (**a**) absorption; (**b**) relaxation; (**c**) radiative recombination and photon emission. *Shaded areas* indicate valence and conduction band

related [9, 10]. The Einstein's relations have been generalized thereafter to the case of broadband spectra of atomic systems dilutely distributed in a dielectric medium, either a liquid, a solid, or a dense gas [11, 12]. A generalized form of these results, which is quite general and very useful in all cases where a detailed theoretical knowledge of a material electronic properties is missing, is the following [13]:

$$R_{sp}(\hbar\omega) = \alpha(\hbar\omega)n^2 e^{-\hbar\omega/k_B T} e^{\Delta F/k_B T} (\hbar\omega)^2 / \left(\pi c^2 \hbar^3\right), \tag{5.10}$$

where n is the material's refractive index and ΔF is the difference between the quasi-Fermi levels of electrons and holes [14].

Photoluminescence, namely, the effect of spontaneous emission in a photoexcited system, can be schematized in terms of a three-step process, as shown in Fig. 5.2:

(a) *Absorption* of a photon of energy $\hbar\omega$ with an ensuing creation of an electron–hole pair.
(b) *Relaxation* of the free electron and hole, respectively, toward the bottom of the CB and the top of the VB (in defect-free samples) by fast scattering with acoustic and optical phonons.
(c) *Radiative recombination* of the electron and hole, during or at the end of the relaxation process, with a return of the system to its ground state by *emission* of a photon of energy lower than $\hbar\omega$.

In a conventional PL experiment, the photons emitted by the sample are spectrally analyzed by varying the detection energy, E_{det}, while keeping the excitation energy, E_{ex}, fixed. The recombination processes more frequently observed in a PL spectrum are related to the excited states of *lowest energy,* as shown in Fig. 5.3. This is due to the fact that the timescale of the relaxation processes ($\sim 10^{-12}$ s) is much faster than that of radiative recombinations ($\sim 10^{-9}$ s), so that photon emission takes place only after carriers have reached the excited states of minimum energy. In *ideal* semiconductors, with no energy level inside the forbidden gap, electrons and holes recombine, respectively, from the extrema of the CB and VB; see fifth panel in the figure. The energy of this *band-to-band* transition is equal to the energy gap of the material whenever the Coulomb interaction in an electron–hole pair, or excitonic interaction, is weak, as in InSb [15]. In most semiconductors, however, this interaction is not negligible and the dominant feature in the PL spectrum is associated with the recombination of *free excitons* (FE); see fourth panel in the figure.

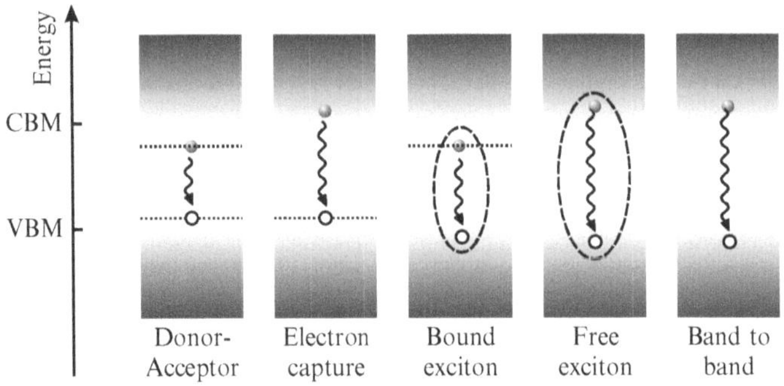

Fig. 5.3 Main recombination processes observed in photoluminescence. The *shaded areas* indicate VB and CB, while the *horizontal dotted lines* mark the energy position of impurity states. Electrons (holes) are indicated by *dark-shaded* (*hollow*) dots. *Undulated arrows* indicate radiative recombination. *Dashed curves* highlight the Coulomb interaction

The energy of the photons emitted in this transition is that of the excitonic ground state, which in turn equals the energy gap minus the exciton binding energy [8]. In addition to these "intrinsic" transitions, PL spectra of real semiconductors are characterized by features related to the presence of *defects* and *impurities*, whose incorporation in the crystal structure is often associated with energy levels within the forbidden gap; see first three panels in the figure. Since the relaxation process greatly favors low-energy states, these impurity levels usually give rise to a significant PL signal. Thus, PL is a powerful tool for the investigation of shallow impurities often characterized by a low volume density and hence difficult to detect by absorption measurements. As an example, we show in Fig. 5.4 the low-temperature spectrum of a GaAs sample where the emission from free excitons [X], excitons bound to neutral donors [D^0,X], to ionized donors [D^+,X], to neutral acceptors [A^0,X], and to ionized acceptors [A^-,X] have been identified on the basis of the paper by Pavesi and Guzzi [16].

Photoluminescence also provides information on less conventional type of impurities, like in the case of complexes made up of "isovalent" impurities in III–V semiconductors such as $GaAs_{1-x}N_x$, $GaAs_{1-x}Bi_x$, or $GaP_{1-x}N_x$. These alloys belong to a class of materials referred to as *highly mismatched alloys* where an element of the host compound is substituted by another isoelectronic element, namely, *sharing the same chemical valence* of the atom being replaced, but having much different size and atomic potential. These alloys exhibit uncommon compositional dependences of their fundamental band structure parameters (e.g., band gap energy and carrier effective mass) thus boosting a number of opportunities for materials engineering [17, 18]. The unusual electronic properties of these semiconductors stem from the impurity-like localized character that the host band edges acquire as a result of the hybridization with the wave function of the isoelectronic impurities. In the case of $GaP_{1-x}N_x$, an indirect band gap semiconductor, this hybridization confers a direct-like character to the material band gap with a large increase in the material

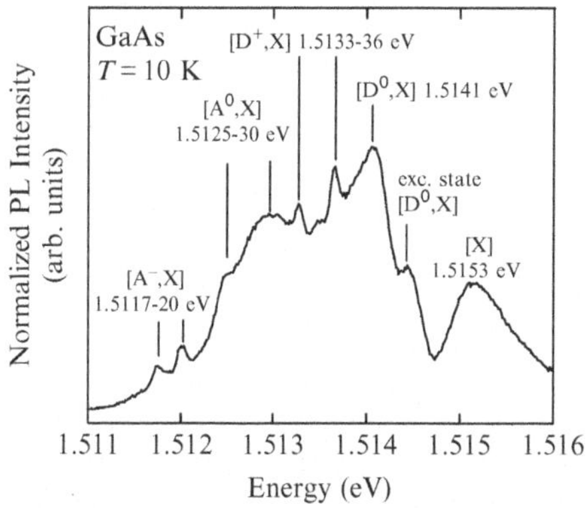

Fig. 5.4 $T = 10$ K photoluminescence spectrum of a GaAs sample. [X] indicates emission from free-excitons, [D^0,X] indicates emission from excitons bound to neutral donors, [D^+,X] indicates emission from excitons bound to ionized donors, [A^0,X] indicates emission from excitons bound to neutral acceptors, [A^-,X] indicates emission from excitons bound to ionized acceptors

radiative efficiency that has been exploited to make a number of optoelectronic devices in particular diodes emitting in the visible energy range. Moreover, the incorporation of N gives rise to a great variety of energy levels in the optical band gap of GaP. These energy levels are due to complexes made from different pairs and triplets of N atoms.

An example of the extremely rich PL spectrum produced by different N pairs and complexes in quite dilute $GaP_{1-x}N_x$ ($x = 0.05\%$) is shown in Fig. 5.5. LO and TO indicate longitudinal and transverse optical-phonon replica of the emission lines, which are labeled according to the classification scheme introduced in [19, 20].

5.2.4 *Photoluminescence Excitation Spectroscopy*

Photoluminescence excitation, PLE, is a technique *complementary* to conventional PL that *mimics absorption* measurements. Conversely to the PL case, in PLE the excitation energy E_{ex} is varied while detection energy E_{det} is kept fixed at the energy of one of the resonances in the PL spectrum of the sample. This technique has become very popular to study thin epilayers and heterostructures grown by modern epitaxial techniques, where standard *transmission* measurements of $\alpha(E_{ex})$ are difficult to perform because of low absorption and/or opaque bulk substrates. The PLE spectrum can be regarded as a good approximation of the absorption spectrum of the material, provided certain conditions are satisfied. The relationship between the intensity of the PL signal, I_{PL}, at energy E_{det} and that of the exciting radiation, I_{ex}, can be written explicitly as [8]

$$I_{PL}(E_{ex}, E_{det}) = P_{abs}(E_{ex}) \cdot P_{rel}(E_{ex}, E_{det}) \cdot P_{em}(E_{det}) \cdot I_{ex}(E_{ex}). \quad (5.11)$$

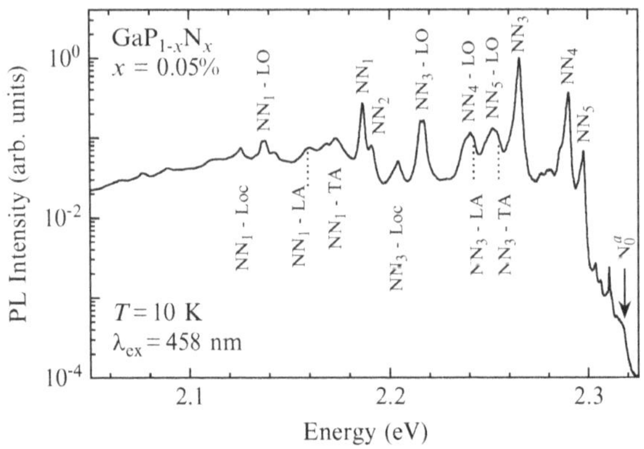

Fig. 5.5 Photoluminescence spectrum at $T = 10\,\text{K}$ of a dilute $\text{GaP}_{1-x}\text{N}_x$ sample ($x = 0.05\%$). The NN_i emission lines, due to exciton recombination on different nitrogen pairs, are labeled according to the classification introduced in [19, 20]. LO stands for longitudinal optical mode, LA and TA for longitudinal and transverse acustical mode, respectively. Loc denotes a local mode. The position of the recombination of an exciton localized on a single nitrogen center, N_0^a, is indicated by an *arrow*

The probability that a photon of energy E_{ex} is *absorbed* by the sample, $P_{\text{abs}}(E_{\text{ex}})$, is proportional to $\alpha(E_{\text{ex}})$. $P_{\text{rel}}(E_{\text{ex}}, E_{\text{det}})$ is the probability that photogenerated e–h pairs will *relax* toward lower energy states giving rise to the emission of photons of energy E_{det}. Finally, $P_{\text{em}}(E_{\text{det}})$ is the probability that photons of energy E_{det} will actually be emitted after relaxation. In a PLE experiment, E_{det} is fixed and $P_{\text{em}}(E_{\text{det}})$ remains constant. Then, according to (5.11), I_{PL} is proportional to $\alpha(E_{\text{ex}})$ only if $P_{\text{rel}}(E_{\text{ex}}, E_{\text{det}})$ is independent of the excitation energy, at least roughly.

This is the case of the PLE spectra [21] shown in Fig. 5.6 and taken in the same $\text{GaP}_{0.0095}\text{N}_{0.0005}$ sample whose PL spectrum has been given in Fig. 5.5. The $T = 10\,\text{K}$ PLE spectra were obtained by detecting the PL signal at the emission energy of the four strongest lines in the PL spectrum, NN_1, NN_3, NN_4, and NN_5 (due to exciton recombination on different N pairs); see Fig. 5.5. Three main features are roughly independent on detection energy and then common to all PLE spectra (1) a sharp resonance at 2.317 eV, the energy of the N_0^a line due to a single-N complex; (2) a rather distinct peak at 2.33 eV, the energy of the free indirect exciton [22, 23]; and (3) a steep edge in the luminescence signal at ~2.85 eV, due to the direct absorption of free carriers from the top of the VB to the CB minimum. The strong reduction of this edge in the spectra relative to the NN_4 and NN_5 lines leads to the uncovering of a fourth feature, a quite pronounced peak at 2.89 eV that has been attributed to an excitonic level introduced near the CB minimum by isolated nitrogen atoms [21].

$P_{\text{rel}}(E_{\text{ex}}, E_{\text{det}})$ depends strongly on E_{ex} and $I_{\text{PL}}(E_{\text{ex}}, E_{\text{det}})$ deviates significantly from a strict proportionality to $\alpha(E_{\text{ex}})$ whenever there are channels of *preferential relaxation* between different states of the crystal. This is the case for detection energies corresponding to emission from impurity-related levels, where phenomena such as phonon-assisted hopping [24] and tunneling from an impurity center to

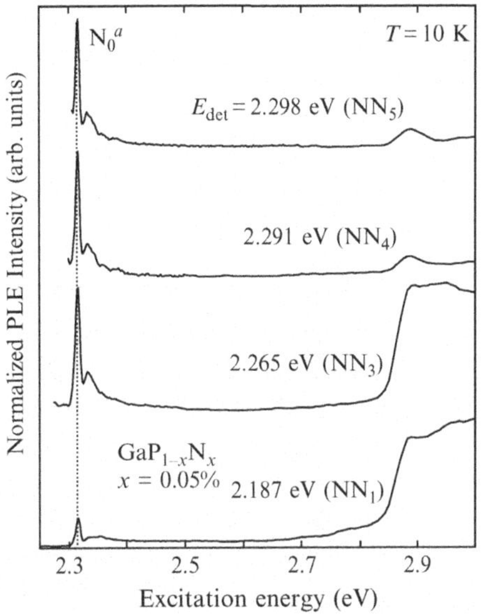

Fig. 5.6 Peak normalized PLE spectra of a $GaP_{1-x}N_x$ sample with $x = 0.05\%$ at $T = 10$ K. Detection energies are in resonance with the four strongest NN_i lines shown in Fig. 5.5

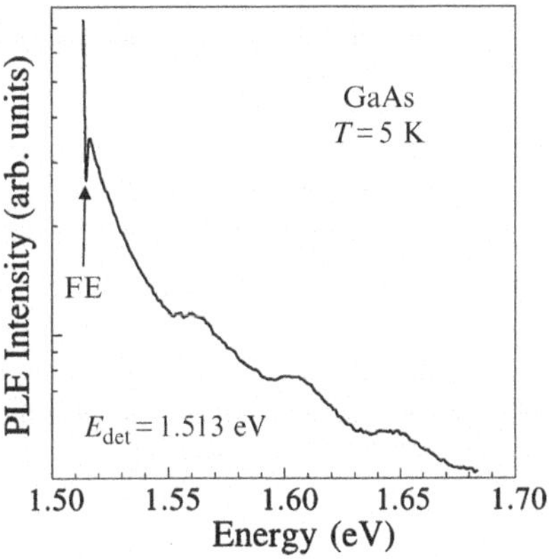

Fig. 5.7 PLE of a bulk GaAs sample. The detection energy, 1.513 eV, is that of an exciton bound to a carbon impurity; see Fig. 5.4

another [21, 25, 26] come into play. As an example, the PLE spectrum of GaAs is shown in Fig. 5.7 (see [27])[2]. The detection energy, ~1.513 eV, is that of the recombination of the [A^0,X] complex in bulk GaAs; see Fig. 5.4. A dip is observed at the FE energy, where a peak in the absorption coefficient should be expected. This shows that the formation of acceptor-bound excitons is due to the separate capture by acceptors of holes and electrons, not to the direct capture of excitons. The intensity oscillations with a 42-meV period, also observed above band gap, indicate, instead, a faster carrier relaxation when the detection and excitation energies differ

[2]This sample has been hydrogenated in order to have a strong PL signal from the well states and make PLE measurements easier

by multiples of $\hbar\omega_{LO}(1 + m_e/m_h)$, where $\hbar\omega_{LO}$ is the energy of the longitudinal optical phonon in GaAs (36 meV) and m_e and m_h are the electron and hole effective mass, respectively [28].

Therefore, it is very important to choose a value of E_{det} that minimizes the dependence of $P_{rel}(E_{ex}, E_{det})$ on the excitation energy when trying to achieve the absorption coefficient of the material from PLE spectra.

5.3 Photoluminescence Spectroscopy: A Tool for Crystalline Disorder Investigation

Photoluminescence spectroscopy provides detailed information about the type of defects with energy levels in the optical band gap and at least qualitative information about their density, as previously exemplified and widely reported in the literature [1–3]. It is also very effective in investigating type and amount of crystalline disorder, both in quantum wells, QWs, and bulk alloys.

5.3.1 *Probing Quantum Well Imperfections*

Photoluminescence is a very convenient, nondestructive tool for measuring deviations from ideality in QWs made up of two different materials, a and b. Because of atomic interdiffusion, islands of the compound b appear on the side a of a real interface, and vice versa, and the well width L fluctuates by δ_1 around its mean value L_0; see inset in Fig. 5.8. For sake of simplicity, let us assume that the island size along the well, δ_2, be independent on the interface side. $C^0_{a,b}$ is the average areal concentration of the islands on the sides (a or b) of the interface. An additional source of disorder is also present when the well and/or the barrier material is made up of an alloy.

When L is comparable with the exciton Bohr radius in bulk, R_{ex}, the excitonic recombination energy in the well strongly depends on L through quantum confinement. If the exciton lateral extension ρ_{ex} is smaller than δ_2, excitons sample different well widths, the excitonic emission is given by multiple peaks, and single monolayer fluctuations in L can be resolved in the emission spectrum. On the contrary, if $\rho_{ex} > \delta_2$, the excitonic recombination is inhomogeneously broadened. Moreover, for $R_{ex} > L$, the exciton wave function penetrates sizably into the barrier and the excitonic linewidth is affected by the barrier quality. As a matter of fact, in $In_xGa_{1-x}As/GaAs$ QWs, the FE linewidth is well described by taking into account all these effects [29]. In particular, for $\rho_{ex} > \delta_2$, interface disorder gives rise to an excitonic Gaussian lineshape with a full width at half maximum, W, given by [30, 31]

$$\gamma_{int} = \sqrt{2}\frac{\delta_2}{\rho_{ex}}(1.4C^0_aC^0_b)^{1/2}(C^0_a\Delta^+ + C^0_b\Delta^-), \qquad (5.12)$$

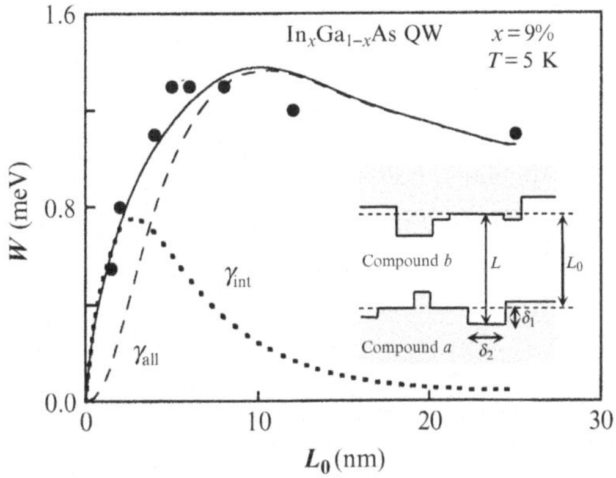

Fig. 5.8 Theoretical full width at half maximum W of the free-exciton recombination line in $In_{0.09}Ga_{0.91}As/GaAs$ QWs (*solid line*) together with experimental results (*dots*) for different well widths L_0. The separate contributions of interface disorder (γ_{int}, *dotted line*) and alloy disorder (γ_{all}, *dashed line*) are also shown. $\delta_2 = 2.4$ nm, $V_{ex} = 4\beta\pi R_{ex}/3$, with $\beta = 6$. A sketch of the atomic interdiffusion-induced islands at the well interfaces is shown in the *inset*; see text

where $\Delta^{\pm} = \left.\frac{\partial E_{ex}}{\partial L}\right|_{L_0 \pm \delta_L} \delta_1$. δ_L is a single monolayer change in thickness, and $E_{ex}(L)$ is the exciton recombination energy, with a well-known dependence on L. For simplicity, it is usually assumed that $C_a^0 = C_b^0 = 0.5$.

For QWs of infinite depth, γ_{int} increases monotonically for decreasing L, becoming infinite for $L = 0$. For QWs of finite depth, instead, γ_{int} reaches a maximum and then decreases to zero as L approaches zero, because of the increasing penetration of the excitonic wave function into the barrier.

As regards alloy compositional fluctuations, E_{ex} fluctuates by $\Delta E_{ex}(x, x_0, V_{ex}) \cong (x - x_0)\Delta_1$ with respect to its value in the virtual crystal approximation, where x_0 is the nominal composition of the alloy, $\Delta_1 = \left.\frac{\partial E_{ex}}{\partial x}\right|_{x_0}$, and V_{ex} is a critical volume of the order of the exciton volume inside which an exciton in its ground state experiences a mean crystal potential [32]. If the probability of finding n indium atoms in a given exciton volume, V_{ex}, is given by a Poisson distribution, the probability distribution function of the exciton energy in the limit of large n and large number of anion sites in V_{ex} is a Gaussian function. The exciton full width at half maximum due to alloy disorder is then given by

$$\gamma_{all} = 2\left(\frac{V_c}{V_{ex}} 1.4 x_0 (1 - x_0)\right)^{1/2} \Delta_1, \tag{5.13}$$

where V_c is the volume per anion site. This formula can be modified in order to take into account the dependence on L of the exciton localization in a QW, whenever the confining potential is finite [29, 32]. If P_{ex} is the probability for the exciton to penetrate into the barrier, one gets [29]

$$\gamma_{\text{all}} = 2\left(\frac{V_c}{V_{\text{ex}}} 1.4 x_0 (1 - x_0)(1 - P_{\text{ex}})\right)^{1/2} \Delta_1. \tag{5.14}$$

The experimental values of the full width at half maximum, W, of the exciton lineshape in a series of $In_xGa_{1-x}As$/GaAs QWs with different well widths, L_0, have been nicely fitted [29] in terms of the calculated values given by $\gamma_{\text{th}} = \sqrt{\gamma_{\text{int}}^2 + \gamma_{\text{all}}^2}$; see Fig. 5.8.

5.3.2 *Localized States in an Alloy*

Photoluminescence has also been used to investigate degree and type of structural disorder in several inhomogeneous alloys, such as $GaAs_{1-x}N_x$, $GaAs_{1-x}Bi_x$, $GaP_{1-x}N_x$, or related quaternary compounds. In $GaAs_{1-x}N_x$, a direct gap semiconductor, arsenic substitution by nitrogen leads to the formation of N pairs, triplets, and other complexes emitting in the band gap of $GaAs_{1-x}N_x$, as in $GaP_{1-x}N_x$ for phosphorous substitution by nitrogen. Moreover, because of the quite large difference in size and electronegativity between N and As, the Virtual Crystal Approximation breaks down and the $GaAs_{1-x}N_x$ energy gap decreases instead of increasing with increasing N concentration [17]. Then, spatial fluctuations in the N complex concentration give rise to a disordered potential characterized by local minima on which excitons are trapped (localized excitons, LE) before recombining. This results in quite broadened asymmetric PL lineshapes skewed on the low energy side, as widely reported in the literature [33–35]. This is shown in Fig. 5.9a that displays PL spectra at different low temperatures in the case of $In_{0.32}Ga_{0.68}As_{0.985}N_{0.015}$, a quaternary compound [36]. Carriers freeze out in local minima and LE recombination dominates the lowest temperature spectra. Only for T's greater than 50 K, recombination from free excitons, FE, appears and increases with T, until it becomes the only recombination at $T \geq 80$ K. This produces the S-shaped dependence of the PL peak energy on temperature shown in panel (b), a typical signature of relevant contributions from localized states to low-temperature PL spectra [36].

The FE recombination energy can be determined by a suitable analysis of the low-T PL lineshape. The low energy tail is accounted for by alloy fluctuations [35] that give rise to an exponential density of localized states [37] $g(h\nu) = g_0 \exp\{-[(h\nu - h\nu_{\text{exc}})/h\nu_0]^{3/2}\}$, where $h\nu_{\text{exc}}$ is the FE energy, $h\nu_0$ is a characteristic energy, and g_0 is a constant. The PL spectrum is then given by [35, 38]

$$I(h\nu) \propto g(h\nu)\tau(h\nu)\exp[f(h\nu)], \tag{5.15}$$

with

$$\tau^{-1}(h\nu) \propto 1 + \exp\{\delta[h\nu_M - (h\nu_{\text{exc}} - h\nu)]/k_B\}. \tag{5.16}$$

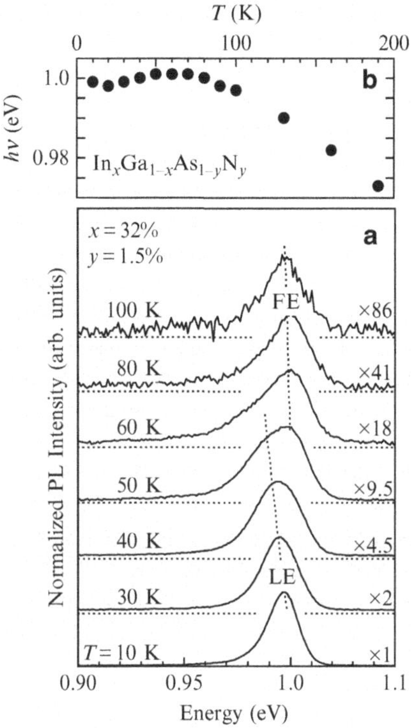

Fig. 5.9 (**a**) Peak normalized PL spectra of an $In_xGa_{1-x}As_{1-y}N_y$ sample ($x = 32\%$; $y = 1.5\%$) for different temperatures. LE and FE indicate recombination from localized and free-exciton states, respectively. (**b**) Temperature dependence of the PL peak energy, $h\nu$

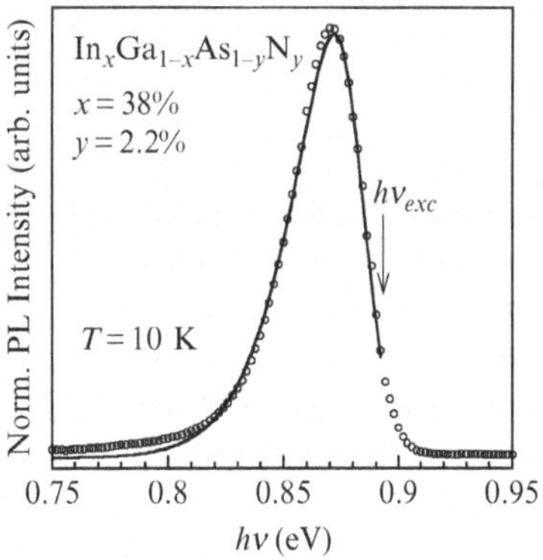

Fig. 5.10 Low-T PL spectrum of an $In_xGa_{1-x}As_{1-y}N_y$ sample ($x = 38\%$; $y = 2.2\%$) (*dots*). The *continuous line* is a simulation to the data, as discussed in the text. $h\nu_{\rm exc}$ indicates the free exciton energy

Here τ is the exciton radiative lifetime, δ is the inverse of an effective temperature, and $h\nu_M$ is the energy at which the radiative recombination probability equals the probability of a transfer toward deeper states [39]. Secondly, the expression of $f(h\nu)$ is given in [35]. The full line in Fig. 5.10 shows the simulation of the PL spectrum of a $In_{0.38}Ga_{0.62}As_{0.978}N_{0.022}$ sample (dots) with $\delta = 77\,\mathrm{K}^{-1}$, $h\nu_{\rm exc} = 0.892\,\mathrm{eV}$, $h\nu_0 = 34\,\mathrm{meV}$, and $h\nu_M = 0.882\,\mathrm{eV}$.

5.3.3 Localized States and Carrier Thermalization

The full width at half maximum of the FE exciton lineshape is a "measure" of the sample quality. The Stokes shift, SS, i.e., the difference between the exciton peak energy as determined by PL and absorption (or PLE) measurements is another quality factor, the origin of which has been long debated. In the pioneering works in $Al_xGa_{1-x}As/GaAs$ single QWs [40], the SS has been attributed to trapping of excitons at interface roughness produced by fluctuations in the QW thickness. In this case, the probability distribution of the exciton energy in the QW plane is assumed to be a Gaussian function [41, 42]. When this topographical disorder extends over a scale much greater than the carrier diffusion length, carriers relax into local minima before recombining as excitons. At low T, the PL spectra reflect the distribution of local minima in the exciton energy distribution and SS $= 0.55\,W$, where W is the full width at half maximum of the exciton *absorption* line [41, 42]. This prediction is supported by results in III–V and II–VI compounds (where W values range from a few meV to about 100 meV) thus suggesting a medium-to-high degree of disorder. With increasing temperature, exciton detrapping increases and SS vanishes.

When thermal effects in the exciton diffusion are added to a Gaussian disorder, only numerical estimate can be made [43]. In this case, SS increases linearly with PL linewidth with a slope smaller (SS $= 0.3\,W$) than that found in absence of thermal effects.

Finally, in the case of a quite small degree of disorder, excitons are in a thermal quasi-equilibrium characterized by a distribution function with an effective carrier temperature T_c greater or equal to the lattice temperature [44]. The ratio between the carrier thermal energy $k_B T_c$ and W measures the degree of carrier thermalization. Whenever this ratio is much smaller than unity, the high energy side of the exciton distribution function is depressed in PL because of population effects and a sizable SS results. If the disorder responsible for the exciton broadening is described by a Gaussian function, the model results into SS $= 0.18 W^2/k_B T_c$, in quite good agreement with experimental data in $Al_xGa_{1-x}As/GaAs$ QWs, where W is of the order of a few meV or less [44]. If this ratio is greater than unity SS vanishes.

Both analytical models described above well account, therefore, for the origin of the Stokes shift provided they are applied to systems where degree and nature of disorder differ by much, as shown in $In_xGa_{1-x}As/GaAs$ single QWs differing in the indium molar fraction and and/or well width L [45]. In this system, indeed, the amount and source of disorder may be tuned and there is a crossover between the topographical and the quasi-equilibrium model for $W \sim 6$ meV. Moreover, for increasing lattice temperature, data turn out to be well described by the quasi-equilibrium model also where the topographical model holds at low T.

In Fig. 5.11, low-T PL (dotted lines) and PLE (full lines) spectra of the (heavy hole) FE are shown for three $In_xGa_{1-x}As/GaAs$ single QWs differing for well width and indium molar fraction. For ease of comparison, PL and PLE spectra (taken instead of absorption because of the low absorbance of QWs) have been shown by taking for each sample the zero of the energy at the PLE peak. Stokes shifts are

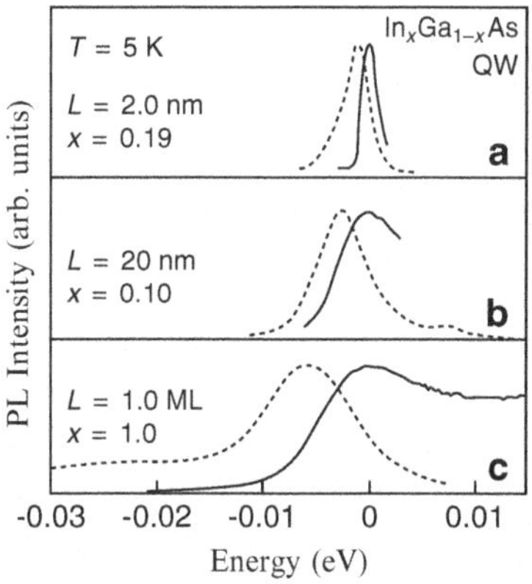

Fig. 5.11 Low-T PL (*dotted lines*) and PLE (*full lines*) of three $In_xGa_{1-x}As/GaAs$ single QWs with different well width (for $x = 1.0$, 1 monolayer, ML, is equal to 3.03 Å) and indium molar fraction. For ease of comparison, the PL and PLE spectra have been shown by taking for each sample the zero of the energy at the PLE peak of FE (at 1.470, 1.400, and 1.471 eV, from the top to the bottom of the figure)

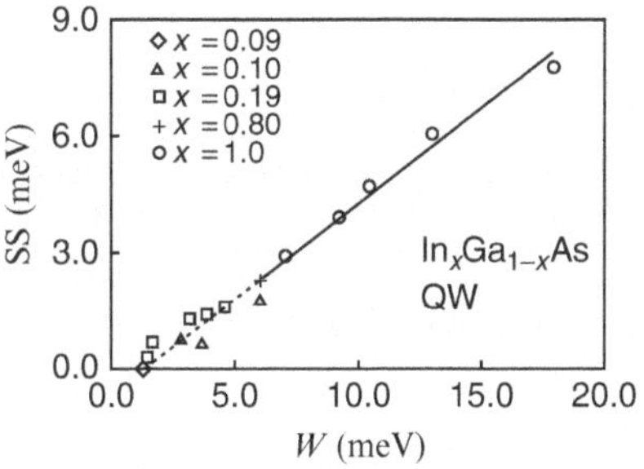

Fig. 5.12 Low-temperature (5 K) Stokes shift, SS, as a function of the full width at half maximum, W, of the FE band measured in PLE for $In_xGa_{1-x}As/GaAs$ single QW samples with different indium molar fraction. The *full line*, $SS = -0.6 + 0.48\,W$ has been obtained by a linear best fit of the data for $W > 6$ meV. The *dashed line* extrapolates the fit to $W < 6$ meV

clearly correlated with recombination and absorption broadenings, all increasing from the top to the bottom of the figure.

The SS values measured at $T = 5$ K in all samples are shown in Fig. 5.12 as a function of W measured in PLE. Where disorder dominates the PL spectra (high W values), data fall on a straight line with slope 0.48, in very good agreement with the topographical model. A closer inspection to the region of small W (≤ 6 meV, or SS ≤ 2 meV), where disorder is small and the topographical model has been questioned [44], shows that the linear extrapolation from the high disorder regime (dashed line in Fig. 5.12) gives a poor fit to the data. Moreover, SS should vanish for finite values of W, a feature inconsistent with the theoretical model.

The experimental results are then compared with the thermalization model, as shown in Fig. 5.13 (T_c has been estimated from the high-energy tail of the PL spectra). This model accounts well for the experimental results in the region of low W values where the thermal energy is about equal to W and a strong communication between the broadened excitonic levels is expected. A quadratic relationship does not hold, instead, for the whole set of data, as shown in the inset where the model

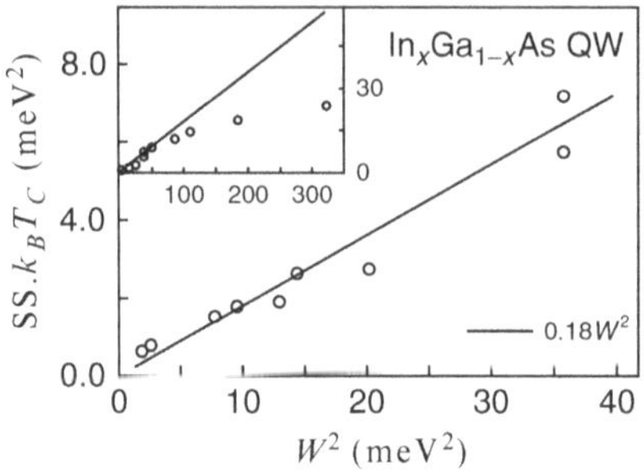

Fig. 5.13 Product of the thermal carrier energy $k_B T_c$ and Stokes shift vs. W^2 for full width at half maximum of the FE band, W, measured in PLE smaller than 6 meV (same data shown in Fig. 5.12). The *full line*, $SS \cdot k_B T_c = 0.18\, W^2$, gives the behavior predicted by the quasi-equilibrium model of [44]. In the *inset*, the comparison between the data and the model of [44] is extended to the full set of data

clearly deviates from the experimental data for $W^2 \geq 36\,\text{meV}^2$. In that case, sample inhomogeneities give rise to local energy minima and potential fluctuations large with respect to the carrier thermal energy. Therefore, the different energy levels do not interact between each other and PL is ruled by the height distribution of local minima, as for model in [41, 42].

In conclusion, data agree better with one model or the other in a same material system according to the value of SS and W, namely, to the sample "quality," and PL and PLE measurements can thus allow to determine nature and amount of structural disorder.

5.3.4 Degenerate Semiconductors

The Coulomb interaction between oppositely charged carriers is cancelled by high electric fields, lattice thermal energies well above the exciton binding energy, or severe carrier–carrier scattering in electron–hole plasmas—obtained in PL experiment at high exciting power densities—or in degenerate semiconductors. In the latter two cases, optical spectra are dominated by band-to-band recombination and band filling effects where the carrier Fermi energy plays a central role.

The $\boldsymbol{k}$-conservation rule requires that vertical transitions take place in the energy–momentum space so that thermalized electrons and holes recombine mainly at $\boldsymbol{k} = 0$, at least in the case of low excitation and direct transitions. This recombination scheme applies to all semiconductors, either intrinsic, lightly, or highly doped. However, when minority carriers, e.g., holes in a degenerate n-type semiconductor, are localized on interface defects or on diluted impurities, the Heisenberg indetermination principle results in a momentum spread that allows holes to recombine with electrons having $\boldsymbol{k}$ vectors ranging between $\boldsymbol{k} = 0$ and $\boldsymbol{k}_F^e$, where $\boldsymbol{k}_F^e = (3\pi^2 n_e)^{1/3}$ is the $\boldsymbol{k}$ value at the electron Fermi energy E_F^e and n_e is the free-electron concentration. This hole localization, together with an enhanced multiple electron–hole scattering rate for electrons close to E_F^e (suppressed by the exclusion principle for electrons at energy $E \ll E_F^e$) gives rise at low T to a strong enhancement in the PL intensity with a resonance-like structure known as Fermi edge singularity [46]. Moreover, because of many body and band tailing effects, the band gap renormalizes [47] and the whole PL emission spectra red-shifts. As far as the absorption is concerned, only the states above the Fermi energy are available for transitions at $T = 0$ and the electrons in n-type degenerate semiconductors populate all states from $\boldsymbol{k} = 0$ to $\boldsymbol{k}_F^e$. Therefore, absorption processes occur only for photon energies higher than that of the band gap. In particular, the absorption edge is blue-shifted by an amount $\Delta E = (1 + m_e/m_h)E_F^e$, not including many-body effects, with respect to the absorption edge in the intrinsic case. This latter effect is known as Moss–Burstein shift by its first discoverers [48, 49].

An underestimate of the Moss–Burstein shift is the most likely origin of the value of 1.89 eV initially reported for the band gap energy of InN (for a review, see [50]), a direct gap compound whose free-electron concentration is well above that of the insulator-to-metal transition even in as-grown samples. This compound has raised an increasing interest since 2002, when it has been found that the actual energy gap of InN is as small as 0.65 meV [51], namely, only one-fifth of the energy of its wide-gap parent compound GaN. Moreover, the free electron concentration in InN can be largely increased by post-growth irradiation with low-energy (10–20 eV) H atoms [52]. These discoveries give to the $In_{1-x}Ga_xN$ system a potential in a variety of applications ranging from photovoltaics to fast and high-power electronics.

Low-temperature ($T = 10$ K) PL spectra of an InN epilayer subjected to different post-growth treatments are shown in Fig. 5.14a (solid lines). From bottom to top: (1) as-grown, a.g.; (2) irradiated by low-energy He atoms, at dose $d_{He} = 3 \times 10^{15}$ ions/cm^2; (3) exposed for 3 min to a H_2 plasma; and (4) and (5) irradiated by a low-energy H ion beam at $d_H = 3 \times 10^{15}$ and $d_H = 5 \times 10^{15}$ ions/cm^2, respectively. All PL spectra exhibit a high energy edge, HEE, whose slope depends on temperature—not shown here [53]—according to a Fermi distribution function. This feature is a clear signature of recombination in a degenerate semiconductor. Irrespective of the H irradiation technique, for increasing d_H the PL HEE rigidly and progressively blue-shifts, the relative intensity of the low-energy side of the PL band increases and broadens, and the overall emission efficiency decreases. No main changes are instead observed in the He irradiated sample, thus excluding that sample damage may account for these results.

A quantitative analysis of PL spectra in degenerate semiconductors requires an improved description of the carrier DOS. Electron–electron interactions narrow the band gap and modify the quasi-particle energies, while electron-random-impurity interactions give rise to conduction and valence band tails. A detailed, rigorous combination of many-body and statistical fluctuation effects is exceedingly complicated, if it exists, and a semiclassical approach has been used [54], namely, the lattice potential is assumed to change by small amounts over an electron wave packet described classically. The semiclassical three-dimensional DOS is then given by

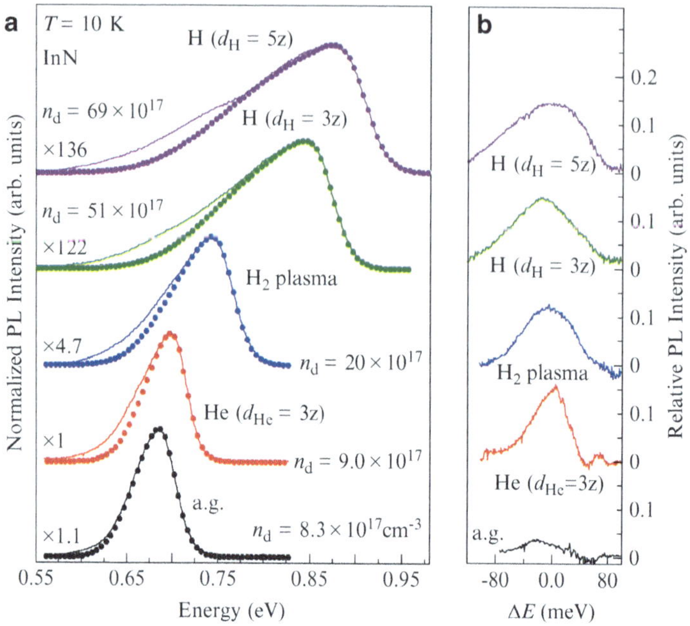

Fig. 5.14 (**a**) Low-temperature ($T = 10$ K) normalized PL spectra (*solid lines*) of an InN sample for different post-growth treatments: As-grown (a.g.), hydrogenated for 3 min in a remote H_2 plasma (H_2 plasma), irradiated at 300°C with different doses $d_{\text{H,He}}$ of low energy (10–20 eV) H or He ions ($z = 1 \times 10^{15}$ ions/cm^2). Normalization factors and carrier concentration related to donors n_d are given in the figure. Laser excitation power density $P_{\text{exc}} = 200\,\text{W/cm}^2$. Simulations of the PL spectra in terms of the semiclassical model described in the text [see (5.18) and (5.19)] are given by *dots*. (**b**) Difference between the experimental data and simulations (misfit bands) as a function of energy for the same samples shown in panel (**a**). Energies are measured with respect to the band gap energy estimated for each PL spectrum

$$\rho_{3D}(E) = \frac{1}{\pi^2} \int_0^\infty k^2 P[E - \varepsilon(\boldsymbol{k})]\mathrm{d}k, \tag{5.17}$$

where $P(V)$ is the distribution function for the potential energy $V(\boldsymbol{r})$ caused by charged impurities and $\varepsilon(\boldsymbol{k}) = \hbar^2 k^2/2m$ in the effective mass approximation. In the impurity density limit, $P(V) = e^{-V^2/2\sigma^2}/(\sqrt{2\pi}\sigma)$ and $\sigma = \sqrt{n_d 8\pi\lambda e^2/\varepsilon}$, where e is the electron charge, ε the dielectric constant, and n_d the impurity concentration. The carrier screening length is given by a Thomas–Fermi approach to a noninteracting Fermi liquid, $1/\lambda = 2k_F/\sqrt{\pi a_B k_F}$ where k_F and a_B are the Fermi wave vector and the effective Bohr radius, respectively. The band gap narrowing due to the many-body interaction is not included in the model and is considered separately in the PL spectra fitting procedure. The resulting DOS is compared in Fig. 5.15 with that of a free electron gas.

The PL intensity $I(\hbar\omega)$ is obtained by integrating the DOS, $\rho_{\text{CB,VB}}(E)$, in the conduction and valence band multiplied by the respective occupation probabilities,

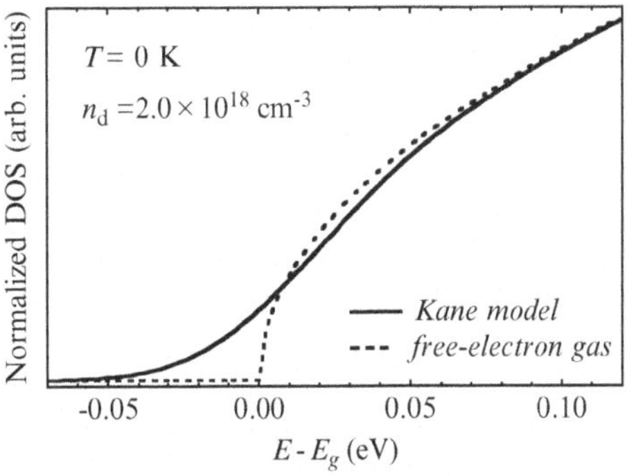

Fig. 5.15 Comparison of the electron density of states as calculated by the model of [54] (*solid line*) with that of an unperturbed free-electron gas (*dashed line*)

namely, by the Fermi–Dirac quasi-distributions functions f_{CB} and f_{VB}. Because of the random potential and strong carrier–carrier scattering, the translational symmetry and the wave vector conservation rule break down (See, e.g., [55]) and

$$I(\hbar\omega) \propto \int \mathrm{d}E_{\mathrm{VB}}\rho_{\mathrm{CB}}^{\mathrm{int}}(\hbar\omega - E_{\mathrm{VB}})\rho_{\mathrm{VB}}(E_{\mathrm{VB}})f_{\mathrm{CB}}(\hbar\omega - E_{\mathrm{VB}})f_{\mathrm{VB}}(E_{\mathrm{VB}}). \quad (5.18)$$

The integrated DOS for the conduction band $\rho_{\mathrm{CB}}^{\mathrm{int}}(E)$, as calculated accordingly to the semiclassical model, is given by

$$\rho_{\mathrm{CB}}^{\mathrm{int}}(E) = \frac{1}{\pi^2}\int_0^{\infty} \mathrm{d}k_{\mathrm{CB}}k_{\mathrm{CB}}^2 P\left(E - \left(E_{\mathrm{g}} + \frac{\hbar^2 k_{\mathrm{CB}}^2}{2m}\right)\right), \quad (5.19)$$

where E_{g} is the renormalized band gap energy. The photoexcited carrier density n_{ph} does not contribute to the statistical broadening of the CB (except for the screening length), while it contributes to the determination of the quasi-Fermi levels. In the present case, where holes can be only photogenerated, the hole quasi-Fermi levels is much smaller than that of electrons and the valence band acts as a δ-like state reservoir in (5.18). Notice that for increasing excitation power, the contribution of photogenerated carriers becomes more important and leads to a blue shift of the electron quasi-Fermi level, thus affecting the high energy edge of the PL spectra. In that case, n_{ph} has to be estimated by simulating spectra recorded under different excitation intensities. Data simulations by (5.18) are shown by dots in Fig. 5.14a; for details see [52, 53]. This model reproduces all PL band lineshapes very well—except for the low-energy side of each PL band—in terms of three parameters: E_{g}, n_{d}, and the carrier temperature T_{c}. The differences between experimental data and lineshape simulations have been then evaluated for each of the PL spectra shown in Fig. 5.14a. All the resulting misfit bands, shown in Fig. 5.14b, have roughly Gaussian lineshapes centered below the band gap energy. The integrated intensity of the misfit band, which is very weak and poorly defined in the as-grown sample, increases with estimated carrier density n_{d}. This band has been then attributed to the (H, h) recombination of a free hole in the valence band with an electron bound to a H donor state 10 ± 5 meV below the conduction band edge [52].

5.4 Magneto-Photoluminescence

In previous sections, we have described the main features and advantages of photoluminescence spectroscopy applied to semiconductor materials and nanostructures. The use of external perturbations in a PL experiment permits to deduce valuable information about the electronic states at the optical band gap. In particular, magnetic fields, B, affect position and lift degeneracy of energy levels involved in spectroscopically observed transitions. The extent of the modifications induced by a magnetic field on the eigenstates of charge carriers in a semiconductor depends on several factors. One of the most relevant is the relative strength of the perturbation provided by the field with respect to the electron–hole Coulomb interaction that determines the optical properties of the most common semiconductors, especially low-dimensional systems. The Coulomb energy scale is given by the effective exciton Rydberg

$$Ry[eV] = \frac{\mu e^4}{2(4\pi\varepsilon_r\varepsilon_0)^2\hbar^2} = \frac{13.6\mu/m_0}{\varepsilon_r^2}, \tag{5.20}$$

where μ is the exciton reduced mass $\mu^{-1} = m_e^{-1} + m_h^{-1}$, m_0 is the electron mass in vacuum, ε_0 is the dielectric constant of free space, and ε_r is the relative static dielectric constant of the material. The energy scale (expressed in eV) provided by the external magnetic field B (expressed in Tesla) can be quantified by

$$\hbar\omega_c\,[\mathrm{eV}] = \frac{\hbar e B}{\mu} = \frac{1.16 \times 10^{-4} B\,[\mathrm{T}]}{\mu/m_0}. \tag{5.21}$$

A dimensionless parameter γ is usually employed to weight the relative strengths of the magnetic field and Coulomb interaction

$$\gamma = \frac{\hbar\omega_c}{2Ry}. \tag{5.22}$$

As it will be shown in detail in the following, depending whether $\gamma \gg 1$ or $\gamma \ll 1$, different approximations can be straightforwardly used to interpret magneto-PL results. Instead, for $\gamma \sim 1$ no analytical formula, though approximate, holds and numerical solutions of the Schrödinger equation are required.

Next, we formulate the general problem of an electron and a hole interacting via the electrostatic force in the presence of a magnetic field. Then, the conditions under which the limit of high ($\gamma \gg 1$), low ($\gamma \ll 1$), and intermediate ($\gamma \sim 1$) field apply are considered and relevant examples from current research in semiconductor physics are discussed. In particular, an emphasis will be posed on a class of semiconductor materials referred to as highly mismatched alloys. The added flexibility of these materials in terms of an engineering of their electronic and lattice properties turns out to be of particular interest in many applicative areas, such as photovoltaics and telecommunications. We will show how magneto-PL spectroscopy allows obtaining important information concerning the physical mechanisms underlying the compositional evolution of the electronic properties of these materials.

5.4.1 Excitons in a Magnetic Field

The effective mass approximation is a most common approach to solve the Schrödinger equation for an electron moving inside a semiconductor [8]

$$(H_0 + U)\Psi(\boldsymbol{r}) = E\Psi(\boldsymbol{r}), \tag{5.23}$$

where H_0 is the one-electron Hamiltonian of the crystal and U represents any external perturbation (e.g., an external magnetic field or the Coulomb attraction due to an impurity). Within this approximation, the wave function $\Psi(\boldsymbol{r})$ can be expanded in terms of Wannier functions [56] $a_n(\boldsymbol{r} - \boldsymbol{R}_i)$ for a given band n and lattice site $\boldsymbol{R}_i$, each weighted by an amplitude $C_n(\boldsymbol{R}_i)$ referred to as envelope wave function

$$\Psi(\boldsymbol{r}) = N^{-1/2} \sum_{n,i} C_n(\boldsymbol{R}_i) a_n(\boldsymbol{r} - \boldsymbol{R}_i), \tag{5.24}$$

where N is the number of unit cells in the crystal. The Wannier functions are related to the Bloch functions $\psi_{nk}(\boldsymbol{r}) = \mathrm{e}^{\mathrm{i}\boldsymbol{k}\cdot\boldsymbol{r}} u_{nk}(\boldsymbol{r})$, solutions of H_0 alone, by

$$a_n(\boldsymbol{r} - \boldsymbol{R}_i) = N^{-1/2} \sum_{\boldsymbol{k}} \mathrm{e}^{-\mathrm{i}\boldsymbol{k}\cdot\boldsymbol{r}_i} \psi_{nk}(\boldsymbol{r}). \tag{5.25}$$

In many cases of interest, e.g., when describing transitions that take place at the band extrema like in PL experiments, the summation in (5.25) can be restricted to $\boldsymbol{k} \sim 0$ and $\Psi(\boldsymbol{r}) \approx u_0(\boldsymbol{r}) C(\boldsymbol{r})$. In general, by taking into account the band structure of the lattice considered, it can be demonstrated that the envelope wave function obeys the Schrödinger equation for a charge carrier with effective mass m subjected to a potential $U(\boldsymbol{r})$ [8]

$$\left[-\left(\frac{\hbar^2}{2m} \right) \frac{\partial^2}{\partial \boldsymbol{r}^2} + U(\boldsymbol{r}) \right] C(\boldsymbol{r}) = E\ C(\boldsymbol{r}). \tag{5.26}$$

The above equation turns out to be a very convenient manner to solve (5.23) for it embeds simply the effect of the crystal potential (namely, the lattice band structure) via the carrier effective mass m and the relative static dielectric constant of the material.

We now consider (5.26) for an electron and a hole, interacting via the Coulomb force, under a uniform stationary magnetic field. The vector potential can then be written as $\boldsymbol{A} = 1/2(\boldsymbol{B} \times \boldsymbol{r})$, where $\boldsymbol{B}$ satisfies the relation $\boldsymbol{B} = \nabla \times \boldsymbol{A}$, with gauge $\nabla \cdot \boldsymbol{A} = 0$. In the presence of $\boldsymbol{B}$, the carrier momentum becomes $\hbar\boldsymbol{k} = \boldsymbol{p} + e\boldsymbol{A}$ and, neglecting the carrier spin, the Hamiltonian is [57–59]

$$\left[-\frac{\hbar^2}{2\mu}\nabla_r^2 - \frac{e^2}{4\pi\varepsilon_0\varepsilon_r r} - i\hbar e\left(\frac{1}{m_e}-\frac{1}{m_h}\right)\boldsymbol{A}(\boldsymbol{r})\cdot\nabla_r + \frac{e^2}{2\mu}\boldsymbol{A}^2(\boldsymbol{r}) + \frac{e\hbar}{m_e+m_h}\boldsymbol{K}\cdot\boldsymbol{A}(\boldsymbol{r})\right]\phi(\boldsymbol{r}) = \left[E - \frac{\hbar^2K^2}{2(m_e+m_h)}\right]\phi(\boldsymbol{r}). \tag{5.27}$$

Here $\boldsymbol{r}$ denotes the electron–hole relative coordinate, and $\boldsymbol{K} = \boldsymbol{k}_e + \boldsymbol{k}_h$ is the center of mass momentum.

We now discuss the various contributions in (5.27).

Using $\boldsymbol{A} = 1/2(\boldsymbol{B}\times\boldsymbol{r})$ and introducing the angular momentum of relative motion $\boldsymbol{L} = -i\hbar(\boldsymbol{r}\times\nabla_r)$, the third addendum on the left-hand side can be written as:

$$H_1 = -i\hbar e\left(\frac{1}{m_e}-\frac{1}{m_h}\right)\boldsymbol{A}(\boldsymbol{r})\cdot\nabla_r = \frac{e}{2}\left(\frac{1}{m_e}-\frac{1}{m_h}\right)\boldsymbol{B}\cdot\boldsymbol{L} \tag{5.28}$$

that accounts for the orbital Zeeman effect. The spin can be effectively included by adding $(\mu_B/\hbar)g_{exc}\boldsymbol{B}\cdot\boldsymbol{S}$, g_{exc} and $\mu_B = e\hbar/2m_0$ being the exciton gyromagnetic ratio and the Bohr magneton, respectively. Assuming $\boldsymbol{B} = (0, 0, B)$, the fourth term becomes

$$H_2 = \frac{e^2}{2\mu}\boldsymbol{A}^2(\boldsymbol{r}) = \frac{e^2B^2}{8\mu}(x^2+y^2). \tag{5.29}$$

The fifth term

$$\frac{e\hbar}{(m_e+m_h)}\boldsymbol{K}\cdot\boldsymbol{A}(\boldsymbol{r}) = \frac{e}{2}(\boldsymbol{V}\times\boldsymbol{B})\cdot\boldsymbol{r}, \tag{5.30}$$

can be regarded as a Stark effect induced by the applied magnetic field and depends on the center of mass velocity $\boldsymbol{V}$ of the exciton [58]. This perturbation represents the effect of the Lorentz force acting on the exciton center of mass moving with velocity $\boldsymbol{V}$. It can be treated as an electric field $\boldsymbol{E} = \boldsymbol{V}\times\boldsymbol{B}$ acting on the pair at rest, as first reported in CdS [60]. The right-hand side of (5.27) contains the kinetic energy of the exciton center of mass.

It is instructive to compare the relative importance of the contributions of H_1 and H_2 [(5.28) and (5.29), respectively] for *a hydrogen atom and an exciton in GaAs* in their fundamental $\boldsymbol{L} = 0$ state. In the two cases, the contribution H_1 [meV] $= (\mu_B/\hbar)g_e\boldsymbol{B}\cdot\boldsymbol{S} \cong \mu_B B = 5.8\times10^{-2}B$ [T] is of the same order of magnitude since the exciton gyromagnetic ratio is close to unity. The contribution of term H_2, instead, differs considerably because it depends on the hydrogenic system considered. In fact, if we write

$$\sqrt{x^2+y^2} \approx a_B = (4\pi\varepsilon_r\varepsilon_0)\hbar^2/(\mu e^2) = [\varepsilon_r/(\mu/m_0)]\times 0.053\,\text{nm}, \tag{5.31}$$

we have that H_2 [meV] $= 6.2\times10^{-8}B^2$ [T^2]$\varepsilon_r^2/(\mu/m_0)^3$ is about 10^6 times larger for an exciton in GaAs (where $\varepsilon_r = 12.5$ and $\mu = 0.054\,m_0$) than for a hydrogen atom.

These estimates of the extent of the perturbation exerted by a magnetic field on *a hydrogenic* system in a semiconductor have also interesting, more general implications. The Hamiltonian in (5.27) has the same form for a H atom (where $a_\mathrm{B} = 0.053\,\mathrm{nm}$ and $Ry = 13.6\,\mathrm{eV}$) and an exciton in GaAs (where $a_\mathrm{B} \sim 12\,\mathrm{nm}$ and $Ry = 4.2\,\mathrm{meV}$). However, for a magnetic field $B = 10\,\mathrm{T}$ (a value typically accessible in laboratory) the parameter $\gamma = (\hbar\omega_\mathrm{c})/(2Ry)$ that quantifies the relative strengths of the magnetic field and Coulomb interaction is equal to about 4×10^{-5} and 2 for a H atom and a GaAs exciton, respectively. Therefore, investigating the properties of excitons in a table-top experiment is equivalent to study the same problem for a H atom in the presence of fields as large as $10^5\,\mathrm{T}$. These field intensities are observable only in astrophysical contexts like in the presence of white dwarfs or neutron stars [61].

In the next sections, we consider three different limits occurring in magneto-PL experiments: High, low, and intermediate magnetic field. For each case, we discuss how to derive quantitative information about the electronic properties of a particular material.

5.4.2 *High Magnetic Field Limit*

In the high magnetic field limit ($\gamma \gg 1$), the Coulomb force acting between the recombining electron–hole pair can be neglected with respect to the cyclotron energy $\hbar\omega_\mathrm{c}$. The charge carrier Hamiltonian H can be written in terms of the effective mass approximation [8, 57]

$$\frac{1}{2m}(\boldsymbol{p} + e\boldsymbol{A})^2. \tag{5.32}$$

The corresponding Schrödinger equation can be easily solved leaving out the spin effects (which we will deal with in the following) and considering nondegenerate conduction and valence bands with parabolic dispersion and spherical symmetry around the Γ point (no interband Coulomb interaction is assumed). In this case, the energy dispersion relations are

$$E_\mathrm{c}(k_z, n) = E_\mathrm{c}^0 + \left(n + \frac{1}{2}\right)\frac{\hbar e B}{m_\mathrm{e}} + \frac{\hbar^2 k_z^2}{2m_\mathrm{e}}, \tag{5.33a}$$

$$E_\mathrm{v}(k_z, n) = E_\mathrm{v}^0 - \left(n + \frac{1}{2}\right)\frac{\hbar e B}{m_\mathrm{h}} - \frac{\hbar^2 k_z^2}{2m_\mathrm{h}}. \tag{5.33b}$$

E_c^0 and E_v^0 are the unperturbed bottom and top of the conduction and valence band, respectively, and m_e and m_h the corresponding mass. The conduction and valence bands are split by the presence of the magnetic field in a series of quantized energy sublevels, sketched in Fig. 5.16, known as *Landau levels* (LLs) and identified by

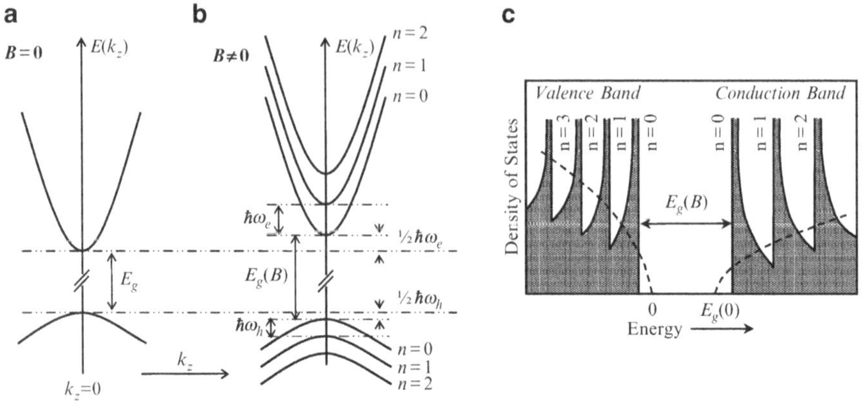

Fig. 5.16 Sketch of the energy bands of a cubic semiconductor in absence [panel (**a**)] and presence [panel (**b**)] of an external magnetic field. k_z indicates the component of the wavevector along the magnetic field direction. (**c**) Sketch of the density of states, DOS, in the conduction and valence band for $B = 0$ (*dashed lines*, DOS $\propto \sqrt{E}$) and $B \neq 0$ [*solid lines*, DOS given in (5.34)]

the quantum number $n = 0, 1, 2, \ldots$. Notice that even the lowest lying conduction (valence) sublevel $n = 0$ is up- (down-) shifted in energy with respect to the zero field case by $\frac{1}{2}\hbar\omega_e$ ($\frac{1}{2}\hbar\omega_h$), namely, the zero-point energy of the harmonic oscillator $\omega_{e,h} = eB/m_{e,h}$. This results in a linear increase of the band-gap energy with the magnetic field B.

Although (5.33) do not depend on k_x or k_y, which can assume arbitrary values, there is not an infinite degeneracy because of the finite crystal volume where carriers are confined. Imposing that the center of the orbit is inside the crystal, the density of states is found to be

$$\rho(E) = \frac{eB}{4\pi^2\hbar}\frac{\sqrt{2m_{e,h}}}{2\hbar}\left|E - E_{c,v}(k_z, n)\right|^{-1/2}, \qquad (5.34)$$

where $E_{c,v}$ (k_z, n) is an energy level with given k_z and belonging to the nth conduction or valence sublevel energy. This expression is satisfied in the conduction band for $E > E_c(0,0)$ and in the valence band for $E < E_v(0,0)$. The density of states is infinite at the edge of each sublevel (see Fig. 5.16), a singularity experimentally not observable because of scattering phenomena and finite temperature. From a classical point of view $\omega_c\tau$, where τ is the carrier scattering lifetime, i.e., the average time between two collisions, should be greater than one so that carriers complete at least one orbit without being scattered and quantization of motion is observed. As an example, for an effective mass equal to 0.1 m_0 and a magnetic field of 1 T, the scattering lifetime has to be larger than 10^{-12} s to observe the Landau quantization. Thus, carriers should be photogenerated in pure crystals at low temperatures.

As already mentioned, all Landau levels are highly degenerate. The degeneracy at a given k_z is the same for all LLs and, if z is the direction of the magnetic field, is given by [8]

$$\xi = \frac{m_i}{2\pi\hbar}\omega_i L_x L_y = \frac{L_x L_y eB}{h} = \frac{L_x L_y}{2\pi l_B^2}, \tag{5.35}$$

where L_x and L_y are the dimensions of the sample in the x and y directions, respectively, and $l_B = \sqrt{\hbar/eB}$ is the so-called Landau magnetic length, equal to the radius of the classical orbit corresponding to the lowest energy Landau level ($n = 0$).

If the electrons are constrained (i.e., confined) to move only in planes perpendicular to the magnetic field and the free-electron-like motion parallel to the magnetic field is suppressed, we can define the number of fully occupied Landau levels or LL *filling factor*

$$\nu = \frac{N_e}{\xi} = \frac{hn_e^{2/3}}{eB}, \tag{5.36}$$

where N_e is the number of electrons inside the crystal. For increasing magnetic fields, more and more electrons fit into a given Landau level and the occupation of the highest LL ranges from completely full to entirely empty. This leads to oscillations in various electronic properties and gives rise, e.g., to the De Haas–van Alphen effect and to the Shubnikov–De Haas effect.

Let us now consider the effect of the electron spin, namely, the *Zeeman effect.* Because of the interaction between the magnetic field and the electron spin, the energy of each level changes by an amount $\Delta E = m_s g \mu_B B$, where m_s is the exciton spin projection along the magnetic field direction, to be added to the second member of the (5.33). The density of states $f(E)$ is modified too by the Zeeman effect

$$f(E) = \sum_{|m_J|}\left[\frac{1}{2}f_0(E + |m_s|g\mu_B B) + \frac{1}{2}f_0(E - |m_s|g\mu_B B)\right], \tag{5.37}$$

where $f_0(E)$, given by (5.34), is the energy density of states calculated without spin. Next we focus on two examples where the approximation of high magnetic field and the Landau level (5.33) hold.

5.4.2.1 Free-Electron to Neutral-Acceptor Transition

First, we consider transitions where the electron is free and the hole is bound to an impurity, as, for instance, in GaAs-based materials where carbon atoms incorporated during the growth process act as traps for holes and give rise to acceptor levels in the host crystal band gap. The PL spectra of a GaAs sample recorded at $T = 20$ K and different magnetic fields are shown in Fig. 5.17, where the (e,C) recombination of a free-electron in the conduction band with a hole localized on a carbon level dominates the spectrum. The longitudinal-optical phonon replica of the (e,C) recombination and the free-exciton, FE, recombination bands are observed too. At zero magnetic field, the (e,C) recombination energy is given by

Fig. 5.17 Low-temperature ($T = 20$ K) normalized PL spectra of a GaAs sample for different values of applied magnetic field. (e,C) and FE indicate the free-electron to neutral-carbon recombination and free-exciton recombination, respectively. The longitudinal optical (LO) phonon replica of the (e,C) recombination is also resolved. The *dashed line* is a guide to the eye

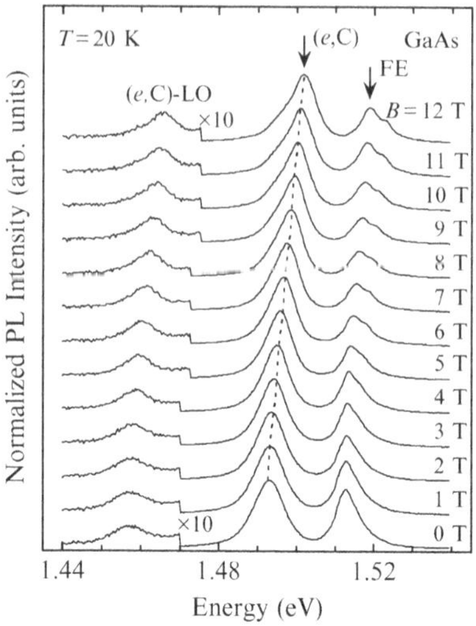

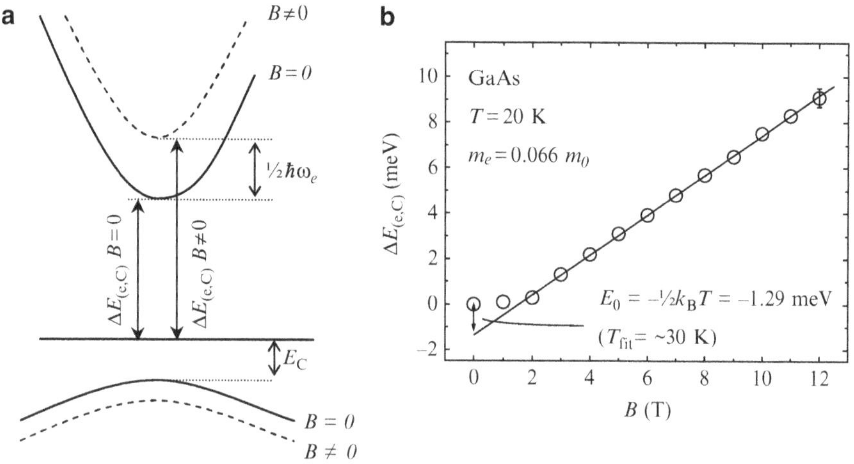

Fig. 5.18 (**a**) Sketch of the levels involved in the (e,C) recombination in the presence (*dashed lines*) and absence (*solid lines*) of an external magnetic field. Under magnetic field, the carbon acceptor level remains fixed in energy whereas the conduction band minimum up-shifts in energy by $1/2\,\hbar\omega_e$. (**b**) Diamagnetic shift of the (e,C) recombination as measured in a GaAs sample (*symbols*). The typical error bar is shown for one data point. The electron effective mass—as estimated by fitting (5.40) to the data for $B > 2\,\mathrm{T}$ (*solid line*) - is indicated in the figure. The extrapolated zero-field energy gives an estimate of the carrier temperature

$$h\nu_{(e,\mathrm{C})}(B = 0) = E^0_{\mathrm{CBM}} - E^0_{\mathrm{C}}, \tag{5.38}$$

where E^0_{CBM} is the energy of the conduction band minimum and E^0_{C} is the energy of the carbon acceptor level. Figure 5.18a sketches the (e,C) recombination in a reciprocal space scheme at $k \sim 0$ when a magnetic field B splits the conduction band into LLs, as discussed in Sect. 5.4.2. In particular, if the (e, C) recombination involves the lowest Landau level ($n = 0$), its energy shifts according to

$$h\nu_{(e,\mathrm{C})}(B) = E^0_{\mathrm{CBM}} - E^0_{\mathrm{C}} + \frac{e\hbar}{2m_{\mathrm{e}}}B - \Delta E_{\mathrm{C}}(B), \tag{5.39}$$

where $\Delta E_{\mathrm{C}}(B)$ indicates the shift of the carbon level induced by the magnetic field. Rossi, Wolfe, and Dimmock [62], followed by many other authors [63–68], first assumed that the C-related level stays fixed in energy because of the dispersionless characteristics (i.e., infinite effective mass) of the C impurity level in k-space. $\Delta E_{\mathrm{C}}(B)$ is very small, indeed, less than 0.075 meV at 19 T ($\sim$0.02 meV at 10 T) in GaAs [64]. Therefore, the slope of the diamagnetic shift $\Delta E_{(e,\mathrm{C})}$ of the (e,C) PL peak energy is proportional to the inverse of the electron effective mass

$$\Delta E_{(e,\mathrm{C})} = h\nu_{(e,\mathrm{C})}(B) - h\nu_{(e,\mathrm{C})}(0) = \frac{e\hbar}{2m_{\mathrm{e}}}B. \tag{5.40}$$

Figure 5.18b shows $\Delta E_{(e,\mathrm{C})}$ as a function of B together with the fit of (5.40) to the experimental data for $B \geq 2$ T (full line). The estimated electron effective mass is 0.066 m_0, in excellent agreement with the value reported in the literature (0.0667 m_0). For low magnetic fields ($B < 2$ T), a deviation from linearity is observed, as already reported in the literature [63,65,66,68]. This deviation is related to a change in the dimensionality of the density of states in the conduction band [see (5.34) and Fig. 5.16c], which determines the PL lineshape of the (e,C) band [63,65]

$$I_{(e,\mathrm{C})}(E) = A\rho_{\mathrm{CB}}(E) \cdot \mathrm{e}^{-E/(k_B T)}. \tag{5.41}$$

The proportionality constant A includes the transition matrix element, $\rho_{\mathrm{CB}}(E)$ is the density of states in the conduction band, and $\mathrm{e}^{-E/(k_B T)}$ is the Boltzmann factor, which determines the high-energy edge of the PL band. In the presence of a magnetic field, the electron density of states is given by (5.34) and it is strongly peaked when the energy E equals the energy of a Landau level. In this case, the energy of the photoluminescence peak for the (e,C) transition almost coincides with the energy distance between the levels involved in the recombination, $E^0_{\mathrm{CBM}}(B) - E^0_{\mathrm{C}}$, namely, the effect of the Boltzmann factor is negligible. On the other hand, when the magnetic field vanishes, the conduction band density of states returns to that of a free-electron gas, which varies as $\sqrt{E - E^0_{\mathrm{CBM}}}$. In this case, the effect of $\mathrm{e}^{-E/(k_B T)}$ is sizable and the (e,C) photoluminescence peak does not coincide with the energy distance between the levels but it is at $(E^0_{\mathrm{CBM}} - E^0_{\mathrm{C}}) + k_B T/2$. As a consequence, a correct evaluation of the electron effective mass can be done only at sufficiently high field or at low temperature.

Magneto-PL has been used in $\mathrm{GaAs}_{1-x}\mathrm{N}_x$ to investigate the effect of the N concentration on the conduction band edge shape, namely, on its curvature as measured by the electron effective mass [69, 70]. Indeed, in highly mismatched alloys such as dilute nitrides, the fundamental band structure parameters, including the carrier effective mass, exhibit uncommon compositional dependences, as already mentioned.

Figure 5.19 displays the magneto-PL spectra of a series of $GaAs_{1-x}N_x$ samples having different N concentrations. Relatively low temperatures ($T \leq 30$ K) and low excitation power densities ($P_{exc} \leq 20\,W/cm^2$) were used to enhance the contribution of the (e,C) transitions with respect to those related to free excitons and excitons trapped on N clusters, NC. Figure 5.20 highlights the different degree of localization that carriers involved in the above-mentioned transitions exhibit. The NC-related transitions concern carriers whose wave functions are strongly localized around the nitrogen atoms, are not perturbed sizably by the magnetic field, and, therefore, do not shift with B. As regards the FE band, it is characterized by a diamagnetic shift smaller and qualitatively different (namely quadratic instead of linear) with respect to that of the (e,C) band. This is due to a Coulomb force acting on the exciton much larger than the force between the free electron and the neutral acceptor, as it will be discussed in more detail in Sect. 5.4.4.

The relative energy shift of the (e,C) transition as a function of B is shown in Fig. 5.21 for a few characteristic N concentrations, while the dependence of m_e on N concentration is shown in Fig. 5.22. After the first abrupt doubling for $x \sim 0.1\%$ ($m_e \sim 0.13\ m_0$), m_e undergoes a second increase for $x \sim 0.35\%$, and finally it fluctuates around $m_e \sim 0.15\ m_0$ for $0.38\% \leq x \leq 1.78\%$. This dependence displays a behavior very different from that expected for a conventional alloy, where the compositional dependence of band structure parameters evolve smoothly; see the dashed line in the figure. Contrarily, the experimental results are remarkably well reproduced up to $x = 0.6\%$ by a *modified* $k \cdot p$ approach (open squares) [17,70,71]. Within this approach, the interaction between the conduction band edge of $GaAs_{1-x}N_x$—that red-shifts with increasing x—and the energy-pinned states due to N clusters in the band gap is duly taken into account. In turn, this explains the data of Fig. 5.22 in terms of on- and off-resonance conditions, driven by composition, between states having a different degree of localization [72].

5.4.2.2 Free Carrier Recombination in Degenerate Semiconductors

In order to study the behavior of single carrier levels in a magneto-PL experiment, the Coulomb interaction between oppositely charged carriers should be close to zero, as in the presence of either high electric fields or a lattice thermal energy well above the exciton binding energy. Exciton dissociation occurs also because of screening effects at high carrier density leading to the formation of an electron–hole plasma [47].

Degenerate semiconductors are another case where free-carrier (or band-to-band) recombination dominates the PL spectrum. It has been already mentioned that InN is a degenerate semiconductor with potential in a variety of applications ranging from photovoltaics to fast and high-power electronics [51]. To this regard, a direct determination of the electron effective mass of InN via cyclotron-like measurements

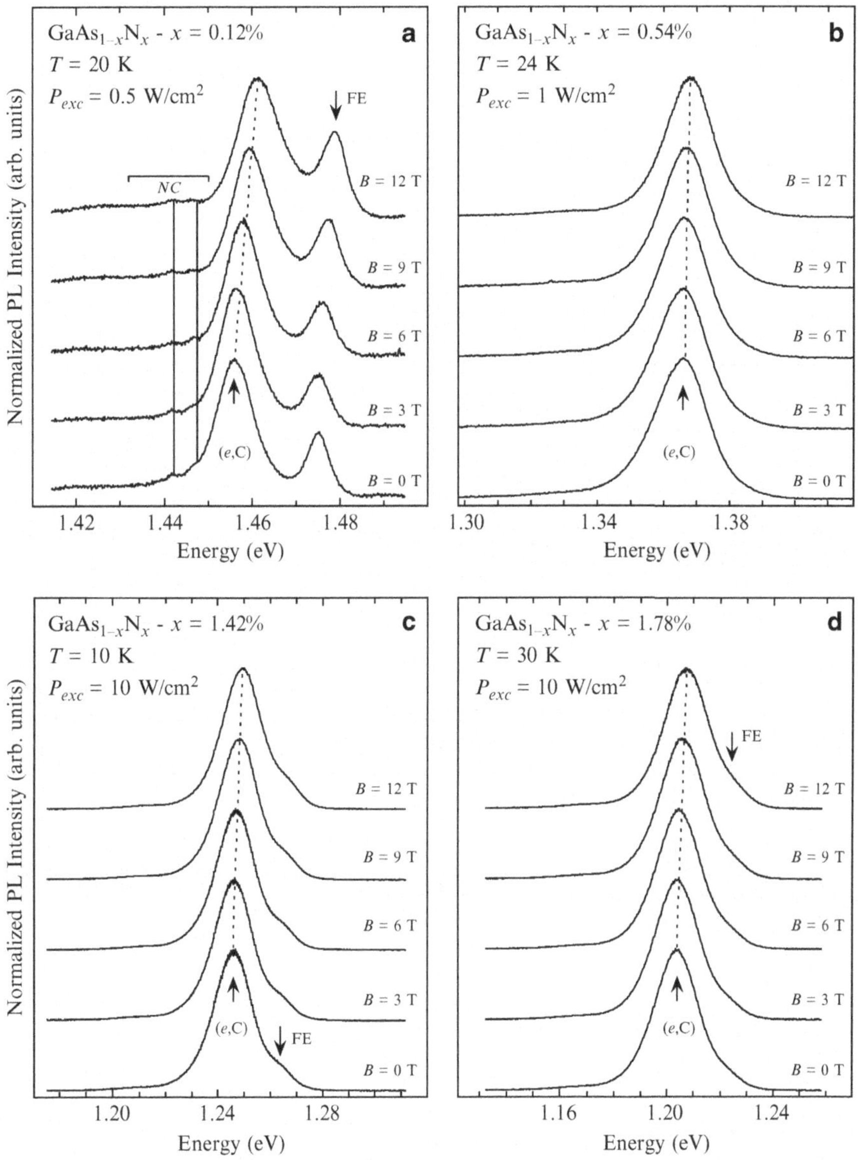

Fig. 5.19 Low-temperature ($T = 10$–30 K) normalized PL spectra of $GaAs_{1-x}N_x$ samples for different nitrogen concentrations [**(a)** $x = 0.12\%$; **(b)** $x = 0.54\%$; **(c)** $x = 1.42\%$; **(d)** $x = 1.78\%$] for different applied magnetic fields. *Downward (upward) arrows* indicate FE [(*e*,C)] recombination. *Dotted lines* are guides to the eye indicating the position of the (*e*,C) peak. *Solid vertical lines* show the energy position of nitrogen-cluster states, NC, which are not perturbed by applied magnetic fields

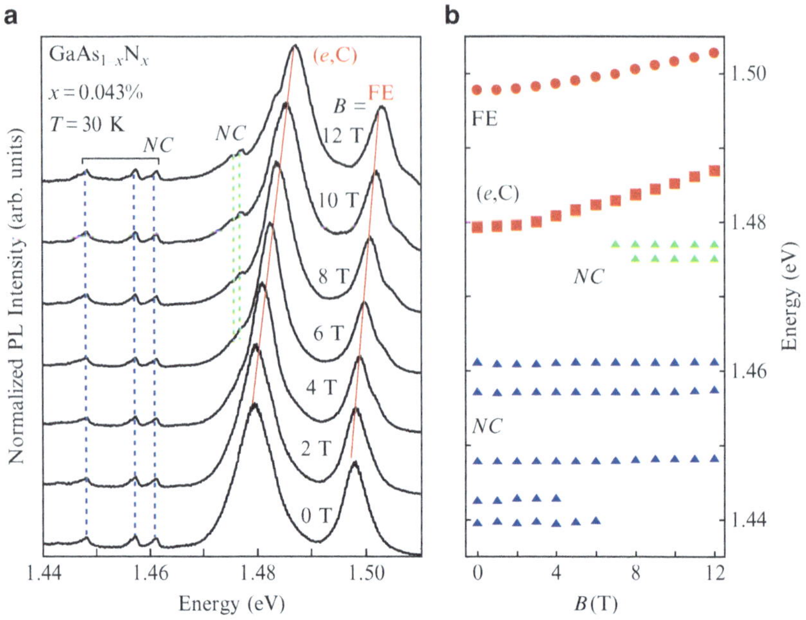

Fig. 5.20 (**a**) Low-temperature ($T = 30$ K) normalized PL spectra of a $GaAs_{1-x}N_x$ ($x = 0.043\%$) epilayer for different applied magnetic fields. (e,C) and FE indicate the free-electron to neutral-carbon recombination and free-exciton recombination, respectively. Other recombination peaks below 1.478 eV are related to N complex states, NC. (**b**) Magnetic field dependence of the peak energy of the different recombination bands shown in panel (**a**). Note that only the (e,C) and FE transitions shift with B

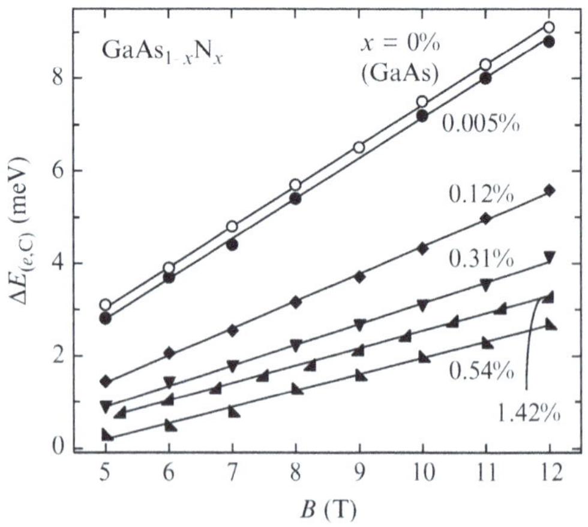

Fig. 5.21 Magnetic field-induced shift of the (e,C) recombination band for some $GaAs_{1-x}N_x$ samples (*full symbols*) with different nitrogen concentrations (indicated in the figure). The shift of the (e,C) recombination band for a GaAs reference sample (*open circles*) is also shown. *Lines* are fits of (5.40) to the data to obtain the m_e value

is of great interest, since most of the m_e values reported have been inferred from infrared reflectivity or ellipsometry measurements. Figure 5.23 shows the magneto-PL spectra recorded at low temperature ($T = 5$ K) on an InN epilayer untreated [panel (a)] and irradiated with two doses of hydrogen atoms: $d_H = 1z$ [panel

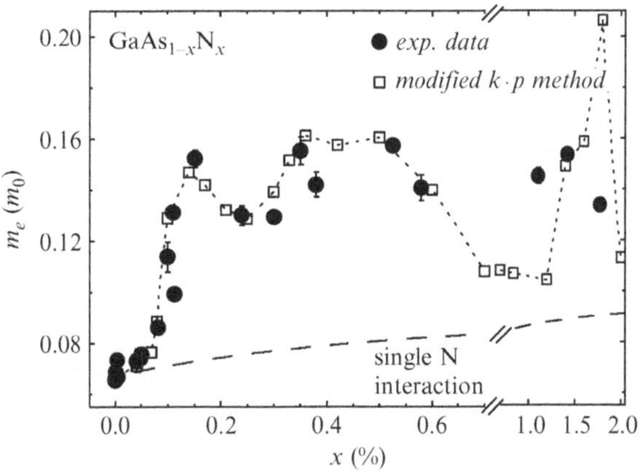

Fig. 5.22 Dependence of the $GaAs_{1-x}N_x$ electron effective mass on nitrogen concentration (*full symbols*). Error bars indicate the uncertainty on the mass values, whenever larger than the symbol size. The *dashed line* is the theoretical expectation based on an anticrossing model for a single N band, whereas the *dotted line* linking open squares is a theoretical estimate based on a modified $k \cdot p$ method that takes into account the effects of the interaction of the CB minimum with N-cluster states (see text and [70, 71])

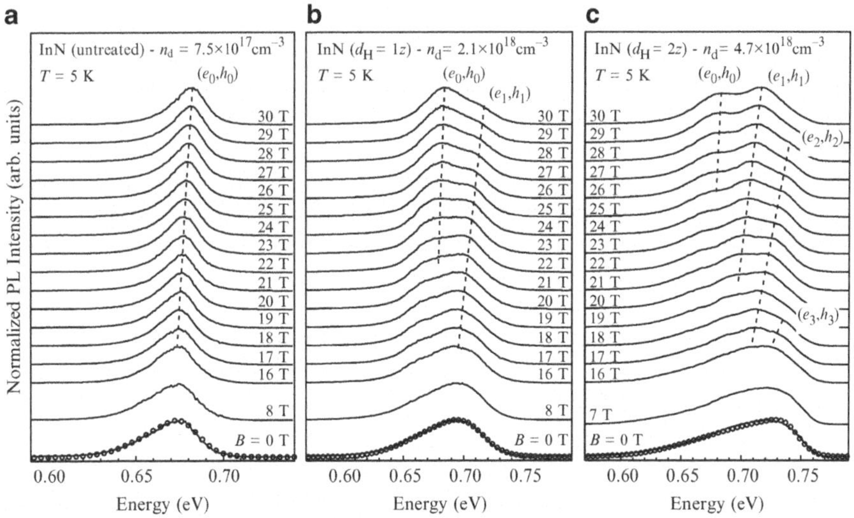

Fig. 5.23 Low-temperature ($T = 5$ K) normalized PL spectra of an InN sample irradiated with different H doses (d_H) to control the free-electron concentration (n_d): (**a**) untreated; (**b**) hydrogenated with $d_H = 1z$; (**c**) hydrogenated with $d_H = 2z$. $z = 1 \times 10^{15}$ ions/cm^2. The spectra were recorded for different magnetic fields, in steps of 1 T. (e_n, h_n) indicates a transition between the nth Landau levels in the conduction and valence band. The *dashed lines* are guides to the eye. PL spectra at $B = 0$ T are well reproduced (*open circles*) by the semiclassical model described in the text; see (5.18) and (5.19). The fitted free-electron concentrations are indicated in figure

(b)] and $d_{\mathrm{H}} = 2z$ [panel (c)], with z equal to 1×10^{15} ions/cm^2. H-irradiation introduces donor levels in n-type InN and leads to an almost linear increase of the electron concentration without changing sizably the material band-gap energy, as discussed in Sect. 5.3.4 and shown in [52].

For increasing B, several bands develop. These bands can be related to Landau levels (LLs) with different indexes n and energies given by

$$E_n(B) = E(0) + \left(n + \frac{1}{2}\right)\left(\frac{1}{m_{\mathrm{e}}} + \frac{1}{m_{\mathrm{h}}}\right)e\hbar B = E(0) + \left(n + \frac{1}{2}\right)\frac{\hbar e B}{\mu}. \quad (5.42)$$

The PL spectrum results from the recombination of both photoexcited and donor-induced electrons in the conduction band and photoexcited holes in the valence band. The open circles superimposed on the zero field spectra in Fig. 5.23 are simulations to the PL lineshape by the Kane model summarized in Sect. 5.3.4. The different number of peaks observable for different samples is related to the changes of the LL degeneracy with magnetic field, and thus to the number of LLs populated. In fact, considering the degeneracy of each spin-resolved LL given by (5.36), $\nu = hn_{\mathrm{e}}^{2/3}/(eB)$ is the number of the *almost* filled LLs ("*almost*" because each LL has an infinite degeneracy in the component of the wavevector along $\boldsymbol{B}$). Therefore, in hydrogenated samples, more Landau levels are visible as a consequence of the greater n_{e} value in those samples. We also note that no sizable variations of the PL band is detected for magnetic field $B < 16$ T. This may be attributed to carrier localization effects prevailing at small B and/or to strong scattering of carriers with other carriers and charged impurities that has to be expected in highly doped materials.

The magnetic field dependence of the peak energy of all transitions observed in the three considered samples is summarized in Fig. 5.24. $(e_{\mathrm{n}}, h_{\mathrm{n}})$ indicates the recombination of an electron and a hole in the nth LL: According to the symmetry properties of the Hermite polynomials, solutions of the Schrödinger equation for a charge carrier in a magnetic field, radiative transitions occur only between Landau levels in the conduction and valence band with the same index n [73]. For each sample, the lines are fits of (5.42) to the whole set of data—where μ and $E(0)$ are fit parameters—accounting for recombinations from different LLs [74]. For all samples and data shown in Fig. 5.24 an excellent agreement is obtained with very similar values of μ and $E(0)$.

The reduced cyclotron mass estimated from the shift of the transitions involving the LLs is related to the cyclotron frequency $\omega_{\mathrm{c}} = eB/\mu$, where $1/\mu = 1/m_{\mathrm{e}} + 1/m_{\mathrm{h}}$. Due to band nonparabolicity effects, the *curvature* mass differs from the *optical* mass, which is involved in optical experiments such as reflectivity or absorption [75–77]. Nevertheless, the two masses coincide in the limit $k \to 0$ and the μ values estimated here provide the *reduced* curvature mass *at the* Γ *point* of the InN conduction and valence band. Indeed, as the magnetic field is turned on, the change of dimensionality of the carrier motion from 3D- to 1D-like [see (5.34) and Fig. 5.16c] implies a change in the energy dependence of the crystal density of states, DOS. Therefore, a much larger carrier occupancy at the band extrema

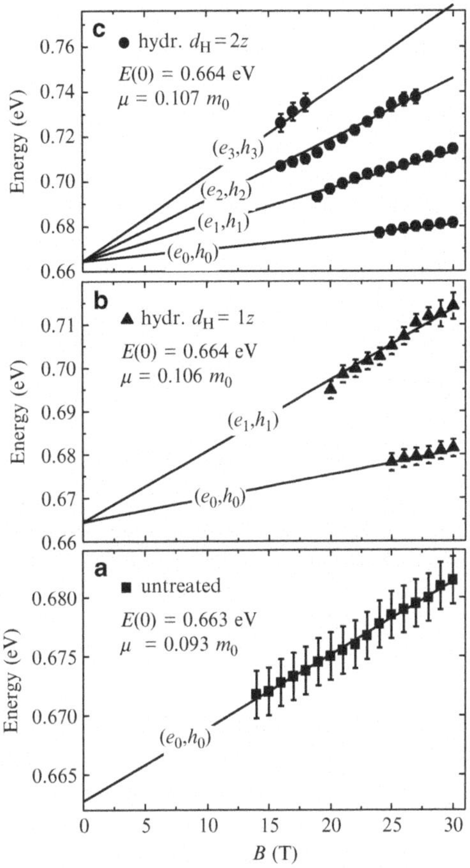

Fig. 5.24 Energies of the (e_n, h_n) transitions as a function of magnetic field, as obtained by the analysis of the magneto-PL spectra shown in Fig. 5.23, for an InN sample irradiated with different hydrogen doses: (**a**) untreated; (**b**) hydrogenated with $d_H = 1z$; (**c**) hydrogenated with $d_H = 2z$. $z = 1 \times 10^{15}$ ions/cm^2. *Lines* are fits of (5.42) to the data. The values of the band gap energy at zero field, $E(0)$, and of the carrier reduced mass, μ, are fitting parameters

is expected for a 1D-DOS ($\propto 1/\sqrt{E}$) than for a 3D-DOS ($\propto \sqrt{E}$). Hence, the PL bands observed in Fig. 5.23 at high magnetic fields can be attributed to carrier recombinations mainly at the Γ point, and the slope of the peak energy with B is ultimately determined by the carrier reduced mass at the conduction band minimum and valence band maximum, *independently on the carrier concentration*. In order to extract the electron effective mass from μ, one should know the value of the hole effective mass. Since a quite large range of hole effective mass values has been reported on the basis of indirect measurements in InN [78–80], only a lower limit of the electron effective mass has been obtained, $m_e \geq 0.106 m_0$.

5.4.3 Low Magnetic Field Limit

When the magnetic field is so small that the Coulomb energy is larger than the magnetic energy, a condition always reached for $B \to 0$, the electrostatic term in (5.27) cannot be neglected. However, in highly polar semiconductors, such as ZnO and ZnS, the exciton binding energy is as high as 40 meV [59,81], and the conditions of low field are satisfied even at the highest B values reachable in a laboratory (45 and 100 T for steady-state and long-pulsed fields, respectively). A high exciton binding energy is observed also in highly confined systems like quantum dots, QDs, with typical dot size of few tens of nanometer, or less. Therein, the strongly confining potential constrains an electron–hole pair to have a relative distance much smaller than that it would have in a bulk material, thus leading to an exciton binding energy of about 40 meV in InGaN/GaN [82] or 20 meV in InAs/InP QDs [83].

Under this condition, the two field-containing terms of the Hamiltonian in (5.27)

$$H_1 = \frac{e}{2}\left(\frac{1}{m_e} - \frac{1}{m_h}\right)\boldsymbol{B}\cdot\boldsymbol{L} + g_{exc}\frac{\mu_B}{\hbar}\boldsymbol{B}\cdot\boldsymbol{S}, \tag{5.43}$$

and

$$H_2 = \frac{e^2}{2\mu}\boldsymbol{A}^2(\boldsymbol{r}) = \frac{e^2B^2}{8\mu}(x^2+y^2), \tag{5.44}$$

can be treated as a perturbation. The energy correction to the exciton ground state $|0\rangle$ as given by perturbation theory keeping only terms up to B^2 is [84, 85]

$$\Delta E = \frac{e^2B^2}{8\mu}\langle 0|x^2+y^2|0\rangle + g_{exc}\mu_B B m_s, \tag{5.45}$$

where the matrix element is related to the exciton Bohr radius a_B by $\langle 0|x^2+y^2|0\rangle = 2a_B^2$.

Equation (5.45) is a very frequently used formula since it renders available important information concerning the wave function extent of an exciton, provided that the exciton reduced mass is known. This turns out to be of relevance in the case of nanostructures. Indeed, the bulk properties of a material are usually known (e.g., dielectric constant, carrier mass), and it would be valuable to estimate the size and symmetry of the exciton wave function, especially when novel nanostructures are formed in a spontaneous process or fabricated.

As an example, a new method was recently developed for the fabrication of III–V semiconductor QDs (the "*hydrogenation and masking*" method) [86, 87]. This method relies on a defect-engineering approach that exploits the hydrogen-induced passivation of N atoms in Ga(AsN), (InGa)(AsN), and Ga(PN) crystals. Therein, the incorporation of small percentages (of the order of 1–5%) of N atoms causes dramatic variations in the electronic properties of the host crystal [17]. These variations comprise a giant reduction of the energy gap with increasing nitrogen incorporation (∼200 meV per N percent) as well as unexpected compositional dependences of the transport and spin properties. The unusual electronic properties of these semiconductors stem from the impurity-like localized character that the host band edges acquire as a result of a hybridization with the isoelectronic-impurity wave function [71]. In dilute nitrides, H atoms form stable N–2H–H complexes that wipe out the N effects on the host lattice thus modifying in a controllable

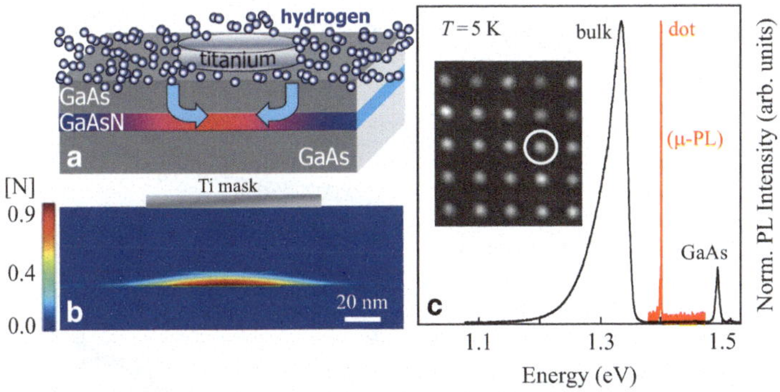

Fig. 5.25 (**a**) Sketch of the "*hydrogenation and masking*" method for the fabrication of QDs in dilute nitrides, see text and [86] and [87]. (**b**) Calculated posthydrogenation 2D distribution of N atoms not passivated by H in a $GaAs_{0.989}N_{0.011}/GaAs$ QW. The sample cross section passes through the Ti-mask dot diameter ($d = 80$ nm) containing the growth axis. Note the formation of a GaAsN QD (lateral size ~50 nm, high ~3 nm) embedded in GaAs-like barriers. (**c**) Low-temperature ($T = 5$ K) normalized PL spectrum of the pristine GaAsN QW and μ-PL spectrum of a single GaAsN QD (highlighted by the circle in the inset) obtained with the "*hydrogenation and masking*" method. The *inset* shows a μ-PL image of an array of dots, that demonstrates the excellent control over the dot position allowed by this fabrication method

manner the electronic (e.g., band gap energy, transport, and spin properties) [69, 70] and optical (refractive index) [88] characteristics of these materials. Therefore, by allowing H incorporation in selected regions of a QW sample, it is possible to attain a spatially tailored modulation of the band gap energy in the growth plane [86]. This is accomplished by prior deposition of H-opaque titanium dots that impede H diffusion beneath them. Therefore, thanks to the extremely short decay length (5 nm/decade) of the hydrogen forefront in bulk GaAsN [89], it is possible to realize site- and size-controlled quantum dots. A sketch of the process leading to the QD formation is shown in panels (a) and (b) of Fig. 5.25. Figure 5.25c shows the μ-PL spectra of a single dot and of the corresponding GaAsN quantum well before the hydrogenation-and-masking processing that leads to the dot formation. On the basis of H-diffusion simulations, the dot size is estimated to be of order of few tens of nm in diameter and few nm in height. Figure 5.26a shows the PL spectra under magnetic field up to 30 T taken in a single dot obtained by the fabrication technique displayed in the previous figure. With increasing B, the exciton peak blue-shifts and splits into Zeeman components. The diamagnetic shift displays a clear quadratic dependence on magnetic field. Panel (b) shows a quantitative analysis of the data by (5.45). If $\mu = 0.095\ m_0$ (as independently determined) [70], we obtain an exciton diameter $d_{exc} = 2a_B = \sqrt{2\ \langle 0|x^2 + y^2|0\rangle} = 4.3$ nm. This indicates a small wave function size caused by carrier quantum confinement in this nanostructure, as found in self-assembled InAs QDs [90]. Panel (c) displays an analysis of the Zeeman-splitting data leading to a value of the exciton g-factor equal to 1.04.

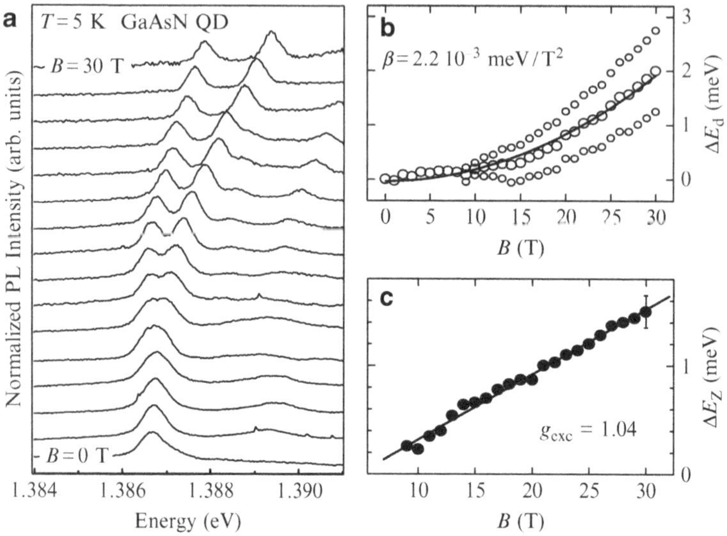

Fig. 5.26 (**a**) Low-temperature ($T = 5\,\mathrm{K}$) μ-PL spectra for different magnetic fields, B, recorded on a single GaAsN QD obtained with the "*hydrogenation and masking*" method depicted in Fig. 5.25. The spectra are displayed in steps of 2 T. (**b**) Diamagnetic shift of the PL peaks shown in panel (**a**) (*large circles* indicate the average over the Zeeman split components). *Solid line* is a fit of $\Delta E_{\mathrm{d}} = \beta B^2$ [see (5.45)] to the data. The error associated to each data point is within the symbol size. (**c**) Zeeman splitting of the PL peak shown in panel (**a**). The *solid line* is a fit of $\Delta E_Z = g_{\mathrm{exc}}\mu_{\mathrm{B}} \cdot B$ to the data, where g_{exc} is the exciton g-factor and μ_{B} is the Bohr's magneton. The error bar is shown for one data point

5.4.4 Intermediate Magnetic Field Limit

In Sect. 5.4, we introduced two energy scales: The magnetic energy $\hbar\omega_{\mathrm{c}}\,[\mathrm{eV}] = (\hbar e B/\mu) = [1.16 \times 10^{-4}/(\mu/m_0)]B\,[\mathrm{T}]$ and the exciton Rydberg $Ry\,[\mathrm{eV}] = (13.6\,\mu/m_0)/\varepsilon_{\mathrm{r}}^2$. The ratio $\gamma = (\hbar\omega_{\mathrm{c}}/2Ry) = 4.26 \times 10^{-6}[\varepsilon_{\mathrm{r}}/(\mu/m_0)]^2 B\,[\mathrm{T}]$ establishes to a first approximation the magnetic field regime to be considered for an electron–hole pair with reduced mass μ and embedded in a medium with static dielectric constant equal to ε_{r}. If we gauge γ from the values of μ and ε_{r} in a typical semiconductor (e.g., GaAs where $\mu \sim 0.054\,m_0$ and $\varepsilon_{\mathrm{r}} \sim 12.5$), we find $\gamma \sim 10$ for $B = 45\,\mathrm{T}$ (the highest steady-state field value attainable with present hybrid—resistive+superconducting—magnet technology). Incidentally, $\gamma \sim 10$ for *a hydrogen* atom would correspond to a magnetic field $B \sim 10^6\,\mathrm{T}$, as already pointed out in Sect. 5.4.1.

Therefore, in most semiconductors, whenever the exciton is not ionized or single carrier levels are not spectroscopically available, an intermediate field regime has to be considered. If we consider an exciton in its ground state (namely, orbital angular momentum $\boldsymbol{L} = 0$) and neglect the Zeeman splitting due to spin, the exciton effective-mass Hamiltonian turns into

$$H\phi(\boldsymbol{r}) = \left[-\frac{\hbar^2}{2\mu}\nabla_r^2 - \frac{e^2}{4\pi\varepsilon_0\varepsilon_r r} + \frac{e^2}{8\mu}(x^2+y^2)B^2\right]\phi(\boldsymbol{r}) = E\phi(\boldsymbol{r}). \tag{5.46}$$

Equation (5.46) is usually rewritten with energy in units of the exciton Rydberg $\mu e^4/\left[2(4\pi\varepsilon_r\varepsilon_0)^2\hbar^2\right]$, positions in units of the exciton Bohr radius $a_\mathrm{B} = (4\pi\varepsilon_0\varepsilon_\mathrm{r}\hbar^2)/(\mu e^2)$ and magnetic field in units of γ [91–93]

$$\left[-\nabla_r^2 - \frac{2}{r} + \frac{\gamma^2}{4}(x^2+y^2)\right]\phi(\boldsymbol{r}) = E\,\phi(\boldsymbol{r}). \tag{5.47}$$

$\boldsymbol{r} \equiv (x, y, z)$ is the electron–hole pair relative coordinate and the magnetic field is applied along the z direction. Since (5.47) has axial symmetry about z and is invariant under space inversion, the envelope function $\phi(\boldsymbol{r})$ has definite parity and z-projection m of angular momentum. Furthermore, allowed electromagnetic dipole transitions for band-edge emission involve an initial state whose envelope wave function does not vanish at the origin ($\boldsymbol{r} = 0$); see [8]. This implies that $\phi(\boldsymbol{r})$ has $m = 0$ and even parity. Following [92–94], $\phi(\boldsymbol{r})$ can be expressed as a combination of spherical harmonics with $m = 0$ and even l

$$\phi(r, \vartheta, \varphi) = \sum_{\text{even } l} f_l(\boldsymbol{r}) Y_{l,0}(\vartheta, \varphi), \tag{5.48}$$

where the radial functions $f_l(\boldsymbol{r})$ are found by substituting (5.48) in (5.47). A detailed description of the numerical methods employed to solve this eigenvalue problem is beyond our scopes. The interested reader can find details in [92–94]. Here, we report the results of the diamagnetic shift ΔE_d of an exciton in its ground state as a function of the magnetic field. Figure 5.27 shows ΔE_d in units of Ry as a function of magnetic field in units of the parameter γ. The data shown in this figure turn out to be particularly useful, since they provide an "universal" solution to the problem of an exciton in a magnetic field holding for any field intensity and for any isotropic semiconductor (provided that static dielectric constant and carrier effective masses are known). Conversely, the exciton reduced mass can be derived from the diamagnetic shift data exploiting the results presented in Fig. 5.27. The solid line in the figure is a fit to the data by a ninth degree polynomial

$$\Delta E_\mathrm{d}(\gamma) = \sum_{i=1}^{9} A_i\,\gamma^i. \tag{5.49}$$

In order to use the above expression for an exciton in a given material, one has to make γ explicit with magnetic field in units of Tesla and express ΔE_d in meV units. Then, (5.49) turns into

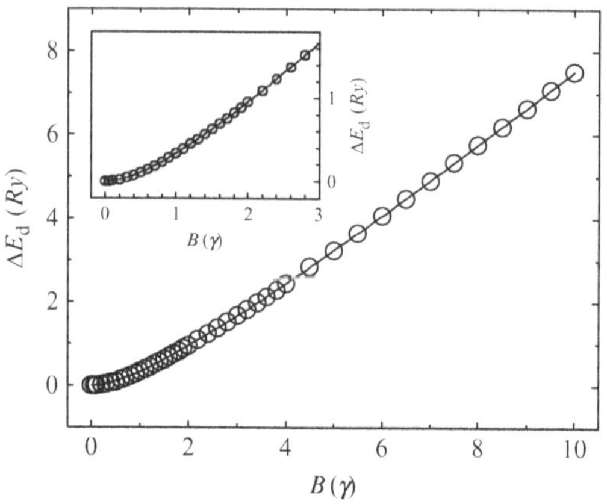

Fig. 5.27 Energy shift (ΔE_d) of the exciton ground state in units of the effective exciton Rydberg [given in (5.20)] as a function of magnetic field in units of the parameter γ [given in (5.22)] as calculated in [91–93]. The *solid line* is a fit of (5.50) to the data ($A_1 = 0.011$, $A_2 = 0.504$, $A_3 = -0.245$, $A_4 = 8.27 \times 10^{-2}$, $A_5 = -1.83 \times 10^{-2}$, $A_6 = 2.59 \times 10^{-3}$, $A_7 = -2.24 \times 10^{-4}$, $A_8 = 1.08 \times 10^{-5}$, and $A_9 = -2.23 \times 10^{-7}$). The *inset* highlights the low-field limit where the energy shift is highly nonlinear

$$\Delta E_d = 13.6 \times 10^3 \left(\frac{(\mu/m_0)}{\varepsilon_r^2} \right) \sum_{i=1}^{9} A_i \left[4.26 \times 10^{-6} \left(\frac{\varepsilon_r}{(\mu/m_0)} \right)^2 B \right]^i . \quad (5.50)$$

To exemplify the use of the model valid for arbitrary fields, we focus on the free exciton in bulk GaAs. Figure 5.28 shows the low temperature ($T = 5$ K) PL spectra of GaAs in a bulk crystal for different magnetic fields up to 30 T. The free-exciton diamagnetic shift (symbols) is shown in Fig. 5.29 together with a fit to the data (solid line) by means of (5.50). The dielectric constant has been set equal to $\varepsilon_r = 12.5$ and the exciton reduced mass used as fit parameter. It is found that $\mu = 0.053\ m_0$, in agreement with data reported in the literature [95]. The numerical solution to (5.47) given in (5.50) agrees very well with the experimental data *over all the range* of B. It is interesting to compare the limit of low (5.45) and high (5.42) field with the present model. The inset of Fig. 5.29 shows an enlargement of the data in the $B = (0 - 9)$ T range displayed in the main part of the figure. The parameter γ is equal to one for $B = 4.3$ T. This sets the field region above which (5.45) should not hold. The dotted line in the inset is a fit to the data by $\Delta E = AB^2$ with $A = 0.102\,\text{meV/T}^2$. A good agreement with the data is found up to about $B = 4$ T, namely, $\gamma \sim 1$. If we assume the exciton reduced mass known, $\mu = 0.053\ m_0$, according to (5.45), then one finds an exciton Bohr radius for GaAs $a_B \sim 12$ nm. This value is very close to an estimation based on *a hydrogenic* approximation for an exciton having a binding energy equal to 4.2 meV. However, above $B = 4$ T, the perturbative model deviates more and more from the experimental data as the field increases.

Let us consider the opposite limit of high field. The experimental data shown in Fig. 5.29 are reproduced in Fig. 5.30 along with a linear fit in the range $B =$

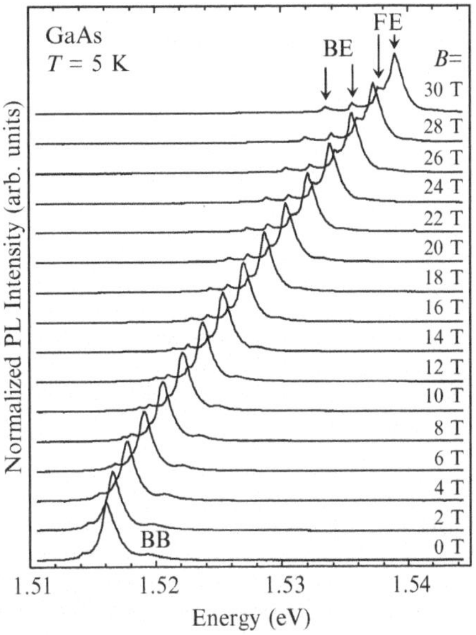

Fig. 5.28 Low-temperature ($T = 5$ K) normalized PL spectra of a bulk GaAs sample for different applied magnetic fields. FE and BE indicate the free-exciton and impurity-bound exciton recombination, respectively, BB indicates band-to-band recombination

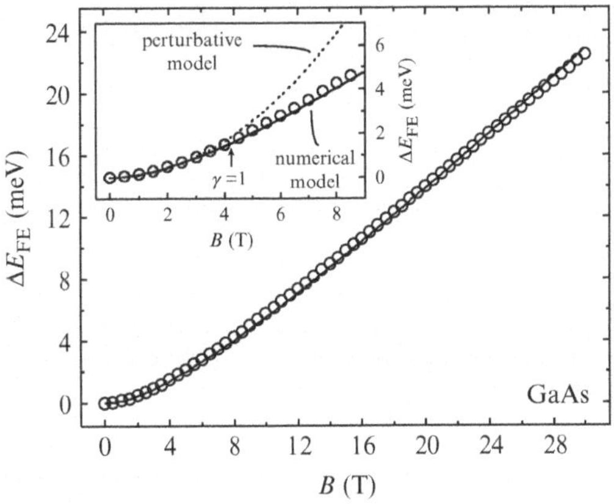

Fig. 5.29 Free-exciton diamagnetic shift (ΔE_{FE}) in bulk GaAs as a function of magnetic field (*open circles*), as obtained from the FE peak shown in Fig. 5.28 (by averaging over the Zeeman split components). The error bar associated to each data point is within the symbol size. *Solid line* is a fit to the data of the variational model described in the text (5.50). The *inset* shows a comparison at low fields ($B < 9$ T) between the numerical model fit [*solid line*, (5.50)] and the perturbation model fit [*dotted line*, (5.45)] restricted to the data satisfying the condition $\gamma \leq 1$

$(15 - 30)$ T. The linear fit agrees very well with the data. Then, one could assume that the electron and hole behave as an uncorrelated pair where the energy level of each carrier follows its Landau level. In this case, the diamagnetic shift would be given by

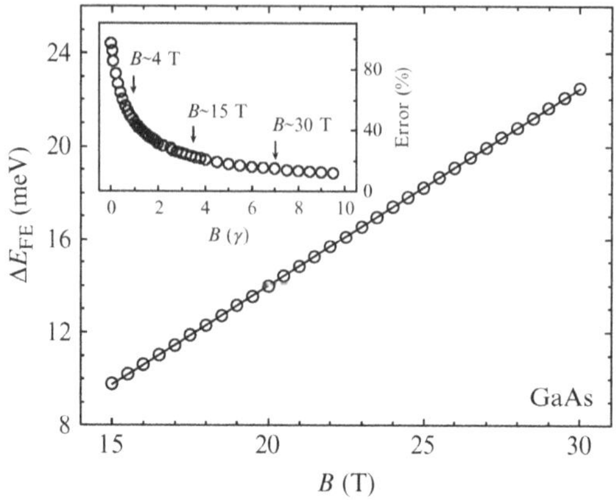

Fig. 5.30 Free-exciton diamagnetic shift (ΔE_{FE}) in bulk GaAs in the region of high magnetic fields, $B \geq 15$ T (*open circles*), as obtained from the FE peak shown in Fig. 5.28 (by averaging over the Zeeman split components). The *error bars* associated to each data point are within the symbol size. The *solid line* is a fit to the data of a Landau level recombination model (5.51). The *inset* shows the percentage error on the estimation of the free-exciton effective mass in terms of Landau level recombination, as a function of magnetic field in units of γ. The results have been derived by an analysis of the data shown in Fig. 5.27 [92, 94]. For GaAs, an analysis of the ΔE_{FE} data in terms of Landau level recombination produces a quite large error (about 20%) even at relatively high fields ($15\,\mathrm{T} \leq B \leq 30\,\mathrm{T}$; namely, $3.5 \leq \gamma \leq 7.1$)

$$\Delta E_{n=0}(B) = \frac{\hbar e B}{2\mu} \tag{5.51}$$

resulting in an exciton reduced mass equal to 0.07 m_0, which is a value sizably larger than the true value ($0.052 \div 0.058\, m_0$) [18]. This reflects the fact that for GaAs γ ($\sim$7 at 30 T) cannot be still considered $\gg$1 at 30 T. Indeed, the curve represented by (5.50) exhibits a first derivative that varies as B increases in such a way that a strictly linear dependence of ΔE_d is not fully reached for the magnetic fields considered here. The inset in Fig. 5.30 shows the percentage error on the estimation of the exciton effective mass that occurs using (5.51). Clearly, the Landau level approximation becomes better as the magnetic field increases.

5.5 Conclusions

First introduced in 1952, photoluminescence has rapidly become a very popular characterization tool in Materials Science after the discovery of the laser in the 1960s. Indeed, photoluminescence allows a nondestructive, quick, simple, and insightful determination of many of the properties that rule the performances of semiconductors and insulators in most modern optoelectronics (and other) devices. In particular, photoluminescence allows to investigate optical properties (the energy edge of fundamental transitions, type, amount and energy levels of impurities, radiative lifetimes, thermalization process, and nonradiative centers) and structural properties (the amount of crystalline disorder in bulk materials as well as interface disorder in heterostructures). Photoluminescence is, therefore, crucial whenever a new material is synthesized, in particular in the form of thin films and nanostructures not suitable for absorption studies. The combination of magnetic fields with photoluminescence permits to determine most important properties of interest for transport, such as carrier mass. Furthermore, magneto-photoluminescence experiments are particularly relevant in the field of Nanotechnology, since the perturbation induced by a magnetic field on the energy levels of a semiconductor can be exploited to estimate the wave function extent (and hence the quantum confinement degree) of a nanostructure whose geometric parameters are not directly accessible.

Acknowledgements We would like to acknowledge many colleagues with whom we collaborated on the topics presented in this chapter. First of all, we are grateful to F. Martelli and A. Patanè for their valuable contribution to the studies presented in the first section of this chapter. We gratefully acknowledge the fruitful collaboration with M. Felici, F. Masia, and R. Trotta on the magneto-PL investigations presented in the second section of this chapter. We also wish to thank S. Rubini, F. Martelli, and V. Lebedev for providing the samples described. High magnetic field (30 T) measurements were performed at the High Field Magnet Laboratory by the support of EuroMagNET under the EU contract 228043 (Nijmegen, The Netherlands) and we appreciate local support from J. C. Maan and P. C. M. Christianen. Finally, it is our pleasure to acknowledge the COST Action MP0805.

References

1. H.B. Bebb, E.W. Williams, *Semiconductors and Semimetals*, vol. 8, ed. by R.K. Willardson, A.C. Beer (Academic, New York, 1966), pp. 181–392
2. M. Gershenzon, *Semiconductors and Semimetals*, vol. 2, ed. by R.K. Willardson, A.C. Beer (Academic, New York, 1966), pp. 316
3. Y. P. Varshni, Phys. Stat. Sol. **19**, 419 (1967)
4. J.I. Pankove, J. Appl. Phys. **39**, 5368 (1968)
5. J.R. Haynes, H.B. Briggs, Phys. Rev. **86**, 647 (1952)
6. G.B. Stringfellow, P.E. Greene, J. Appl. Phys. **40**, 502 (1969)
7. P.Y. Yu, M. Cardona, Solid State Commun. **9**, 1421 (1971)
8. P.Y. Yu, M. Cardona, *Fundamentals of Semiconductors – Physics and Materials Properties* (Springer, Berlin, 2005)
9. A. Einstein, Z. Physik **18**, 121 (1917)
10. A. Einstein, P. Ehrenfest, Z. Physik **19**, 301 (1923)
11. W. van Roosbroeck, W. Shockley, Phys. Rev. **94**, 1558 (1954)
12. D.E. Mc Cumber, Phys. Rev. **136**, A954 (1964)
13. G. Laher, F. Stern, Phys. Rev. **133**, A553 (1964)
14. J.S. Blakemore, *Semiconductor statistics* (Pergamon Press, New York, 1962), Ch. 4
15. G.W. Gobeli, H.Y. Fan, *Semiconductor Research, Second Quarterly Rept.* (Purdue University, Lafayette, 1956); see also Ref. [6]
16. L. Pavesi, M. Guzzi, J. Appl. Phys. **75**, 4779 (1994)
17. E.P. O'Reilly, A. Lindsay, P.J. Klar, A. Polimeni, M. Capizzi, Semicond. Sci. Technol. **24**, 033001 (2009)

18. G. Pettinari, A. Polimeni, J.H. Blokland, R. Trotta, P.C.M. Christianen, M. Capizzi, J.C. Maan, X. Lu, E.C. Young, T. Tiedje, Phys. Rev. B **81**, 235211 (2010)
19. D.G. Thomas, J.J. Hopfield, C.J. Frosch, Phys. Rev. Lett. **15**, 857 (1965)
20. D.G. Thomas, J.J. Hopfield, Phys. Rev. **150**, 680 (1966)
21. M. Felici, A. Polimeni, A. Miriametro, M. Capizzi, H.P. Xin, C.W. Tu, Phys. Rev. B **71**, 045209 (2005)
22. J.J. Hopfield, P.J. Dean, D.G. Thomas, Phys. Rev. **158**, 748 (1967)
23. E. Cohen, M. Sturge, Phys. Rev. B **15**, 1039 (1977)
24. N.F. Mott, E.A. Davis, *Electronic Processes in Noncrystalline Materials* (Oxford University Press, Oxford, 1971)
25. T. Niebling, O. Rubel, W Heimbrodt, W. Stolz, S.D. Baranovskii, P.J. Klar, J.F. Geisz, J. Phys. Condens. Matter **20**, 015217 (2008)
26. P.J. Wiesner, R.A. Street, H.D. Wolf, Phys. Rev. Lett. **35**, 1366 (1975)
27. A. Polimeni, D. Marangio, M. Capizzi, A. Frova, F. Martelli, Appl. Phys. Lett. **65**, 1254 (1994)
28. C. Weisbuch, Solid-State Electron. **21**, 179 (1978)
29. A. Patanè, A. Polimeni, M. Capizzi, F. Martelli, Phys. Rev. B **52**, 2784 (1995)
30. J. Singh, K.K. Bajaj, S. Chauduri, Appl. Phys. Lett. **44**, 805 (1994)
31. S. Hong, J. Singh, Appl. Phys. Lett. **49**, 331 (1986)
32. J. Singh, K.K. Bajaj, J. Appl. Phys. **57**, 5433 (1985); and references therein
33. H.P. Xin, C.W. Tu, Appl. Phys. Lett. **72**, 2442 (1998)
34. M. Hetterich, M.D. Dawson, A.Yu. Egorov, D. Bernklau, H. Riechert, Appl. Phys. Lett. **76**, 1030 (2000)
35. D. Ouadjaout, Y. Marfaing, Phys. Rev. B **41**, 12096 (1990)
36. A. Polimeni, M. Capizzi, M. Geddo, M. Fischer, M. Reinhardt, A. Forchel, Appl. Phys. Lett. **77**, 2870 (2000)
37. B. Halperin, M. Lax, Phys. Rev. **148**, 722 (1966)
38. A. Aít-Ouali, R.-F. Yip, J.L. Brebner, R.A. Masut, J. Appl. Phys. **83**, 3153 (1998)
39. M. Oueslati, C. Benoit à la Guillaume, M. Zouaghi, Phys. Rev. B **37**, 3037 (1988)
40. G. Bastard, C. Delalande, M.H. Meynadier, P.M. Frijlink, M. Voos, Phys. Rev. B **29**, 7042 (1984)
41. F. Yang, M. Wilkinson, E.J. Austin, K.P. O'Donnell, Phys. Rev. Lett. **70**, 323 (1993)
42. M. Wilkinson, F. Yang, E.J. Austin, K.P. O'Donnell, J. Phys. Condens. Matter **4**, 8863 (1992)
43. E.-X. Ping, V. Dalal, J. Appl. Phys. **74**, 5349 (1993)
44. M. Gurioli, A. Vinattieri, J. Martinez-Pator, M. Colocci, Phys. Rev. B **50**, 11817 (1994)
45. A. Polimeni, A. Patanè, M. Grassi Alessi, M. Capizzi, F. Martelli, A. Bosacchi, S. Franchi, Phys. Rev. **54**, 16389 (1996)
46. M.S. Skolnick, J.M. Rorison, K.J. Nash, D.J. Mowbray, P.R. Tapster, S.J. Bass, A.D. Pitt, Phys. Rev. Lett. **58**, 2130 (1987)
47. A. Selloni, S. Modesti, M. Capizzi, Phys. Rev. B **30**, 821 (1984); and references therein
48. T.S. Moss, Proc. Phys. Soc. B **67**, 775 (1954)
49. E. Burstein, Phys. Rev. **93**, 632 (1954)
50. W. Walukiewicz, S.X. Li, J. Wu, K.M. Yu, J.W. Ager III, E.E. Haller, Hai Lu, William J. Schaff, J. Cryst. Growth **269**, 119 (2004)
51. W. Walukiewicz, J. W. Ager III, K. M. Yu, Z. Liliental-Weber, J. Wu, S.X. Li, R.E. Jones, J.D. Denlinger, J. Phys. D **39** R83 (2006); and references therein
52. G. Pettinari, F. Masia, M. Capizzi, A. Polimeni, M. Losurdo, G. Bruno, T.H. Kim, S. Choi, A. Brown, V. Lebedev, V. Cimalla, O. Ambacher, Phys. Rev. B **77**, 125207 (2008)
53. G. Pettinari, Electronic properties and response to hydrogen incorporation in novel semiconductors materials: GaAsN, GaAsBi, and InN, Ph. D. thesis, Sapienza University of Rome (2008)
54. E.O. Kane, Phys. Rev. **131**, 79 (1963)
55. A.R. Goñi, K. Syassen, *Optical properties of semiconductors under pressure, High Pressure in Semiconductor Physics I*, ed. by T. Suski, W. Paul Semiconductors and Semimetals, vol. 54, R.K. Willardson, E.R. Weber Series eds. (Academic, San Diego, 1998), p. 324

56. G. Wannier, *Elements of Solid State Theory* (Cambridge University Press, Cambridge, 1959)
57. F. Bassani, G. Pastori Parravicini, *Electronic States and Optical Transitions in Solids* (Pergamon Press, Oxford, 1975)
58. D.L. Dexter, R.S. Knox, *Excitons* (Interscience, New York, 1965)
59. R.S. Knox, *Theory of Excitons, Solid State Physics*, ed.by F. Seitz, D. Turnbull (Academic, New York, 1963)
60. D.G. Thomas, J.J. Hopfield, Phys. Rev. **124**, 657 (1961)
61. W. Rosner, G. Wunner, H. Herold, H. Ruder, J. Phys. B: At. Mol. Phys. **17**, 29 (1984)
62. J.A. Rossi, C.M. Wolfe, J.O. Dimmock, Phys. Rev. Lett. **25**, 1614 (1970)
63. W. Rühle, E. Göbel, Phys. Stat. Sol. (b) **78**, 311 (1976)
64. D. Bimberg, Phys. Rev. B **18**, 1794 (1978)
65. P.J. Dean, H. Venghaus, P.E. Simmonds, Phys. Rev. B **18**, 6813 (1978)
66. S. Zemon, P. Norris, E. S. Koteles, G. Lambert, J. Appl. Phys. **59**, 2828 (1986)
67. X.L. Zheng, D. Heiman, B. Lax, F.A. Chambers, K.A. Stair, Appl. Phys. Lett. **52**, 984 (1988)
68. B.J. Skromme, R. Bhat, M.A. Koza, S.A. Schwarz, T.S. Ravi, D.M. Hwang, Phys. Rev. Lett. **65**, 2050 (1990)
69. F. Masia, A. Polimeni, G. Baldassarri Höger von Högersthal, M. Bissiri, M. Capizzi, P.J. Klar, W. Stolz, Appl. Phys. Lett. **82**, 4474 (2003)
70. F. Masia, G. Pettinari, A. Polimeni, M. Felici, A. Miriametro, M. Capizzi, A. Lindsay, S.B. Healy, E.P. O'Reilly, A. Cristofoli, G. Bais, M. Piccin, S. Rubini, F. Martelli, A. Franciosi, P.J. Klar, K. Volz, W. Stolz, Phys. Rev. B **73**, 073201 (2006)
71. A. Lindsay, E.P. O'Reilly, Phys. Rev. Lett. **93**, 196402 (2004)
72. G. Pettinari, A. Polimeni, F. Masia, R. Trotta, M. Felici, M. Capizzi, T. Niebling, W. Stolz, P.J. Klar, Phys. Rev. Lett. **98**, 146402 (2007)
73. L.M. Roth, B. Lax, S. Zwerdling, Phys. Rev. **114**, 90 (1959)
74. G. Pettinari, A. Polimeni, M. Capizzi, J.H. Blokland, P.C.M. Christianen, J.C. Maan, V. Lebedev, V. Cimalla, O. Ambacher, Phys. Rev. B **79**, 165207 (2009)
75. J.S. Blakemore, J. Appl. Phys. **53**, R123 (1982)
76. L.R. Bailey, T.D. Veal, P.D.C. King, C.F. McConville, J. Pereiro, J. Grandal, M.A. Sánchez-Garcìa, E. Muñoz, E. Calleja, J. Appl. Phys. **104**, 113716 (2008)
77. M. Millot, N. Ubrig, J.-M. Poumirol, I. Gherasoiu, W. Walukiewicz, S. George, O. Portugall, J. Léotin, M. Goiran, J.-M. Broto, Phys. Rev. B **83**, 125204 (2011)
78. A.A. Klochikhin, V.Y. Davydov, V.V. Emstev, A.V. Sakharov, V.A. Kapitonov, B.A. Andreev, H. Lu, W.J. Shaff, Phys. Rev. B **71**, 195207 (2005)
79. X. Wang, S.-B. Che, Y. Ishitani, A. Yoshikawa, Appl. Phys. Lett. **90**, 201903 (2007)
80. X. Wang, S.-B. Che, Y. Ishitani, A. Yoshikawa, Appl. Phys. Lett.**92**, 132108 (2008)
81. Ü. Özgür, Y.I. Alivov, C. Liu, A. Teke, M.A. Reshchikov, S. Doan, V. Avrutin, S.-J. Cho, H. Morkoç, J. Appl. Phys. **98**, 041301 (2005)
82. J.-J. Shi, C.-X. Xia, S.-Y. Wei, Z.-X. Liu, J. Appl. Phys. **97**, 083705 (2005)
83. H. Pettersson, L. Landin, R. Liu, W. Seifert, M.-E. Pistol, L. Samuelson, Appl. Phys. Lett. **85**, 5046 (2004)
84. K.J. Nash, M.S. Skolnick, P.A. Claxton, J.S. Roberts, Phys. Rev. B **39**, 10943 (1989)
85. S.N. Walck, T.L. Reinecke, Phys. Rev. B **57**, 9088 (1998)
86. M. Felici, A. Polimeni, G. Salviati, L. Lazzarini, N. Armani, F. Masia, M. Capizzi, F. Martelli, M. Lazzarino, G. Bais, M. Piccin, S. Rubini, A. Franciosi, Adv. Mater. **18**, 1993 (2006)
87. R. Trotta, A. Polimeni, F. Martelli, G. Pettinari, M. Capizzi, L. Felisari, S. Rubini, M. Francardi, A. Gerardino, P.C.M. Christianen, J.C. Maan, Adv. Mater. **23**, 2706 (2011)
88. M. Geddo, T. Ciabattoni, G. Guizzetti, M. Galli, M. Patrini, A. Polimeni, R. Trotta, M. Capizzi, G. Bais, M. Piccin, S. Rubini, F. Martelli, A. Franciosi, Appl. Phys. Lett. **90**, 091907 (2007)
89. R. Trotta, D. Giubertoni, A. Polimeni, M. Bersani, M. Capizzi, F. Martelli, S. Rubini, G. Bisognin, M. Berti, Phys. Rev. B **80**, 195206 (2009)
90. I.E. Itskevich, M. Henini, H.A. Carmona, L. Eaves, P. Main, D.K. Maude, J.C. Portal, Appl. Phys. Lett. **70**, 505 (1997)

91. A.H. MacDonald, D.S. Ritchie, Phys. Rev. B **33**, 8336 (1986)
92. D. Cabib, E. Fabri, G. Fiorio, Solid State Commun. **9**, 1517 (1971)
93. D. Cabib, E. Fabri, G. Fiorio, Il Nuovo Cimento **10B**, 185 (1972)
94. Yong Zhang, A. Mascarenhas, E.D. Jones, J. Appl. Phys. **83**, 448 (1998)
95. S. Shokhovets, O. Ambacher, G. Gobsch, Phys. Rev. B **76**, 125203 (2007)

Chapter 6
Pressure Studies

Andrew Prins, Alf Adams and Stephen Sweeney

Abstract High pressure is one of the most valuable characterization tools available in semiconductor research. In this chapter, we show how it has been used directly to measure some of the fundamental properties of materials and also to investigate electronic and optoelectronic devices, resulting in improved theoretical models of material properties and optimized device performance. We hope that the examples we give and the experimental methods we describe will inspire others to use this powerful but underused research technique.

6.1 Introduction

Starting his PhD at Harvard University in 1905, Percy Williams Bridgman (1882–1961) began the unlikely process of squeezing materials to see the effect of pressure on their physical properties. Undaunted by the experimental limits at the time, he soon developed new methods of sealing fluids at high pressure and was able to reach previously unobtainable hydrostatic pressures to reveal new and exciting properties. He developed piston-in-cylinder and also opposing anvil high pressure equipment, both still the basis of most modern techniques, and was able to perform electrical measurements up 10 GPa (100 kbar) [1]. To set this in perspective, the hydrostatic pressure experienced at the bottom of the deepest ocean on Earth is approximately 0.1 GPa and the pressure at the center of the Earth is around 340 GPa. In 1946, Bridgman received the Nobel Prize in physics for his pioneering work and he is considered the father of high pressure physics. His collected works are available in [2] and his book [3] is also an outstanding source of information.

A. Adams (✉)
Advanced Technology Institute, University of Surrey, Guildford, Surrey, GU2 7XH, UK
e-mail: Alf.Adams@surrey.ac.uk

A. Patanè and N. Balkan (eds.), *Semiconductor Research*, Springer Series in Materials Science 150, DOI 10.1007/978-3-642-23351-7_6,

With the advent of transistors in 1947, it became clear that an understanding of the electronic band structure of semiconductors was essential and pressure proved to be a powerful probe in its investigation; high pressure allows us to turn the lattice constant of materials into a variable which can often lead to a better physical interpretation than the much more widely used thermodynamic variable, temperature. A full and complete review of high pressure semiconductor physics up to 1998 can be found in two excellent volumes [4]. We also recommend books by Spain and Paauwe [5] and Sherman [6] which cover the experimental details of the equipment and materials used in their construction. There are a multitude of aspects of and types of semiconductors studied with pressure, but we will principally concern ourselves with the properties of those with diamond or zincblende lattices and see how pressure has proved itself a most valuable tool in manipulating and understanding their properties. For a consideration of some future possibilities in different and new materials under pressure, we would also point the reader to the article by Yu [7] and importantly, their own imagination.

6.2 The Effect of Pressure on Electronic Band Structure

The electrical and optical properties of elemental diamond-type cubic lattices (e.g., Si and Ge) and the related zincblende compound semiconductors (e.g., GaAs and InP) at ambient and high pressure can be understood from Fig. 6.1a. In this diagram, we illustrate the band structure (energy–momentum dispersion relationship) of GaAs. The effective mass of a particle at any point is inversely proportional to the curvature of this energy–momentum curve. In these materials, there are three distinct sets of conduction band minima as seen at the Γ, L, and X (or nearby Δ) points in the Brillouin zone. The exact energy positions of these minima play a key

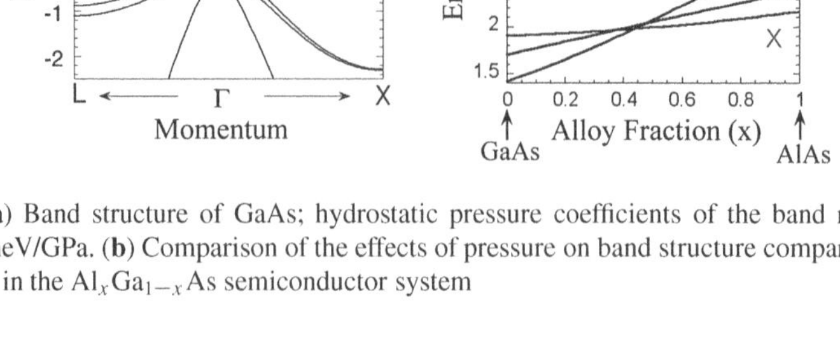

Fig. 6.1 (**a**) Band structure of GaAs; hydrostatic pressure coefficients of the band minima are shown in meV/GPa. (**b**) Comparison of the effects of pressure on band structure compared to those of alloying in the $Al_xGa_{1-x}As$ semiconductor system

role in determining the electrical and optical characteristics of each semiconductor. See [8–10] for a review of these parameters in the most commonly studied III–V and group IV materials. The fact that these energies can be varied in a smooth and reversible manner by the application of hydrostatic pressure is the single most important reason why pressure is used to study semiconductor materials and devices. Indeed it should be noted that for many experiments, the effect of pressure on material or device properties is best related directly to the resulting changes in their band structure rather than an absolute measurement of the pressure itself. Since the top of the valence band is at the Γ point in all diamond and zincblende materials, when Γ is the lowest conduction band minimum the material has a direct band gap (e.g., GaAs and InP). From pressure measurements on Ge, Si, and a number of III–V materials, Paul [11] was able to form the basis of an *Empirical Rule* concerning minima behavior; the pressure coefficient of the direct gap at Γ is in the range 100–150 meV/GPa while those of the indirect gaps of L and X are near 50 meV/GPa, and between -10 and -20 meV/GPa, respectively. Therefore, high pressure allows us to change band ordering from direct (e.g., GaAs) to X or Δ-like (e.g., Si) or L-like (e.g., Ge). This is a very important result since an understanding of this means that the band structure at ambient conditions can be determined by extrapolating back the results of pressure experiments. It can also be used to manipulate properties in much the same way as alloying, while avoiding unwanted changes such as doping density or thickness variations. This results in cost savings since a single sample can be studied rather than having to grow multiple structures with varying alloy fractions. An example is clearly shown in Fig. 6.1b where the effects of pressure and alloy fraction are compared. Other comparisons will occur later in this chapter. While collected data is now available for a range of III–V and group IV semiconductors, the exact position of the band structure minima was not always clear. For example, it was originally believed that GaAs possessed a Γ–X–L arrangement in ascending order of their energies [12], but evidence showing that the L minima were situated below that of the X resulted from high pressure measurements on the Gunn effect as discussed below.

6.2.1 The Gunn Effect

The Gunn diode played an important role in the history of semiconductor devices. Before its discovery, semiconductors had only been used as passive detectors of microwaves but the Gunn diode was able to replace bulky microwave vacuum tube sources with cheap and simple solid-state oscillators, just as the transistor had replaced low frequency vacuum tubes in the 1950s. The possibility of such devices based on transferred electrons was first predicted by Ridley et al. [13]. They considered an n-type semiconductor such as GaAs where the electrons normally occupy the low-mass, high-mobility Γ minimum. In such a case, they pointed out that if an electric field is applied, which is sufficiently large to cause the electrons to heat up, then the electrons could gain enough energy to scatter across to the higher lying satellite minima. Here their mobility would be much lower both because of the increased effective mass and because strong intervalley scattering would occur. This would result in a decreasing current with increasing applied voltage and so to a negative differential resistance resulting in current instability. In 1963, John Battiscombe Gunn discovered that if the applied electric field across certain bulk III–V crystals (GaAs and InP) exceeded a critical value, then pulses of current at microwave frequencies were produced [14]. Kroemer [15] proposed that electron transfer was the mechanism behind the effect even though Gunn himself had considered and dismissed it, and there was intense speculation about the underlying mechanism. The crucial experiment to demonstrate that the transfer model was correct was carried out on GaAs under pressure by Hutson et al. [16] who found that the Gunn effect disappeared at high pressure as would be expected if it is due to the transferred electron effect. Careful measurements of the pressure variation of the threshold field for the Gunn effect were undertaken and compared with the results of Monte Carlo simulations of the transferred electron effect [17, 18], and it was concluded that agreement could be obtained but only if it is assumed that the L minima are below the X minima in GaAs. This ordering was confirmed by the uniaxial stress measurements of Pickering et al. [19] which showed that the threshold field for the Gunn effect in GaAs is much more sensitive to stress in the $\langle 111 \rangle$ direction along which the L minima lie, than stress in the $\langle 100 \rangle$ direction where the X minima are. This agreed also with a reassessment of other previous experimental data to that point carried out by Aspnes et al. [20]. A similar result was later observed in an analogous experiment by measuring threshold as a function of phosphorous composition in a range of $GaAs_xP_{1-x}$ alloys [21].

In semiconductor materials that do not have suitable band structure for the Gunn effect, impact ionization of electron-hole pairs across the band gap occurs before intervalley transfer (see our later discussion of avalanche photodiodes in Sect. 6.4.3), but in some cases pressure can be used to tune a material to the correct band configuration. InAs is such a material, with a narrow direct band gap of 0.35 eV at room temperature. This was studied by Pickering et al. [22] up to 5 GPa. A piston-in-cylinder (see Sect. 6.4. on optoelectronic device measurements) was used up to 1.5 GPa and beyond this opposing anvil pressure apparatus with diodes potted in epoxy resin at the center of an MgO-loaded epoxy Bridgman ring [23]. At atmospheric pressure, the current voltage plot for this material was linear up until the onset of impact ionization, which was marked by a large increase in the current (see Fig. 6.2 example marked in *blue*), and this onset (the avalanche voltage) was measured as a function of pressure. The avalanche voltage increased due to the increasing band gap until at $\approx$3.3 GPa, violent oscillations were observed with a definite current drop-back (Fig. 6.2 marked in *red*). These oscillations were similar to those seen in GaAs and InP and associated with the negative differential conductance of the Gunn effect. The oscillations continued until 4.8 GPa and beyond this, the current saturated (Fig. 6.2 marked in *green*) in the same way as observed in GaAs and InP under pressure. This showed that InAs could be tuned by pressure from a band structure where the direct band gap is smaller than the separation to the higher lying minima resulting first in avalanche breakdown, through the

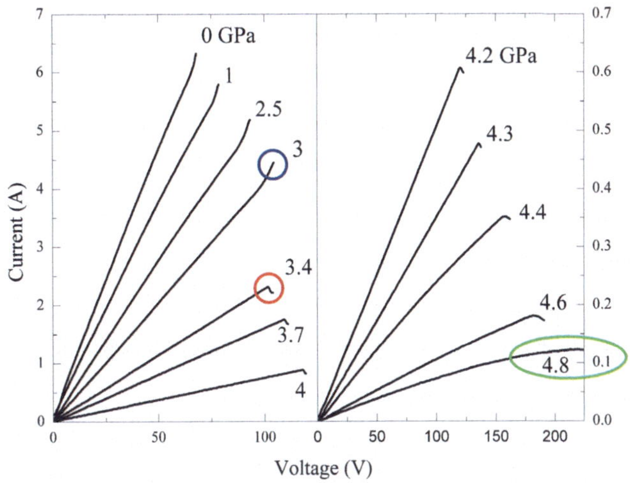

Fig. 6.2 I/V characteristics of InAs up to 4.8 GPa of pressure [19]. Examples of the onset of impact ionization and the Gunn effect are marked in *blue and red*, respectively. Ohmic like behavior is marked in *green*

Gunn region and into a region of almost ohmic behavior where the satellite minima are so close that they are already occupied by most of the electrons simply due to thermal excitation. Analysis of the Gunn effect measured in $InAs_{1-x}P_x$ alloys under pressure was also able to give a good indication of the L minima position in this material at ambient pressure [24].

6.2.2 Anvil Cells and Optical Work

Bridgman's opposed-anvil based high-pressure cells used tungsten carbide and allowed electrical measurements only. The introduction of diamond as an anvil material in the 1950s opened up the possibility of optical work. A schematic of a diamond anvil cell (DAC) is shown in the inset of Fig. 6.3; a drilled metal gasket is compressed between the small flat faces of two opposing cut diamonds and the hole contains the sample and a ruby pressure calibrant (see [25, 26]) in a hydrostatic pressure transmitting medium. Photoluminescence (PL) is well established for characterizing semiconductors and is the most straightforward technique in the DAC. PL uses optical excitation above the band gap energy to create electron-hole pairs which thermalize down to recombine across the minimum energy gap possible, emitting a photon at that energy and giving rise to an emission peak characteristic of the material. The linewidth, shape, and intensity of the PL peaks yield information pertaining to crystal purity, band structure, and recombination paths along with alloy disorder and interface roughness in quantum wells. The reader is referred to Chap. 5 of this book for further information. The results of a PL experiment [27] are used to illustrate beautifully the *Empirical Rule* in Fig. 6.3, in a material

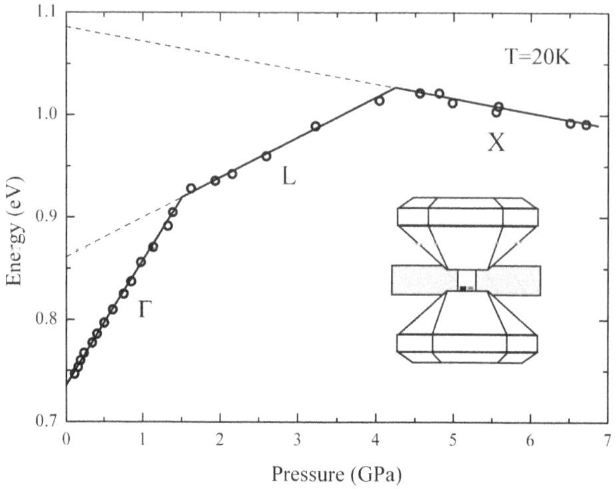

Fig. 6.3 Photoluminescence peak energy against pressure for a $Ga_{1-x}In_xSb/GaSb$ quantum-well [27]. Transition changes from Γ to L to X are clearly seen. The *inset* shows a schematic diamond anvil cell with gasket, sample and ruby pressure calibrant

system where pressure clearly reveals all three types of minima and their behavior as outlined above.

Modern semiconductor devices are complex and consist of many layers of these constituent materials doped, undoped, and often in thin layers called quantum wells (QWs), and this gives rise to other important questions: how do the energy levels within different materials align with each other? How do they change with doping? While model solid theory [28] and less direct measurement methods can give some estimate of band alignment, results have varied greatly and a direct method of measurement is required. With a knowledge of the *Empirical Rule*, Wolford et al. [29] was able to show that it was possible directly to measure band offsets by high pressure spectroscopy in heterostructures where the lowest X minima were in the barrier material, as shown in the inset of Fig. 6.4. This method has been extended to other material systems [30] and the results of such an experiment carried out in the DAC are shown in Fig. 6.4. In this figure, after correcting for any quantum confinement effects, the energy difference shown as ΔE_1 corresponds to the difference in material band gaps (ΔE_g) and, after extrapolation back to ambient pressure, ΔE_2 corresponds to the valence band offset (ΔE_v). Thus, from such an experiment, the band offsets are fully determined:

$$\Delta E_g = \Delta E_c + \Delta E_v. \tag{6.1}$$

For the most accurate determination, good samples are needed that unambiguously show both the barrier and quantum well transitions in the same experiment above the X crossover pressures. Low temperature identification of phonon lines can help rule out defect related emissions and give reliable data above crossover for extrapolation, allowing any changes in the valence band offset with pressure ($\mathrm{d}\Delta E_v/\mathrm{d}P$) to be

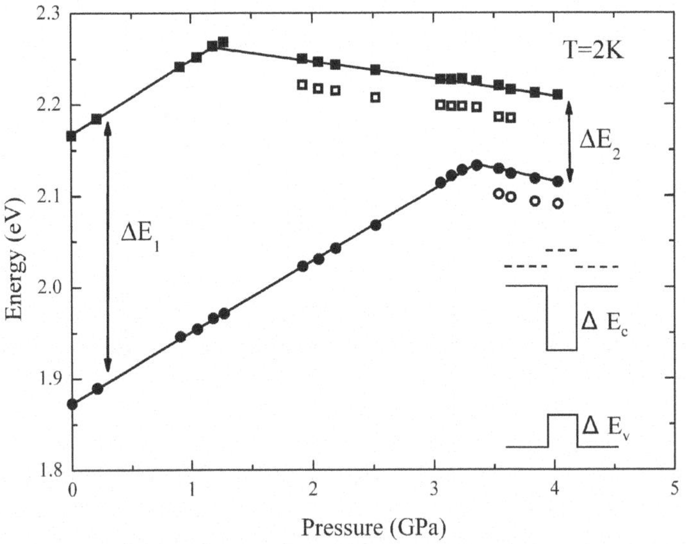

Fig. 6.4 Photoluminescence peak energies against pressure for a strained GaInP quantum well in $(Al_{0.3}Ga_{0.7})_{0.5}In_{0.5}P$ barriers [34]. Above crossover phonon replicas are observed. The *inset* shows a schematic of the sample structure including the X band edge (*broken line*) and the Γ conduction and valence bands (*solid lines*) at ambient pressure (see text for explanation)

measured accurately as well as the offset ratio at ambient pressure. Lambkin et al. [31] found that in the $GaAs/Al_xGa_{1-x}As$ system, $d\Delta E_v/dP = 1.1x$ (meV/kbar) while it was independent of pressure in GaInAs/InP [32]. Other systems specific for optoelectronic applications have also been measured and can be found in the literature, while Whitaker et al. [33] proposed an extension of this pressure method between a larger set of III–V materials.

From the simple pieces of information discussed so far and with the correct samples, the band structure and band offsets for a new material system can be fully determined using pressure and used to help model and improve devices [34].

The DAC lends itself well to these kind of optical measurements but requires small thinned samples (the gasket hole is typically around 300 μm diameter and indented to about 100 μm thickness), and while electrical measurements on specially prepared devices are possible [35], the DAC is not well suited to this purpose. A good review of the standard use of DACs can be found in [36] and their operation in [37]; we also recommend the book by Eremets [38].

6.2.3 *Transport Properties*

As mentioned earlier, the curvature around the band minima is related to the effective mass (m^*) at that point and this largely determines the transport, carrier mobility, and optical properties of semiconductor materials and devices. The $k \cdot p$

method developed by Kane [39] can be used to calculate the band structure at these minima and in its simplest form for electrons in the Γ conduction band minimum is expressed as:

$$\frac{m_0}{m^*} = 1 + \frac{2p^2}{3m_0}\left(\frac{2}{E_0} + \frac{1}{E_0 + \Delta_0}\right), \tag{6.2}$$

where E_0 is the fundamental gap, p the momentum operator, and Δ_0 is the spin–orbit splitting. Interactions with higher lying bands can be treated as further expanded perturbations and in this manner the theory can be used across the entire Brillouin zone. To calculate the effective masses, the only data required are the energy gaps and the momentum-matrix elements at these points. Pressure is the ideal tool for investigating this relationship since the fundamental gap, as we know, can be altered using pressure. An increase in band gap will result in less curvature and a higher effective mass. Shantharama et al. [40] were able to adapt a Bridgman anvil system to include optical fiber access and simultaneously measured the change in energy gap and effective mass with pressure up to 4 GPa in n-type InP and GaAs using photoconductivity and magnetophonon oscillations. They showed that the contribution to the effective mass from higher lying bands was small and that the simplified $k \cdot p$ theory of (6.2) modeled the results well. Recently, magnetophotoluminescence in the DAC up to about 0.1 GPa has been used to study the effective mass of dilute nitrides with up to 0.2%N [41–43] and showed increases larger than those seen in conventional alloys. Endicott et al. [41] were able to fit the unusually large increase in effective mass fairly well using a two-level band anticrossing (BAC) model (see Sect. 6.2.5 on BAC) but only had three pressure data points and one sample concentration. Pettinari et al. [42] measured 15 pressure points in two well-defined sample concentrations and discovered large variations (up to 60%/0.1 GPa) with increasing pressure and marked changes in behavior between the N concentrations. However, despite these variations, both [42,43] were able to show that their results were qualitatively consistent with a $k \cdot p$ model if modified so that the conduction band interacted with specific N complex levels, thus helping in identifying which of these levels contribute most to modify the band structure under ambient conditions.

Measurement of free-carrier mobility and concentration can be determined using the Hall effect and this was commonly used to characterize material by assuming a standard scattering mechanism at room temperature. The effective mass of electrons in the conduction band increases with band gap, resulting in decreased mobility, and this means that scattering mechanisms in semiconductors can be studied as deviations from the expected behavior with pressure. A number of scattering mechanisms can affect conduction in semiconductors, including ionized impurity and acoustic or optical phonon scattering. Each electron scattering mechanism has its own characteristic dependence on the carrier effective mass and so scattering of electrons in n-type, direct-gap semiconductors can be investigated by measuring the pressure dependence of their mobility. From Hall measurements under pressure, Lancefield et al. [44] found that the standard theories of ionized impurity scattering were inadequate to explain the behavior in GaAs and InP unless correlation effects were taken into account for the impurity distribution.

6.2.4 Defects Under Pressure

Doping is used in semiconductors to incorporate controlled levels of impurities to produce free charge carriers for controlling conductivity. These impurities can have electronic levels within the band gap either close to the band edges (shallow defects) or in the middle (deep defects). Their pressure dependence is one of the most reliable ways to distinguish between them since it will depend either on the nearest minima in the case of shallow levels or a large region of the Brillouin zone and many bands in the case of deep levels. The DX center has been highly studied [45] for very good reasons; it is a deep defect found in many n-type compound semiconductors and their alloys. It has been extensively studied in n-doped (Si) $Al_xGa_{1-x}As$ with $x > 0.22$ since it dominates the transport properties of the alloy, affecting its use in high-speed electronic and optoelectronic devices. It gives rise to persistent photoconductivity (PPC) and also causes a drastic decrease in carrier concentration at cryogenic temperatures. Extrapolating the DX center energy against alloy content back to $x = 0$ showed that it should be above the conduction band in GaAs by around 0.17 eV, and Mizutu et al. [46] were able to show its existence in GaAs by applying around 2.4 GPa of hydrostatic pressure to push it below the conduction band edge. This use of high pressure has separated band structure changes from alloying effects and shown that certain substitutional donors in III–V compound semiconductors transform into DX centers; for example S in InP at over 8.2 GPa [47]. These experiments demonstrated that DX centers are substitutional shallow donors that undergo a structural change to form a deep level under certain conditions of alloying or hydrostatic pressure.

Techniques used in the DAC to study the DX center were: deep level transient spectroscopy (DLTS), Hall measurements, and optical absorption. Where electrical feedthrough wires were required (DLTS and Hall) the gaskets were coated with Al_2O_3 and samples specially thinned, soft powders such as $CaSO_4$ were then used as pressure transmitting media.

Introducing impurities into semiconductor substrate materials such as InP and GaAs during growth has also been shown to decrease dislocations and provide what is sometimes called "lattice hardening" [48, 49]; this is important for improving the lifetime and performance of electronic and optoelectronic devices and is discussed later on in relation to the study of high pressure phase transitions (see Sect. 6.3).

Another area of interest is the use of rare-earth ions doped into semiconductors which greatly change the host band structure [50]. There are potential optoelectronic applications around the 1 μm wavelength for InP doped with Yb from intra-4f-shell luminescence. Experimental work by Kaminska et al. [51] has shown that up to 6 GPa, these transitions are sensitive to pressure but above this they remarkably change behavior and become almost independent of pressure. Such studies help enhance our understanding of these materials.

6.2.5 Band Anticrossing

Nitrogen complexes in the indirect band gap material GaP are of technological importance since they give rise to radiative recombination suitable for LEDs, from green to orange, and were originally believed to be a shallow donor. Diamond anvil pressure studies and theory, however, showed that they failed to follow the nearby conduction band [52] and that they were only accidentally nearby (less than 0.1 eV below). It was not until 1990, also with the aid of pressure, that nitrogen pair complexes were finally observed in GaAs [53] and shown to be around 0.3 eV above the conduction band edge. The highly localized nature of the N states suggests that they will interact only weakly with extended states of these materials at low concentrations. Alloying low concentrations of nitrogen into III–V compounds has however been shown to drastically modify the electronic band structure [54] and is very important for device applications. This perturbation gives rise to a huge band gap decrease and greater electron effective mass. Initial attempts to explain this behavior used various unsuccessful models, but these were abandoned for the two-level BAC model which was able to explain the pressure and composition dependencies of the band gap; the interaction of the N level with the conduction band edge splits the conduction band into two highly nonparabolic subbands E_- and E_+ and the position of these sub-bands depends on the alloy concentration, a coupling parameter, and the location of E_N in relation to the conduction band edge (see Fig. 6.5 for a representation of the BAC model). The band gap reduction is accounted for by the downward shift of the conduction band due to the anticrossing interaction and gives rise to E_-. Photoreflectance (PR) measurements clearly show the E_- and E_+ transitions and the anticrossing character manifests itself in their pressure dependencies as seen in Fig. 6.6 for the case of $GaAs_{0.985}N_{0.015}$ [55]. Pressure measurements have been essential in these materials to reach an understanding of the electronic band structure.

6.3 Phase Transitions in Bulk, Superlattices and Nanoparticles

The work we have discussed so far has investigated the electrical and optical properties of materials at pressures under 10 GPa and has neglected structural properties. Under large pressures, it may become energetically favorable for a material to change its atomic arrangement and undergo a sequence of structural phase transitions. The properties of the resulting materials can be very different from those at ambient pressures and pressure vs. temperature phase diagrams are often plotted to show phase boundaries of these different states. Assuming a fixed ambient temperature, most fourfold semiconductor materials (diamond, zincblende, or wurtzite) undergo an initial phase transition into sixfold structures (rocksalt, Cmcm, or β-Sn). The Cmcm structure is a distortion of the NaCl (rocksalt) structure and was

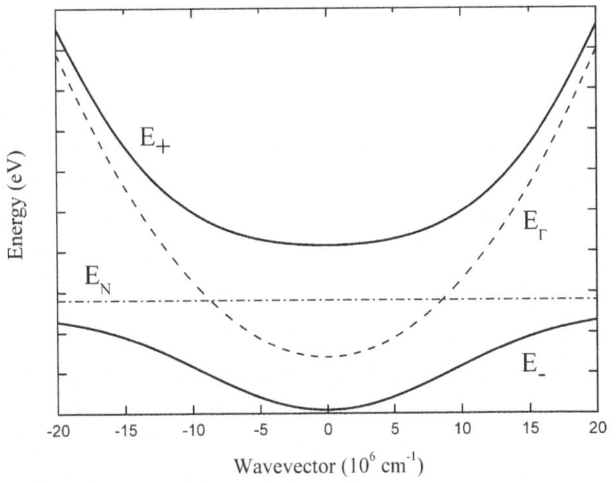

Fig. 6.5 Energy momentum diagram of a dilute-N band structure calculated using the BAC approach. The effect of the nitrogen level on the conduction band E_Γ produces two repelled bands E_+ and E_- and as a result reduces the band gap of the material

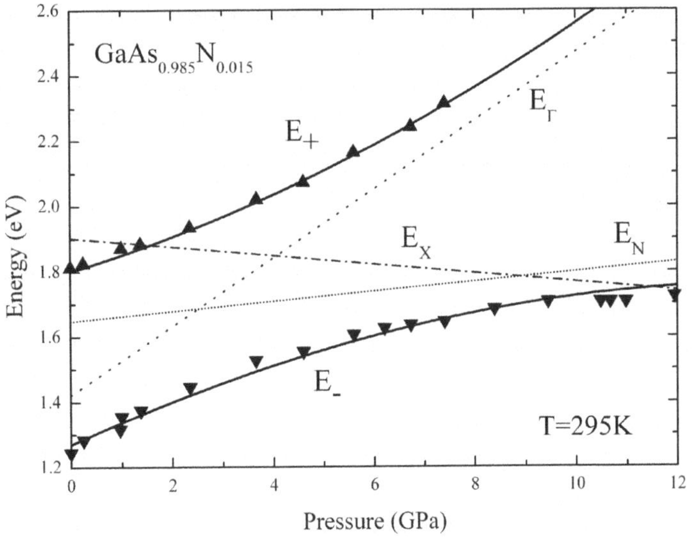

Fig. 6.6 Photoreflectance transition energies against pressure for $GaAs_{0.985}N_{0.015}$ [55]

frequently misinterpreted in many early measurements since it had not been seen previously in the binary materials [10]. For structural observations, the measurement of choice is X-ray diffraction, but simultaneously using a range of techniques such as optical transmission, X-ray diffraction, and Raman scattering greatly improves the interpretation of results under pressure and allows comparison with other work, see, for example the DAC studies of GaAs by Besson et al. [56] or the analysis of results for InP by Whitaker et al. [57]. It is important to note that computational methods play an important part in predicting and interpreting experimental results. The whole area of structural transitions is complex and changing and the reader is referred to [4, 58] for full and detailed reviews. Care should be taken as many texts are not in agreement and, with this in mind, we will discuss some of the pressure results of InP in relation to its physical and electronic band structure.

The DAC allows pressures in excess of 100 GPa to be generated and measurements are carried out using powder diffraction techniques due to the large volume changes involved (see Fig. 6.7 as an example of this). The ruby pressure calibration method can be used with an accuracy of around 1% up to 10 GPa and 3% around 100 GPa when referenced to primary pressure scales based on the change in lattice constant of known materials (see [25, 26]). The first phase transition in Si and Ge occurs at around 10 GPa and in GaAs at around 17 GPa. For InP 10.8 GPa is usually cited [59], although there has been significant reported variation outside the expected pressure calibration error. An analysis of this experimental data [57] proposed that the transition pressure from zincblende to rocksalt was variable (they quote from 9 GPa to more than 10.5 GPa) and this rocksalt structure was shown to be stable only up to 19 GPa [59] and later up to 28 GPa [60]. Figure 6.7 shows the variation of volume with pressure for heavily sulfur-doped InP [61] compared to that of pure InP [59]. Lin et al. [61] observed the rocksalt to Cmcm phase transition at an even higher pressure (35.8 GPa) and proposed that the dopant increased the ionicity of the rocksalt phase and enhanced its bulk modulus and stability but slightly reduced the stability of the zincblende crystal phase. In dilute nitrides, no significant change in the first phase transition pressure was seen between GaP and the dilute nitride $GaP_{0.979}N_{0.021}$ through the examination of Raman modes and the N local vibrational modes (LVMs) [62]. Please refer to Chap. 9 of this book for a

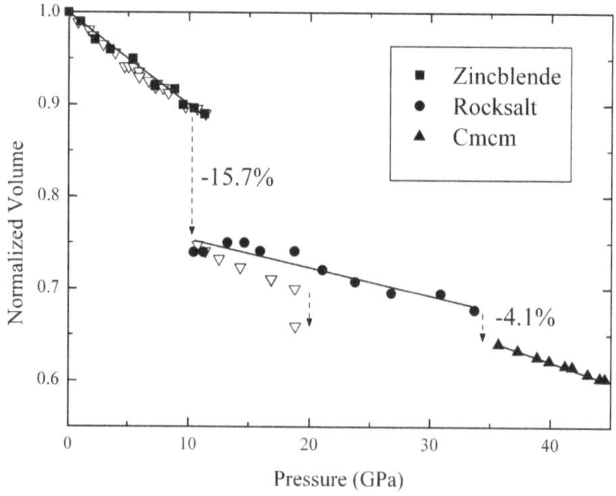

Fig. 6.7 Volume vs. pressure data for InP:S [61] (*solid points and lines*) and InP [59] (*open triangles*) under compression at ambient temperature. Volume changes associated with phase changes are clearly seen

detailed description of Raman spectroscopy and specifically to Sect. 9.4.2 Impurities and Alloys, concerning LVMs. No real in-depth study of the effects of doping on semiconductor phase transitions and structure seems to have ever been undertaken, although Bose et al. [63] suggested a decrease in the first phase transition pressure with increased doping in heavily selenium-doped InP. There would seem to be scope for future work in this area.

The $\Gamma - X$ band minima crossing pressure for InP has also been disputed, but we believe it to be above 12 GPa from the extrapolation of pressure data for GaInP alloys studied by Tozer et al. [64, 65], and as a result the expected $\Gamma - X$ crossover would not be observed in bulk InP under pressure due to the first phase transition.

An interesting effect related to phase transitions and little discussed is superpressing in which a material in a superlattice will persist in its zincblende structure above its normal phase transition pressure; this has been seen in AlAs/GaAs [66] and in CdTe/ZnTe [67] and has been used to measure the deformation of a semimetal [68], and these effects can easily suppress phase transitions by up to 10 GPa.

There has, however, been much interest in nanoscale materials and their associated properties [69] and in the study of phenomena in nanocrystalline semiconductor systems [70]. The earliest pioneering pressure studies [71] were on CdSe and related to phase stability, showing that with decreasing particle radius the first phase transition pressure was systematically higher and was as much as 13 GPa higher than in the bulk material. Large elevations in phase transition pressure, normally attributed to an increase in surface energy, have also been observed in Si nanocrystals [72], in Ge [73], and in a range of other materials [70]. However, GaAs is reported as showing no pressure difference in its bulk and nanostate phase transitions [74]. There also appears to be no current measurement of the pressure for the phase transition of InP nanocrystals in the literature although the work of [75] cites a value of 14 GPa. With this elevated transition pressure and as a result of quantum confinement effects, it would be expected that the $\Gamma - X$ band minima crossing discussed earlier should be seen in InP nanoparticles under pressure. The experimental results however are not clear since Menoni et al. [76] observed transitions that were L and X like in nature that changed with particle size, while Lee et al. [75] saw no real difference from the bulk. The analysis by Díaz et al. [77] goes some way to explain these results in terms of L and X mixing of internal and external surface states. To complicate matters further doping can also affect surface energies and phase transitions of nanocrystals [78]. In conclusion, it is again clear that pressure has an important part to play in understanding these complex semiconductor nanocrystalline materials and that further work is required.

6.4 Optoelectronic Device Measurements Under Pressure

Figure 6.8 shows two types of high pressure systems used in the study of optoelectronic devices at the University of Surrey, both are capable of reaching pressures of around 1.5 GPa and have fairly large sample spaces allowing simple mounting and

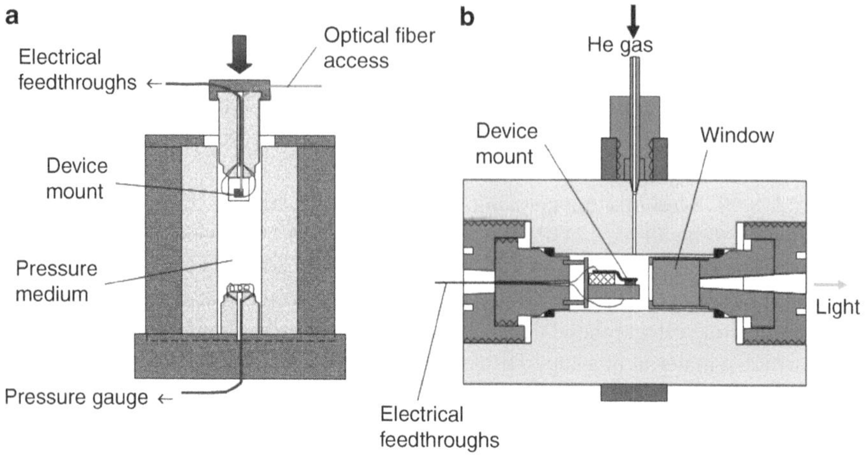

Fig. 6.8 High pressure piston-in-cylinder apparatus and the helium gas high pressure system sample chamber (University of Surrey)

contacting of devices. The first (Fig. 6.8a) is a piston-in-cylinder, which consists of two opposing pistons made from hardened steel within a compound double cylinder. A load is applied using a large hydraulic ram to the upper moveable piston which is sealed against the liquid pressure media with a neoprene O-ring and phosphor-bronze ring. Electrical leads and an optical fiber are both fed through this piston to supply current and to couple light output into a spectrometer. The addition of an optical fiber in this way for the first time [79], essentially as an extended high pressure optical window, allowed direct optical access to devices in the high pressure volume. This improved optical coupling allows the studies of light output not just from a laser facet but also from windows milled into the substrate contacts of devices so that pure spontaneous emission can be observed [80]. The use of optical fibers as high-pressure windows has proved successful because their very small cross-sectional area (πr^2), where r is the fiber radius, minimizes the force (F) exerted by the pressure media ($F = \pi r^2 P$) while the large length (l) distributes and supports the total shear stress ($F = 2\pi r l \tau$) along the fiber's length, the shear strength of Stycast-type epoxy is $\tau \approx 0.1$ GPa and as a result the glued fiber is able to sustain high pressures. This has allowed the investigation of a large number of optoelectronic devices working over a wide range of operating wavelengths from the blue into the mid-infrared (see for example [81–83]). Pressure is calibrated in this system by measuring the resistance change in a manganin wire coil attached to the lower fixed piston. This is a large bulky system which cannot be cooled to low temperatures, but raising its temperature is quite straightforward using heating coils.

Figure 6.8b shows the sample chamber of the second pressure system used at the University Surrey in which compressed helium gas, via a capillary, is supplied from a large three stage piston-in-cylinder compressor (Unipress, Poland). Each stage has its own control valves and its operation is quite complex. It does, however, have advantages; the chamber can be cooled to low temperature and the helium pressure media and sapphire windows have better transparency properties, helium is also less reactive with device structures and coatings, has a better matched refractive index to air and does not freeze under typical operating temperature and pressures, unlike liquid pressure media. This type of gas apparatus was first used by Langer and Warschauer [84] and is described in great detail there.

For both systems, safety is a major consideration since the energy stored by compressing the media into relatively large volumes can be explosive if released suddenly and special shielding around both setups is required.

6.4.1 Semiconductor Lasers

In terms of technological impact, semiconductor lasers are second only to the transistor and integrated circuit. They are essential to communications, data storage, and a multitude of other applications too numerous to discuss here. Unlike group IV materials such as Si, the direct band gap of III–Vs has led to them becoming the materials of choice for optoelectronic applications. The commercial importance of certain wavelengths and advances in growth techniques have driven much of the technology. The earliest semiconductor lasers were simply GaAs homojunctions, later clad by AlGaAs to form double heterostructures providing electrical and optical confinement for wave guiding. The long-wavelength fiber-optic transmission windows at 1.3 and 1.55 μm meant that other materials and heterojunctions were more suitable. GaAs/AlGaAs would later return for CD players and laser printers and the ability to grow QW structures [85] would reduce laser threshold currents through improved carrier confinement effects. The threshold current is the current where lasing begins and is determined for a device by extrapolating the linear regions of the light output vs. current graph as shown in Fig. 6.9 for a modern 1.55 μm InGaAsP/InP laser. For data storage, wavelengths as short as possible (currently around 400 nm for Blu-ray) are needed to maximize storage capacity while wavelengths in the mid-infrared are suitable for gas detection due to strong absorption resonances in gas molecules. For all operating wavelengths, it is crucial to reduce the threshold current, improve the efficiency, and decrease the temperature sensitivity. In general, carrier recombination processes in semiconductor lasers are strongly band gap dependent and pressure has proved to be an ideal technique to investigate recombination mechanisms especially when coupled with temperature dependence observations. With this in mind, experimental results are often plotted directly against wavelength or energy rather than pressure itself. The first reported pressure tuning of lasers was carried out by Feinlab et al. [86] on GaAs diodes up to 0.8 GPa at room and liquid nitrogen temperatures and was used to investigate the electronic transitions involved. Since then, pressure has been an essential tool for understanding loss mechanisms in lasers so that designs can be optimized to achieve their lowest threshold currents. The incorporation of strain into lasers to enhance their performance was first suggested by Adams [87] and Yablonovitch

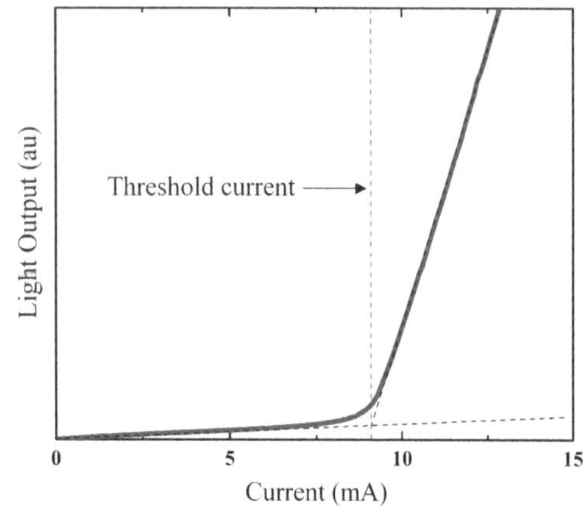

Fig. 6.9 Light-current characteristic for a semiconductor laser showing how the threshold current is determined

and Kane [88] and is now the basis of most modern semiconductor lasers. From our knowledge of band structure under pressure, the most obvious effect to be seen when squeezing a III–V-based laser is a reduction in the operating wavelength as the direct band gap increases, but other important results are now discussed.

The threshold current of a semiconductor laser can be expressed approximately in terms of carrier density (n) dependent terms:

$$I_{\mathrm{thr}} = eV(An + Bn^2 + Cn^3) + I_{\mathrm{leak}}, \tag{6.3}$$

where e is the electronic charge; V is the volume of active region; and A, B, and C are monomolecular (defect-related), radiative (spontaneous), and Auger recombination coefficients, respectively; while I_{leak} accounts for leakage from the active region into the barrier or cladding layers. In good material, An is negligible and in ideal QW lasers the threshold current would be expected to increase with pressure at a rate approximately equal to E_{g}^2 (the direct band gap squared) [89], and this is indeed the case for GaAs/AlGaAs lasers which are dominated by spontaneous emission. The third term in this equation corresponds to Auger recombination, a nonradiative process whereby the energy and momentum of the recombining electron-hole pair are given to a third carrier, either an electron or hole, lifting it to a higher energy state. This energy is subsequently dissipated by phonon emission and so it is lost from the lasing process and results in lattice heating.

The threshold current of lasers typically increases with temperature; in Fig. 6.10a, b a visible AlGaInP laser (672 nm) and an InGaAs IR laser (1.5 μm) are shown for comparison. The threshold current of these devices increases much more rapidly with temperature than those based on GaAs. Their temperature dependence looks fairly similar and it is only when compared using pressure (Fig. 6.10c, d) that we see that very different loss mechanisms are at work. In visible lasers, especially as they approach shorter wavelengths, leakage current is the dominant mechanism which limits performance [34]; this can be seen in Fig. 6.10c where theoretical fitting based on the Γ and X minima states approaching is shown as a solid line

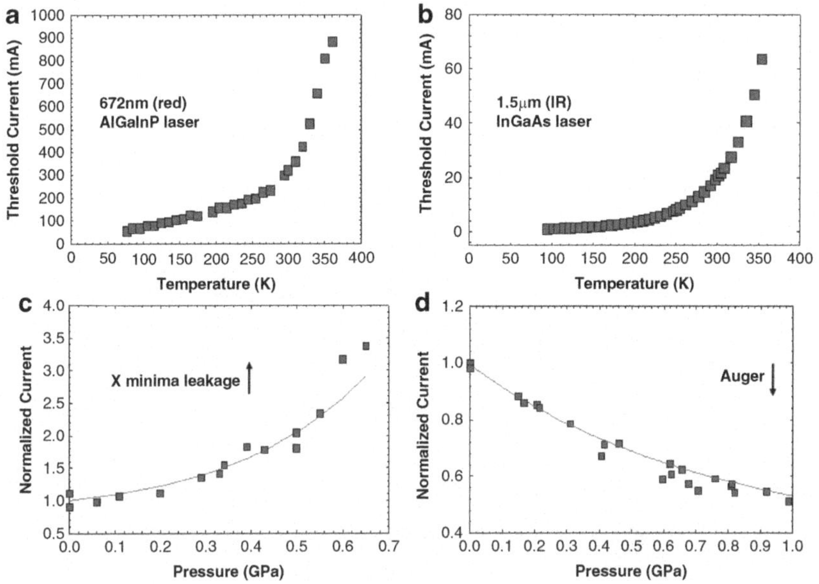

Fig. 6.10 Comparison of the effects of temperature and pressure on threshold current for visible and long wavelength lasers

Fig. 6.11 Schematic of a type-I QW where injected carriers can leak out of the well and into the barrier X level without recombining in the well

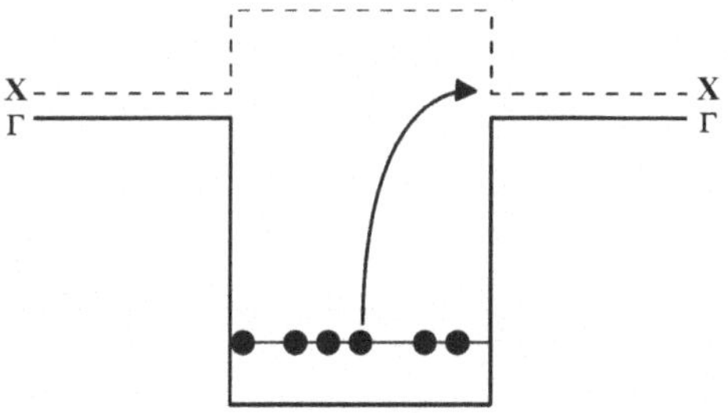

[90], Fig. 6.11 illustrates this leakage process into the X minima in the barrier. The long wavelength laser, however, behaves very differently; instead of the threshold current increasing, as with radiative recombination or leakage, there is a decrease that slows with pressure (Fig. 6.10d). The pressure dependence therefore shows that these lasers are dominated by Auger recombination. Increasing pressure gives rise to an increase in band gap meaning that the Auger transitions move further out in k-space where the probability of finding a state decreases, hence resulting in a decrease in the Auger recombination rate, see Fig. 6.12 for illustrations of two types of Auger: production of a hot conduction electron high into the conduction band (CHCC) and a hole deep into the spin–orbit valence band (CHSH). Measuring the pressure dependence of the Auger process can show clearly whether the particle absorbing the energy is an electron or a hole which allows the device to be designed to minimize the effect [91]. Figure 6.13 shows the normalized threshold current

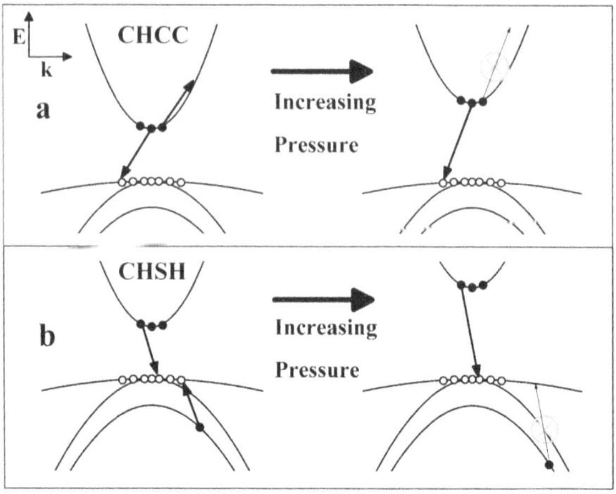

Fig. 6.12 The main Auger processes in long wavelength material and the effect on these of applied pressure

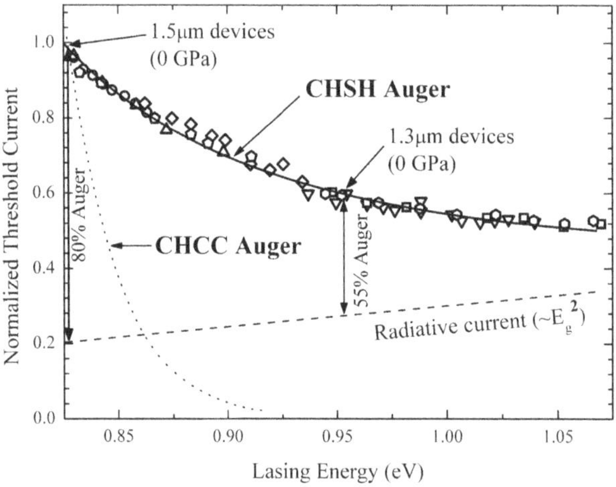

Fig. 6.13 Variation of the normalized threshold current with lasing energy for a range of lasers, along with theoretical curves. Normalization is against a 1.5 μm laser. The theoretical radiative current is shown as a *dashed line* and the CHSH and CHCC Auger curves with *solid and dotted lines*, respectively

results, from high pressure experiments, for a range of lasers plotted against lasing energy as well as theoretical fits based on the two Auger processes discussed. High pressure clearly identifies CHSH as the dominant loss mechanism.

6.4.2 Quantum Cascade Lasers

We have seen that as we move toward the mid-infrared, there is an increase in the effect of Auger recombination which limits laser performance. Quantum cascade lasers (QCLs) are very different in their operation from conventional QW lasers. Instead of the optical emissions being the result of interband transitions (recombination of electrons in the conduction band with holes in the valence band), they are the result of intersubband transitions, i.e., transitions between the electron states in the conduction band QW only. As a result, they are termed a "unipolar" device (usually electrons) rather than a "bipolar" device (electrons and holes). These extremely complex devices typically contain ~25–75 QWs and over 1,000 layers in total and operate at high electric fields. The field tilts the energy band diagram and after the electrons emit a photon, they tunnel through the barrier to be captured in the next QW (which is at a lower energy due to the field) and once there emit another photon; a cascade effect results, see Fig. 6.14 inset. Clearly the emission wavelength of these devices does not depend on the band gap of the heterostructure materials and the shortest wavelengths are in theory limited only by the conduction band offset. This type of laser was first demonstrated in 1994 by Faist et al. [92] operating at 4.2 μm. High power and temperature stable operation have been achieved over a wide range of wavelengths and provide a route to making THz sources. There is also considerable effort to produce QCL devices operating in the 2–4 μm region, where interband lasers are Auger-limited. Modeling has shown that leakage into the indirect minima may effect their operation. Again pressure can be used to tune the positions of these minima and identify limiting loss mechanisms. Figure 6.14 shows the normalized threshold current as a function of pressure for two InAs/AlSb QCLs operating at 3.3 and 2.9 μm, respectively [93]. It can be seen that the threshold current of the 3.3 μm laser is almost pressure independent, while that of the 2.9 μm shows a considerable increase with pressure and temperature. These results can be fitted theoretically [93] and show that at 2.9 μm, the principle loss mechanism in

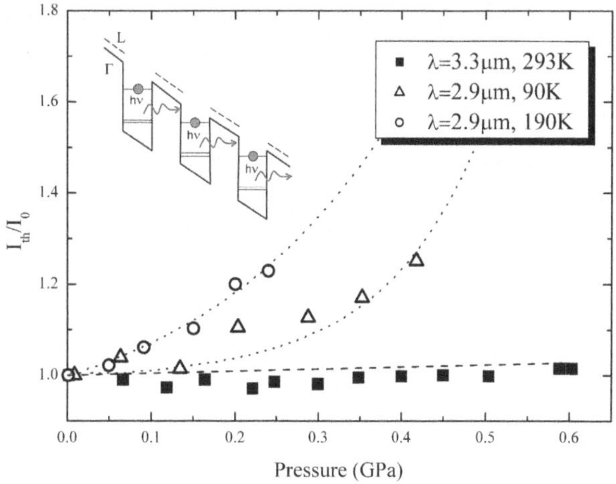

Fig. 6.14 Effect of pressure and temperature on the threshold current of two QCLs showing the effect of carrier leakage into the L minima for the shorter wavelength laser [93]. *Inset* shows a schematic of the QCL process

these lasers is leakage from the upper electron level into the L minima. Using this experimental data, Cathabard et al. [94] were able to optimize the design of lasers operating at 2.6 μm by weakening losses to the L minima and were able to operate them up to 175 K.

6.4.3 Avalanche Photodiodes

Avalanche photodiodes (APDs) are the highly sensitive low-noise solid-state optical detector equivalent of photo-multiplier tubes. Under high electric fields, photogenerated carriers gain energy and impact ionize (the reverse of Auger recombination) leading to a multiplication of photocurrent. This is an important process in semiconductors and pressure measurements have yielded its fundamental dependence on band structure. For good APDs, a material is required with a ratio of electron to hole ionization coefficients far from unity (either much smaller or much higher than unity is ideal). Sze and Gibbons [95] attempted to relate the band structure of experimentally measured ionization coefficients in terms of band gap, however their formulae failed in some important cases such as InP. From the pressure measurements of avalanche breakdown voltage V_b of p–n junction photodiodes in a range of materials, Allam [96] proposed a universal dependence. Figure 6.15 shows the experimental results for GaAs under pressure with calculated pressure coefficients equal to those of Γ (107 meV/GPa), L (38 meV/GPa) and X (−13 meV/GPa) conduction band minima [97]. The best fit to the experimental data is obtained using a threshold value of 27 meV/GPa and this is close to that of a Brillouin-zone averaged energy

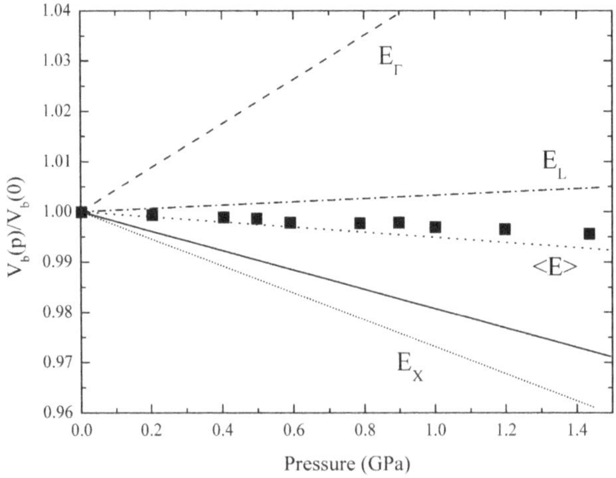

Fig. 6.15 Experimental pressure dependence of the breakdown voltage for GaAs compared to the calculated dependence for threshold energy independent of pressure (*solid line*) or scaling with E_Γ, E_X, E_L, or "Universal" scaling ($\langle E \rangle$) [97]

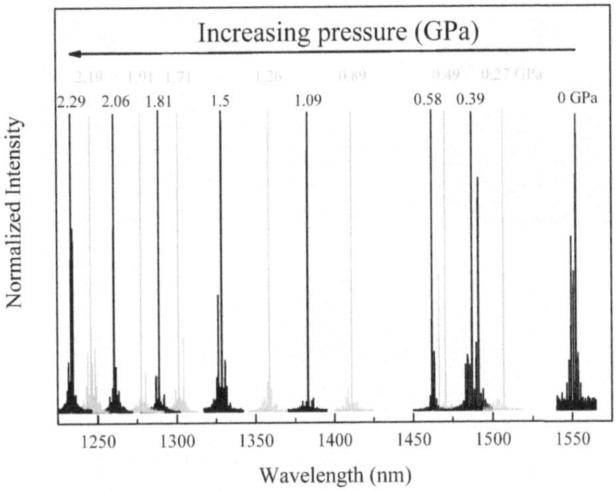

Fig. 6.16 Tuning a 1.55 μm laser over 300 nm using pressure. Pressure values are marked in alternating *black and gray* text for clarity

gap $\langle E \rangle$ approximated as:

$$\langle E \rangle = \tfrac{1}{8}(E_{\Gamma} + 4E_L + 3E_X). \tag{6.4}$$

This scaling effect with $\langle E \rangle$ implies that final electron states of the process are distributed between conduction band valleys according to their prevalence. This extended to show clear trends; for wide band gap materials the weighted average band gap of the Γ, X, and L points dominates while the value of the direct gap is the controlling parameter if it is less than about half of the lowest indirect band gap. This "universal model" gives an accurate prediction of the onset of impact ionization in all semiconductor devices where large electric fields are employed.

6.5 Commercial Equipment and Applications

With access to a good machine shop the production of high pressure anvil cells is well within the scope of most research facilities [98]. Easylab (http://www.easylab.co.uk) supply a range of DAC products for optical, X-ray magnetic, and electrical investigations. While Unipress, the high pressure equipment laboratory in Poland (http://www.unipress.waw.pl) provides a range of equipment for "the high pressure advancement of multidisciplinary research and practical applications". A spin out from this is their "Widely tunable laser diodes group" (WTLAB) (http://w3.unipress.waw.pl/wtlab) who aims to produce compact pressure systems working up to 2 GPa and low temperatures [99]. Such tunable lasers are needed in spectroscopy, telecommunications, gas detection, and medicine. At the Advanced

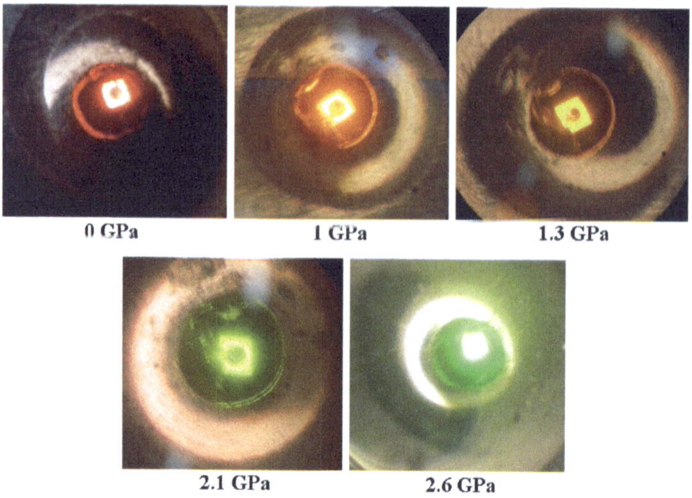

Fig. 6.17 Close up photographs of a commercial AlGaInP red LED device chip operating at 0, 1, 1.3, 2.1, and 2.6 GPa

Technology Institute, University of Surrey (http://www.ati.surrey.ac.uk), work is also being carried out on a microtunable laser system using pressure. Figure 6.16 shows a telecoms 1.55 μm laser reversibly tuned by over 300 nm using 2.3 GPa of pressure, while in Fig. 6.17 an AlGaInP red LED is tuned into the green by around 2.6 GPa of pressure. Both results are from a new design of an optical anvil based system with electrical feedthroughs [100].

Acknowledgments The authors would like to thank I.P. Marko and S.R. Jin for their contributions to this chapter and would also like to thank the Engineering and Physical Sciences Research Council (EPSRC) for their support of high pressure activities at the University of Surrey.

References

1. P.W. Bridgman, Proc. Am. Acad. Arts Sci. **79**, 167 (1952)
2. P.W. Bridgman, *Collected Experimental Papers*, vol. I–VII, (Harvard University Press, Cambridge, 1964)
3. P.W. Bridgman, *The Physics of High Pressure*, (G. Bell and Sons, Ltd., London, 1931) (reprinted by Dover Publications, 1970)
4. T. Suski, W. Paul (eds.), in *High Pressure in Semiconductor Physics*. Semiconductors and Semimetals, vol. 54 and 55 (Academic, New York, 1998)
5. I.L. Spain, J. Paauwe, *High Pressure Technology: Volume 1: Equipment Design, Materials, and Properties* (CRC Press, London, 1977)
6. W.F. Sherman, *Experimental Techniques in High Pressure Research* (Wiley, New York, 1988)
7. P. Yu, Phys. Stat. Sol. (b) **248**, 1077 (2011)

8. O. Madelung (ed.) *Data in Science and Technology*, Semiconductors Group-IV Elements and III-V Compounds, (Springer, Berlin, 1991)
9. I. Vurgaftman, J.R. Meyer, L.R. Ram-Mohan, J. Appl. Phys. **89**, 5815 (2001)
10. O. Madelung, *Semiconductors: Data Handbook* (Springer, Heidelberg, 2004)
11. W. Paul, J. Appl. Phys. **32**, 2082 (1961)
12. H. Ehrenreich, Phys. Rev. **120**, 1951 (1960)
13. B.K. Ridley, T.B. Watkins, Proc. Phys. Soc. **78**, 293 (1961)
14. J.B. Gunn, Solid State Commun. **1**, 88 (1963)
15. H. Kroemer, Proc. IEEE **52**, 1736 (1964)
16. A.R. Hutson, A. Jayaraman, A.G. Chynoweth, A.S. Coriell, W.L. Feldman, Phys. Rev. Lett. **14**, 639 (1965)
17. P.J. Vinson, C. Pickering, A.R. Adams, W. Fawcett, G.D. Pitt, in *Proceeding of the 13th International Conference on Physics of Semiconductors*, ed. by F.G. Fumi (Rome, Italy, 1976) p. 1243
18. A.R. Adams, P.J. Vinson, C. Pickering, G.D. Pitt, W. Fawcett, Electron Lett. **13**, 46 (1977)
19. C. Pickering, A.R. Adams, J. Phys. C **10**, 3115 (1977)
20. D.E. Aspnes, Phys. Rev. B **14**, 5331 (1976)
21. J.W. Allen, M. Shyam, Y.S. Chen, G.L. Pearson, Appl. Phys. Lett. **7**, 78 (1965)
22. C. Pickering, A.R. Adams, G.D. Pitt, Solid State Commun. **16**, 1359 (1975)
23. G.D. Pitt, J. Phys. E **1**, 915 (1968)
24. A. El-Sabbahy, A.R. Adams, M.L. Young, Solid State Electron. **21**, 83 (1978)
25. J.D. Barnett, S. Block, G.J. Piermarini, Rev. Sci. Instrum. **44**, 1 (1973)
26. K. Syassen, High Pressure Res. **28**, 75 (2008)
27. R.J. Warburton, R.J. Nicholas, N.J. Mason, P.J. Walker, A.D. Prins, D.J. Dunstan, Phys. Rev. **543**, 4994 (1991)
28. C.G. Van de Walle, Phys. Rev. B **39**, 1871 (1989)
29. D.J. Wolford, T.F. Keuch, J.A. Bradley, M.A. Gell, D. Ninno, M. Jaros, J. Vac. Sci. Technol. **B4**, 1043 (1986)
30. T. Suski, W. Paul (eds.), in *High Pressure in Semiconductor Physics*. Semiconductors and Semimetals, vol. 54 (Academic, New York, 1998), pp. 25 and 334
31. J.D. Lambkin, A.R. Adams, D.J. Dunstan, P. Dawson, C.T. Foxon, Phys. Rev. B **39**, 5546 (1989)
32. J.D. Lambkin, D.J. Dunstan, E.P. O'Reilly, B.R. Butler, J. Cryst. Growth **93**, 323 (1988)
33. M.F. Whitaker, D.J. Dunstan, M. Missous, L. Gonzalez, Phys. Stat. Sol. (b) **198**, 349 (1996)
34. A.T. Meney, A.D. Prins, A.F. Phillips, J.L. Sly, E.P. O'Reilly, D.J. Dunstan, A.R. Adams, A. Valster, IEEE J. Sel. Top. Quant. Elect. **1**(2), 697 (1995)
35. D. Patel, C.S. Menoni, H. Temkin, R.A. Logan, D. Coblenz, Appl. Phys. Lett. **62**, 2459 (1993)
36. A. Jayaraman, Rev. Sci. Instrum. **57**, 1013 (1986)
37. D.J. Dunstan, I.L. Spain, J. Phys. E: Sci. Instrum. **22**, 923 (1989)
38. M.I. Eremets, *High Pressure Experimental Methods* (Oxford University Press, Oxford, 1987)
39. E.O. Kane, J. Phys. Chem. Solids **1**, 249 (1957)
40. L.G Shantharama, A.R. Adams, C.N. Ahmad, R.J. Nicholas, J. Phys. C: Solid State Phys. **17**, 4429 (1984)
41. J. Endicott, A. Patanè, D. Maude, L. Eaves, M. Hopkinson, G. Hill, Phys. Rev. B **72**, 041306(R) (2005)
42. G. Pettinari, A. Polimeni, F. Masia, R. Trotta, M. Felici, M. Capizzi, T. Niebling, W. Stolz, P.J. Klar, Phys. Rev. Lett. **98**, 146402 (2007)
43. A. Polimeni, G. Pettinari, R. Trotta, F. Masia, M. Felici, M. Capizzi, A. Lindsay, E.P. O'Reilly, T. Niebling, W. Stolz, P.J. Klar, F. Martelli, S. Rubini, Phys. Stat. Sol. (a) **205**, 107 (2008)
44. D. Lancefield, A.R. Adams, M.A. Fisher, J. Appl. Phys. **62**, 2343 (1987)
45. P.M. Mooney, J. Appl. Phys. **67**, R1 (1990)
46. M. Mizuta, M. Tachikawa, H. Kukimoto, Jpn. J. Appl. Phys. **24**, L143 (1985)
47. J.A. Wolk, W. Walukiewicz, M.L.W. Thewalt, E.E. Haller, Phys. Rev. Lett. **68**, 3619 (1992)
48. Y. Seki, H. Watanabe, J. Matsui, J. Appl. Phys. **49**, 822 (1978)

49. A.S. Jordan, A.R. Vonneida, R. Caruso, J. Cryst. Growth **70**, 555 (1984)
50. M.G. Brik, A. Kaminska, A. Suchocki, J. Appl. Phys. **108**, 103520 (2010)
51. A. Kaminska, A. Kozanecki, S. Trushkin, A. Suchocki, Phys. Rev. B **81**, 165209 (2010)
52. H.P. Hjalmarson, P. Vogl, D.J. Wolford, J.D. Dow, Phys. Rev. Lett. **44**, 810 (1980)
53. X. Liu, M.-E. Pistol, L. Samuelson, S. Schwetlick, W. Seifert, Appl. Phys. Lett. **56**, 1451 (1990)
54. W. Shan, W. Walukiewicz, J.W. Ager, E.E. Haller, J.F. Geisz, D.J. Friedman, J.M. Olson, S.R. Kurtz, Phys. Rev. Lett. **82**, 1221 (1999)
55. J. Wu, W Shan, W Walukiewicz, Semicond. Sci. Technol. **17**, 860 (2002)
56. J.M. Besson, J.P. Itié, A. Polian, G. Weill, J.L. Mansot, J. Gonzalez Phys. Rev. B **44**, 4214 (1991)
57. M.F. Whitaker, S.J. Webb, D.J. Dunstan, J. Phys. Condens. Mater. **10**, 8611 (1998)
58. A. Mujica, A. Rubio, A. Munoz, R.J. Needs, Rev. Mod. Phys. **75**, 863 (2003)
59. C.S. Menoni, I.L. Spain, Phys. Rev. B **35**, 7520 (1987)
60. R.J. Nelmes, M.I. McMahon, N.G. Wright, D.R. Allan, H. Liu, J.S. Loveday, Phys. Rev. Lett. **79**, 3668 (1997)
61. C.M. Lin, H.S. Sheu, M.H. Tsai, B.R. Wu, S.R. Jian, Solid State Commun. **149**, 136 (2009)
62. M.P. Jackson, M.P. Halsall, M. Güngerich, P.J. Klar, W. Heimbrodt, J.F. Geisz, Phys. Stat. Sol. (b) **244**, 336 (2007)
63. D.N. Bose, B. Seishu, G. Parthasarsthy, E.S.R. Gopal, Proc. R. Soc. London. Ser. A, Math. Phys. Sci. **405** (1829), 345 (1986)
64. S.W. Tozer, D.J. Wolford, J.A. Bradley, D. Bour, G.B. Stringfellow, in *Proceedings of the 19th ICPS*, Poland, 1988, p. 881
65. S.W. Tozer, D.J. Wolford, J.A. Bradley, D. Bour, J. Electron. Mater. **S35** (1988)
66. B.A. Weinstein, S.K. Hark, R.D. Burnham, R.M. Martin, Phys. Rev. Lett. **58**, 781 (1987)
67. D.J. Dunstan, A.D. Prins, B. Gil, J.P. Faurie, Phys. Rev. B **44**, 4017 (1991)
68. V. Latussek, C.R. Becker, G. Landwehr, R. Bini, L. Ulivi, Phys. Rev. B **71**, 125305 (2005)
69. C. Burda, X. Chen, R. Narayanan, M.A. El-Sayed, Chem. Rev. **105**, 1025 (2005)
70. B.A. Weinstein, Phys. Stat. Sol. (b) **244**, 368 (2007)
71. S.H. Tolbert, A.P. Alivisatos, Science **265**, 373 (1994)
72. S.H. Tolbert, A.B. Herhold, L.E. Brus, A.P. Alivisatos, Phys. Rev. Lett. **76**, 4384 (1996)
73. H. Wang, J.F. Liu, Y. He, Y. Wang, W. Chen, J. Z. Jiang, J. Staun Olsen, L. Gerward, J. Phys.: Condens. Mater **19**, 156217 (2007)
74. J.Z. Jiang, J. Staun Olsen, L. Gerward, S. Steenstrup, High Pressure Res. **22**, 395 (2002)
75. C.J. Lee, A. Mizel, U. Banin, M.L. Cohen, A.P. Alivisatos, J. Chem. Phys. **113**, 2016 (2000)
76. C.S. Menoni, L. Miao, D. Patel, O.I. Mic'ic', A.J. Nozik, Phys. Rev. B **84**, 4168 (2000)
77. J.G. Díaz, G.W. Bryant, W. Jaskólski, M. Zieliski, Phys. Rev. B **75**, 245433 (2007)
78. Z. Zong, Y. Ma, T. Hu, G. Cui, Q. Cui, M. Zhang, G. Zou, Solid State Commun. **151**, 607 (2011)
79. K.C. Heasman, PhD thesis, University of Surrey, 1985
80. S.J. Sweeney, A.F. Phillips, A.R. Adams, E.P. O'Reilly, P.J.A. Thijs, IEEE Phot. Tech. Lett. **10**, 1076 (1998)
81. A.F. Phillips, S.J. Sweeney, A.R. Adams, P.J.A. Thijs, IEEE J. Sel. Top. Quant. Elect. **5**, 401 (1999)
82. S.J. Sweeney, G. Knowles, T.E. Sale, Appl. Phys. Lett. **78**, 865 (2001)
83. N.F. Masse, E. Homeyer, I.P. Marko, A.R. Adams, S.J. Sweeney, O. Dehaese, R. Piron, F. Grillot, S. Loualiche, Appl. Phys. Lett. **91**, 131113 (2007)
84. D. Langer, D.M. Warschauer, Rev. Sci. Instrum. **32**, 32 (1961)
85. R. Dingle, W. Wiegmann, C.H. Henry, Phys. Rev. Lett. **33**, 827 (1974)
86. J.F. Feinlab, S. Groves, W. Paul, R. Zallen, Phys. Rev. **131**, 2070 (1963)
87. A.R. Adams, Electron. Lett. **22**(5), 249 (1986)
88. E. Yablonovitch, E.O. Kane, J. Lightwave Technol. **T-4**(5), 504 (1986)
89. T. Suski, W. Paul (eds.), in *High Pressure in Semiconductor Physics*. Semiconductors and Semimetals, vol. 55 (Academic, New York, 1998), p. 311

90. S.J. Sweeney, G. Knowles, T.E. Sale, A.R. Adams, Phys. Stat. Sol. (b) **223**, 567 (2001)
91. S.J. Sweeney, A.R. Adams, M. Silver, E.P. O'Reilly, J.R. Watling, A.B. Walker, P.J.A. Thijs, Phys. Stat. Sol. (b) **211**, 525 (1999)
92. J. Faist, F. Capasso, D.L. Sivco, C. Sirtori, A.L. Hutchinson, A.Y. Cho, Science **264** (5158), 553 (1994)
93. I.P. Marko, A.R. Adams, S.J. Sweeney, R. Teissier, A.N. Baranov, S. Tomic, Phys. Stat. Sol. (b) **246**, 512 (2009)
94. O. Cathabard, R. Teissier, J. Devenson, J.C. Moreno, A.N. Baranov, Appl. Phys. Lett. **96**, 141110 (2010)
95. S.M. Sze, G. Gibbons, Appl. Phys. Lett. **8**, 111 (1966)
96. J. Allam, Jpn. J. Appl. Phys. **36**, 1529 (1997)
97. J. Allam, A.R. Adams, M.A. Pate, J.S. Roberts, Appl. Phys. Lett. **67**, 22, 3304 (1995)
98. D.J. Dunstan, I.L. Spain, J. Phys. E: Sci. Instrum. **22**, 913 (1989)
99. http://w3.unipress.waw.pl/wtlab see publications list or for example: P. Adamiec, A. Salhi, R. Bohdan, A. Bercha, F. Dybala, W. Trzeciakowski, Y. Rouillard, A. Joullié, Appl. Phys. Lett. **85**, 4292 (2004)
100. A.D. Prins, A.R. Adams, S.J. Sweeney, to be published

Chapter 7
Spatially Resolved Luminescence Spectroscopy

Gintautas Tamulaitis

Abstract Spatially resolved luminescence spectroscopy is a useful tool for the study of semiconductors with inhomogeneities of their properties on submicrometer scale and semiconductor nanostructures. In this chapter, basic operation principles, instrumentation, and advantages and disadvantages of micro-photoluminescence (μ-PL), confocal microscopy, scanning near-field optical microscopy (SNOM), and cathodoluminescence (CL) are discussed.

7.1 Introduction

Luminescence spectroscopy with spatial resolution provides an informative tool for the study of semiconductors and their nanostructures. Two principal approaches can be followed in the study of the spatial distribution of the photoluminescence intensity on the sample surface: direct imaging of the sample area under investigation and point-by-point scanning of the area.

According to the first approach, the spatially resolved information can be obtained by homogeneous illumination of the area, optical formation of the image of the light emitted by the sample, and recording the image for storing and inspection (e.g., by using photographic film, two-dimensional CCD or CMOS matrix, etc.). The technical capabilities of optical image magnification boosted and saturated encountering technological and physical challenges at the end of the nineteenth century.

First of all, the focusing ability of a lens is limited by its aberrations. There are two types of on-axis aberrations (chromatic and spherical) and four types of off-axis

G. Tamulaitis (✉)
Semiconductor Physics Department and Institute of Applied Physics, Vilnius University, Vilnius, Sauletekio al. 9-III, Lithuania
e-mail: gintautas.tamulaitis@ff.vu.lt

A. Patanè and N. Balkan (eds.), *Semiconductor Research*, Springer Series in Materials Science 150, DOI 10.1007/978-3-642-23351-7_7,

aberrations (coma, astigmatism, curvature of field, and distortion). The aberration-induced deformation of the image in optical instruments can be compensated to a rather high degree by designing sophisticated optical systems consisting of many different lenses made of appropriate optical materials.

However, even in optical systems with negligible aberrations, the spatial resolution of the image formation is limited by light diffraction. Due to diffraction, the light hitting the edge of an object bends around the edge into the geometric shadow of the object. In particular, propagation of the light through a round aperture results in the formation of a central spot called Airy disc (after Sir George Airy, 1801–1892) and concentric bright and dark rings around the spot. The Airy pattern remains also intact when a light beam is focused by a lens into its focal plane, which acts as an aperture and limits the smallest resolvable spot size on the image. According to geometric optics, the light with a planar wave front should be focused by an aberration-free lens into a point. Diffraction, however, transforms the point into the Airy disc. The diameter of the Airy disc d depends on the light wavelength λ and the aperture angle of the lens. For a planar wave front, the aperture angle can be described by the focal ratio, i.e., the ratio between the focal length f and the diameter D of the lens. Thus, d can be expressed as

$$d = 1.22\lambda(f/D). \tag{7.1}$$

The expression demonstrates that the spot size d can be diminished by increasing the lens diameter D or using a lens with a shorter focus length f. For $f/D < 1$, (7.1) has to be replaced by a more accurate expression:

$$d_{\mathrm{R}} = 0.61\lambda/\mathrm{NA}. \tag{7.2}$$

Here, $\mathrm{NA} = n \sin\theta$ is the numerical aperture, and θ is the half angle between the marginal converging rays of the light cone. The numerical aperture also takes into account the refraction index n of the medium between the lens and the focal plane.

Expression (7.2) defines also the spatial resolution when the lens is used to collect light from the surface under study. The light passing through the lens interferes with itself creating a ring-shaped Airy pattern. Expression (7.2) reflects the Rayleigh criterion, stating that two closely located equally bright points are distinguishable from each other if the center of the Airy disc of the first point coincides with the first zero response ring of the second point. According to the alternative Sparrow criterion, two points of equal brightness can be distinguished if the intensity at the midway point is lower or equal to the intensity at the points. The corresponding spatial resolution

$$d_{\mathrm{S}} = 0.51\lambda/\mathrm{NA} \tag{7.3}$$

is similar to that according to the Rayleigh criterion (7.2).

The lenses can be replaced by optical systems (objectives), but none of them can overcome the diffraction limit, unless nonlinear phenomena (e.g., saturation effects, multiphoton absorption, and stimulated depletion) are utilized. The optical systems for linear microscopy are designed to increase the angle θ, while immersion liquids help in increasing the refraction index n. For modern microscope objectives working in high-index media such as water ($n = 1.33$) or oil ($n = 1.56$), the numerical aperture can be as high as 1.3–1.4. Thus, the limit of the spatial resolution obtainable with high-quality objectives is equal to approximately half the wavelength of the light analyzed.

According to the second, currently more common approach, the spatial distribution of the photoluminescence (PL) intensity on the sample surface can be accomplished by scanning the sample surface point by point. To obtain the PL spatial distribution with high resolution, either the excitation light has to be focused into a sufficiently small spot or the luminescence light has to be selectively collected from a small spot.

In this chapter, the basic operation principles of micro-photoluminescence (μ-PL) and other spectroscopic techniques using confocal microscopy, scanning near-field optical microscopy (SNOM), and cathodoluminescence (CL) are discussed.

Note that the volume occupied by nonequilibrium carriers in semiconductors can be considerably larger than the volume of primary excitation. Excitation peculiarities and spreading of the nonequilibrium carriers outside the volume, which is initially excited by absorbed light or electron beam, are discussed in Sect. 7.6. The chapter ends with a short list of advantages and disadvantages of the main techniques for spatially resolved luminescence spectroscopy of semiconductors and semiconductor nanostructures.

7.2 Micro-photoluminescence (μ-PL)

The term μ-PL is slightly ambiguous. Sometimes the term is used to describe any type of PL measurement system with the spatial resolution of approximately 1 μm. The traditional and still more common use of the term μ-PL implies a technique to measure PL by using a setup where the luminescence is excited with a tightly focused light beam employing an optical microscope or a similar optical system. We follow the latter option thereafter.

Instrumentally, the spatial resolution in μ-PL is usually limited to 1–2 μm but might be lower due to properties of the object under study (see Sect. 7.6). Two approaches allow mapping the sample surface: scanning the surface by moving the excitation laser beam or moving the sample on the beam focus plane at a fixed beam position. The second approach is more prevalent, but the first one is advantageous for experiments performed at low temperatures when the sample has to be placed inside a cryostat. In any case, the cryostat windows inflict optical distortions. Since the excitation light in the μ-PL experiments is focused on the sample surface with short-focus objectives, the distortions are important and special measures have to be taken to compensate them.

Monochromatic PL intensity mapping, i.e., the PL intensity mapping performed at a fixed wavelength, can be obtained using band-pass filters, while spectrometers are used to obtain the luminescence spectra of the light collected from the single spots of the surface scan. Using this experimental information, mappings of PL intensity, PL band peak position, or bandwidth can easily be generated using the appropriate software. To add the time resolution, picosecond or femtosecond lasers are used for pulsed excitation, and the PL response is analyzed by exploiting streak cameras, up-conversion systems, or optical Kerr shutters, as in any other time-resolved photoluminescence experiments (see review of time-resolved PL techniques in Chap. 8).

Being comparatively simple, the μ-PL systems are commonly used in semiconductor industry for a fast automated inspection of wafers to evaluate quality, get information for the optimization of fabrication or processing, etc. The μ-PL systems are also used for a deeper study of the properties of semiconductors and their heterostructures. The comparatively low spatial resolution sometimes limits the applicability of the μ-PL systems for the study of the semiconductor nanostructures where the distance between the neighboring nano-objects is so small that the $\sim$ 1-μm spot used in μ-PL covers several nano-objects. Sometimes mesa structures are etched with a single nano-object on top of it. However, preparation of such a sample is quite difficult and destructive.

7.3 Scanning Confocal Microscopy (SCM)

The principle of confocal microscope was suggested and patented in 1957 by Minsky. The commercial instruments appeared in 1987 and gradually became a quite conventional tool in the study of biological objects and semiconductor structures. Capability of scanning the object luminescence in three dimensions at comparatively good spatial resolution and user-friendly operation are the most attractive features of this device.

The typical configuration of a confocal microscope is schematically depicted in Fig. 7.1. Excitation light is focused by an objective to a small, nearly diffraction-limited spot on the sample surface. The same objective is also used to collect the light emitted by the photoexcited sample. The plane of the image obtained by the objective coincides with a screen containing a pinhole confocally aligned with the excited spot under study. Thus, the light coming out of the spot passes the pinhole, while the light from outside the spot is focused off the pinhole and is blocked by the screen from entering the detection system. Moreover, the light emitted off the focal plane along the optical axis is focused in front of the pinhole or behind it. This emission is defocused on a large area of the screen with the pinhole. Thus, only a small fraction of this emission passes the pinhole. This on-axis spatial selectivity enables scanning the sample in direction z perpendicular to the sample surface by shifting the focal plane of the objective.

Various lasers are used as excitation sources and their radiation is usually delivered to the confocal microscope by optical fibers. Selection of the laser is primarily based on the wavelength required for excitation.

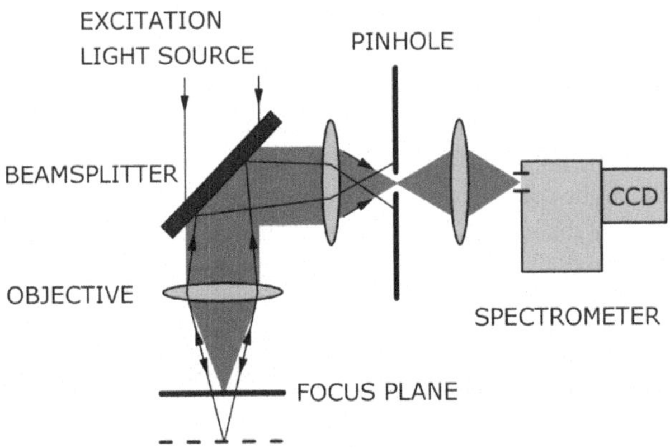

Fig. 7.1 Schematic configuration of a confocal microscope with spectrometer. Paths of the luminescence emitted from the focus plane and off the focus plane are depicted by *shaded area and solid lines*, respectively

A spectrally integrated spatial PL distribution can be obtained by delivering the luminescence, which passed though the pinhole, directly to a photodetector (photodiode, photomultiplier tube). To get a monochromatic spatial PL distribution, band-pass filters are used. The most informative analysis of the luminescence is accomplished by delivering the selectively collected luminescence into a spectrometer.

Two methods are used to accomplish the in-plane (*xy*) scanning of the sample surface (1) moving the excitation beam and (2) shifting the stage with the sample on top of it. The position of the excitation spot is changed using two mirrors deflecting the focused light beam in two perpendicular directions. However, piezoelectric sample stage for accurate shifting of the sample in the *xy* plane is currently more convenient. Tilting of the sample surface with respect to the *xy* plane of the stage might sometimes be an issue but can be, at least partially, compensated using corresponding software.

The spatial resolution of a confocal microscope depends on the excitation and luminescence wavelengths, which are inherent to the sample under study and the desirable conditions of its excitation. Instrumentally, the resolution mainly depends on the numerical aperture of the objective. The numerical aperture determines the excitation spot size and the size of the spot image at the pinhole plane. The pinhole diameter is a subject of an unavoidable trade-off: the aperture should be small enough to ensure high spatial resolution and possibly large to let more light into the detection system. The aperture of a multimode optical fiber (typically 25 μm in diameter) delivering the luminescence light to detection system is often used as a pinhole. At the objective magnification of $\sim 100\times$, the aperture picks up the central spot of the Airy disc for a single diffraction-limited point source on the sample surface and ensures the spatial resolution of ~ 200 nm, which is typical for conventional confocal microscopes.

The spatial resolution of a confocal microscope has certain peculiarities in respect to that of a standard optical microscope [1]. In standard microscopes, a large sample is uniformly illuminated using an incoherent light source. Thus, the illumination is spatially incoherent. An ideal aberration-free objective images every single point on the sample surface as a diffraction-limited spot on the observation plane. Since the light source is spatially incoherent, the spatial resolution of the system is limited by the spatial distribution of the square of the amplitude response of the objective. Meanwhile in confocal microscope, only one point on the sample surface is illuminated and serves as a spatially coherent light source. As a result, the resolution is limited by the spatial distribution of the amplitude response of the objective. The smallest resolvable distance between two points in the xy plane can be approximated as [2]

$$d_{xy} \approx 0.4\lambda/\mathrm{NA}. \tag{7.4}$$

Thus, under appropriate conditions, the xy spatial resolution of a confocal microscope can exceed that obtainable with wide-field optics by a factor of ~ 1.4 [3]. This improvement is sometimes called superresolution. The term also covers other techniques for increasing the resolution of optical imaging systems: Pendry lens, Kino's solid immersion lens, and Toraldo di Francia's apodizing masks [4].

The spatial resolution along the z-axis can be expressed as [2]

$$d_z \approx 1.4\lambda n/\mathrm{NA}^2. \tag{7.5}$$

Here, n is the refraction index of the medium between the sample and the objective (air or immersion fluid). Note that n is also encompassed in the numerical aperture $\mathrm{NA} = n\sin\theta$. The effective wavelength λ has to be calculated using the wavelengths of excitation (λ_{exc}) and luminescence (λ_{lum}) as $1/\lambda = 1/\lambda_{\mathrm{exc}} + 1/\lambda_{\mathrm{lum}}$. For typical values $\lambda = 0.5\,\mu\mathrm{m}$, $n = 1$, and $\mathrm{NA} = 0.9$, the resolution of $0.86\,\mu\mathrm{m}$ can be expected. Thus, the spatial resolution in z direction is lower and has a stronger dependence on the numerical aperture than that in xy plane.

The axial spatial resolution (in z direction) can be increased by up to the factor of five by using a technique called $4-\pi$ confocal. The name implies 4π steradians of a complete sphere to indicate a large solid angle employed for excitation and detection in $4-\pi$ confocal microscope. The microscope contains two objectives. There are three types of $4-\pi$ confocal microscopes. Configuration of the simplest and the most widely used type-A device is presented in Fig. 7.2. In this device, a two-side excitation and a one-side detection configuration are used. Laser excitation beam is split into two equal parts by a beam splitter BS, and the two beams are focused into the same point in the sample after being reflected by mirrors M1–M2–M3 and M4–M5, respectively. The luminescence is collected through lens L2 and directed to the pinhole via dichroic mirrors DM2 and DM1. In type-B device, a one-side excitation and a two-side detection configuration is used, while a two-side excitation and a two-side detection configuration is employed in type-C device. The principle of resolution improvement is illustrated in Fig. 7.3. Here, the z response of an infinitely thin fluorescent layer, which reflects the spatial distribution of the

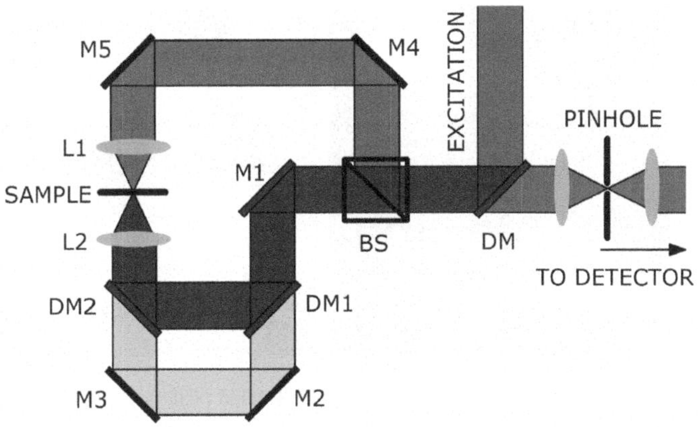

Fig. 7.2 The simplest configuration of a $4 - \pi$ confocal microscope (type-A) containing two objective lenses (L), mirrors (M), and dichroic mirrors (DM) to ensure two excitation paths (M1–M2–M3–L2 and M1–M4–M5–L1) and one detection path (L2–DM2–DM1–M1). After [5]

Fig. 7.3 Z-response of an infinitely thin fluorescent layer for (**a**) a conventional confocal microscope and (**b**) a type-A $4 - \pi$ confocal microscope, calculated for objective lenses with NA = 1.4, and excitation and emission wavelengths of 633 and 725 nm, respectively. After [6]

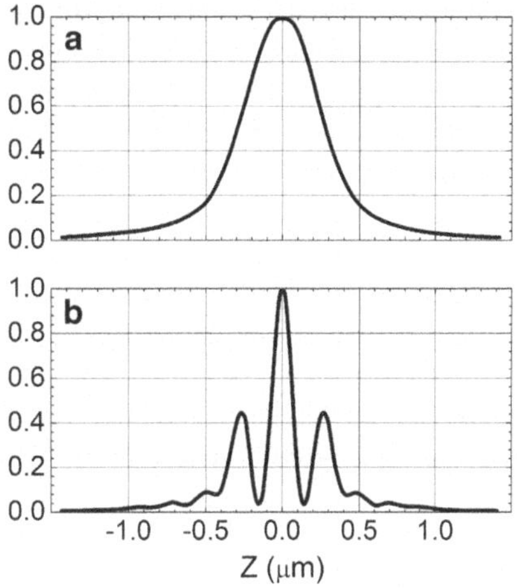

detected signal intensity along the z axis, is presented for a conventional confocal microscope (Fig. 7.3a) and a type-A $4 - \pi$ confocal microscope (Fig. 7.3b) [6]. In conventional confocal microscope, the finite-sized excitation volume and imperfect pinhole discrimination result in nearly Gaussian-shaped z response. The width of this response function determines the axial spatial resolution. In $4 - \pi$ confocal microscope, the counter-propagating excitation light beams create an interference pattern. As a result, the z response consists of a central peak, which is considerably narrower than the nearly Gaussian band in conventional confocal microscope, and side lobes (see Fig. 7.3b). Several techniques using hardware and software are being applied to filter out these side lobes (see, e.g., [5] and references therein). The most effective gain in spatial resolution is obtained in type-C $4 - \pi$ confocal microscope, which is also the most complicated when compared with type-A and -B devices.

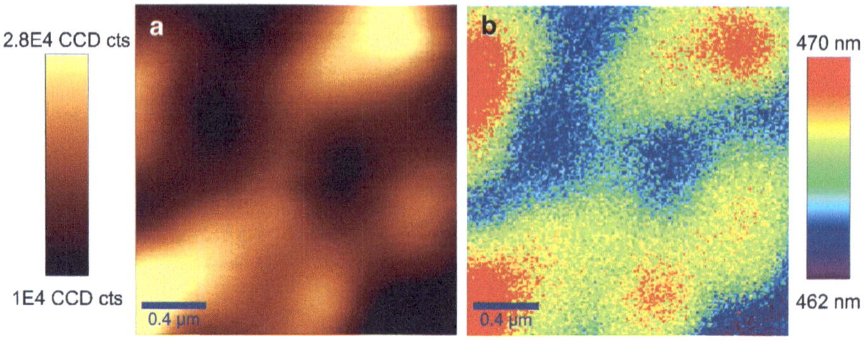

Fig. 7.4 Typical confocal images of PL intensity and PL band peak wavelength mappings of an InGaN epilayer

Four-π confocal microscopy requires samples that are transparent to the excitation light. This is often the case in study of biological objects. Four-π confocal technique is usually not suitable for investigation of bulk semiconductors; however, it might turn out to be a very convenient tool to study semiconductor quantum wells and nanostructures, provided that the substrates and all layers embedding the structures under study have a wider band gap.

Typical mapping images of PL intensity and band peak wavelength are presented in Fig. 7.4 for InGaN. Island-like bright areas of 0.1–1 μm in size are observed. Usually, PL bands in these bright areas are red-shifted in respect of those in the dark background. This is an indication that the effective band gap in the bright areas is smaller than that in the background, and the nonequilibrium carriers accumulate there. Bright islands with blue-shifted PL bands have also been observed in certain InGaN-based structures and interpreted by assuming the existence of regions, where the nonequilibrium carriers accumulate and screen the built-in field [7], or by inhomogeneous distribution of non-radiative recombination centers [8].

7.4 Scanning Near-Field Optical Microscopy

The limit of the spatial resolution due to self-diffraction is an inherent property of the light in the far field, i.e., at a distance from the light source, which is considerably longer than the wavelength. However, the limit can be substantially decreased in the near field. Historically, two names are used for the same microscopy technique (actually, the same group of related techniques) exploiting the near field: scanning near-field optical microscopy (SNOM) and near-field scanning optical microscopy (NSOM).

The idea of the technique has been first suggested by Synge [9, 10] in the third decade of the twentieth century. The idea is based on light propagation through an aperture with a diameter considerably smaller than the wavelength λ (see Fig. 7.5).

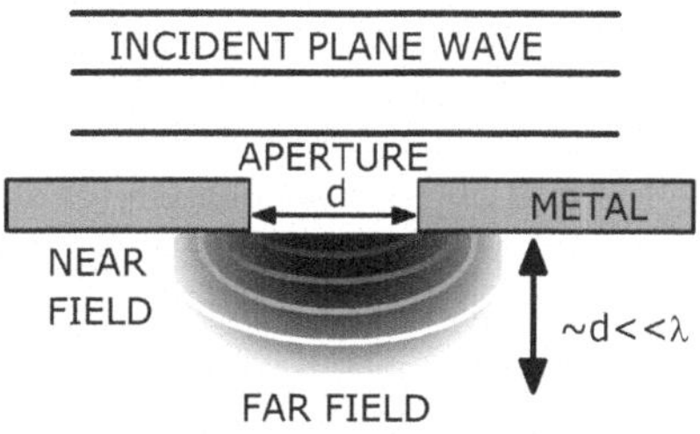

Fig. 7.5 Illustration of a plane light wave transforming into an evanescent wave by passing a sub-wavelength-size aperture in infinitesimally thin metal screen

The light wave vector $k \equiv 2\pi/\lambda$ might be expressed as

$$k^2 = k_t^2 + k_l^2, \tag{7.6}$$

where k_t and k_l are transverse and longitudinal components of k, respectively. Under the simple assumption that the field is strictly confined within the dimensions of the aperture, the transverse component k_t is defined by the aperture diameter d as $k_t = j(\pi/d)$, with $j = 1, 2, \ldots$ Thus, for d smaller than a certain value, the following inequality holds:

$$k_t = \frac{\pi}{d} > \frac{2\pi}{\lambda} \equiv k. \tag{7.7}$$

According to this inequality and (7.6), the longitudinal component of the wave vector in these conditions has to be imaginary, which corresponds to an evanescent wave. Thus, the light propagates through the aperture with $d \ll \lambda$ only in the near field (distance of the order of d). Consequently, placing a sample under study in close proximity of such an aperture would result in excitation of a spot with a diameter considerably smaller than the light wavelength. Moreover, moving the aperture and sample parallel to each other enables scanning the sample surface with a very high spatial resolution.

Realization of this idea had to wait for decades due to technical problems. In 1972, the idea was implemented in practical sub-wavelength microscopy in the microwave region ($\lambda = 3$ cm) [11]. The breakthrough in application of the SNOM technique in the optical region, which requires the aperture diameter and probe–sample distance in the sub-micrometer range, occurred in the mid-1980s and has been led by two research groups at Cornel University, USA, and IBM Zurich, Switzerland.

In parallel with the development of experimental capabilities, the theory of the field distribution in the near field of the aperture has been developed. The distribution of electromagnetic field on the opposite side of a circular sub-wavelength-size aperture in a perfectly conducting screen of infinitesimally small thickness and infinite lateral extend, which is illuminated by a plane wave, was calculated by solving Maxwell's equations by Bethe [12] and Bouwkamp [13, 14]. These calculations revealed that the light throughput through the aperture scales down as the sixth power of the aperture diameter, and the light polarization is strongly affected by the aperture. Calculations according to the aperture models, which take into account the real geometry of the SNOM probes, become quite difficult (see, e.g., [15] for review). Moreover, real metals have a finite skin depth (e.g., ~ 13 nm at 500-nm wavelength for aluminum), so the field strength decays within this distance into the metal. It has been claimed that the penetration of the field into the metal limits the SNOM resolution to ~ 12 nm (i.e., to $\sim \lambda/43$) [16], though further experiments in microwave range showed that the SNOM resolution can be improved down to the aperture diameter in spite of the skin depth being larger than the diameter [17].

In addition, the sample may significantly influence the field near the probe. Thus, the properties of the samples, especially of those containing areas with strong contrast in conductivity, change excitation conditions and might significantly distort the SNOM image, as calculated, e.g., in [18].

The sub-wavelength-size aperture in a conducting screen can serve well as a simplistic model for the SNOM probe; however, it is inconvenient for both scanning and collecting the luminescence light. Thus, tapered tips with an aperture on its apex are practically used instead. In early SNOM experiments, glass pipettes and etched quartz crystals have been used as probes (see, e.g., [19] for review). Nowadays, majority of SNOM systems use tapered single-mode optical fibers. The tips of the tapered fibers are coated with metal (usually with aluminum) to prevent escaping of light when the fiber diameter approaches the waveguide mode cutoff (at $\sim 0.4\lambda$). These probes gained popularity for comparatively high efficiency in delivering light to the aperture and for the relative simplicity of their production. Currently, most probes for SNOM in visible are produced from fused-silica single-mode optical fibers. Tapering the fiber is usually performed by heating it above the melting temperature under CO_2 laser irradiation in commercially available pipette pullers. Microprocessor-controlled technological parameters during this process enable one to obtain a rather high reproducibility of geometry and optical and mechanical properties of the tip. As mentioned above, the throughput of the probe decreases as the sixth power of the aperture diameter. At fixed aperture size, the throughput can be enhanced by increasing steepness of the final taper (converging angles between 30° and 70° are used) as well as by optimizing the shape of the taper [20]. The tapered fiber probes can also be produced using chemical etching (see, e.g., [21]) and nanophotolithography [22]. Hollow pyramid-shaped silicon probes and transparent silicon carbide probes [23] can also be used in SNOM.

Another approach to use the near field in spatially resolved spectroscopy of semiconductors is aperture-less SNOM. In this case, a metal probe in close proximity of the sample surface is used. The tip and the sample area under the tip are illuminated by a light source in the far field. The light induces electron oscillations in the probe tip, and the light field directly below the tip is enhanced by many orders of magnitude in respect to the field outside the tip area. Thus, the sample is effectively excited only directly below the metal tip. Consequently, the excitation spot size depends mainly on the tip diameter. As a result, aperture-less probes enable smaller excitation spots and ensure a higher spatial resolution than that in the SNOM systems with probes containing apertures.

Though being a simpler system, the aperture-less device has certain technical and physical drawbacks. First of all, the signal in these devices is collected in the far field. Thus, a large part of the luminescence from areas outside the spot directly under the tip is also collected into the detector. Though excited at considerably lower excitation power density, this luminescence has a significant contribution due to the large sample surface area, which is illuminated in the far field. Thus, the luminescence under study, which is emitted from the area directly below the tip, is distorted by the luminescence collected from a broad area around the tip. This effect is strongly reduced by using two-photon excitation, which enhances the excitation intensity dependence of the density of photoexcited electron–hole pairs and, consequently, diminishes the influence of the undesirable luminescence component. Another important physical drawback of the aperture-less device is a danger that the close proximity of the metal tip has a significant influence on properties of the object under study. For example, the surface plasmons in the tip might enhance the radiative recombination rate in the semiconductor under study. This effect is desirable in light-emitting devices and is exploited by introduction of metal nanoparticles in the vicinity of InGaN quantum wells [24, 25]. However, the effect distorts the measurement results in aperture-less SNOM devices. In general, the aperture-less SNOM provides opportunities for reaching a higher spatial resolution, but a closer study of the tip–sample system is often necessary to correctly exploit the system and interpret the experimental results. This type of SNOM is currently under intense study and development. Theoretical calculations of the field distribution under the probe and implementation of differently shaped probes are being carried out.

The SNOM technique is suitable for the study of surface plasmon excitations occurring in metallic nanoparticles [26] and might turn out to be an informative tool in plasmonics – a field of studying localization and guiding of electromagnetic energy at the interfaces between metals and dielectrics or semiconductors [27, 28]. The probe can be also exploited in a scattering mode when the signal from the area under the tip is obtained at the wavelength of the incident light [29].

Spatial resolution in SNOM devices containing apertures is basically determined by the aperture diameter. However, the probe transmission efficiency decreases approximately as the fourth [30] or sixth [12–14] power of the aperture diameter. Thus, a trade-off between the spatial resolution and the intensity of luminescence available for detection is important. Though resolution of $\lambda/40$ was demonstrated two decades ago [31], majority of the SNOM systems currently in use exploit apertures of ~ 100 nm (corresponding to resolution of $\sim \lambda/5$).

Deviations in aperture diameters of different tips, roughness of the aperture edges or metallic coating of the tip, and the slight damage gradually or abruptly occurring to the tip during experiment deteriorate the reproducibility of the experimental results and are considerable problems in SNOM applications. Thus, improvement of the probe design and fabrication is still a serious challenge for producers of SNOM equipment.

A SNOM device basically contains a probe, a feedback mechanism for probe positioning, a piezoelectric sample stage, a light source, and a detector. A simplified outline of the main SNOM device configuration modes is presented in Fig. 7.6.

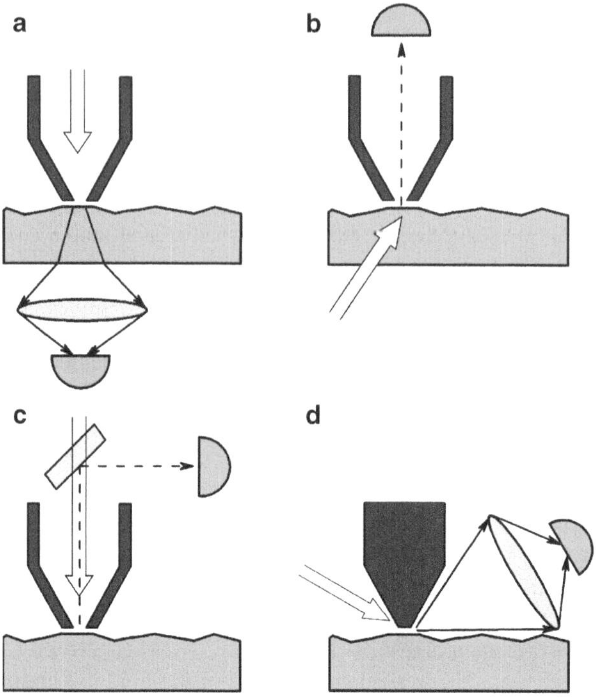

Fig. 7.6 Operation modes of a scanning near-field optical microscope: (**a**) illumination, (**b**) collection, (**c**) illumination-collection, (**d**) aperture-less probe. *White and narrow arrows* depict excitation and collected luminescence light, respectively

In illumination (I) mode, the sample is photoexcited through the probe in the near field, and the luminescence is detected in the far field (Fig. 7.6a). In collection mode (C), the sample surface is excited in the far field, while the luminescence light is collected in the near field via the probe (Fig. 7.6b). In illumination–collection (I–C) mode, the sample is excited and the luminescence is collected through the probe (Fig. 7.6c). In SNOM devices with aperture-less probe (Fig. 7.6d), the sample surface is illuminated and the luminescence signal is collected in the far field; however, the excitation light field is strongly enhanced only directly under the probe, thus enabling prevalence of the PL signal from the small spot under the probe in the signal detected. Selection of the configuration depends, first of all, on sample transparency.

Benefits of the SNOM images taken in I–C and I modes are illustrated in Fig. 7.7, where the results of a study of an InGaN single quantum well (SQW) [32] are presented. In the areas exhibiting longer diffusion lengths (encircled by solid white lines), the PL intensity in I–C mode is weaker than that in I mode, since carriers diffuse beyond the probe aperture into the areas with lower effective band gap, where an intense emission is observed in I mode. Meanwhile in the areas where the diffusion length is shorter (encircled by dashed white lines), a larger part of the photoexcited carriers recombine non-radiatively before reaching the areas with the lower band gap. The emission spectra in Fig. 7.7c support this conclusion.

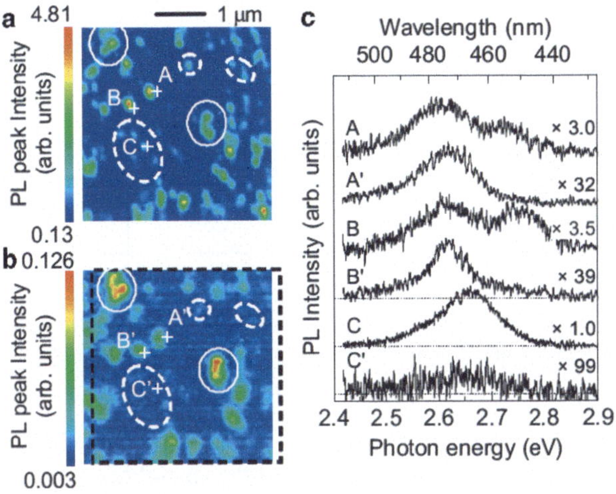

Fig. 7.7 SNOM images of an InGaN single quantum well in the I–C mode (**a**) and the I mode (**b**), and PL spectra at spots labeled in the images (**c**). Reproduced from [32]

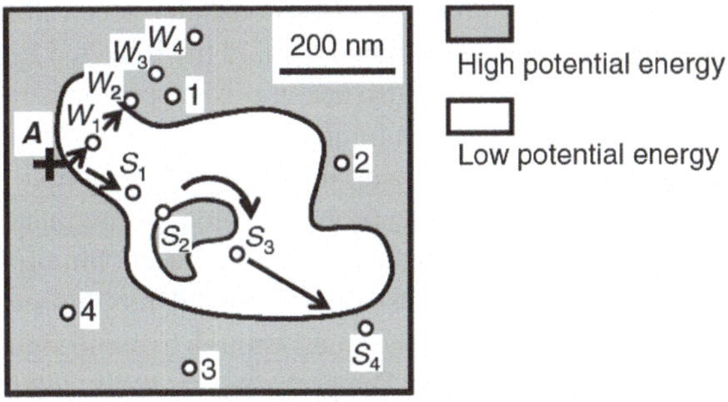

Fig. 7.8 Schematic image of the potential profile and carrier diffusion routes elucidated using a dual-probe SNOM system with one probe fixed at point A and the second probe at different positions. The *shaded and white* domains correspond to high and low potential energy regions, respectively. Reproduced from [33]

The carrier diffusion in InGaN has been recently studied by a dual-probe SNOM system independently controlling the distances between probe and sample and between the probes in real time [33]. This technique enables reconstruction of the potential profile as illustrated for an InGaN SQW in Fig. 7.8.

To keep the probe tip in close proximity of the sample surface, an accurate position-sensitive feedback has to be established. Feedback methods and instrumentation employed in SNOM are similar to those used in other scanning probe microscopes, including the currently common atomic force microscopes (AFM) with the most mature feedback systems. Two methods for sensing the probe–sample distance are basically used in SNOM: shear-force and tapping mode feedbacks.

In the shear-force feedback, the probe is dithered (vibrated) parallel to the sample surface. Piezoelectric devices are used to provide the driving force to dither the

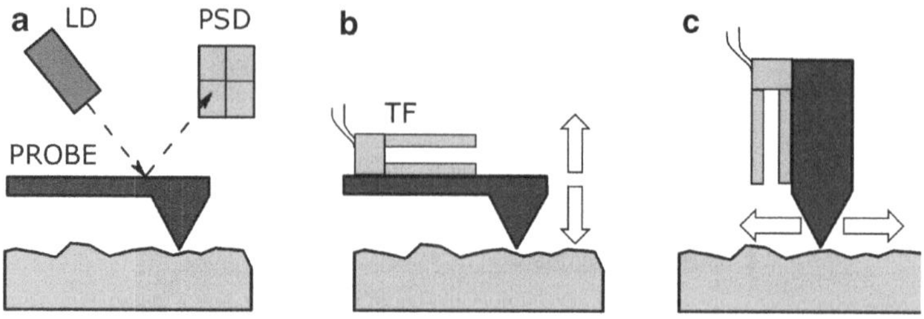

Fig. 7.9 Types of topography feedback used in SNOM systems: (**a**) laser-bounce feedback with position-sensitive photodiode (PSD), (**b**) tuning fork (TF) feedback with normal force, and (**c**) tuning fork feedback with shear force

probe at one of its mechanical resonances. The dither amplitude is usually maintained below 10 nm to prevent deterioration of the spatial resolution. Lateral forces of interaction between the probe and the sample change the amplitude and phase of the probe vibrations. These changes serve as a measure of the probe–sample proximity and are employed for maintaining the constant distance of the probe to the sample surface. Detection of the dither characteristics is usually accomplished by optical means: a tiny laser beam is focused on the probe and the reflected light is employed for the detection. The detection can be also accomplished piezoelectrically, by using a tuning fork (see Fig. 7.9). The shift in tuning fork resonance frequency or quality coefficient serves as an indication of the tip proximity to the sample surface.

In the tapping mode feedback, the probe position is modulated in direction normal to the sample surface. Piezoelectrically activated bimorphs (cantilevers consisting of two active layers) are usually employed for this purpose, and the modulation amplitude is maintained to be small enough to avoid significant changes in the near field on the sample surface. The probe makes weak, intermittent contact with the sample surface. The vertical movement of the probe is driven at a mechanical resonance and is monitored optoelectronically by using a laser beam, which is reflected from the probe, or piezoelectrically by using a tuning fork. The forces employed in the tapping mode (typically from nanonewtons down to piconewtons) are smaller than those used in the shear-force feedback mode. Moreover, the tapping can be used for additional measurements of mechanical properties of the sample or for intentional mechanical modification of the sample surface.

Since transmittance of the tapered waveguide is usually as low as 10^{-8} to 10^{-6}, laser sources are usually employed for excitation. Though high excitation intensity is favorable for better signal-to-noise ratio in SNOM experiments, the intensity of the light delivered to the probe should be kept low enough to avoid significant probe heating, which can cause reversible or even irreversible deterioration of the device operation. The external lasers are usually coupled to the SNOM system via optical fibers. Excitation wavelength is basically limited only by transparency of the optics used.

Single photomultiplier tubes (PMT) or photodiodes (PD) can be used for detection of the spectrally integrated luminescence, while spectrometers equipped with PMT, PD, or coupled charge device (CCD) cameras can be employed for measuring luminescence spectra. Small sample volume probed in SNOM experiments results in weak optical signals to be detected and degrades the accuracy of the spectroscopic results. Nevertheless, pulsed laser excitation and luminescence detection in time-resolved mode by exploiting streak camera are also used in SNOM devices [34]. Transient lens technique [35] exploiting focusing and defocusing of a CW probe beam by a lens created in the sample by changes in the refractive index due to the spatially distributed nonequilibrium carrier density generated by a pulsed pump beam has been also exploited for time-resolved PL measurement with a twin SNOM system [36]. In this system, both beams were coupled into a metal-coated optical fiber tip with the aperture size of 200 nm. The photoluminescence and the transient lens signals were separated by a dichroic mirror and a colored glass filter. They were detected with a PMT and a highly sensitive photodetector, respectively, using a lock-in amplifier system.

Scanning of the sample surface is accomplished at a fixed probe position by using a piezoelectric sample stage to shift the sample laterally. Commercial SNOM devices are usually designed for use at room temperature. However, a higher quantum yield, lower emission bandwidths, and the ability to influence the carrier dynamics by variable temperature encourage attempts to carry out SNOM experiments at low temperatures. The most straightforward approach is to cool down the entire microscope setup, including sample holder, near-field probe, piezo-positioning stage, and positioning control system, by using gaseous or liquid helium [37]. This approach has significant technical difficulties. Meanwhile, cooling the sample by attaching it to the cold finger of a helium flow cryostat, while keeping the rest of the system in the vacuum chamber [38], is another approach viable up to the recent SNOM applications [39].

Capability of SNOM technique to map the PL spatial distribution with high resolution is indispensable for study of materials with content fluctuations or/and inhomogeneous defect and strain distribution. Figure 7.10 illustrates that application of SNOM with a 30-nm aperture probe enables observation of PL bands in InGaN single quantum well structures, which are narrower than the spatially integrated

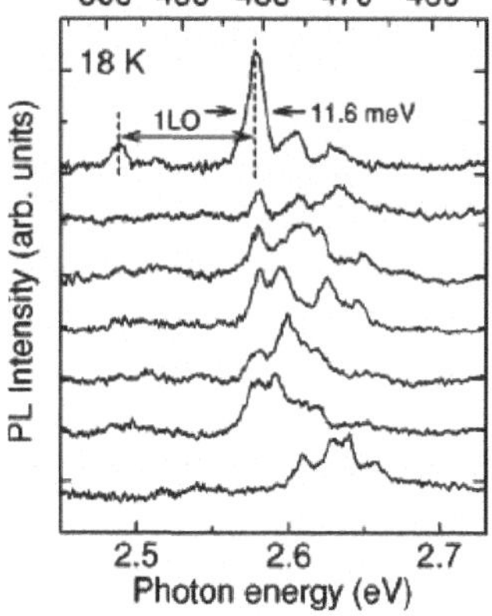

Fig. 7.10 PL spectra obtained using SNOM with 30-nm aperture probe at different positions of an InGaN single quantum well structure at 18 K. Reproduced from [34]

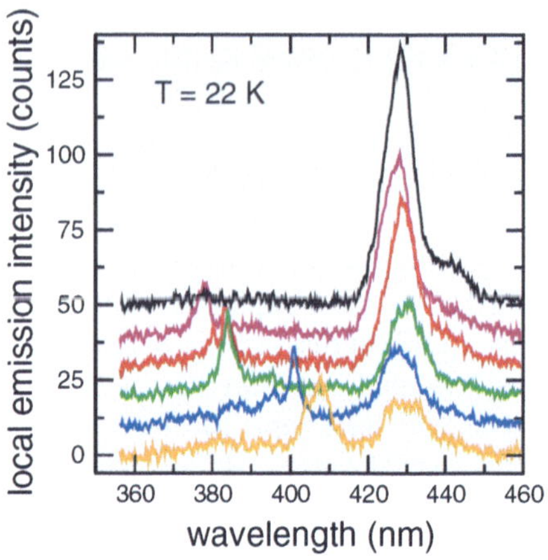

Fig. 7.11 Low-temperature PL spectra obtained using SNOM from nanometer-size areas at different locations close to a threading dislocation. Reproduced from [40]

band by a factor of five [34]. SNOM studies of high-efficiency GaInN/GaN quantum wells revealed an additional high-energy PL band caused by emission from the narrow band-gap sidewall regions of a V-shaped pit formed by a dislocation crossing the quantum well (see Fig. 7.11). This observation provided a sound evidence supporting the assumption that such V-shaped pits prevent non-equilibrium carriers from non-radiative recombination at the dislocations [40].

7.5 Cathodoluminescence

Cathodoluminescence is the process of light emission caused by the interaction of an electron beam with a material. Scattering of the high-energy primary electrons having energies of 0.1–30 keV creates backscattered electrons, secondary electrons, Auger electrons (when an electron in the excited atom relaxes from an outer shell to the inner shell by transferring its energy to another electron), X-rays (due to transition of an electron from the higher state down to the vacancy in the lower electron shell), and visible, UV, or IR light. In semiconductors, the high-energy electrons excite electrons from the valence band to the conduction band. The effective ionization energy E_i, i.e., the average energy required to generate one electron–hole pair, depends on the band gap E_g as $E_i = 2.8E_g + M$, where M is a material-dependent constant ranging from 0 to 1 eV [41]. The number of electron–hole pairs generated per one primary electron G depends on E_i and acceleration energy E_{acc}: $G = (1-\gamma)(E_{acc}/E_i)$. Here, γ is the fraction of energy scattered back or transmitted through the sample under excitation.

After relaxing to the bottom of the conduction band by a cascade emission of electron and phonon excitations, the excited electrons can radiatively recombine with holes left in the valence band and emit a photon with energy, which is close to that of the band gap. The mechanisms of radiative recombination (band-to-band, excitonic, and defect-related) and non-radiative recombination are the same for excitation by light or by electron beam. Just the excess energy, i.e., the initial energy of the excited electron (hole) with respect to the bottom of the conduction (valence) band, is considerably larger under the electron beam excitation. Consequently, heating of the lattice or the electron–hole system at the same density of the excited nonequilibrium electron–hole pairs is more probable in CL than in PL experiments.

CL spectroscopy is usually performed using an electron beam produced by a scanning electron microscope (SEM) or a scanning transmission electron microscope (STEM). The electron beam of adjustable current and energy is provided by the electron gun. In the gun, electrons are emitted by an electron source and accelerated toward the anode (accelerating voltage of $\sim 5\,\mathrm{kV}$ and beam current of $\sim 200\,\mathrm{pA}$ are typical values for CL experiments). The majority of the classic electron guns consist of a tungsten filament bent in the shape of a V-shaped hairpin and heated to $\sim 2,700\,\mathrm{K}$. Other types of electron guns are also used and exhibit longer lifetimes. The emission area of the V-shaped hairpin tip is typically of $\sim 100\,\mu\mathrm{m}$. The converging beam is focused and the first crossover in the electron gun is demagnified on the sample surface using apertures and electron lenses. A rotationally symmetric electromagnetic lens consists of a coil of wire inside an iron pole piece. The magnetic field is weaker in the center of the gap and forms a bell-shaped field distribution focusing the electron beam via Lorentz force, changing its magnitude and direction along the bore radius. The diameter of the focused electron beam is limited not by physical constraints, but rather by aberrations of electron lenses. The real volume excited in semiconductor usually exceeds the size of the focus spot on the semiconductor sample due to elastic and inelastic scattering and spreading of primary and secondary electrons (see Sect. 7.6).

Since the scanning is performed by manipulating the electron beam while the sample is maintained at a fixed position, it is comparatively easy to equip the CL systems with a cold stage to perform measurements at variable temperatures down to the liquid helium temperature (4.2 K) when using a flow cryosystem, or 8 K when a closed-cycle helium cryosystem is used.

The luminescence of the sample under electron beam excitation is collected and analyzed using the same equipment as in PL.

CL images of InGaN epilayers and QWs (see Fig. 7.12) served as the first direct evidence of exciton localization in this material [42].

In CL experiments performed using SEM or STEM equipment, there is a possibility to complement the results obtained by CL spectroscopy with the information provided by SEM or STEM techniques. SEM can provide the topographical information by detection of secondary electrons, the atomic number contrast revealed by detection of backscattered electrons, or the elemental analysis obtained using energy-dispersive X-ray spectroscopy (denoted by three synonymous acronyms EDS, EDX, or EDXS). An example of matching information obtained using CL and SEM techniques is presented in Fig. 7.13, where CL mapping images of a single InGaN pyramid are displayed. STEM facilitates visualization of dislocations; determination of Burger's vector; direct lattice imaging; getting information on

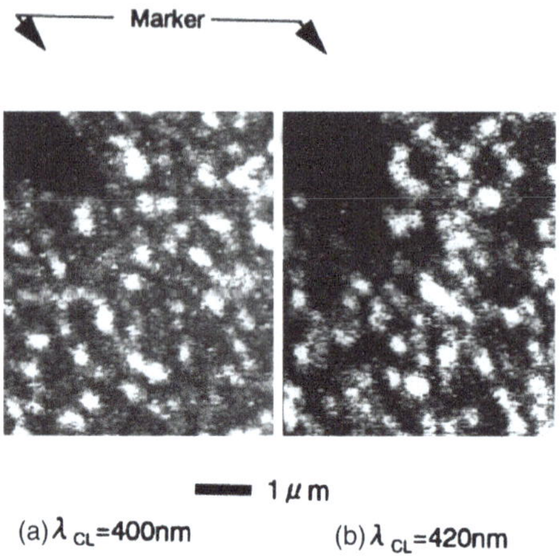

Fig. 7.12 Monochromated scanning CL images of an $In_{0.2}Ga_{0.8}N$ SQW at wavelengths of 400 nm (**a**) and 420 nm (**b**). Reproduced from [42]

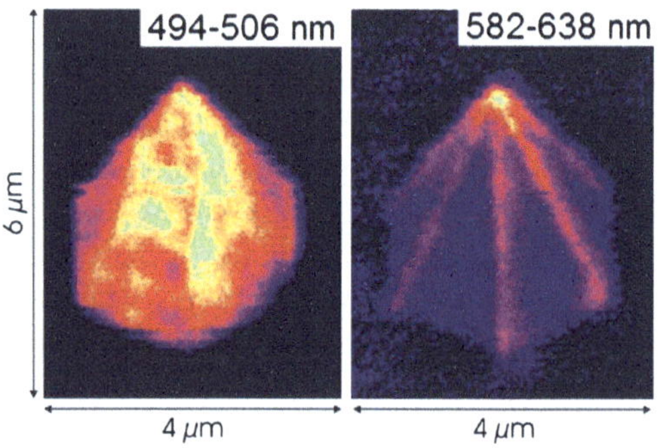

Fig. 7.13 CL intensity mapping of a single InGaN pyramid for two different wavelength ranges. Reproduced from [43]

atomic composition, chemical bonding, valence and conduction band electronic properties, surface properties and enables determination of element-specific pair distance distribution functions by using electron energy loss spectroscopy (EELS).

High spatial resolution in CL spectroscopy can be combined with time resolution. The first attempts were based on the fast deflection of the electron beam away from the spot under study using deflection plates [44]. This technique suffered undesirable sweeping of the sample by the electron beam and low time resolution limited by a few hundred picoseconds. Currently, beam blankers are successfully used in experiments where no high time resolution is necessary. The study of donor–acceptor pair recombination in Mg-doped GaN by using 100-ns pulses produced by a beam blanker operating at a frequency of 100 kHz can serve as an example [45].

A high time resolution of 10 ps at 50-nm spatial resolution was achieved using a different approach: an optically driven electron gun [46]. The picosecond CL setup is depicted in Fig. 7.14. The usual hairpin tungsten electron gun is replaced by a 20-nm-thick gold photocathode deposited on a quartz plate. The pulsed electron beam is generated by illuminating the photocathode with 200-fs pulses of a UV (266-nm) mode-locked laser. The pulsed electron beam is bright enough to record the surface

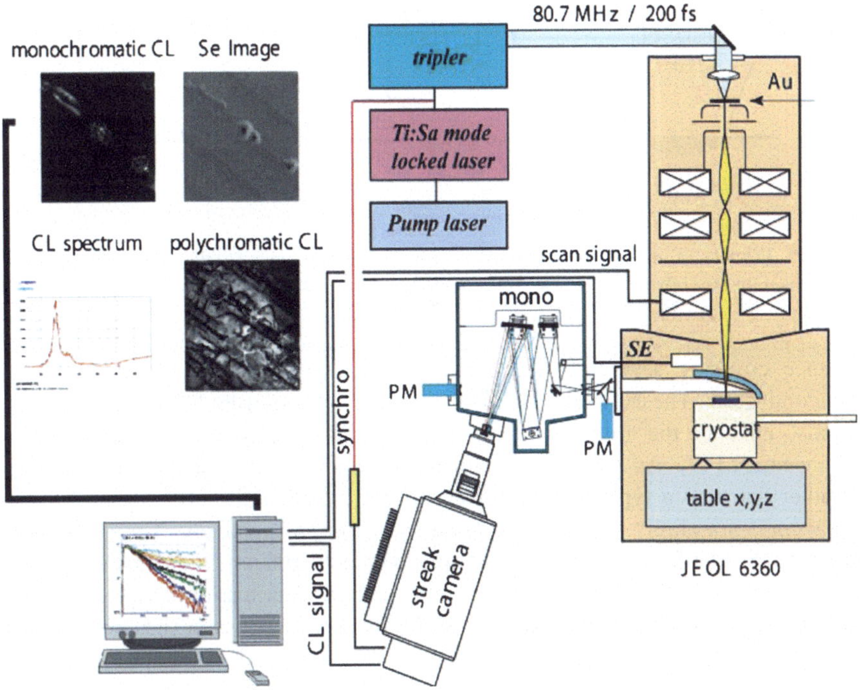

Fig. 7.14 Picosecond cathodoluminescence setup with examples of secondary electron and CL images. Reproduced from [47]

morphology by detection of secondary emission and to use a streak camera for time-resolved luminescence spectroscopy.

7.6 Excitation of Luminescence in Semiconductors

To interpret correctly the results in spatially resolved luminescence experiments, the real spatial distribution of excitation in the sample has to be taken into account. The excited volume might be influenced by light or electron beam absorption, diffusion of photoexcited carriers, photon recycling, phonon wind, Fermi pressure, etc.

In the simplest model of semiconductors, the coefficient of the linear optical absorption is zero below the band gap and increases in proportion to the density of states above the band gap. Simultaneous absorption of phonons enables absorption of photons with energies below the band gap and results in the Urbach tail in the semiconductor absorption [48], i.e., in an approximately exponential decrease of the absorption coefficient below the band gap. This tail can also be influenced by impurities. In addition, Coulomb interaction between electron and hole results in the formation of excitons. Wannier–Mott excitons (i.e., excitons with hole–electron

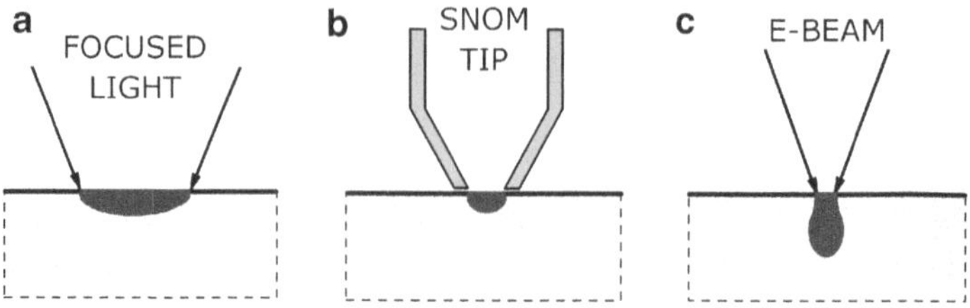

Fig. 7.15 Primary excitation volume in (**a**) microphotoluminescence (μ-PL), (**b**) scanning near-field optical microscopy (SNOM), and (**c**) cathodoluminescence (CL) experiments

distance covering several lattice constants) are typically formed in inorganic semiconductors. The excitons not only have a discrete spectrum below the band gap but also enhance the above-band-gap absorption due to the Coulomb interaction enhancement [49]. As a result, the absorption spectrum of α above the band gap is rather flat, and a typical value for absorption coefficient is 10^4–$10^5\,\mathrm{cm}^{-1}$ and 1–$10^3\,\mathrm{cm}^{-1}$ for direct and indirect semiconductors, respectively. The absorption length $l_\alpha = 1/\alpha$ for $\alpha = 10^5\,\mathrm{cm}^{-1}$ is approximately 100 nm. Thus, the excitation volume at band-to-band excitation in μ-PL experiments has the shape of a disc with a diameter equal to the spot size of the focused light beam (usually not less than $\sim 1\,\mu$m, though the diffraction limit is ~ 200 nm), and a height of ~ 100 nm (see Fig. 7.15a). Semiconductor excitation by light with photon energy significantly lower than the band gap is also possible, though considerably less probable, due to two-photon absorption (via virtual states in the band gap) and two-step absorption (via real defect-related states in the band gap), but is rarely used in spatially resolved PL study, since the absorption length, which is excitation power density dependent, is of the order of centimeters and more.

In SNOM experiments, the size of the excited spot approximately equals the diameter of the tip aperture, both in parallel and perpendicular directions to the surface (see Fig. 7.15b).

In CL experiments, the electrons impinging the sample encounter elastic and inelastic scattering and spread in the volume, which is considerably larger in size than that of the electron beam focus spot on the surface. The shape of the excited volume in this case is sketched in Fig. 7.15c. The depth of the excited area depends on the incident electron energy and might vary from hundreds of nanometers to several micrometers.

Excitation creates strong carrier density gradients at the boundaries of the excited volume. Consequently, the photoexcited carriers might spread beyond the excited volume via diffusion. The diffusion length L_D depends on the diffusion coefficient D and the carrier lifetime τ as $L_D = \sqrt{D\tau}$. At low density of nonequilibrium carriers, the diffusion is governed by minority carriers. Meanwhile, at high excitation power densities, when the density of nonequilibrium carriers exceeds that of the equilibrium carriers, the densities of nonequilibrium electrons and holes are equal, and ambipolar diffusion takes place. In the vast majority of inorganic semiconductors, the diffusion coefficient of electrons is considerably larger than that of holes. However, the diffusion of electrons is inhibited via Coulomb interaction by holes dragging behind. The ambipolar diffusion coefficient can be expressed as a function of electron and hole mobilities μ_{n} and μ_{p}, respectively:

$$D_{\mathrm{a}} = \frac{2\mu_{\mathrm{n}}\mu_{\mathrm{p}}}{\mu_{\mathrm{n}} + \mu_{\mathrm{p}}} \frac{kT}{q}. \tag{7.8}$$

If $\mu_{\mathrm{n}} \gg \mu_{\mathrm{p}}$ (as usual for most inorganic semiconductors), $D_{\mathrm{a}} \approx 2D_{\mathrm{p}}$, where D_{p} is the hole diffusion coefficient related to the hole mobility through the Einstein relation. A typical value for ambipolar diffusion coefficient of GaN at room temperature is $\sim 2\,\mathrm{cm}^2/\mathrm{s}$, which, for $\tau = 2\,\mathrm{ns}$, results in $L_D \approx 0.6\,\mu\mathrm{m}$ [50]. Similar diffusion lengths have been observed in other direct band-gap semiconductors (see review in [51]). The diffusion length depends on temperature and can be strongly modified by carrier localization. Giant ambipolar diffusion coefficients exceeding the usual values by orders of magnitude can occur in certain semiconductor structures [52]. The carrier lifetime in indirect band-gap semiconductors is considerably larger than that in direct band-gap semiconductors, and the diffusion length in semiconductors such as Si and SiC can be of the order of millimeters.

The volume containing the nonequilibrium carriers can considerably exceed that of the direct photoexcitation due to photon recycling. A photon emitted via electron–hole recombination can be absorbed outside the volume of direct photoexcitation, thus causing extension of the excited volume. The importance of this effect depends on recombination mechanisms, the steepness of absorption edge, etc., and might be rather high in semiconductors and structures with significant absorption involving defects or localized states. In addition to the increased spatial spreading of excitation, the photon recycling results in a red shift of the emission band due to more effective reabsorption at the short-wavelength side of the band and in an increase of carrier lifetimes with respect to those expected without recycling. Note that the photon recycling might be important even for semiconductor quantum dots and nanocrystals, where the excitation volume is restricted by the natural boundaries of the nanostructure, but reabsorption in the neighboring nanostructures is possible. The photon recycling, which increases the role of non-radiative recombination, is very important in light extraction out of light-emitting diode (LED) structures. Ray tracing calculations to take this effect into account are currently a routine procedure in LED structure design (SimuLED from STR group or LightTools by ORA are examples of the corresponding software). The photon recycling is also important for light conversion in photovoltaic cells.

The excitation of high density of electron–hole pairs in semiconductor materials might trigger additional mechanisms of spatial expansion of the photoexcited carriers. Phonon wind [53] at intense photoexcitation of a thin semiconductor surface layer by photons with large excess energy can cause a significant drag of photoexcited electron–hole pairs out from the surface and, consequently, increase the penetration depth of nonequilibrium carriers. Phonons emitted by the photoexcited carriers cascading down to the band bottom propagate predominantly into the depth of the sample. Interaction of these phonons with electrons and holes provides the nonequilibrium carriers with momentum directed outward the excited volume and increases their penetration depth. Such a phonon wind-based transport has been observed, e.g., in InGaAs-InP quantum well by using the μ-PL technique [54]. At excitation of dense electron–hole plasma above a critical value corresponding to the minimum of free energy per electron–hole pair, the Fermi pressure [55] can cause spreading of the plasma at velocities exceeding the equilibrium diffusion velocity by orders of magnitude [56].

Good understanding of a semiconductor photoresponse can serve for pushing the spatial resolution well beyond the diffraction limit. In continuous luminescent objects, there are two fundamental requirements for sub-diffraction resolution in the far field: (1) spatially nonuniform illumination and (2) a nonlinear photoresponse [57]. Sub-diffraction resolution can also be achieved in studies of single fluorescence point sources such as single molecule labels or semiconductor quantum dots [58]. Digital image processing enables considerably higher determination accuracy for the "peak position" than for the extent of a single emitter [59].

7.7 Comparison of Techniques for Spatially Resolved Spectroscopy of Semiconductors

The major advantages and disadvantages of the main techniques for spatially resolved PL of semiconductors and semiconductor structures can be shortly summarized as follows.

7.7.1 *Micro-photoluminescence (μ-PL)*

Advantages: The simplest and least expensive technique for spatially resolved PL study.
Disadvantages: Comparatively low spatial resolution. No capability for matching of PL mapping with surface morphology or mapping of other material parameters.

7.7.2 *Scanning Confocal Microscopy*

Advantages: Scanning capability in all three dimensions, in contrast to all other techniques discussed in this chapter. Flexibility in choosing photon energy for excitation when compared with electron beam excitation in CL.
Disadvantages: Lower spatial resolution in comparison with SNOM and CL, though higher than that in μ-PL. No capability for matching of PL mapping with surface morphology or mapping of other material parameters.

7.7.3 Scanning Near-Field Optical Microscopy

Advantages: High spatial resolution. Capability of simultaneous mapping of surface morphology and PL characteristics.
Disadvantages: Small detected signal due to low light throughput of the probe. Problems with repeatability due to variations in geometrical parameters of intentionally identical probes and probe wear during experiment.

7.7.4 Cathodoluminescence

Advantages: Possibility to complement the results obtained by CL spectroscopy with data on SEM or STEM mapping. Capability of the electron beam to excite semiconductors with any band gap including wide-band-gap semiconductors such as high-Al-content AlGaN, AlN, BN, and diamond.
Disadvantages: Dissipation of large amount of energy accumulated in the beam of fast electrons in a small sample volume that enhances the danger of reversible sample heating or heat-induced damage.

References

1. T.R. Corle, G.S. Kino, *Confocal Scanning Optical Mocroscopy and Related Imaging Systems*, p. 39 (Academic, San Diego, 1996)
2. J.B. Pawley, *Handbook of BiologicalConfocal Microscopy,*2nd edn. (Plenum, New York, 1995)
3. D.B. Murphy, *Fundamentals of Light Microscopy and Electronic Imaging* (Wiley, New York, 2001)
4. R. Pike, D. Chana, P. Neocleous, Sh. Jiang, in *Optical Imaging and Microscopy*, ed. by P. Török, F.-J Kao (Springer, Heidelberg, 2007)
5. P.L. Lu, in *Handbook of Microscopy for Nanotechnology*, ed. by N. Yao, Z.L. Wang (Kluwer, Boston, 2005)
6. S. Hell, E.H.K. Stelzer, J. Opt. Soc. Am. A, **9**, 2159 (1992)
7. K. Okamoto, A. Kaneta, Y. Kawakami, S. Fujita, J. Choi, M. Terazima, T. Mukai, J. Appl. Phys. **98**, 064503 (2005)
8. D. Dobrovolskas, J. Mickevičius, E. Kuokštis, G. Tamulaitis, M. Shur, M. Shatalov, J. Yang, R. Gaska, J. Phys. D: Appl. Phys. **44** 135104 (2011)
9. E.H. Synge, Philos. Mag. **6**, 356 (1928)
10. E.H. Synge, Philos. Mag. **13**, 297 (1932)
11. E.A. Ash, G. Nicholls, Nature **237**, 510 (1972)
12. H.A. Bethe, Phys. Rev. **66**, 163 (1944)
13. C.J. Bouwkamp, Philips Res. Rep. **5**, 321 (1950)
14. C.J. Bouwkamp, Philips Res. Rep. **5**, 401 (1950)
15. L. Novotny, D.W. Pohl, in O. Marti, R. Möler (eds), *Photons and Local Probes* (Kluwer, The Netherlands, 1996)
16. E. Betzig, J.K. Trautmann, T.D. Harris, J.S. Weiner, R.L. Kostelak, Science **251**, 1468 (1991)
17. B. Knoll, F. Keilmann, A. Kramer, R. Guckenberger, Appl. Phys. Lett. **70**, 2667 (1997)

18. L. Novotny, D.W. Pohl, P. Regli, J. Opt. Soc. Am. A **11**, 1768 (1994)
19. D.W. Pohl, in *Advances in Optical and Electron Microscopy*, ed. by C.J.R. Sheppard, T. Mulvey (Academic, London, 1990)
20. G.A. Valaskovic, M. Holton, G.H. Morrison, Appl. Opt. **34**, 1215 (1995)
21. T. Saisaki, K. Matsuda, Appl. Phys. Lett. **74**, 2773 (1999)
22. T. Matsumoto, M. Ohtsu, J. Lightwave Technol. **14**, 2224 (1996)
23. N.F. van Hulst, M.H.P. Moers, O.F.J. Noordman, R.G. Tack, F.B. Segerink, B. Bölger, Appl. Phys. Lett. **62**, 461 (1993)
24. D.-M. Yeh, C.-F. Huang, C.-Y. Chen, Y.-C. Lu, C.C. Yang, Appl. Phys. Lett. **91**, 171103 (2007)
25. C.-F. Lu, C.-H. Liao, C.-Y. Chen, C. Hsieh, Y.-W. Kiang, C.C. Yang, Appl. Phys. Lett. **96**, 261104 (2010)
26. A. Klar, M. Perner, S. Grosse, G. von Plessen, W. Spirkl, J. Feldmann, Phys. Rev. Lett. **80**, 4249 (1998)
27. S.A. Maier, H.A. Atwater, Appl. Phys. Lett. **98**, 011101 (2005)
28. E. Ozbay, Science **311**, 189 (2006)
29. L. Novotny, S.J. Stranick, Annu. Rev. Phys. Chem. **57**, 303 (2006)
30. R.D. Grober, T.D. Harris, J.K. Trautman, E. Betzig, Rev. Sci. Instrum. **65**, 626 (1994)
31. E. Betzig, J.K. Trautman, T.D. Harris, J.S. Weiner, R.L. Kostelak, Science **251**, 1468 (1991)
32. A. Kaneta, M. Funato, Y. Kawakami, Phys. Rev. B **78**, 125317 (2008)
33. A. Kaneta, T. Hashimoto, K. Nishimura, M. Funato, Y. Kawakami, Appl. Phys. Express **3**, 102102 (2010)
34. A. Kaneta, K. Okamoto, Y. Kawakami, Sh. Fujita, G. Marutsuki, Y. Narukawa, T. Mukai, Appl. Phys. Lett. **81**, 4353 (2002)
35. K. Okamoto, K. Inoue, Y. Kawakami, Sh. Fujita, M. Terazima, A. Tsujimura, I. Kidoguchi, Rev. Sci. Instrum. **74**, 575 (2003)
36. K. Okamoto, A. Scherer, Y. Kawakami, Appl. Phys. Lett. **87**, 161104 (2005)
37. R.D. Grober, T.D. Harris, J.K. Trautman, E. Betzig, Rev. Sci. Instrum. **65**, 626 (1994)
38. G. Behme, A. Richter, M. Süptitz, Ch. Lienau, Rev. Sci. Instrum. **68**, 3458 (1997)
39. V. Liuolia, A. Pinos, S. Marcinkeviius, Y.D. Lin, H. Ohta, S.P. DenBaars, S. Nakamura, Appl. Phys. Lett. **97**, 151106 (2010)
40. A. Hangleiter, F. Hitzel, C. Netzel, D. Fuhrman, U. Rossow, G. Ade, P. Hinze, Phys. Rev. Lett. **95**, 127402 (2005)
41. B.G. Yacobi, D.B. Holt, *Cathodoluminescence Microscopy of Inorganic Solids* (Plenum, New York, 1990)
42. S. Chichibu, K. Wada, S. Nakamura, Appl. Phys. Lett., **71**, 2346 (1997)
43. V. Pérez-Solórzano, A. Gröning, M. Jetter, T. Riemann, J. Christen, Appl. Phys. Lett. **87**, 163121 (2005)
44. A. Steckenborn, H. Munzel, D. Bimberg, J. Lumin. **24–25**, 351 (1981)
45. M. Fischer, S. Srinivasan, F.A. Ponce, B. Monemar, F. Bertram, J. Christen, Appl. Phys. Lett. **93**, 151901 (2008)
46. M. Merano, S. Sonderegger, A. Crottini, S. Collin, P. Renucci, E. Pelucchi, A. Malko, M. H. Baier, E. Kapon, B. Deveaud, J.-D. Ganie‘re, Nature, **438**, 479 (2005)
47. M. Merano, S. Sonderegger, A. Crottini, S. Collin, E. Pelucchi, P. Renucci, A. Malko, M.H. Baier, E. Kapon, J.D. Gganière, B. Deveaud, Appl. Phys. B **84**, 343 (2006)
48. F. Urbach, Phys. Rev. **92**, 1324 (1953)
49. J. Nunnenkamp, J.H. Collet, J. Klebiczki, J. Kuhl, K. Ploog, Phys. Rev. B, **43**, 14047 (1991)
50. J. Mickevičius, M.S. Shur, R.S. Qhalid Fareed, J.P. Zhang, R. Gaska, G. Tamulaitis, Appl. Phys. Lett. **87**, 241918 (2005)
51. A. Gustafsson, M.-E. Pistol, L. Montelius, L. Samuelson, J. Appl. Phys. **84**, 1715 (1998)
52. K.H. Gulden, H.Lin, P. Kiesel, P. Riel, G.H. Dohler, K.J. Eberling, Phys. Rev. Lett. **66**, 373 (1991)
53. A.E. Bulatov, S.G. Tikhodeev, Phys. Rev. B, **46**, 15058 (1992)
54. F.G. Monte, S.W. Da Silva, J.M.R. Cruz, P.M. Morais, A.S. Chaves, Phys. Rev. B, **62**, 6924 (2000)

55. W.F. Brinkman, T.M. Rice, Phys. Rev. B **7**, 1508 (1973)
56. A. Cornet, M. Pugnet, J. Collet, T. Amand, M. Brousseau, J. Phys. Colloques **42** C7–471(1981)
57. R. Heintzmann, M.G.L. Gustafsson, Nat. Photonics **3**, 362 (2009)
58. K.A. Lidke, B. Riegel, T.M. Jovin, R. Heintzmann, Opt. Express **13**, 7052 (2005)
59. W.E. Moerner, Nat. Methods **3**, 781 (2006)

Chapter 8
Time-Resolved Optical Spectroscopy

Andrea Balocchi, Thierry Amand, and Xavier Marie

Abstract This chapter presents an overview of different complementary techniques used in ultra-fast time-resolved optical spectroscopy experiments in the picosecond and sub-picosecond time range, with particular emphasis on their application to the investigation of the optical and electronic properties of semiconductors.

8.1 Introduction

The development of lasers providing ultra-short pulses [1] has sparked the growth of a new and interdisciplinary research field, the time-resolved optical spectroscopy, investigating the optical phenomena occurring on picosecond to millisecond time scales. One of the main applications in semiconductor physics offered by the availability of such ultra-short lasers is the analysis of transient phenomena induced on a sample by the excitation light pulse. For example, after a laser pulse has excited charges in a semiconductor conduction band, the carriers can thermalize, diffuse, or recombine radiatively or not. Or, again, a light pulse can be used to impress a transient modification of the optical, electronic, or magnetic properties of a sample. It is of fundamental importance, not only to the semiconductor physics field, to understand these mechanisms whose temporal evolution brings an invaluable insight into the underlying physics. The role of the time-resolved optical spectroscopy is precisely to quantify the amount and the duration of these modifications.

Basically, two kind of analysis can give insight into the temporal processes occurring in a given physical system: the spectral approach and the temporal approach. In the former, the temporal evolutions are contained in the width of the

X. Marie (✉)
Physics Department - LPCNO, Université de Toulouse, INSA-CNRS-UPS, LPCNO, 135, Av. de Rangueil, 31077 Toulouse, France
e-mail: marie@insa-toulouse.fr

A. Patanè and N. Balkan (eds.), *Semiconductor Research*, Springer Series in Materials Science 150, DOI 10.1007/978-3-642-23351-7_8,

measured spectral line, provided the spectral resolution is high enough.[1] However, two strong limitations appear here: (1) the system under investigation may contain many micro-systems which in turn can be inhomogeneous, thus masking the homogeneous line broadening due to intrinsic processes. (2) The evolution may correspond to a nonlinear process making the interpretation of the linewidth difficult and non-univocal. On the contrary, time-resolved spectroscopy gives a direct access to the time evolution of a given system (through radiative recombinations or the time evolution of the complex index of refraction) provided the time resolution of the system is high enough. The latter is basically limited by the convolution of the excitation temporal width with the temporal response function of the detection. For instance, in time-resolved photoluminescence spectroscopy, the two limiting times are the excitation laser pulse duration τ_p and the time resolution of the detection τ_d, yielding an overall time resolution of the order of $\Delta t \approx \sqrt{\tau_\mathrm{p}^2 + \tau_\mathrm{d}^2}$. Thus fast excitation sources are required, and the detection strategy should be optimized with respect to the characteristic evolution time of the observed process. The counterpart of a high temporal resolution analysis is the loss of spectral selectivity in the excitation process. As a matter of facts, a short excitation pulse consists in a coherent superposition of many Fourier components expanding on a range of the order of $\Delta\omega_\mathrm{p} \geq 1/\tau_\mathrm{p}$, preventing the selective excitation of two optical transitions lying closer than $\Delta\omega_\mathrm{p}$ in the spectral domain. On the other side, coherent superposition of quantum states can be achieved in this case, and the temporal observation of subsequent beats between the two states yields additional information, provided that the beating period occurs within the temporal window of observation (for instance, the radiative lifetime in time-resolved photoluminescence) [2]. High spectral resolution can thus be achieved in the temporal domain.[2]

In this general context, ultra-fast lasers (providing picosecond or sub-picosecond pulses) are the sources of choice for ultimate temporal resolution spectroscopy. The availability of these light sources has consequently stimulated the advance of new measurement devices and experimental techniques capable of resolving ultra-fast light phenomena [1]. All-optical techniques or electro-optical methods have been developed and the purpose of this chapter is to present an overview of some of these techniques for time-resolved optical spectroscopy in the picosecond or sub-picosecond time range with emphasis, when possible, on III–V-N semiconductors when presenting application examples.

[1]The spectral resolution of a system is obtained by the convolution of the excitation spectral function with the spectral response function of the detection.

[2]The spectral resolution achieved is then $\Delta\omega \cong 1/\tau_\mathrm{obs}$, where τ_obs is the duration of the temporal window of observation.

8.2 Streak Cameras for Time-Resolved Photoluminescence Spectroscopy

A technologically complex, although very versatile and fast photoelectric instrument for the measurements of time-resolved ultra-fast phenomena is the streak camera [3, 4]. The streak camera is a multichannel device allowing for the simultaneous measurement of the dynamics of light events with picosecond or sub-picosecond resolution, as a function of either their wavelength components (if coupled to a spectrometer) or their spatial position (if coupled to an imaging optical system). The fundamental principle is the conversion of a light phenomenon temporal profile into a spatial one. This is achieved in several steps. The light signal is first converted by photoemission into electrons whose emission time is temporally resolved by spatial dispersion, and finally back-converted into photons for optical detection. The instrument name finds its origin from the early development of ultra-fast recording devices where a rotating drum would reflect light onto a photographic film, leaving an impression "streak", spatially reproducing the temporal event.

8.2.1 Working Principle

The basic principle of a streak camera can be summarized as follows (Fig. 8.1): the train of emitted light pulses to be analyzed is imaged, using a system of lenses, onto a photocathode where a slit, reducing the vertical extension of the beam, is used to limit the time resolution degradation (see Sect. 8.2.5). Light is converted into electrons in a vacuum tube, by photoelectric effect, spatially and temporally separated according to the arrival time and horizontal (spectral) position of the light pulses onto the photocathode. The resulting flux of electrons, an electronic replica of the light pulse, will then be focused, accelerated, and sent into a deflection region constituted by two deflection plates. A very fast sweeping voltage is applied in this region, vertically deflecting the traversing electrons proportionally to their arrival time. A temporal sequence is thus linearly converted into a spatial one. The horizontal profile, corresponding to different wavelengths (or positions), is left unchanged. The final process consists in the impact of the time-resolved electron flux onto a phosphorous screen where the signal is converted back to photons to be captured, typically, by a CCD camera. The result of the measurement is, therefore, a tri-dimensional image where the horizontal axis represents the space coordinate (wavelength or position), the vertical axis the time coordinate, and the signal intensity is recovered from the phosphorous screen intensity (usually converted into a color code by the CCD software). An electron amplifier, a micro-channel plate delivering up to 10^3 gain [5], is inserted before the phosphorous screen and after the temporal resolution has been performed in order to increase the brightness of the recorded image. This allows also the limitation of the space charge effects between the photoelectrons, minimizing repulsion effects hampering the temporal resolution.

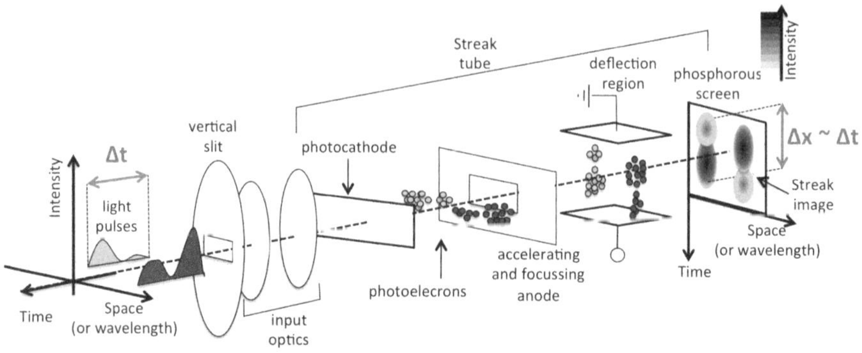

Fig. 8.1 Sketch of the working principle of a streak camera

Compared to fast photomultipliers or avalanche photodiodes, which can provide time resolutions down to a few tens of picoseconds, the key advantage of a streak camera is the recording of three-dimensional images thanks to its multichannel character.

The interpretation of a streak camera image produced on the phosphorous screen relies on three main aspects:

- The photon arrival time is related to the deflection angle imposed to the photoelectrons in the sweep stage and, therefore, to the photoelectrons' vertical impact position on the phosphorous screen.
- The intensity of the light pulse is related to the flux of photoelectrons created and, therefore, to the luminous intensity on the phosphorous screen.
- Different horizontal incident points of photons on the photocathode, reflecting different signal wavelengths or spatial positions, are replicated by the photoelectrons and, therefore, will appear at different horizontal positions on the phosphorous screen.

8.2.2 Synchronization and Sweep Methods

The key element performing the temporal resolution of a streak camera is the sweep stage in the deflection region. In order to detect transient phenomena correctly with high time resolution, two main different deflecting techniques, namely the "triggered mode" and "Synchroscan sweep", are employed depending on the type of measurement and laser used. A synchronization system (Fig. 8.2a) between the deflection stage and arrival time of the photoluminescence is obtained by sending a portion of the excitation pulse intensity usually to a PIN photodiode serving as a synchronization reference for the sweep voltage circuit. A delay stage is then used to adjust the trigger signal properly to the PL pulse arrival time.

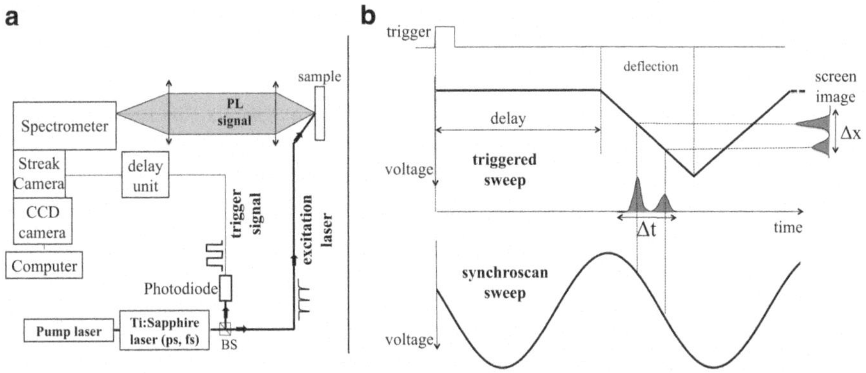

Fig. 8.2 (**a**) Experimental setup for time-resolved photoluminescence spectroscopy using a streak camera with a photodiode synchronization system (BS = beam splitter). (**b**) Principle of time operation of a streak camera in single sweep (*top*) and synchroscan (*bottom*) mode

Triggered mode. This operation mode [6, 7] is used with low repetition rate phenomena from single pulses (from where it derives its name) up to a few MHz and can cover temporal dynamics from a few tens of picoseconds to milliseconds. The sweep voltage (Fig. 8.2b) consists in a simple linear ramp and it is ideally suited for a single exposure. If multiple images have to be averaged, trigger jitter correction techniques are necessary to minimize the temporal resolution degradation. In the latter case, time resolutions of a few tens of picoseconds are typically achievable with picosecond systems.

Synchroscan sweep. This sweep method [4] takes advantage of the high repetition rate of mode-locked lasers (typically 80 MHz) to average multiple photoluminescence (PL) traces onto the same image. Very weak signals, un-recordable with triggered mode cameras, can be measured with high signal-to-noise ratio and high temporal resolution. Although the same electronic signal provided by the PIN diode is used, the synchronization technique is here fundamentally different. The electronic signal generated by a sequence of laser pulses detected by the PIN photodiode is filtered, and the frequency of the fundamental harmonic extracted in order to drive an oscillator and amplification circuit. The phase between the first harmonics of the photodiode signal and the streak local oscillator is locked using a phase comparator (Fig. 8.2b). For this reason, a very accurate tuning of the frequencies of the streak oscillator and the mode locked laser must be achieved before the experiment to bring them into resonance. The resulting high voltage sine wave is then applied to the sweeping region. Its central part is used to produce, with sufficient linearity, the deflection electrical field to convert the photoelectrons' temporal profile into a vertical spatial one for repetitive streak imaging. Phase-locking techniques are employed to minimize the electronic jitter and track the laser pulses dephasing with high precision. The same light signal issued from the same experiment repeated at the laser frequency is then accumulated on the phosphorous screen, each pulse arriving exactly at the same position as the previous one. Greater time resolution ($\sim$2 ps) is achievable for signal integration with the synchroscan compared to the triggered mode technique ($\sim$20 ps). However, short temporal windows (<2 ns) and a very narrow laser repetition frequency bandwidth, with respect to the designed one, are the constraints imposed by this high precision sweep technique. In addition to a single fast synchroscan vertical sweep, on certain systems it is possible to apply a slower perpendicular voltage sweep thanks to supplementary horizontal deflection plates, shifting the pulses in the horizontal direction [8]. Two temporal axes (dual time-base camera) with different time scales are available, allowing measuring the slow variation phenomena of a fast signal.

8.2.3 Measurement Methods

The measurement method described in the preceding section consists in the record of one or several consecutive images produced on the phosphorous screen. The individual light pulses are integrated and the average intensity of the phosphorous screen is then recorded. For this reason, this acquisition mode is called "analog integration." As stray light and electronic noise of different origins add up to each image, a background subtraction of the raw data is unavoidable to recover the "real" dynamics. This operation allows, in state-of-the-art systems, the recording of signal with $\sim$200 fs time resolution in single sweep mode and a dynamic range of about 10^3, the latter mainly limited by phosphorous screen saturation phenomena.

Thanks to the high sensitivity of streak cameras, the system can be operated also in "photon-counting" mode. This technique allows for the measurements of single photon events and is well suited for very weak signals. The principle is the following (Fig. 8.3a): a single photon event ejecting a single electron on the photocathode

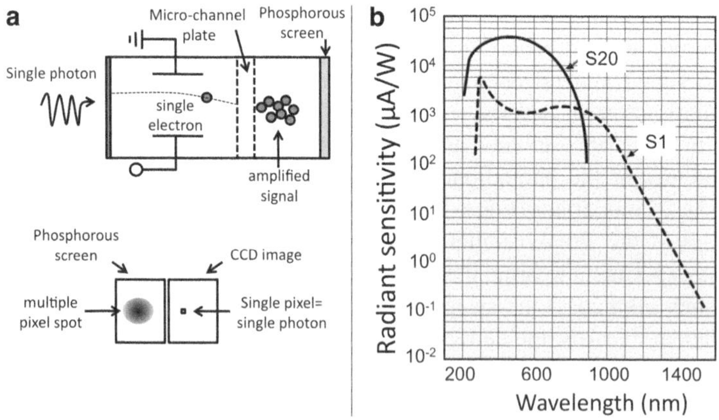

Fig. 8.3 (**a**) Schematic drawing of the single photon-counting method in streak cameras and (**b**) the S1 and S20 photocathode spectral dependence of the radiant sensitivity [8]

will cause a measureable light intensity on the phosphorous screen thanks to the electron amplification achieved by the micro-channel plate. Different from the analog integration, the light intensity on the phosphorous screen will be interpreted and recorded as a single event (a single pixel) by the CCD software, corresponding to the arrival of a single photon. By suitably establishing a threshold value for the measurement, it is possible to discriminate between noise and the signal created by a single photon event, eliminating the need for background subtraction. The multichannel character of the measurement is conserved, namely, the simultaneous spatial (spectral) and time resolution of a light event. As for any photon-counting techniques, the condition for single photon event, namely, very weak intensities, has to be fulfilled to avoid multiple event distortions or saturation of the recorded dynamics. As each event is stored in the CCD buffer memory, a higher dynamic range can be achieved than with the analogue integration mode.

8.2.4 Photocathode Type and Sensitivity

The efficiency of the photo-conversion of light into electrons, relying on the photoelectric effect with a photocathode with negative affinity, is a strong function of the signal wavelength. Photocathodes optimized to the experiment wavelength range have to be properly chosen. Similarly to photomultipliers, this is one of the major drawbacks of these systems, although the spectral band covered by each type of photocathode is relatively large. The spectral response of different photocathodes is usually identified by the "S" convention, S1 being the system of choice for near-infrared detection [9, 10] and S20–25 [11] the most widely used photocathode types for the UV to visible spectral range (see Fig. 8.3b). Commercially available systems do not extend to wavelengths above $\sim 1.5\,\mu m$ with their quantum efficiencies dropping dramatically already above $1\,\mu m$ (approximately an order of magnitude every $0.1\,\mu m$) and requiring photocathode cooling for lowering the noise level. If detection of longer wavelengths is needed, different techniques, such as the up-conversion, have to be employed (see Sect. 8.3). The strong wavelength dependence of the photocathodes' detectivity, especially at their extreme spectral regions, requires a careful intensity correction of the recorded signals in order to recover the real spectral features.

8.2.5 Time Resolution

Temporal resolutions as low as 2 ps for synchroscan mode (200 fs for femtosecond single sweep-triggered cameras) are achievable with commercial systems. The temporal resolution is intrinsically limited by several factors. Electrons transit time spread in the acceleration region and synchronization jitter, for instance, play an important role. Clearly, the time resolution is also limited by the vertical slit width proportionally to the deflection speed (see Fig. 8.1). Apart from these aspects, the experimental conditions can have an important influence. This is especially true when the system is used coupled to a spectrometer. Diffraction gratings unavoidably introduce time broadening to the light pulse proportionally to the grating width and groove density. This time broadening can sensibly limit the time resolution of an experimental system. The remedy to this problem consists in using a double monochromator in the subtractive mode. While recovering the full temporal resolution of the streak camera, the multichannel character of the system is lost as only a single spectral component temporal evolution can be recorded in a broad emission spectrum. In an experiment, although the intrinsic streak camera time resolution is not affected, it is inevitably the whole system resolution (including the jitter of the mode-locked laser) that has to be taken into account for the interpretation of a measurement. Despite the few limitations listed above, streak cameras remain an extremely versatile (although expensive) time-resolved luminescence spectroscopy tool in the picosecond and sub-picosecond time scale.

8.2.6 Application Examples

Figure 8.4 presents the result of a typical time-resolved photoluminescence spectroscopy experiment applied to the study of the photoluminescence dynamics of dilute nitride GaAsN semiconductors [12]. The sample contains two 8-nm quantum wells grown under exactly the same conditions by molecular beam epitaxy on a (001) semi-insulating GaAs substrate and separated by a 130-nm GaAs barrier. The first QW is a $In_{0.34}Ga_{0.66}As$ layer (nitrogen free), whereas the second is a $In_{0.34}Ga_{0.66}As_{0.994}N_{0.006}$ layer. The excitation source is a mode-locked Ti:Sa laser with a 1.2-ps pulse width (FWHM) and a repetition frequency of 80 MHz. The resulting PL intensity is dispersed by an imaging spectrometer and recorded by an S1 photocathode streak camera. The overall system temporal resolution is ~8 ps. Figure 8.4a reproduces the room temperature time-integrated spectrum of the two quantum wells, showing that a small fraction of N (less than 1%) considerably reduces the band-gap energy. Figure 8.4b presents the corresponding photoluminescence dynamics taken at the well peak emission. The photoluminescence decay time is greatly reduced for the second quantum well compared to the N-free one due to the formation non-radiative defects caused by the N incorporation [12].

8.3 Up-Conversion Technique for Time-Resolved Photoluminescence Spectroscopy

The up-conversion represents the technique of choice for time-resolved luminescence spectroscopy in the infrared spectral region where other techniques employing detectors such as photodiodes or photomultiplier tubes show no or very weak

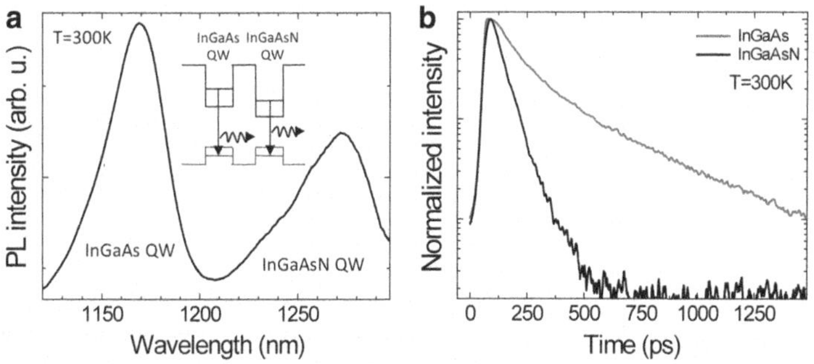

Fig. 8.4 (**a**) Room temperature time-integrated luminescence spectra of an $In_xGa_{1-x}As$ and an $In_xGa_{1-x}As_{1-y}N_y$ quantum well with $x = 0.34$ and $y = 0.006$. (**b**) The corresponding normalized photoluminescence dynamics measured at the emission peaks [12]. Reprinted with permission from H. M. Zhao, L. Lombez, B. L. Liu, B. Q. Sun, Q. K. Xue, D. M. Chen, and X. Marie, Applied Physics Letters, Vol. 95, page 041911 (2009). Copyright 2009, American Institute of Physics.

detectivity with poor signal-to-noise ratio [13–15]. The up-conversion is also the time-resolved technique that can yield the best time resolution, down to 5 fs [16]. The up-conversion relies on the nonlinear properties of dielectric crystals and it is particularly suited for the measurements of the dynamics of PL signals excited by the ultra-short pulses of mode-locked lasers with high repetition frequencies. This time-resolution technique is essentially limited by the laser pulse width and can, therefore, be typically of the order of 100 fs or less.

The basic principles behind the up-conversion techniques are presented in Fig. 8.5 [13–15, 17]. The typical source is a mode-locked Titanium–Sapphire (or a synchronously pumped dye) laser in the picosecond or femtosecond regime. The laser is split into two separate beams by a beam splitter (BS). The first one (the excitation laser) is used to excite the sample whose emitted PL is then collected by a system of lenses (or parabolic mirrors) and focused onto the nonlinear crystal ($LiIO_3$ in the examples used here; less dispersive BBO crystals are for instance more suitable for higher time resolutions). The second beam (delayed laser) is sent through a variable-length optics path (delay line) and then focused on the same spot as the PL signal onto the crystal. The nonlinear properties of the crystal[3] allow then for the frequency mixing (up-conversion) of the PL and pump signals if they are simultaneously present and satisfy definite polarization and incidence angle conditions (phase-matching conditions). It is this up-converted signal, of frequency equal to the sum of the laser and PL frequencies, that will be finally dispersed by a monochromator and detected by a photomultiplier tube linked to a photo-counting system or a CCD camera [18]. As the simultaneous presence of the photoluminescence and laser signals is necessary to produce the frequency mixing,

[3] It is the $\chi^{(2)}$ contribution of the crystal nonlinear susceptibility which is used here (see (8.12)), which is non-zero in non-centrosymmetric crystals.

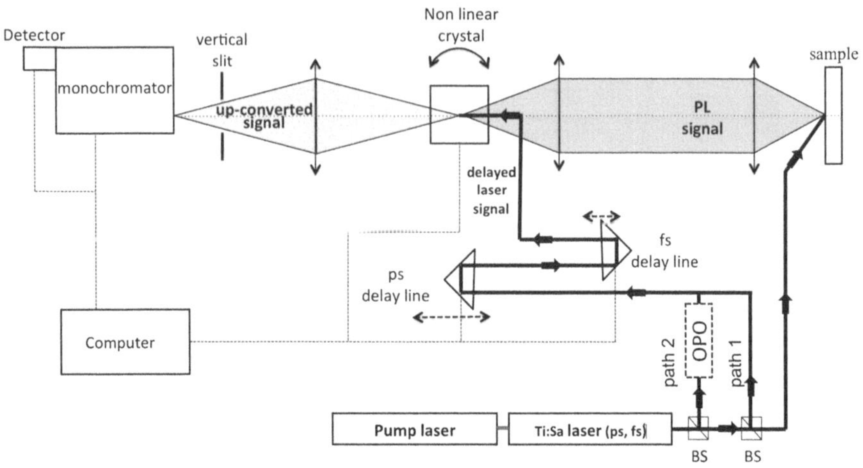

Fig. 8.5 Up-conversion setup. Path "1": one-color up-conversion technique. The same Ti:sapphire laser signal is used for excitation and opening the optical gate. Path "2": two-color up-conversion. The Ti:sapphire laser is used to excite the sample and to synchronously pump the optical parametric oscillator (OPO), while the signal from the OPO is used for the gate opening. Note that the part played by the two lasers can be swapped (see Fig. 8.7c, d)

the nonlinear crystal works as an optical gate opened by the retarded laser: temporal sampling of photoluminescence dynamics, with time resolutions comparable to the laser pulse width, is therefore possible by simply varying the delay between excitation and delayed laser signals. As the PL signal is up-converted to a higher energy by mixing it with a laser of suitable energy, the detected signal can be made to fall in a spectral domain (typically the visible) for which very sensitive detectors are available (photomultiplier tubes, CCD cameras). This characteristic, allowing for the analysis of photoluminescence in a wider infrared domain compared to photocathodes (up to $\cong$ 10 μm [17]), represents one of the key advantages of this technique. Historically, this technique was conceived to detect far infra-red radiation in astronomy [17].

8.3.1 Phase-Matching and Polarization Conditions

In addition to the time and spatial overlap of the luminescence and laser signals in the nonlinear crystal, the frequency up-conversion needs to satisfy phase-matching and polarization conditions in order to maximize the conversion efficiency. The former condition translates the conservation of both momentum and frequency of photons during the process and read [19] (see Fig. 8.6a):

$$\mathbf{k}_{Sum} = \mathbf{k}_{\mathrm{Lum}} + \mathbf{k}_{\mathrm{Laser}}, \tag{8.1}$$

$$\omega_{\mathrm{Sum}} = \omega_{\mathrm{Lum}} + \omega_{\mathrm{Laser}}, \tag{8.2}$$

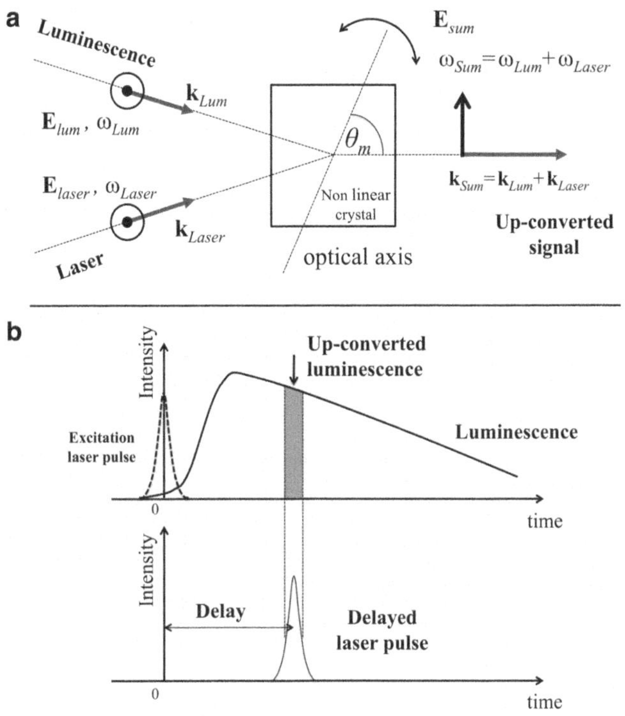

Fig. 8.6 (**a**) Type I phase-matching conditions. (**b**) Sketch of the all-optical sampling of the photoluminescence by the up-conversion technique

where the indices Sum, Lum, and Laser refer, respectively, to the up-converted, luminescence and delayed laser signals. These conditions can be satisfied in a uniaxial crystal with negative birefringence (being the extraordinary refractive index n_e smaller than the ordinary one n_o) such as $LiIO_3$, BBO, LBO, and KDP, whose choice essentially depends on the transparency range, the group velocity dispersion, the corresponding magnitude of the nonlinear coefficient, and the temporal resolution needed. For a given nonlinear crystal, these conditions impose, as well, restrictions on the polarization state of the light signals to be mixed. For sake of simplicity and illustration purposes, we will suppose quasi-collinear incident beams on a Lithium Iodate crystal ($LiIO_3$), one of the currently used nonlinear materials.

The phase-matching conditions (Fig. 8.6a) and the crystal symmetry dictate that the polarization of the up-converted beam will be extraordinary for ordinary polarized incoming beams (o + o → e, type I up-conversion[4]) [15]. For such a crystal, the effective refractive index n_{eff} seen by the extraordinary signal (the up-converted one) traveling at an angle θ with respect to the crystal optical axis is

[4]The up-conversion is named "type I" or "type II," respectively, if the incoming beams (PL and laser) are polarized parallel (either ordinary or extraordinary) or orthogonal to each other [15].

$$\frac{1}{n_{\mathrm{eff}}^2(\theta, \lambda_{\mathrm{Sum}})} = \frac{\cos^2(\theta)}{n_0^2(\lambda_{\mathrm{Sum}})} + \frac{\sin^2(\theta)}{n_{\mathrm{e}}^2(\lambda_{\mathrm{Sum}})}. \tag{8.3}$$

In this case, the phase-matching conditions (8.1) and (8.2) can be rewritten, respectively, as

$$\frac{n_{\mathrm{eff}}(\theta, \lambda_{\mathrm{Sum}})}{\lambda_{\mathrm{Sum}}} = \frac{n_0(\lambda_{\mathrm{Lum}})}{\lambda_{\mathrm{Lum}}} + \frac{n_0(\lambda_{\mathrm{Laser}})}{\lambda_{\mathrm{Laser}}}, \tag{8.4}$$

$$\frac{1}{\lambda_{\mathrm{Sum}}} = \frac{1}{\lambda_{\mathrm{Lum}}} + \frac{1}{\lambda_{\mathrm{Laser}}}, \tag{8.5}$$

where λ_{Sum}, λ_{Lum}, and λ_{Laser} are the respective signals' wavelengths in vacuum. The phase-matching conditions can be satisfied, for instance, by a fine angle-tuning of the crystal to the matching angle $\theta_{\mathrm{m}} \equiv (\vec{k}_{\mathrm{sum}}, \vec{c})$, being $\vec{c}$ the direction of the crystal optical axis, defined by [15]:

$$\sin^2 \theta_{\mathrm{m}} = \frac{n_{\mathrm{eff}}^{-2}(\theta_{\mathrm{m}}, \lambda_{\mathrm{Sum}}) - n_0^{-2}(\lambda_{\mathrm{Sum}})}{n_{\mathrm{e}}^{-2}(\lambda_{\mathrm{Sum}}) - n_0^{-2}(\lambda_{\mathrm{Sum}})}, \tag{8.6}$$

where $n_{\mathrm{eff}}(\theta, \lambda_{\mathrm{Sum}})$ is given by (8.4).

For a fixed θ_{m} and laser frequency, the up-conversion is realized for one single PL wavelength. The nonlinear crystal plays here the role of a monochromator by selecting only one spectral component of the sample emission to be detected. The variation of the delay line will provide with the measurement of the luminescence dynamics. Conversely, keeping fixed the delay line and synchronously varying the crystal angle and monochromator central wavelength allow for the measurement of the PL spectrum at a fixed time delay after the excitation. Finally, it is interesting to note that the crystal acts also as a very efficient linear polarization analyzer since only the component of the PL linearly polarized along the ordinary axis is up-converted. This property is very useful when analyzing the polarized components of the photoluminescence.

8.3.2 *Two-Color Up-Conversion*

Time-resolved photoluminescence spectroscopy is a powerful tool for the fine investigation of the carrier dynamics and the coherent properties in semiconductors. For these purposes, it is often necessary to excite the system quasi- or strictly resonantly. In general, being the intensity of the photoluminescence several orders of magnitude weaker than the excitation light, resonant excitation is often impossible. The up-conversion technique makes no exception, as under resonant excitation, the phase-matching condition will be satisfied for the second harmonic generation of the incident laser backscattered in the PL analyzed solid angle and also for direct frequency doubling of the "gate-opening" pulse. This limitation can be overcome by the use of two synchronous lasers at two different wavelengths. One of them is used to excite the sample, whereas the other is used as the delayed beam to be mixed with the PL signal ("open the optical gate" in the up-conversion terminology). As the wavelengths of the PL and pump beam will be now different, no direct second harmonic generation of the "gate opening" laser can be achieved, and strictly resonant excitation can be performed (Fig. 8.7c, d). In practice, the two sources can be two synchronized Titanium–Sapphire lasers or an optical parametric oscillator (OPO) synchronously pumped by a fraction of the Titanium–Sapphire laser intensity used for the excitation. The latter technique allows avoiding the complexity of the synchronization techniques of two independent lasers and generates no jitter between the two sources [20]. Nevertheless, during strictly resonant experiments, the mixing of the OPO signal with the scattered light of the exciting Titanium–Sapphire laser

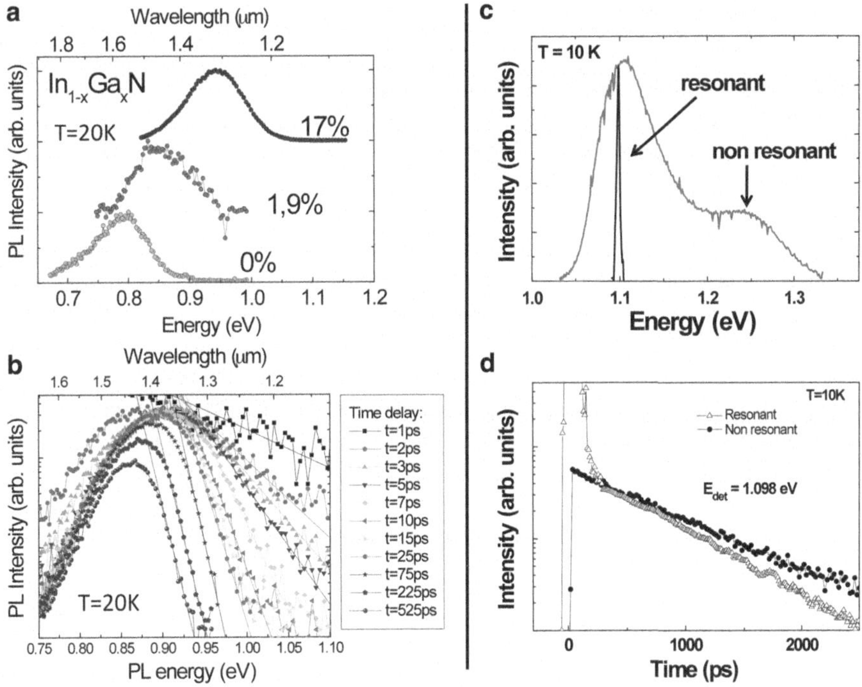

Fig. 8.7 Application examples of the up-conversion technique. One-color up-conversion: (**a**) Time-integrated photoluminescence of different InGaN alloys emitting up to 1.8 μm. (**b**) The photoluminescence spectra of the $In_{0.83}Ga_{0.17}N$ sample recorded at different delay times [21]. Two-color up-conversion: The comparison of the photoluminescence spectra (**c**) and the decay time (**d**) of InAs quantum dots under strictly resonant and non-resonant excitations [22]. Here, part of the Ti:sapphire laser is used to pump an OPO which opens the optical gate. Reprinted with permission from M. Paillard, X. Marie, E. Vanelle, T. Amand, V. K. Kalevich, A. R. Kovsh, A. E. Zhukov, and V. M. Ustinov Applied Physics Letters, Vol. 76, page 76 (2000). Copyright 2000, American Institute of Physics

can mask the first instants of the PL dynamics (see Fig. 8.7d). This problem can be minimized by a careful preparation of the sample surface and aligning the excitation beam so that its mirror reflection does not enter into the detection solid angle.

8.3.3 Quantum Efficiency

If phase-matching conditions are satisfied, the quantum efficiency of the up-conversion can be calculated by solving the system of coupled equations for the PL, laser, and up-converted beams [17, 19]. In the approximation of plane waves and negligible depletion of the laser intensity, the efficiency η_q for a perfect phase matching, $\Delta\mathbf{k} = \mathbf{k}_{\mathrm{Sum}} - (\mathbf{k}_{\mathrm{Lum}} + \mathbf{k}_{\mathrm{pump}}) = 0$, reads [15]

$$\eta_q(\Delta\mathbf{k} = 0) = \frac{2\pi d_{\mathrm{eff}} L^2 I_{\mathrm{Laser}}}{c\varepsilon_0 \lambda_{\mathrm{Lum}} \lambda_{\mathrm{Sum}} n_0(\lambda_{\mathrm{Lum}}) n_0(\lambda_{\mathrm{Laser}}) n_{\mathrm{eff}}(\theta_{\mathrm{m}}, \lambda_{\mathrm{Sum}})}, \tag{8.7}$$

where d_{eff} is the effective nonlinear susceptibility of the crystal,[5] L the interaction length in the nonlinear crystal, I_{Laser} the intensity of the delayed laser, c the speed of light, and ε_0 the vacuum permittivity. Typical values obtained for the quantum efficiency in the case of a picosecond excitation laser under the experimental conditions $I_{Laser} = 400\,\mathrm{mW}$ average power, $\lambda_{\mathrm{Laser}} = 0.8\,\mu\mathrm{m}$, $\lambda_{\mathrm{Lum}} = 1\,\mu\mathrm{m}$, and $L = 1\,\mathrm{cm}$ are $\eta_q \cong 7\%$. This small value is the major drawback of the up-conversion technique and it is mainly due to the very small acceptance angle (mrad) inherent to the technique. In case of sub-picosecond pulses, the nonlinear crystal will induce a stronger group velocity dispersion (GVD), and thinner nonlinear crystals of the order of 1–2 mm are employed limiting the conversion efficiency. However, the peak power is increased in this case, compensating the loss of efficiency to some extent. In a practical situation, however, the hypothesis underpinning the expression (8.7) is not strictly satisfied. Focusing and the non-co-linearity of the incident beams reduce the interaction length to a value much smaller than the actual length of the crystals.

In the event that a strict phase-matching condition cannot be satisfied ($\Delta\mathbf{k} \neq 0$), the conversion efficiency drops according to [19]:

$$\eta_q(\Delta\mathbf{k}) = \eta_q(0) \cdot \mathrm{sinc}^2\left(\frac{\Delta\mathbf{k}L}{2}\right). \tag{8.8}$$

In this case, the coherence length falls down according to $L_{\mathrm{coh}} \cong \pi/|\Delta\mathbf{k}|$.

[5] $d_{\mathrm{eff}} = d_{15}\sin\theta$ and $d_{15} = 4.87 \times 10^{-23} A \cdot s \cdot V^{-2}$ for $\mathrm{LiIO_3}$.

8.3.4 Spectral Resolution

For a definite angle between the crystal and the incoming beams, at a given laser wavelength satisfying the phase-matching condition, the frequency mixing is efficient only for a narrow spectral band. Considering the spectral bandwidth as the interval in energy for which the quantum efficiency drops to 50% from its maximum, the spectral band ΔE of converted luminescence is given by [15]:

$$\Delta E = \frac{3.66 \times 10^{-12}}{L((\partial k_{\mathrm{Sum}}/\partial \omega) - (\partial k_{\mathrm{Lum}}/\partial \omega))}, \tag{8.9}$$

where the energy is expressed in meV and the lengths in cm. The omnidirectional character of the PL signal can contribute to limiting the spectral resolution of a measurement. Slightly different PL wavelengths at slightly different incidence angles, aside from θ_{m}, will satisfy the phase-matching conditions. An up-converted signal of different wavelength and emerging angle with respect to the central one will be produced. In order to improve the spectral resolution of the detected signal, a supplementary vertical slit (in addition to the monochromator horizontal slit) is employed along the detection optical path, setting the spectral resolution of the experiment. In this case, the phase-matching conditions with respect to the nominal ones are approximately fulfilled for all the laser pulse bandwidth, while maintaining the photon energy condition (8.5), setting the spectral resolution to the laser bandwidth. Under these conditions, the experimentally measured spectral resolution in picosecond regime by up-converting a broad PL spectrum is in the order of $\Delta E = 2$–4 meV, close to the laser energy width.

8.3.5 Time Resolution

The time resolution achievable by the system is mainly determined by the spectral width of the laser and by the GVD induced by the different optical elements (lenses, mirrors) composing the setup. In the case of the type I up-conversion presented here, the temporal response broadening induced by the GVD in the nonlinear crystal is given by

$$\Delta t = L\left(\frac{\partial k_{\mathrm{Laser}}}{\partial \omega_{\mathrm{Laser}}} - \frac{\partial k_{\mathrm{Lum}}}{\partial \omega_{\mathrm{Lum}}}\right), \tag{8.10}$$

and can be particularly severe in the femtosecond domain where thinner nonlinear crystals have to be employed and parabolic mirrors should replace the PL collection lens to minimize the experimental system-induced GVD. Using picosecond lasers, the GVD becomes negligible, and the time resolution is in first approximation, directly determined by the laser temporal pulse width. The accuracy of the PL time sampling step is instead determined by the precision of the mechanical delay line. Using micro-translation stages or high precision stepping motors, accuracy of a few femtoseconds can be easily achieved (3.3 fs/μm).

8.3.6 Acceptance Angle

The strict phase-matching conditions of the frequency mixing process lead to the up-conversion of a tiny fraction of the photoluminescence intensity emitted by the sample despite its omnidirectional character. The acceptance angle of the nonlinear crystal is in fact defined by

$$\Delta\phi = \frac{2,78\, n_o(\lambda_{\text{Laser}})\lambda_{\text{Laser}}}{L[1 - (n_o(\lambda_{\text{Laser}})\lambda_{\text{Lum}})/(n_{\text{eff}}(\theta_m, \lambda_{\text{Sum}})\lambda_{\text{Laser}})]}, \tag{8.11}$$

giving only $\Delta\phi \approx 10^{-3}$ rad for the picosecond system presented here.

8.3.7 Calibration Procedures

Temporal calibration is required in order to define the time origin of the experimental setup. This corresponds to the condition of equal optical path between the excitation and the delay line. This position is determined by measuring the autocorrelation (or cross-correlation in case of two-color up-conversion) between the backscattered excitation laser light and the delayed laser signal. This calibration provides as well with an estimation of the laser beam pulse width and the system response time. This information is retrieved from the measured correlation curve fitted using a laser time intensity profile modeled by $\text{sech}^2(1.76t/\tau)$ [1], τ being the temporal full width at half maximum.

Angular/spectral calibration is necessary in order to determine the value of the nonlinear crystal angle θ_m, satisfying the phase-matching condition for given PL and pump laser wavelengths. The calibration is obtained by frequency-mixing a broad light source (Tungsten–Halogen lamp, for instance) with the laser, thus providing the value of the up-converted wavelength $\lambda_{\text{Sum}} = \lambda_{\text{Sum}}(\theta, \lambda_{\text{Laser}}, \lambda_{\text{PL}})$ as a function of the crystal angle (θ), the excitation laser (λ_{Laser}), and PL (λ_{PL}) wavelengths. These data are then stored on the computer controlling the movement of the crystal and the monochromator. The use of a CCD camera yields less stringent calibration procedures [18].

8.3.8 Streak Camera and Up-conversion: A Comparison

After having presented the main features of the streak camera and up-conversion technique, it can be interesting to evidence their differences and complementarities. The type of experimental information that can be extracted from both systems, namely, the photoluminescence dynamics, is in practice, similar. Nevertheless, the precise experimental conditions will eventually determine the more suitable system for a particular need. The most striking difference between the two techniques concerns the demand in preparation and operation of the systems. It is undeniable that the up-conversion technique requires both lengthy and careful calibration procedures in addition to long measurement times linked to its mono-channel character and small effective quantum efficiency. Individual acquisitions are needed for each point of a PL dynamics at a single wavelength. The reconstruction of a full dynamics of an entire spectrum can be extremely time consuming. Conversely, all this information is obtained with a single exposure with a streak camera. On the other side, the up-conversion technique can offer both superior temporal resolution, being limited in principle by the sole laser pulse width, and a much greater dynamic range. High electron densities, and phosphorous screen saturation phenomena represent instead the main limitation affecting the time resolution and dynamic range in streak cameras. Moreover, if photoluminescence signals of wavelengths longer than the available photocathodes' detection limits have to be measured, or strictly resonant excitation condition has to be used, the up-conversion technique remains, up to now, the only possible solution (for instance, with the LiO_3 system presented here, no drop of the sensitivity is observed up to $\lambda_{Lum} \sim 2\,\mu m$). Moreover, when polarization analysis is required, the up-conversion system provides polarization resolution with a very high rejection, comparable to the best Glan-Taylor analyzer, without any additional signal loss. Finally, financial arguments might also play a non-negligible role considering the much heavier investment demanded for a streak camera system.

8.3.9 Application Examples

Figure 8.7 presents a few typical experimental results obtained with the up-conversion technique. Figure 8.7a, b shows the application of this technique to the study of InGaN semiconductors grown by MBE, characterized by emission wavelengths up to 1.8 μm. Although InGaAs photodiodes can be used for detecting time-integrated PL, the up-conversion represents the only available solution to acquire time-resolved data with picosecond time resolutions. The carrier dynamics was characterized using for sample excitation the 1.2-ps excitation pulses generated by a mode-locked Ti-doped sapphire laser with a repetition frequency of 80 MHz with an excitation wavelength set to $\lambda_{exc} \sim 800$ nm. The detection was performed with the up-conversion technique using LiO_3 as the nonlinear crystal. The sum frequency signal was dispersed by a monochromator and detected by a photomultiplier tube linked to a photon-counting system. Figure 8.7a reproduces the time-integrated spectra obtained at $T = 20$ K of samples containing a variable composition of Ga, showing a band-gap wavelength for pure InN, centered on $\lambda \sim 1.55\,\mu m$. Figure 8.7b reports the PL spectra obtained at different time delays for the sample containing 17% of Ga. This experiment allows the observation of the thermalization of the hot carriers generated by the 800-nm laser pulse [21]. The energy peak of the spectra during the first picoseconds shows a pronounced blue shift with respect to the time-integrated spectrum. As the delay time increases, we observe a progressive red shift of the instantaneous PL spectral peak, reflecting the hot carrier relaxation dynamics and the progressive reduction of the band-filling effect as the photoexcited carriers recombine. By fitting the high-energy tail of each delayed spectrum with an exponential function, it is possible to extract the instantaneous carrier temperature and evaluate the time necessary for the hot carriers to reach thermal equilibrium at $T = 20\,\mathrm{K}$ [21].

Figure 8.7c, d reports an application example of the two-color up-conversion technique used for the investigation of InAs Stranski—Krastanov quantum dots under strictly resonant excitation conditions at cryogenic temperatures. The sample presented was grown by MBE and consists of ten layers of quantum dots separated by 8-nm GaAs barriers. They are located in a Bragg mirror microcavity in order to better extract the emitted photons and reduce the emission solid angle [22]. The strictly resonant conditions allow the selective excitation of the fundamental energy level of a quantum dots family. The observed PL spectrum (Fig. 8.7c) shows a much narrower spectral emission compared to the non-resonant excitation, and coincides with the exciting laser spectral width. These observations demonstrate that at low temperature, no charge transfer occurs between different dots belonging to either the same or different planes. Figure 8.7d reports the correspondent PL dynamics of the quantum dots fundamental states under non-resonant and strictly resonant excitation conditions. The intensity peak observed during the first picoseconds of the resonant dynamics is due to the laser scattering from the sample surface. Under resonant conditions, the carriers are directly photogenerated on the dot fundamental energy level, where they remain confined. The dynamics allows the direct determination of the exciton radiative lifetime. The slightly longer PL decay time observed under non-resonant conditions reveals the contribution to the PL dynamics of the carrier relaxation processes in the quantum dots [22].

8.4 Pump and Probe Time-Resolved Spectroscopies

The pump and probe technique is a straightforward and one of the most widely used ultra-fast spectroscopic techniques offering high temporal resolution, large sensitivity and signal-to-noise ratio. Under the same name fall several different spectroscopic techniques linked by a common fundamental principle. The basic idea is to use two ultra-fast light pulses, the pump and the probe one, to respectively first perturb the sample, and second to sense the perturbation effects by monitoring the changes induced on the probe beam interacting with the sample (Fig. 8.8). The arrival of the pump beam is used to define the origin of the time creating the perturbation and its effect can then be monitored at different pump–probe time delays by sending the probe beam through an optical delay line. The temporal variation of the perturbation effects can, therefore, be reconstructed with a temporal resolution defined by the probe pulse width and a sampling accuracy determined by the precision of the delay line step.

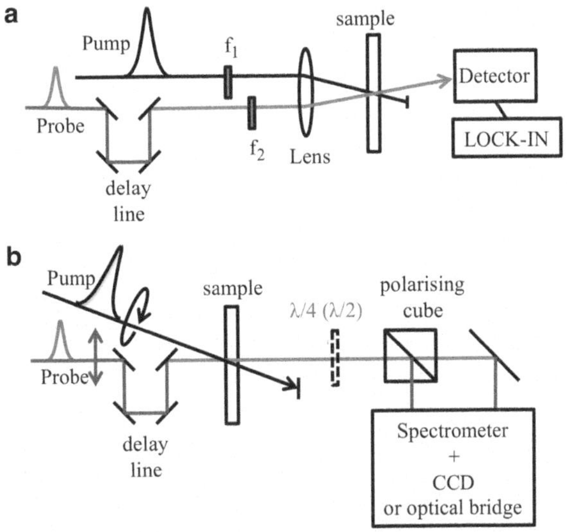

Fig. 8.8 Typical realizations of the pump–probe experiment: (**a**) Differential transmission setup with lock-in detection in double modulation. (**b**) Experimental realization for pump–probe dichroism or Faraday rotation measurements

The first fundamental assumption to the interpretation of pump and probe results is the hypothesis that the intensity of the probe beam is low enough not to influence the measurement: the result has to be verified to be independent from the probe intensity while still conserving a reasonable signal-to-noise ratio. This means that the excited system response is still linear with respect to the probe. Second, it is important to achieve a correct spatial overlap of the pump and probe beam in order to ensure that the probe senses a homogeneously perturbed region (since the sample response is in general non-linear). To this purpose, spatially narrower probe spot sizes, about half in diameter, compared to the pump ones are typically used. A variation to this experiment consists in the measurement of the spatial diffusion of the pump-induced perturbation in addition to its temporal evolution. In this case, the probe beam optics is mounted on a precision x–y translation stage, allowing the record of a two-dimensional map of the perturbation spatial extent. Finally, it is worth noting that although the pump and probe technique is an ultra-fast spectroscopic tool, it does not require fast detectors. Similarly to the up-conversion technique (Sect. 8.3), the experiment is repeated for any delay several times (millions times per second in the case of high-frequency mode-locked lasers) and the averaged response of the detector is then recorded.

The pump–probe interaction is, in a classical description, a nonlinear optical process whose resulting signal is proportional to the third-order optical susceptibility of the material: the experiment involves the evaluation of the probe signal variation induced by the pump beam intensity. It is assumed usually that the material polarization $\vec{P} = \varepsilon_0 \chi \vec{E}$ can be expressed in terms of the Taylor series with respect to the electric field as

$$\vec{P} = \varepsilon_0\left(\chi^{(1)}\vec{E} + \chi^{(2)}\vec{E}\vec{E} + \chi^{(3)}\vec{E}\vec{E}\vec{E} + \ldots\right), \tag{8.12}$$

where $\chi^{(1,2,3)}$ represent, respectively, the linear, second-order and third-order susceptibility of the material and ε_0 the vacuum dielectric constant. If we consider, for the sake of simplicity, only centrosymmetric materials, the second-order susceptibility $\chi^{(2)}$ disappears for symmetry considerations. The total electric field writes $E_{\text{total}} = \text{Re}\{E_{\text{p}}(\omega_{\text{p}})e^{i\omega_{\text{p}}t} + E_{\text{pr}}(\omega_{\text{pr}})e^{i\omega_{\text{pr}}t}\}$, where the pump and probe fields are labeled E_{p} and E_{pr}, respectively. By injecting E_{total} into the preceding equation and considering only the terms radiating at the probe frequency, it is straightforward to identify that the term describing the probe variation under the influence of the pump signal is the third-order polarization $\vec{P}^{(3)} \propto \chi^{(3)} I_{\text{p}} \vec{E}_{\text{pr}}$, where I_{p} is the pump intensity. Since the material complex refractive index $\tilde{n}$ is linked in a dielectric material to the susceptibility through $\varepsilon(\omega) = 1 + \chi(\omega) = \tilde{n}^2(\omega) = [n(\omega) + i\kappa(\omega)]^2$ (where n and κ represent, respectively, the real and imaginary parts of the complex refractive index $\tilde{n}$), the nonlinear polarization at the probe frequency induces a corresponding variation on $\tilde{n}$. The real and imaginary parts of the refractive index, being related through the Kramers–Krönig relations, are mutually influenced, and the pump–probe experiment results depend on the whole $\tilde{n}$. Nevertheless, if the probe beam frequency ω_{pr} is coincident or close to a material transition ($\omega_{\text{pr}} \approx \omega_0$ in Fig. 8.9), the effects on the probe signal will be dominated by the variation of the absorption coefficient $\Delta\alpha \propto \Delta\kappa$. On the contrary, if ω_{pr} senses a transparency region of the material ($\omega_{\text{pr}} \approx \omega_1$ in Fig. 8.9), the measurement will be dominated by the variation Δn of the real part of the refractive index.

If the pump and probe experiment measures the pump-induced variation of the material absorption in reflection or transmission, the experiment is respectively termed differential transmission and differential reflection (see Sect. 8.4.2). If the polarization of the pump and probe beams is taken into account, the experiment may also reveal induced dichroism or birefringence (see Sect. 8.4.3). Depending on the symmetry of the global system, the dichroism can be either linear or circular, signifying, respectively, a different absorption coefficient for orthogonally linearly ($\kappa_x \neq \kappa_y$) or orthogonally circular ($\kappa_+ \neq \kappa_-$) polarized probe beams. Correlatively, linear or circular birefringence phenomena ($n_x \neq n_y$ and $n_+ \neq n_-$, respectively) can be as well induced [23, 24]. Often time-resolved pump and probe experiments make use of sub-picosecond laser pulses, characterized by a considerable spectral width ($\Delta E \approx 20$ meV for 100-fs pulses), which unavoidably covers

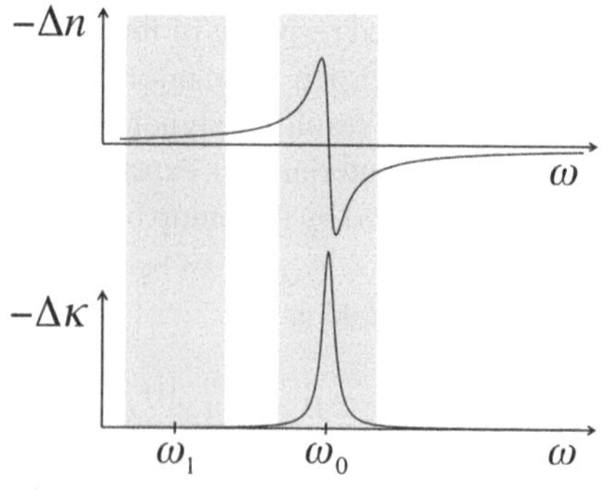

Fig. 8.9 Pump-induced changes of the real and imaginary parts of the refractive index for a homogeneously broadened transition described by a Lorentz oscillator. The shaded regions correspond to the position and spectral width of the probe signal

both transparency and absorption regions: absorption, dichroism, and birefringence phenomena coexist, and it is the detection method which has to be adapted to the demanded analysis [24–26].

8.4.1 Probe Characteristics and Detection Techniques

The spectral characteristics of the probe beam are chosen according to the experimental needs and can be either narrow or have a broad spectral width. The former kind of probe beam is used if a particularly narrow discrete spectral level has to be selectively probed avoiding the contribution of near states. Typical examples are studies of the absorption or the induced dichroism of quantum dots [27] or defect-bound states. The same frequency for pump and probe beam is often used ("one-color" pump and probe experiments). In this case, the same laser source is split in appropriate proportion in the two beams and a delay line allows the tuning of the time delay between the pump and the probe. In the cases where the analysis is needed on a wide spectral region, spectrally broad signals with relatively flat intensity distribution can be generated by sending high peak-power laser pulses on liquids, glasses, or non-linear optical fibers [28]. The generated signal, the so-called continuum, typically appears as white light; it can be spectrally filtered to the desired wavelength range and makes possible the record of transient absorption phenomena simultaneously at different wavelengths using, in general, a CCD camera. A few typical examples of applications in semiconductor physics are the measurement of carrier distribution, relaxation dynamics, and exciton screening or formation times under non-resonant excitation [29–31].

In order to achieve the best sensitivities, lock-in synchronous detection is often used along with an optical chopper. The chopper normally modulates the pump beam and is used as synchronization signal for the lock-in amplifier. The probe signal is measured with a photodiode or a photomultiplier tube linked to the lock-in input. The result is directly proportional to the transmission (reflection) difference of the probe with and without the presence of the pump. For superior signal-to-noise ratio, a double frequency modulation technique is usually employed [32]. Both pump and probe beams are modulated by a mechanical light chopper at two different frequencies, respectively, f_{pump} and f_{probe}. The resulting probe variations, detected by a photodiode or a photomultiplier tube, will bear the signal $S(t)$ at both the sum and difference frequencies $S(t) \propto \cos[2\pi(f_{\mathrm{probe}} - f_{\mathrm{pump}})] + \cos[2\pi(f_{\mathrm{probe}} + f_{\mathrm{pump}})]$, and the demodulation to retrieve the signal can be performed on either of the two frequencies. This yields (1) an improved sensitivity and (2) the elimination of measurement of the pump stray light scattered into the probe propagation direction: only the cross-correlated signal between pump and probe is measured. Care has to be taken, like in any lock-in detection experiment, to avoid the demodulation frequency to correspond to a multiple or divisor of the main frequency: the f_{pump} and f_{probe} frequencies are chosen to be incommensurable. Signal variation of the order of 10^{-5} can thus be routinely detected. As the frequencies achievable by mechanical choppers are at most a few kilohertz, acousto-optic modulators (photo-elastic modulators in case of polarization-dependent experiments), with modulation frequencies of up to the MHz region, can be additionally employed to improve the signal-to-noise ratio. This third light frequency modulation f_{MHz} can be added at the laser output before generating the pump and probe signal. Then a two-step demodulation protocol is used: the first stage demodulates the high-frequency modulation, while the second stage demodulates the remaining signal at $f_{\mathrm{pump}} \pm f_{\mathrm{probe}}$. By shifting the signal at higher frequencies, laser intensity instabilities, mechanical vibrations, and the $1/f$ component of the noise are considerably reduced.

Balanced photo-detection technique (optical photodiode bridge) is another detection scheme commonly adopted if great sensitivity and low signal-to-noise ratio are required. This technique is implemented by using two photodiodes connected in such a way that their photocurrents cancel if submitted to the same signal intensity. Usually, in order to compensate the non-identical sensitivity of the detector or the non-perfectly equal intensities of the signals, signal compensation means such as variable neutral density filters are utilized (nulling the optical bridge). The output of the balanced bridge is, therefore, zero until a mechanism modifies one beam (the signal one) with respect to the reference one. A net signal appears then on the output. This scheme is particularly useful, for instance, in polarization-dependent pump–probe experiment (dichroism or birefringence-induced polarization rotation): the polarization components of the probe beam are separated and simultaneously sent to the optical bridge. A polarization-dependent differential absorption will unbalance the bridge and appear as a signal. One of the advantages of the method is that the background and electronic noise are automatically subtracted, provided that the two bridge photodetectors are selected to be nearly identical.

These sensitive lock-in-based measurement techniques are, however, bound to be limited to a narrow wavelength range as the detection is performed by photodiodes or photomultiplier tubes. It is often of crucial importance to accompany the dynamical information with a spectral analysis of the probe modification by using a spectrometer coupled to a CCD camera or a photodiode array. This technique has the non-negligible advantage of offering the simultaneous measurement of the transient at different wavelengths.

It is known that in a pump–probe experiment, measurement artifacts can appear due to coherent interactions between the pump and probe pulses for small pump–probe relative delays. In practice, these artifacts overlap with the signal, making extraction of the desired information difficult. A simple solution consists in the comparison between co-polarized and cross-polarized experiment results. This is, however, not possible in the case where polarization-dependent phenomena have to be investigated as the researched effects cannot be isolated. Another more complex technique consists in the separate measurement of the isolated coherent artifact emerging at a different angle [33], which is then subtracted from the raw. Other artifacts may arise from the Fabry–Perot interferences between the front and back faces of the sample [31]. In this case, antireflection coatings can be employed to minimize the interference fringes. On the contrary, putting a sample in an optical cavity can enhance the sensitivity of the pump–probe experiment due to the probe multiple reflections inside the sample [34].

8.4.2 Time-Resolved Differential Absorption and Dichroism Experiments

In the case of semiconductor physics, the differential absorption technique relies on the absorption modification by the pump-generated carriers due to the Pauli exclusion principle. The experiment usually consists in exciting the material with laser pulses, creating a nonequilibrium carrier distribution reflecting the energy profile of the pump pulse. The excess carriers affect, in turn, the optical properties of the material through the complex index of refraction determining the absorption coefficient [30, 31, 35, 36].

In a simple approach, neglecting excitonic contributions and considering a direct band-gap semiconductor, the pump-induced change in the absorption coefficient α can be expressed as $-\Delta\alpha(\Delta t, \lambda) = \alpha_0(f_e + f_h)$, where α_0 represents the absorption coefficient of the unperturbed sample, and f_e and f_h are the electron and hole population distributions. The pump-generated carriers subsequently modify their energy distribution and will eventually thermalize with the lattice as the time progresses due to different phenomena such as carrier–carrier interactions or phonon emissions. Therefore, for a given pump wavelength of spectral width $\Delta\lambda$, the pump perturbation may also have repercussions on a broader energy range. By choosing a suitably large spectral width of the probe beam and by varying the pump–probe delay Δt, the population dynamics can be thus reconstructed. In practice, it is the amplitude of the transmitted (or reflected) probe, as a function of the arrival time after the pump, which is measured providing insight into the lifetime and recombination dynamics of the photogenerated carriers.

For instance, in the case of transmission geometry, the differential absorption $\Delta\alpha = \alpha_p - \alpha_0$ (where α_p and α_0 are, respectively, the sample absorption coefficient with and without the pump) can be related to the corresponding transmission ΔT by[6] $\Delta\alpha = -1/L[\ln(1 + \Delta T/T)]$, where L is the thickness of the layer of interest [29–31]. Figure 8.10 illustrates the use of the pump and probe technique to measure the fast relaxation times of non-thermal photoexcited carriers in GaAs quantum wells [29]. In the experiment, the well absorption changes are monitored with a few tens of femtosecond resolution over a wide energy range, which allows a temporal and spectral analysis of the relaxation of hot photogenerated carriers. The sample consists of 65 9.6-nm wide GaAs quantum wells separated by 9.6-nm AlGaAs barriers. The pump and probe beams are derived from a white-

[6] For small signal variations, this equation can be approximated to $\Delta\alpha \cong -1/L(\Delta T/T_0)$, stating that a measurement of the differential transmission is directly related to the absorption coefficient variation.

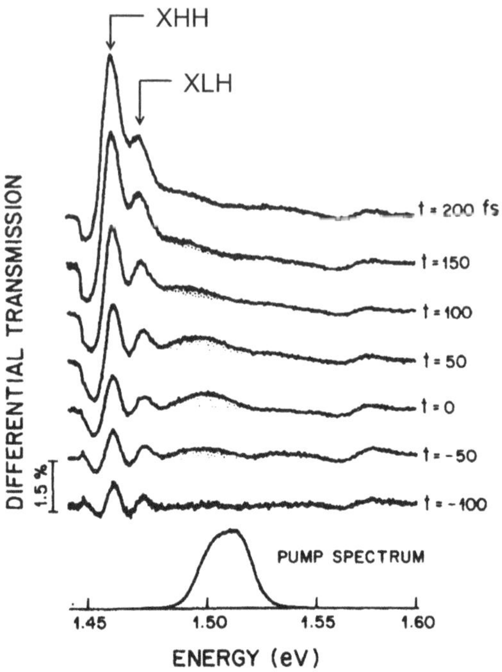

Fig. 8.10 Room temperature differential transmission spectra of GaAs/AlGaAs quantum wells taken at different pump–probe delays. The detail of the pump signal is reported at the bottom of the graph [29]. XHH and XLH correspond, respectively, to the fundament heavy and light hole exciton transitions. Reprinted figure with permission from W. H. Knox, C. Hirlimann, D. A. B. Miller, J. Shah, D. S. Chemla, C. V. Shank, Phys. Rev. Lett., 56, 1191 (1986). Copyright (1986) by the American Physical Society. http://link.aps.org/doi/10.1103/PhysRevLett.56.1191

light continuum. The former is passed through an interference filter to tailor its spectral width and energy position. The pulse is centered on 1.51 eV, i.e., above the lowest heavy- and light-hole transition energies, and focused to a $\sim 20 - \mu$m spot size. A white-light continuum of $\sim 50 -$ fs pulse width is used to probe the absorption dynamics over a 150-meV spectral range. Figure 8.10 presents the differential transmission spectra, from negative to positive pump–probe delays, showing snapshots of the thermalization process: the transmission increase near the fundamental excitonic transition is delayed compared to the variation observed near the pump energy, proving that the exciton absorption bleaching [31] occurs after the carrier thermalization and the filling of the lower-lying band states [29].

Differential absorption experiments can be performed as well as a function of the pump and probe beam polarizations providing information on the transient dichroism induced by the pump in the sample [37–39]. In this case, information on the carrier spin dynamics in addition to simple population dynamics can be extracted by monitoring the absorption variations as a function of the relative pump and probe beam polarizations. As an application example, we show in Fig. 8.11a the time-resolved differential absorption technique used to measure the heavy-hole exciton spin relaxation in GaAs/AlGaAs superlattices. The circularly polarized pump and probe beams are generated by a 1-ps mode-locked dye laser and are respectively tuned to and slightly below the heavy-hole transition. Figure 8.11a shows the temporal evolution of the right (σ^+) and left (σ^-) circularly polarized differential transmission probe signals after a right circularly polarized pump beam. The

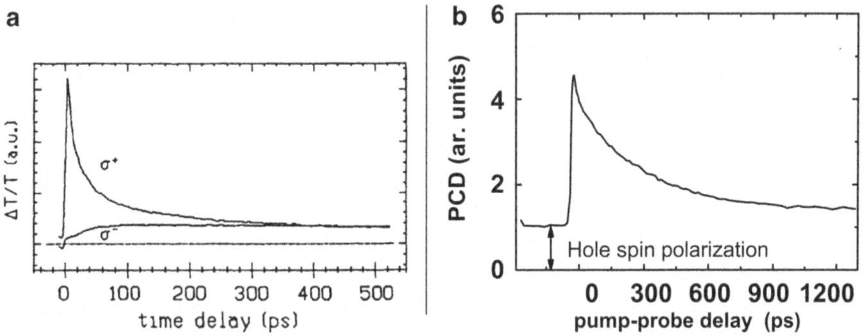

Fig. 8.11 (**a**) Temporal evolution of the polarization-dependent differential transmission signal probe slightly below the heavy hole transition in a GaAs/AlGaAs superlattice at $T = 4\,\text{K}$ [37]. The pump is σ^+ polarized. Reprinted figure with permission from S. Bar-Ad, I. Bar-Joseph, Phys. Rev. Lett., 68, 349, (1992). Copyright (1992) by the American Physical Society. http://link.aps.org/doi/10.1103/PhysRevLett.68.349 The effect relies on the polarization-dependent exciton non-linear absorption saturation under resonant excitation. (**b**) Photo-induced dichroism experiment: Experimental measurement at $T = 2\,\text{K}$ of the temporal evolution of InAs/GaAs quantum dots dichroism revealing a nonzero value after the trion recombination time. The repetition frequency of the experiment is ~80 MHz so that the resident hole spin is not fully relaxed before the next pulse excites the sample. This measurement reflects the long spin lifetime of the resident hole [27]. Reprinted figure with permission from B. Eble, C. Testelin, P. Desfonds, F. Bernardot, A. Balocchi, T. Amand, A. Miard, A. Lemaître, X. Marie, and M. Chamarro, Phys. Rev. Lett., **102**, 146601 (2009). Copyright (2009) by the American Physical Society. http://link.aps.org/doi/10.1103/PhysRevLett.102.146601

analysis of the temporal evolution of the intensity difference of the two orthogonally polarized probe beams, which is proportional to the population difference between right and left circular excitons, gives insight into the spin dynamics governing the heavy-hole exciton transition [37].

These types of differential absorption experiments monitor the transient dynamics of the pump photogenerated carriers while still present in the sample. In other words, the probe senses the dynamics of externally created carriers in the sample "environment." A very useful feature offered by the pump–probe technique is the possibility of sensing the *modifications* imparted on the sample under the action of the pump, and as such, also after the pump-generated carriers have disappeared: the probe senses here the result of the previous presence of the pump on the sample. For instance, information can be imprinted on the sample through the photogenerated carriers, encoded, for example, in the form of carriers spin through optical orientation experiments. This information can be transferred to the intrinsically present charges, or to nuclear or magnetic impurity spins, and might then persist long after the photogenerated carriers have recombined [26, 27]. The role of the probe beam is here to determine the lifetime and the dynamics of the written information.

A typical example of this application is the photo-induced dichroism experiment presented in Fig. 8.11b [27]. Here, the pump beam prepares the system, a p-doped quantum dot ensemble, with a preferential carrier spin orientation. The linear probe beam experiences here a different absorption for its circular polarization components. The measurement of the intensity imbalance is proportional to the net spin polarization created in the system. This photo-induced dichroism measurement is applied to the study of the resident hole spin relaxation due to hole–nuclei interaction in InAs quantum dots. The sample consist of 30 planes of InAs/GaAs quantum dots p-modulation doped with carbon delta doping ensuring, on average, the presence of a single resident hole per dot of random spin orientation. The pump and probe beams are provided by a mode-locked Ti:sapphire laser. A photoelastic modulator oscillating at 42 kHz is used to modulate the polarization of the pump beam between circular right and left polarizations. The probe beam is instead linearly polarized. The probe circular polarization components' intensities are measured after the transmission through the sample with a balanced optical bridge. Due to the optical selection rules, the pump beam selectively generates positive trions with a preferential orientation of the electron spin. Due to the fast electron–nuclei hyperfine interaction, the electron flips its spin before the trion radiative recombination: this effectively leaves the quantum dot with a resident hole with an opposite spin orientation compared to that before the excitation. This spin-cooling preparation mechanism effectively unbalances the population of spin-up to spin-down resident hole quantum dots. Due to the Pauli principle, the resonant linearly polarized probe will, therefore, experience a difference in the absorption coefficients of its left and right circularly polarized components. The measurement of the persistence of this absorption unbalance after the trion recombination time is a direct estimation of the resident hole spin dephasing time due to the hyperfine interaction with the nuclei, which occurs on a time scale of $\sim$10 ns.

The pump and probe technique is, therefore, a fundamentally different spectroscopic tool compared to photoluminescence techniques. The latter relies on the presence of photogenerated carriers and the results are principally sensitive to the energetically lower-lying states. The pump and probe technique can instead sense any of the available states falling in the probe energy profile and, relying on the absorption, is sensitive to the available density of states. Finally, as just mentioned, the probe beam can be used to analyze the persistence of pump-imprinted features even for times longer than the photogenerated carriers' lifetimes (which is seldom achievable with photoluminescence techniques).

8.4.3 Faraday and Kerr Rotation Spectroscopy

"Optically active" materials exposed to a magnetic field cause linearly polarized light traversing them to rotate its axis of polarization. This phenomenon, known under the name of Faraday effect, is the consequence of an induced inequality of the material refractive index for left (n_+) and right (n_-) circularly polarized light, termed circular birefringence. This difference is wavelength dependent and proportional to the material magnetization and its thickness. The material magnetization is usually induced by external magnetic fields. In the case of optical orientation experiments in non-magnetic semiconductors however, the polarization state of the incident light can photogenerate carriers with a preferential spin orientation thanks to the optical selection rules [40] which, in turns, creates a net magnetic moment within the sample. This will similarly cause a dissymmetry on the sample refractive indices n_+ and n_-. The opposite helicity components of a linearly polarized light signal will experience, therefore, a relative phase shift which is at the origin of the rotation of the linear polarization direction according to

$$\theta_{\mathrm{F}}(\omega) = \frac{\omega L}{2c}(n_- - n_+). \tag{8.13}$$

The amount of the probe polarization rotation is proportional to the carrier spin orientation in the sample: the spin-polarized photogenerated carriers reduce the optical transitions preferentially for one circularly polarized component of a linear probe. The difference in the absorption causes, through Kramers–Krönig relations, a difference in the respective refractive index at the origin of the rotation of the linear polarization direction. A measurement of the time evolution of the polarization rotation angle proportionally reflects, therefore, the temporal evolution of the carrier spin polarization degree [25]. If the variation of the probe polarization state is monitored in reflection or transmission spectroscopy (Fig. 8.8b), the experiment is termed Kerr and Faraday rotation, respectively [23, 41, 42]. This last terminology is used whenever a polarization change is experienced by the probe signal, even without the presence of an external magnetic field as, for instance, in optical orientation experiments.

The light-induced Faraday (Kerr) rotation is an extremely sensitive spectroscopy tool for the detection of very small carrier spin polarization. The typical experimental setup for Faraday rotation is, in its main aspects, identical to the polarization-dependent differential absorption, as presented in Fig. 8.8b. Circularly polarized pump and linearly polarized probe beams are sent onto the sample. After transmission through the sample, the probe beam is projected on two orthogonal polarization directions having equal intensity without the presence of the pump. The difference in their intensities is measured typically in an optical bridge or simultaneously spectrally resolved on a CCD camera on different pixel lines. Typical magnitudes of the rotational angles for semiconductors are generally between 10^{-4} and 10^{-3} degrees, with resolution up to 10^{-6} degrees using the balanced photodiode bridge technique [26].

As an application example, we show in Fig. 8.12a the use of the time-resolved Kerr rotation technique applied to the measurement of the electron Landé g-factor in thin films of dilute nitride GaAsN [43]. The sample consists of a 190-nm-thick $GaAs_{0.991}N_{0.009}$ layer grown on a semi-insulating (001)-oriented GaAs substrate by molecular beam epitaxy. The time-resolved Kerr rotation experiment was carried out at room temperature and the sample was excited at near-normal incidence with a degenerate pump and delayed probe pulses from a Coherent mode-locked Ti:sapphire laser delivering 120-fs pulses at 76-MHz repetition frequency. The

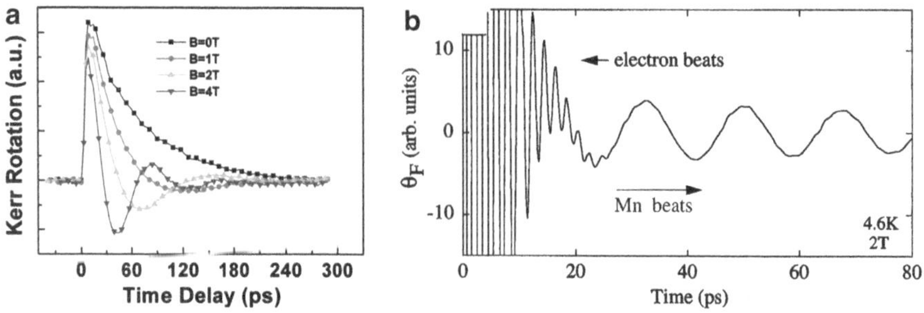

Fig. 8.12 (**a**) Conduction band electron Quantum Beats measured by Kerr rotation on a thin $GaAs_{0.991}N_{0.009}$ layer at room temperature at different external magnetic fields perpendicular to the excitation and sample growth direction [43]. At $B = 2$ T, we deduce that the electron Zeeman splitting is about 4.4 μeV. (**b**) Mn^{2+} ion magnetic moment Quantum Beats measured at $T = 4.6$ K in Mn doped ZnCdSe/ZnSe quantum wells under an external magnetic field $B = 2$ T applied perpendicular to the excitation and sample growth direction [44]. Reprinted figure with permission from S. A. Crooker, J. J. Baumberg, F. Flack, N. Samarth, and D. D. Awschalom, Phys. Rev. Lett., **77**, 2814 (1996). Copyright (1996) by the American Physical Society. http://link.aps.org/doi/10.1103/PhysRevLett.77.2814. The conduction electron beats are also clearly visible for time delays below 25 ps

pump beam is modulated between left and right circular polarization by a 50-kHz photoelastic modulator, and the probe linear polarization rotation is detected through a lock-in amplifier. The pump pulse creates a net conduction band electron spin polarization thanks to the optical selection rules [40] inducing a circular birefringence. The probe polarization rotation θ_K at different pump–probe delays Δt is a direct measurement of the conduction band electron spin dynamics. The application of an external magnetic field B perpendicular to the laser excitation direction induces the Larmor precession of the conduction band electron spin reflected by the oscillation of the Kerr signal (Quantum Beats). The temporal variation of the Kerr signal provides a measurement of the Landé g-factor and the conduction band electron spin dephasing time τ_s through the equation

$$\theta_K(\Delta t) = A e^{-\Delta t/\tau_s} \cos(2\pi \nu_L \Delta t), \tag{8.14}$$

where A is a constant and $\nu_L = g\mu_B B/h$ is the Larmor precession frequency (μ_B being the Bohr magneton and h the Planck constant). The electron Zeeman splitting is thus measured with high accuracy, and the g factor deduced.

Figure 8.12b [44] presents an example of a Faraday rotation experiment applied to the measurement of the electron spin resonance of Manganese ions in ZnCdSe/ZnSe multiple quantum wells containing Mn^{2+} ions. The experimental technique is similar to the Kerr rotation previously described. A circularly polarized 120-fs pulse-width pump beam, traveling along the sample growth axis z and tuned to the electron heavy-hole transition, injects spin-polarized electrons and holes into the sample. The application of an external magnetic field perpendicularly to the sample growth causes the precession of the photogenerated carriers according to their respective Larmor frequencies. The Larmor precession is measured through the period of the oscillations of the rotation imparted to the direction of the polarization of a linearly polarized probe in transmission geometry (Quantum Beats measured in Faraday rotation). The observation of the probe rotation oscillations after the recombination of the photogenerated carriers shows a characteristic period corresponding to the Mn^{2+} ion g-factor $g = 2.0$. This result is interpreted as the Larmor precession of the Mn^{2+} ions magnetic moment around the direction of the externally applied magnetic field. The Mn^{2+} ions magnetic moment, misaligned from its original direction parallel to the external magnetic field, thanks to the spin-torque induced along the magnetic field created by the photogenerated carriers, freely precesses for several hundreds of picoseconds after the carriers' recombination.

To conclude this non-exhaustive overview, let us briefly mention multiple-beam pump–probe experiments frequently employed in semiconductor physics:

1. In spin-grating experiments [45–47], two synchronous pump pulses create a transient carrier density grating on the sample, by which the probe is subsequently diffracted. These experiments are sensitive tools for the probe of the carrier population or spin diffusion in the sample.
2. In four-wave mixing experiments [48, 49], the two pump pulses impinge on the sample at different time delays. These experiments allow measuring the polarization decay of the sample once excited by the first pulse. Moreover, time-resolving the differential signal by up-converting it gives additional information on the nature of the physical process (free induction decay versus photon echo corresponds, respectively, to homogeneous versus inhomogeneous system) [50]. All these techniques rely on the nonlinearity, induced in particular by $\chi^{(3)}$, of the optical response of the sample.

8.5 Time-Correlated Single-Photon Spectroscopy

A well-established technique for the measurement of luminescence dynamics is the time-correlated single-photon counting (TCSPC). It represents a very versatile tool suitable for the time-resolved spectroscopy studies from the ultra-fast domain (picosecond scale) up to the millisecond regime. This measurement method, based on the counting of single photon events, is statistical in nature. In order to acquire a statistically significant number of events in a short time, TCSPC is ideally coupled to the high-frequency mode-locked lasers as excitation source. The working principle can be schematically outlined as follows when operating with a pulsed laser (see Fig. 8.13): for each single pulse, the START signal can be triggered by an auxiliary photodiode which detects some fraction of the excitation laser pulse. The first luminescence photon arriving at the detector triggers the STOP signal. For the detected photon, the time elapsed between the reference signal and the photon arrival instant is measured. The histogram of the number of counts is, therefore, progressively constructed. One count is added in the histogram to the memory

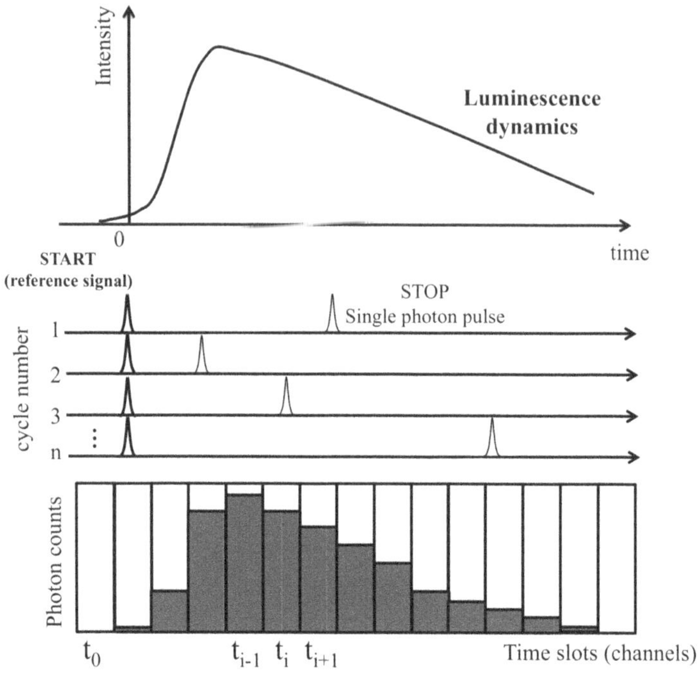

Fig. 8.13 Schematic representation of the START–STOP technique operation under pulsed excitation

corresponding to the time slot (the channel) determined by the elapsed time. These same steps are repeated a very large number of times (millions per second in case of mode-locked lasers). Due to the random character of the spontaneous emission, the obtained histogram (number of photons counted as a function of the START–STOP time interval) is a reconstruction of the luminescence dynamics. As the technique relies on single photon measurements, it is very sensitive. However, it is of capital importance to ensure that the signal intensity is low enough to avoid multiple photon events. A distorted dynamics will otherwise be measured due to the "pulse pile-up" leading to a saturation of the number of counts per channel. As a practical rule, the photon detection rate is kept to a few percent of the reference signal rate. Photodetectors with high gain and fast response times, such as micro-channel plates, photomultipliers, and avalanche photodiodes, are therefore necessary to feed a detectable electrical signal to the START–STOP electronics and to achieve high temporal resolution.

8.5.1 TCSPC Setup and Electronics Components

A schematic diagram of a typical TCSPC setup is presented in Fig. 8.14a. The excitation source is separated into two beams by a beam splitter (BS). The first

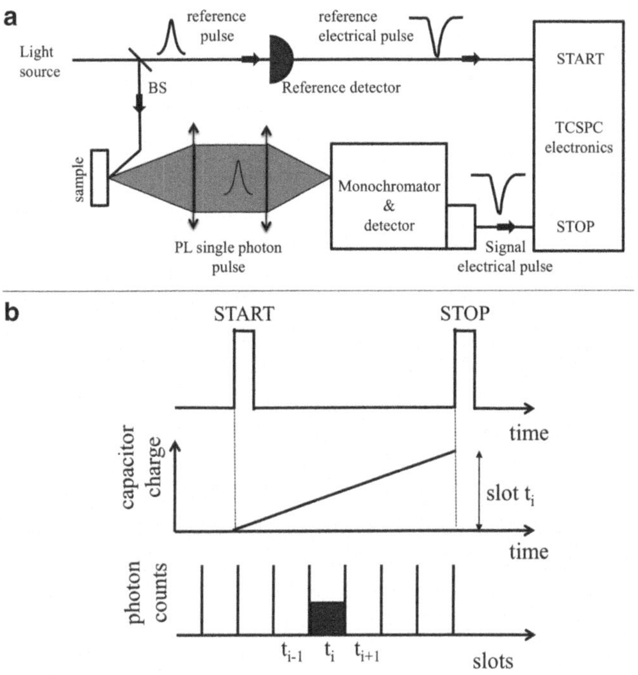

Fig. 8.14 Setup and detection principles of the START–STOP technique. (**a**) Implementation of the experiment. (**b**) Principle of the TCSPC

beam constitutes the reference signal, detected by a fast photodiode. The second is used as excitation for the photoluminescence, which is then collected by a system of lenses and dispersed by a monochromator. Finally, the light pulse will reach the detector with a certain delay compared to the reference signal. At this point, the light signals are converted into their electrical counterparts and fed to the START–STOP electronics. The reference and luminescence electrical signals generated by the detectors are sent to the start and stop inputs, respectively, where a constant fraction discriminator (CFD) evaluates their intensity with respect to set noise and multiple event levels. Only signals larger than the minimum and lower than the estimated double photon occurrence levels proceed further: this effectively removes low-intensity electrical noise from the signal. At this point, the electrical pulses are shaped into standard short rectangular signals ($\leq$ 10 ns TTL pulses). A delay stage is then used to shift the entire measurement into the detection temporal window, as the latter can be considerably smaller than the laser repetition period. Both signals subsequently reach the time-to-amplitude converter (TAC). The TAC (Fig. 8.14b) can be viewed as a high precision stopwatch: the reference pulse provides the start signal, whereas the luminescence pulse associated with one single photon emission gives the stop time. It is important to note that reference and luminescence signals are not simultaneous. At the start pulse arrival, the TAC begins charging a capacitor until the stop pulse reaches itself the TAC. Thus, the time elapsed between the start and stop is proportionally converted into the amplitude of an analog voltage. A single count (corresponding to the detected photon) will then be assigned to the memory channel corresponding to the time slot t_i determined by the TAC-generated amplitude digitalized by an analog-to-digital converter. The repetition of this sequence generates the histogram of the photon counts versus arrival time, effectively reproducing the luminescence dynamics.

This counting method constitutes the forward measuring method. In case of very low signals, it may occur that no luminescence signal reaches the detector for several start signals. In order to decrease the operating charge of the electronics, the counting system can be operated in the reverse mode, namely, in the STOP–START method. The role of the reference and luminescence signals is interchanged: even if a very low photon flux is generated as luminescence, the TCSPC electronics is started only when a photon issued from the sample reaches the detector, and it is stopped by the subsequent reference signal. The TCSPC software usually automatically adjusts the representation of this inverted histogram to show the luminescence dynamics correctly. The advantage of the reversed mode is that it considerably reduces the dead time of the electronics. In the normal operation mode, this corresponds to the accumulation of scans where no PL photons are detected, up to the capacitor maximum charge where the system is then reset, fixing the upper temporal limit.

A few drawbacks of this technique have to be highlighted. In addition to the modest time resolution (a few tens of picoseconds typically, see Sect. 8.5.2), the low count rate demanded to fulfill the single photon event condition often imposes long acquisition times: care has to be taken on the excitation laser drifts among other factors reducing the accuracy of the measurement. Moreover, as for the up-conversion technique, the single channel nature of the method requires multiple acquisitions at different wavelengths if the full dynamics of a spectrally large signal has to be reconstructed.

8.5.2 TCSPC Temporal Resolution and Sensitivity

TCSPC is a versatile technique capable of resolutions of a few tens of picoseconds and suitable for measuring dynamics in a very vast temporal window, limited in the higher range by the laser repetition frequency. Its characteristics make the TCSPC a valuable and relatively low cost technique for ultra-fast time-resolved photoluminescence. Its temporal resolution is determined by the convolution of the response time of the main elements constituting the setup: the light source pulse width, the TCSPC electronic jitter, and the detector. It is often the latter element that sets the temporal resolution. The minimum Transit Time Spread for micro-channel plates or the response time for avalanche photodiodes is often a few tens of picoseconds, at least an order of magnitude greater than picosecond laser system pulses and TCSPC electronics jitter. As said before, a histogram of the number of detected photons (the events) as a function of time memory channel is built. The frequency of events in each channel t_i must be sufficient to reduce the statistical noise due to the Poisson distribution: $\Delta N_i/\bar{N}_i \cong 1/\sqrt{\bar{N}_i}$, where $\bar{N}_i$ represents the average event number in the t_i time slot. On the contrary, the photodetector dark current adds up to the signal so that the sensitivity is eventually limited by the detector detectivity, electronics, and statistical noises.

8.5.3 Application Examples

Figure 8.15a presents an example of a photoluminescence decay time of a single InGaAs/GaAs quantum dot isolated by use of a confocal microscope [51], measured by the TCSPC technique. The sample is excited by a 80-MHz mode-locked Ti:sapphire laser delivering 1.2-ps (FWHM) pulses at 867 nm, and the luminescence intensity is dynamically recorded by a single-photon avalance photodiode (SPAD) with $\tau_R = 350$ ps time resolution. The electrical signal generated by the SPAD is then sent to a TCC900 Edinburgh Instrument card performing the time-correlated single-photon detection providing 4,096 time slots (channels) and less than 25 ps electronics jitter. The whole system is sensitive enough to detect the luminescence dynamics of a single quantum dot filled with a single electron-hole pair. It is clear here that the temporal resolution is limited by the SPAD constituting the "slow" response time element of the chain. The measured lifetime is here about $\tau_r = 1$ ns.

Another application of such method is currently used to record the correlation function $g^{(2)}(\tau)$ of single photon sources. Here, the source can be either pulsed (Fig. 8.15b) or continuous wave (Fig. 8.15c). A confocal microscope isolates some single radiant nano-object. The detection follows the Hanbury Brown–

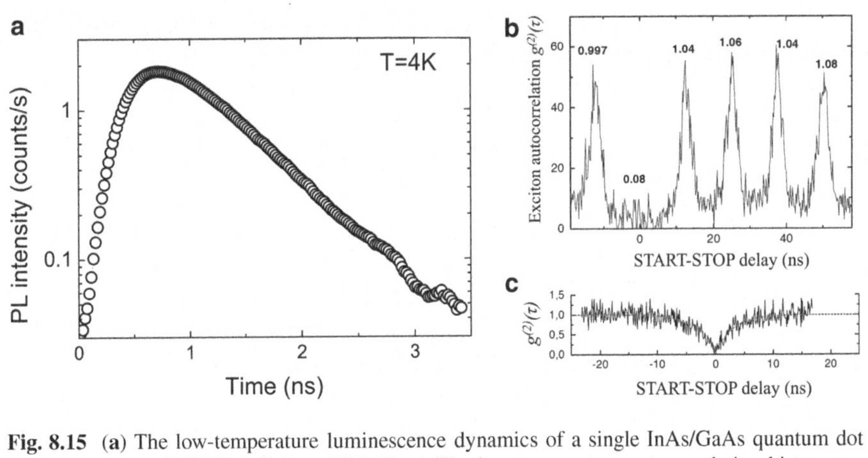

Fig. 8.15 (**a**) The low-temperature luminescence dynamics of a single InAs/GaAs quantum dot measured by the TCSPC technique [51]. (**b**, **c**) The low-temperature autocorrelation histograms of the exciton line in a single InAs/GaAs quantum dot under pulsed (**b**) and continuous wave (**c**) excitation, respectively [52]. The curve (**c**) clearly displays a photon antibunching behavior, characteristic of single photon sources

Twiss scheme [52]. Two examples of $g^{(2)}(\tau)$ correlation curves are shown in Fig. 8.15b, c for, respectively, a pulsed and a continuous wave excitation applied to the investigation of the luminescence properties of a single InAs/GaAs quantum dot. The correlation time of the exciton line τ_c can be extracted and the obtained value of $g^{(2)}(0)$ demonstrates the single photon source character of these quantum dots. Here, under low excitation power, $\tau_c \cong \tau_r \cong 1$ ns, the quantum dot radiative exciton lifetime.

8.6 Time-Resolved Optical Spectroscopy: A Synoptic Comparison of the Different Techniques

To conclude this chapter, we propose a synoptic comparison of the characteristics of the different time-resolved optical spectroscopic techniques presented here. In the table below, the goodness of a characteristic is proportionally coded to the number of star used: one star indicates "good," whereas three stars denote "very-good."

Technique/Property	Sensitivity	Time resolution	Spectral domain	Intensity dynamics	Temporal range	System cost
Streak camera	**	**	*	*[a]	**[b]	*
Up-conversion	*[c]	***	***	***	**	**
Pump–probe	**	***	*[d]	***	**[e]	***
Start–Stop	***	*	*	***	***	***

[a]Except in single photon counting mode

[b]In triggered or dual-time base modes. In synchroscan mode it is limited to ~2 ns

[c]Except for the near infrared region (above ~1 μm) where its sensitivity is the highest compared to other detection techniques

[d]Limited by the available pulsed laser wavelengths unless a white light continuum is used

[e]Depending on the pump laser repetition rate

References

1. C. Rullière, *Femtosecond Laser Pulses* (Springer, Berlin, 2004)
2. A.P. Heberle, J.J. Baumberg, K. Köhler, Phys. Rev. Lett. **75**, 2598 (1995)
3. M.C. Adams, W. Sibbett, D.L. Bradley, Opt. Commun. **273**, 26 (1978)
4. D.J. Bradley, W. Sibbett, Appl. Phys. Lett. **27**, 382 (1975)
5. M. Yamashita, T. Honda, T. Sato, K. Aizawa, IEEE Trans. Instrum. Measur. **IM-32**, 124 (1983)
6. F. Heisel, J.A. Miehé, B. Sipp, Rev. Sci. Instrum. **52**, 992 (1981)
7. P. Geist, F. Heisel, A. Martz, J.A. Miehé, Opt. Commun. **45**, 992 (1983)
8. Hamamatsu Photonics K. K., Japan
9. C.W. Bates, Phys. Rev. Lett. **47**, 204 (1981)

10. F. Gex, R. Alexandre, D. Horville, C. Cavailler, N. Fleurot, M. Nail, D. Mazataud, E. Mazataud, Rev. Sci. Instrum. **54**, 161 (1983)
11. B.P. Varma, C. Ghosh, J. Phys. D: Appl. Phys. **6** 628, (1973)
12. L. Lombez et al., Appl. Phys. Lett. **87**, 252115 (2005)
13. J.E. Midwinter, Appl. Phys. Lett. **14**, 29 (1969)
14. J.E. Midwinter, J. Warner, J. Appl. Phys. **38**, 519 (1967)
15. J. Shah, IEEE J. Quant. Electron. **24**, 276 (1988)
16. W.A. Hugel, M.F. Heinrich, M.U. Wehner, M. Wegener, QELS '99 Technical Digest, Quantum Electronics and Laser Science Conference, (1999)
17. F. Zernike, J.E. Midwinter, *Applied Nonlinear Optics* (Wiley, New York, 1973)
18. S. Haacke, R.A. Taylor, I. Bar-Joseph, M.J.S.P. Brasil, M. Hartig, B. Deveaud, JOSA B **15**, 1410 (1998)
19. A. Yariv, *Optical electronics* (Saunders Collège, USA, 1991)
20. X. Marie, P. Le Jeune, T. Amand, M. Brousseau, J. Barrau, M. Paillard, R. Planel, Phys. Rev. Lett. **79**, 3222 (1997)
21. D. Lagarde, H. Carrère, P.-M. Chassaing, A. Balocchi, X. Marie, N. Balkan, W.J. Schaff, Phys. Stat. Sol. B **248**, 1180–1182 (2011)
22. M. Paillard, X. Marie, E. Vanelle, T. Amand, V.K. Kalevich, A.R. Kovsh, A.E. Zhukov, V.M. Ustinov, Appl. Phys. Lett.**76**, 76 (2000)
23. R.E. Worsley, N.J. Traynor, T. Grevatt, R.T. Harley, Phys. Rev. Lett. **76**, 3224 (1996)
24. A. Malinowski, R.S. Britton, T. Grevatt, R.T. Harley, D.A. Ritchie, M.Y. Simmons, Phys. Rev. B **62**, 13034 (2000)
25. J.J. Baumberg, D.D. Awschalom, N. Samarth, H. Luo, J.K. Furdyna, Rev. Lett. **72**, 717 (1994)
26. S.A. Croocker, D.D. Awschalom, N. Samarth, IEEE J. Select. Top. Quant. Electron. **1**, 1082 (1995)
27. B. Eble, C. Testelin, P. Desfonds, F. Bernardot, A. Balocchi, T. Amand, A. Miard, A. Lemaître, X. Marie, M. Chamarro, Phys. Rev. Lett. **102**, 146601 (2009)
28. R.R. Alfano, S.L. Shapiro, Phys. Rev. Lett. **24**, 592 (1970)
29. W.H. Knox, C. Hirlimann, D.A.B. Miller, J. Shah, D.S. Chemla, C.V. Shank, Phys. Rev. Lett. **56**, 1191 (1986)
30. J. Collet, J.L. Oudar, T. Amand, Phys. Rev. B **34**, 5443 (1986)
31. T. Amand, J. Collet, B.S. Razbirin, Phys. Rev. B **34**, 2718 (1986)
32. S.V. Frolov, Z.V. Vardeny, Rev. Sci. Instrum. **69**, 1257 (1998)
33. P. Borri, F. Romstad, W. Langbein, A.E. Kelly, J. Mørk, J.M. Hvam, Opt. Expr. **7**, 107 (2000)
34. M. Haddad, R. André, R. Frey, C. Flytzanis, Solid State Commun. **111**, 61 (1999)
35. C.V. Shank, R.L. Fork, R.F. Leheny, J. Shah, Phys. Rev. Lett. **42**, 112 (1979)
36. J.L. Oudar, D. Hulin, A. Mingus, A. Antonetti, F. Alexandre, Phys. Rev. Lett. **55**, 2074, (1985)
37. S. Bar-Ad, I. Bar-Joseph, Phys. Rev. Lett. **68**, 349, (1992)
38. T. Kuroda, T. Yabushita, T. Kosuge, K. Taniguchi, T. Chinone, N. Horio, A. Tackeuchi, Appl. Phys. Lett. **85**, 3116 (2004)
39. C. Brimont, M. Gallart, O. Crégut, B. Hönerlage, P. Gilliot, Phys. Rev. B **77**, 125201 (2008)
40. F. Meier, B.P. Zakharchenya (eds), *Optical Orientation* (North-Holland, Amsterdam, 1984)
41. C. Camilleri, F. Teppe, D. Scalbert, Y.G. Semenov, M. Nawrocki, M. Dyakonov, J. Cibert, S. Tatarenko, T. Wojtowicz, Phys. Rev. B **64**, 085331 (2001)
42. J.M. Kikkawa, D.D. Awschalom, Nature **397**, 139 (1999)
43. H.M. Zhao, L. Lombez, B.L. Liu, B.Q. Sun, Q.K. Xue, D.M. Chen, X. Marie, Appl. Phys. Lett. **95**, 041911 (2009)
44. S.A. Crooker, J.J. Baumberg, F. Flack, N. Samarth, D.D. Awschalom, Phys. Rev. Lett. **77**, 2814 (1996)
45. J. Hegarty, L. Goldner, M.D. Sturge, Phys. Rev. B **30**, 7346 (1984)
46. J.R. Salcedo, A.E. Siegman, D.D. Dlot, M.D. Fayer, Phys. Rev. Lett. **41**, 131 (1978)
47. A.R. Cameron, P. Riblet, A. Miller, Phys. Rev. Lett. **76**, 4793 (1996)
48. P.D. Maker, R.W. Terhune, Phys. Rev. **137**, A801 (1965)

49. D.A.B. Miller, D.S. Chemla, D.J. Eilenberger, P.W. Smith, A.C. Gossard, W. Wiegmann, Appl. Phys. Lett. **42**, 925 (1983)
50. S. Weiss, M.-A. Mycek, J.-Y. Bigot, S. Schmitt-Rink, D.S. Chemla, Phys. Rev. Lett. **69**, 2685 (1992)
51. Pierre François Braun, Ph.D. Thesis, INSA Toulouse, 2007
52. Emmanuel Moreau, Ph.D. Thesis, Université de Paris VI, 2002

Chapter 9
Raman Spectroscopy of Compound Semiconductors

Jordi Ibáñez and Ramon Cuscó

Abstract Raman spectroscopy has become a widely used characterization tool in today's semiconductor research. In this chapter, we provide an introductory background to the physics of Raman scattering and discuss present-day applications of Raman spectroscopy in the field of compound semiconductor physics. Illustrative examples of Raman studies are given on a variety of topics such as crystal quality assessment, strain determination, alloy composition, impurities, and free-charge characterization in doped semiconductors.

9.1 Introduction

Raman spectroscopy is a powerful spectroscopic tool for the investigation of materials. It is widely applied in many different areas of science and technology, from materials science, chemistry, or geology, to archeology, art identification, or forensics. Although the Raman effect (i.e., the inelastic scattering of light by elementary excitations in condensed matter systems) was discovered more than 80 years ago,[1] Raman spectroscopy has only become widely extended after the commercialization of inexpensive laser sources and the development of multichannel detectors in the 1980s and 1990s. Raman spectroscopy has found widespread use for the study and characterization of semiconductor materials and structures,

[1]Although independently discovered by Raman and Krishnan and by Landsberg and Mandelstam (see [1, 2]), the Raman effect was named after the Indian scientist Sir C. V. Raman, who was awarded the Nobel Prize in Physics in 1930 for this discovery.

J. Ibáñez (✉) · R. Cuscó
Institut Jaume Almera, Consell Superior d'Investigacions Científiques (CSIC), Lluís Solé i Sabarís s/n, 08028 Barcelona, Catalonia, Spain
e-mail: jibanez@ictja.csic.es; rcusco@ictja.csic.es

A. Patanè and N. Balkan (eds.), *Semiconductor Research*, Springer Series in Materials Science 150, DOI 10.1007/978-3-642-23351-7_9,

as it provides valuable information about the crystal quality, the strain state, or the composition of the samples in a nondestructive and contactless way. From a fundamental point of view, Raman scattering is highly attractive as it allows one to investigate the lattice dynamics of semiconductors and also to probe their electronic structure, the electron–phonon interaction, or free carrier effects.

The constant evolution of the available instrumentation, together with an ever-increasing theoretical and analytical insight, has led to several different approaches of Raman spectroscopy that are gaining considerable attention among researchers. These include surface-enhanced Raman scattering, nonlinear Raman spectroscopies (e.g., stimulated Raman scattering), scanning near-field Raman microscopy (SNOM Raman), tip-enhanced Raman spectroscopy, or inelastic X-ray scattering. These nonstandard methodologies will not be discussed here.

The main objective of this chapter is to present a brief overview of standard (spontaneous) Raman spectroscopy for the investigation of compound semiconductors. The text is mainly aimed at students or researchers with no or little experience in Raman spectroscopy whose investigations might benefit from this technique. In Sect. 9.2, we briefly introduce the theory of Raman scattering. The selection rules for zinc-blende and wurtzite compounds are discussed, as these are the most relevant crystalline structures of III–V and II–VI semiconductors. After briefly describing the most common, commercially available instrumentation (Sect. 9.3), we present several examples of Raman spectroscopy to investigate semiconductor materials (Sect. 9.4). The text is not restricted to any particular material system, although special attention is paid to III–V–N materials, since these are currently attracting a considerable deal of attention. Examples about the characterization of the crystal quality and strain of as-grown epilayers and ion beam implanted material, impurity incorporation, alloying effects, and the study of LO-plasmon coupling to determine the free-carrier concentration in polar semiconductors are given.

9.2 Raman Scattering by Phonons

9.2.1 The Raman Effect

As is well known, light scatters elastically when it encounters imperfections within a medium (single atoms, molecules, or dust particles in air, or surface roughness and dislocations in crystalline materials). Elastic scattering phenomena, also known as Rayleigh scattering, change the direction of propagation of light, but leave its frequency and wavelength unchanged [3, 4]. A very small fraction of light (around 1 per 10^6 photons depending on the wavelength and the electronic structure of the material) may also undergo inelastic scattering processes, generically known as Raman scattering. Although the term Raman scattering may in principle refer to inelastic light scattering by any elementary excitation in a medium (phonons, plasmons, magnons, electronic and spin excitations, etc.) [3–5], here we mainly focus on vibrational normal modes (i.e., phonons in crystalline materials and local vibrational modes associated with impurities).

In a first-order Raman process, a photon of energy $h\nu_i$ and wave vector $\boldsymbol{k}_i$ is scattered by one phonon of energy $E_q = \hbar\omega_q$ and wave vector $\boldsymbol{q}$, which is excited or absorbed in the process. Energy and momentum are conserved, and as a consequence, the energy and momentum of the scattered photon are given by

$$\hbar\omega_s = \hbar\omega_i \pm \hbar\omega_q, \tag{9.1a}$$

$$\hbar\boldsymbol{k}_s = \hbar\boldsymbol{k}_i \pm \hbar\boldsymbol{q}. \tag{9.1b}$$

In the previous expressions, the positive sign corresponds to phonon absorption (the so-called anti-Stokes component), while the negative sign refers to phonon excitation (Stokes component). In a standard Raman experiment, a monochromatic light source (a laser) is focused on the sample; light that has undergone inelastic scattering is collected and analyzed with the appropriate instrumentation (see Sect. 9.3).

Typical Raman spectra show the intensity of the collected light, usually backscattered radiation, as a function of the Raman shift, i.e., the difference between the wavenumber of the incoming and the scattered radiation, $(1/\lambda_i)$–$(1/\lambda_s)$, customarily expressed in cm^{-1} units. Given that photon wavenumbers, energies, and frequencies are all proportional quantities on account of Planck's equation ($E = \hbar\omega = hc/\lambda$), it is clear from energy conservation (9.1a) that the Raman shift is proportional to the energy and frequency of the excited/absorbed phonons. Thus, in Raman scattering jargon, phonon energies and phonon frequencies are commonly expressed in cm^{-1}.

In conventional Raman experiments, one most often detects the Stokes component because the probability of anti-Stokes processes is much smaller (a phonon population must exist before the phonon can be absorbed by the incoming photon). Since the phonon population in the lattice depends on the temperature of the sample, the intensity ratio between the Stokes and anti-Stokes components can be used to evaluate the temperature of layers, devices, or nanostructures [6].

One important consideration directly emerges from energy and wave vector conservation (9.1a, b). First, note that the largest phonon wave vector accessible in a Raman experiment is attained in a backscattering configuration, for which $k_s \sim k_i = 2\pi n/\lambda_i$, where n is the refractive index of the medium. Consequently, the wave vector of the excited or absorbed phonon in a one-phonon process is $q \lesssim 4\pi n/\lambda_i$, which is much smaller than the wave vector of the phonons at the edge of the first Brillouin zone (i.e., $4\pi n/\lambda_i \ll 2\pi/a_0$, where a_0 is the lattice constant). Thus, it can be assumed that the wave vector of the phonons that participate in first-order Raman scattering processes is equal to zero. This implies that in a crystal with N atoms (and, therefore, $3N$ normal modes of vibration), only first-order features arising from zone-center optical modes may be expected in the Raman spectrum. Note that zone-center acoustic phonons correspond to rigid translations of the crystal and, therefore, the frequency of acoustic phonons with $q \sim 0$ is vanishingly small.

Higher-order Raman processes, in which two or more phonons are excited or absorbed, are also possible. Since such processes are less likely, the intensity of higher-order Raman bands is usually low. In this case, however, phonons with non-zero wave vector can participate in the Raman scattering process provided that the sum of the wave vectors of all phonons involved is equal to zero.

9.2.2 Macroscopic Theory and Selection Rules

Group theory allows one to classify the vibrational modes of molecules and crystals in terms of their symmetry properties [3, 4, 7–9]. By using group theoretical arguments, it can be shown that the zone-center optical modes of zinc-blende crystals belong to the triply degenerate Γ_{15} irreducible representation (also denoted Γ_4 in Koster notation), which, as a consequence of the cubic symmetry of the zinc-blende structure, transforms like the three components of a vector [3,4,7,9]. While cubic symmetry imposes that these modes be degenerate at $q = 0$ in zinc-blende crystals, for $q \gtrsim 0$ the longitudinal optical (LO) modes exhibit higher frequencies than the transverse optical (TO) modes because of the additional restoring force originated by the macroscopic electric fields associated with the longitudinal vibrations in the polar crystal [7]. In the case of wurtzite crystals, the zone-center optical modes can be decomposed as $\Gamma_{\text{opt}} = 2E_2 + 2B_1 + A_1 + E_1$ (in molecular notation, see for instance [3,4]). Phonons belonging to the E_2 and B_1 representations are nonpolar, while those with E_1 and A_1 symmetry are polar modes. In the case of A_1 modes (one-dimensional irreducible representation), the atomic displacement takes place along the c axis of wurtzite, while in E_1 modes (two dimensional), the atomic displacements are perpendicular to the c axis. Thus, for phonon propagation parallel (perpendicular) to the c axis, the A_1 modes are longitudinal (transverse), while those with E_1 symmetry are transverse (longitudinal). One of the modes with E_2 symmetry mainly involves motion of the heavy atom, and therefore it is a low-frequency mode (E_{2l}). Conversely, the remaining mode of E_2 symmetry involves motion of the lighter atom, and as a consequence, it is a high-frequency mode (E_{2h}).

Raman spectroscopy is a very powerful tool to study the symmetry of zone-center phonons. The intensity of a given first-order Raman peak depends on the polarization of the incoming and scattered light through the so-called Raman tensor, the form of which is given by the symmetry of the phonon involved in the process (i.e., it depends on the crystal structure of the material). For a detailed discussion on the theory of Raman scattering in crystals, see for instance [3, 4]. The textbook *Fundamentals of Semiconductors* by Yu and Cardona [7] offers an excellent introductory discussion of this topic.

The dependence of the Raman intensity on the polarization of the incident and scattered light can be understood within the framework of a semiclassical macroscopic theory of light scattering. Within this framework, the elastic scattering of light by a material can be viewed as the emission of electromagnetic radiation by a large ensemble of dipoles that are induced to oscillate by the incident light, i.e., by an oscillating electric field $\boldsymbol{E} = E_0 \cos(\boldsymbol{k}_i \cdot \boldsymbol{r} - \omega_i t)\hat{\boldsymbol{e}}_i$. Here, $\hat{\boldsymbol{e}}_i$ is an unitary vector indicating the polarization direction of the incident radiation. The intensity of the light emitted by the oscillating dipoles polarized in a given direction $\hat{\boldsymbol{e}}_s$ is proportional to $|\hat{\boldsymbol{e}}_s \cdot \boldsymbol{P}|^2$, where $\boldsymbol{P} = \chi(\omega_i, \boldsymbol{k}_i)\boldsymbol{E}$ is the polarization induced by the incident radiation and $\chi(\omega_i, \boldsymbol{k}_i)$ is the electric susceptibility of the crystal [3, 4, 10]. The form of $\chi(\omega_i, \boldsymbol{k}_i)$, which in general is a second-rank tensor, is determined by the electronic structure of the material [4].

The inelastic scattering of light by vibrational excitations can be included in the previous treatment by expanding the unperturbed susceptibility, $\chi(\omega_i, \boldsymbol{k}_i)$, in terms of the normal coordinate of vibration of the ions, $\boldsymbol{\xi} = \xi_0 \cos(\boldsymbol{q} \cdot \boldsymbol{r} - \omega_q t)\hat{\boldsymbol{\xi}}$, where $\hat{\boldsymbol{\xi}}$ is an unitary vector parallel to the atomic displacements. Thus,

$$\chi(\omega_i, \boldsymbol{k}_i, \boldsymbol{\xi}) = \chi(\omega_i, \boldsymbol{k}_i) + \frac{\mathrm{d}\chi(\omega_i, \boldsymbol{k}_i)}{\mathrm{d}\boldsymbol{\xi}} \cdot \boldsymbol{\xi} + \cdots . \tag{9.2}$$

By introducing this expression in the definition of $\boldsymbol{P}$, it is found that in addition to the elastic component that oscillates in phase with the incident radiation, the induced polarization exhibits two different sinusoidal terms that oscillate with frequencies $\omega = \omega_i \pm \omega_q$ and wave vectors $\boldsymbol{k} = \boldsymbol{k}_i \pm \boldsymbol{q}$, as in (9.1). These two terms correspond to the Stokes and anti-Stokes components of the inelastically scattered light. Within this approach, higher-order Raman processes are described by higher-order terms in the expansion of $\chi(\omega_i, \boldsymbol{k}_i)$.

The total intensity of light scattered by the medium and polarized in the direction defined by the vector $\hat{\boldsymbol{e}}_s$ can be obtained by calculating the time average of the power radiated by the oscillating dipoles. By using the last term of (9.2), this leads to

$$I \propto \left| \hat{\boldsymbol{e}}_s \cdot \overleftrightarrow{\mathcal{R}} \cdot \hat{\boldsymbol{e}}_i \right|^2 \langle \xi \xi^* \rangle , \tag{9.3}$$

where $\overleftrightarrow{\mathcal{R}} = (\mathrm{d}\chi/\mathrm{d}\boldsymbol{\xi})\hat{\boldsymbol{\xi}}$ is a second-rank tensor known as the Raman tensor. The brackets represent averages over time. Equation (9.3) holds for the Stokes component; a similar expression with $\langle \xi^* \xi \rangle$ instead of $\langle \xi \xi^* \rangle$ applies in the case of the anti-Stokes component. Within the formalism of second quantization, these two quantities can be evaluated by substituting the normal displacements ξ and ξ^* by phonon creation and annihilation operators, ξ and $\xi^\dagger$ so that $\langle \xi \xi^\dagger \rangle = \hbar[n(\omega) + 1]/2\omega_q$ and $\langle \xi^\dagger \xi \rangle = \hbar n(\omega)/2\omega_q$, where $n(\omega)$ is the Bose–Einstein occupation factor. These expressions, together with (9.3), can be employed to evaluate the temperature of a sample if the intensities of the Stokes and anti-Stokes components are determined experimentally.

From the definition of $\overleftrightarrow{\mathcal{R}}$, it follows that the Raman tensor transforms like the irreducible representation of the phonon involved in the first-order Raman scattering process. Consequently, Raman tensors for the different modes can be constructed by using group theory techniques. In [3, 4], the tensors for the Raman-active modes corresponding to the 32 crystal classes are listed. By using (9.3) and

Table 9.1 Selection rules for several backscattering polarization configurations in zinc-blende crystals

Scattering geometry	TO mode	LO mode
$z(xy)\bar{z}$		A
$z(xx)\bar{z}$		
$z(y'y')z$		A
$z(x'y')\bar{z}$		
$y'(z'z')\bar{y}'$	A	
$x''(z''z'')\bar{x}''$	A	A

Symmetry allowed modes are indicated with the letter A. The scattering geometry is expressed by using Porto's notation: in a given $a(bc)d$ configuration, a and d denote the direction of the incident and scattered light, respectively; b and c denote the polarization direction of the incident and scattered light, respectively. $x = [100]$, $y = [010]$, $z = [001]$, $\bar{x} = [\bar{1}00]$, $\bar{y} = [0\bar{1}0]$, $\bar{z} = [00\bar{1}]$, $x' = [110]$, $y' = [1\bar{1}0]$, $x'' = [111]$, $y'' = [1\bar{1}0]$, $z'' = [11\bar{2}]$, etc. In the cubic lattice, the x, y, and z directions are equivalent

Table 9.2 Symmetry allowed (A) Raman modes for several scattering geometries in wurtzite crystals

Scattering geometry	Raman-active modes in wurtzite crystals				
	E_2	A_1(TO)	E_1(TO)	A_1(LO)	E_1(LO)
$z(xx)\bar{z}$	A			A	
$z(xy)\bar{z}$	A				
$x(yy)\bar{x}$	A	A			
$x(zz)\bar{x}$		A			
$x(zy)\bar{x}$			A		
$x(yz)y$			A		A

Porto's notation is used (see Table 9.1). Here, $z = [001]$ direction is parallel to the c axis of the wurtzite structure. In the hexagonal lattice, the x and y directions, perpendicular to the c axis, are equivalent and can be arbitrarily defined

the appropriate tensors for zinc-blende and wurtzite crystals, selection rules for polarized Raman scattering measurements can be obtained. Tables 9.1 and 9.2 show the selection rules for different scattering geometries in zinc-blende and wurtzite crystals, respectively. According to the selection rules, the TO mode of zinc-blende compounds is forbidden in backscattering experiments with incidence on (001) faces, whereas the TO and LO modes are simultaneously observed on (111) faces. For a more detailed discussion of the selection rules in zinc-blende materials, see for instance [4, 7]. In the case of wurtzite compounds, it is found that the B_1 modes are not Raman active. These modes are referred to as silent modes, since they are neither Raman nor infrared active. The E_2 and A_1(LO) modes can be detected in a typical backscattering configuration with incidence along the c axis of wurtzite and parallel polarizations. The A_1(TO) and E_1(TO) modes can be observed for incidence along a direction perpendicular to the c axis, while the E_1(LO) mode is allowed in a 90° configuration.

In order to illustrate how the selection rules are applied to identify the different modes of a crystalline material, we show in Fig. 9.1 room-temperature Raman spectra of a wurtzite ZnO crystal acquired in different polarization configurations [11].

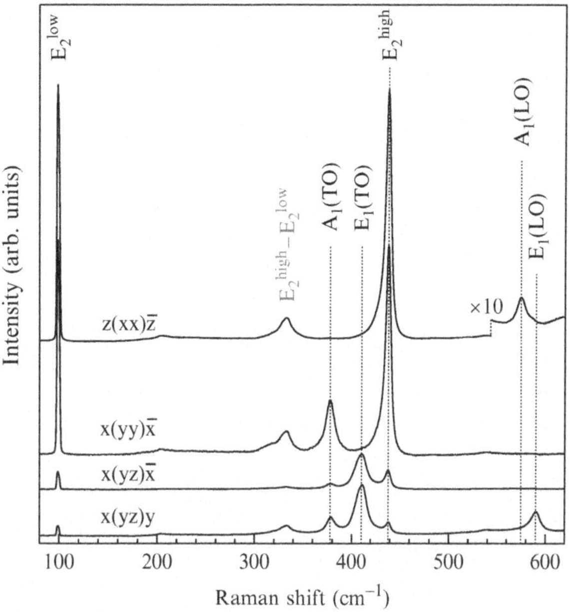

Fig. 9.1 Room-temperature first-order Raman spectra of ZnO in different scattering geometries [11]

As expected from the selection rules, the $z(xx)\bar{z}$ spectrum is dominated by the two intense, sharp E_2 modes at 99 and 438 cm^{-1}. The low-frequency E_2 mode, involving mainly Zn motion, displays an extremely narrow Raman peak whose linewidth is basically determined by the spectral bandwidth of the experimental setup. The A_1(LO) mode is observed as a weak band at 574 cm^{-1}. The prominent feature at 333 cm^{-1} corresponds to second-order scattering [11]. In the $x(yy)\bar{x}$ spectrum, in addition to the strong E_2 modes, the A_1(TO) mode appears at 378 cm^{-1}. An additional peak emerges at 410 cm^{-1} in the $x(yz)\bar{x}$ and $x(yz)y$ spectra, which is assigned to the E_1(TO) mode. Finally, the E_1(LO) mode is observed at 590 cm^{-1} in the $x(yz)y$ spectrum. Note that a residual intensity of the forbidden E_2 modes is seen in the $x(yz)\bar{x}$ and $x(yz)y$ spectra. Residual forbidden signals may arise from a relaxation of the selection rules due, for instance, to lattice disorder or from a slight misalignment of the sample surface and/or of the optical path.

9.2.3 Resonant Raman Scattering

The energy of the excitation radiation used in conventional Raman experiments is much larger than the energy of the phonons involved, which is typically of a few tens of meV. Thus, it is clear that direct photon–phonon interaction is not responsible for the creation/annihilation of phonons in the inelastic scattering of light; electrons and holes must mediate the process. From a quantum mechanical point of view, a first-order Raman event can be viewed as a three-step process in which (1) an incoming photon excites an electron–hole pair; (2) the (virtual) electron or hole excites or absorbs a phonon; and (3) the electron–hole pair recombines and the outgoing photon is emitted. The probability of first-order Raman scattering can be calculated with time-dependent perturbation theory as a function of the strength of the electron–photon and electron–phonon coupling [3, 7]. The electron–phonon interaction takes place through the so-called deformation potential mechanism, by which the electronic bands are modulated by the bond distortions associated with the phonons. This is just the quantum mechanical analog of the modulation of the classical electronic susceptibility introduced in (9.2) to describe the *allowed* Raman scattering in continuous media.

For the present discussion, it suffices to keep in mind that the probability obtained from perturbation theory contains denominators of the form $E_g - \hbar\omega_i$ and $E_g - \hbar\omega_s$, where E_g refers to any direct band gap of the material (the numerators basically contain electron–photon and electron–phonon matrix elements). Thus, when the energy of the excitation radiation is close to a direct electronic transition, a strong enhancement of the Raman signal is observed. This resonance effect is important from an experimental point of view since it may allow the detection of otherwise very weak Raman modes. Resonant Raman scattering also enables one to probe the electronic structure of semiconductors by tuning the excitation wavelength. The observation of a resonance enhancement can be linked to critical points in the electronic structure of the material. On the other hand, since real electron–hole pairs are created under resonant excitation, these may also mediate the Raman scattering processes. Thus, additional electron–phonon interaction channels different from the deformation potential mechanism may contribute to the resonant enhancement of the inelastic light scattering. Such scattering mechanisms exhibit selection rules that differ from those of the symmetry-allowed scattering. They are usually referred to as *forbidden* scattering. For instance, electrons and LO phonons can directly interact through Coulomb interaction by means of the long-range macroscopic electric field associated with the LO modes (Fröhlich interaction). The Raman tensor is in this case diagonal [3, 4] and, as a consequence, LO modes excited through the Fröhlich scattering mechanism may only be observed when the polarization of the incident and the scattered light has the same direction.

9.3 Raman Instrumentation

Next, we briefly discuss the most common configurations used to perform conventional Raman scattering experiments. For a more detailed discussion, see for instance Chap. 2 of [5].

Today, most of the commercial, state-of-the-art Raman systems rely on single or triple dispersive spectrometers equipped with holographic gratings and multichannel detectors (liquid nitrogen-cooled or Peltier-cooled charge coupled devices). These systems can be used for IR, visible, and UV Raman experiments.

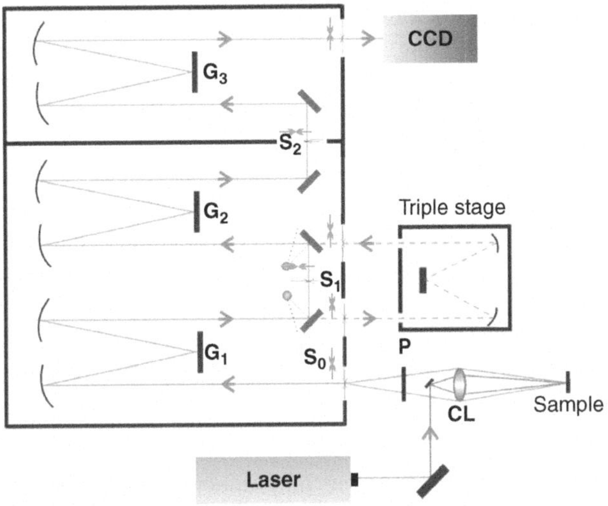

Fig. 9.2 Sketch of a triple spectrometer. The following elements are represented: collecting lens (CL), polarization analyzer (P), entrance slit (S_0), first-intermediate (S_1) and second-intermediate (S_2) slits, gratings G_i ($i = 1, 2, 3$), and coupled-charge device multichannel detector (CCD)

Since the wavelength of the Raman scattered light is usually very close to the excitation wavelength, the Rayleigh radiation needs to be eliminated. In single spectrometers, this is most often achieved with a holographic notch filter or a dielectric edge filter. These filters have a cut-off frequency around 100–200 cm^{-1}, and as a consequence, single spectrometers do not allow one to detect low-frequency modes. Filters with a much lower cut-off frequency are currently being developed. This is expected to overcome in a near future the limitation of single spectrometers for ultra-low-frequency measurements.

The resolution of single spectrometers depends on the focal length of the spectrometer, the groove density of the grating, and the aperture of the entrance slit, and is ultimately limited by the size of the pixels of the multichannel detector. In triple spectrometers, Rayleigh rejection can be achieved by using the so-called subtractive mode. Figure 9.2 shows a typical sketch of a triple spectrometer, with gratings represented by $G_i (i = 1, 2, 3)$. S_0 is the entrance slit, S_1 is the first intermediate slit, and S_2 is the second intermediate slit. In the subtractive mode, the first two gratings (G_1 and G_2) and the first intermediate slit (S_1) are set to block the stray light, while the third grating (G_3) acts as a single spectrometer. This mode of operation permits one to detect Raman signals as low as $\sim 10\,cm^{-1}$. Triple spectrometers can also be operated in an additive configuration, in which the three gratings disperse the collected light for high-resolution measurements. In this case, the optical path is modified by means of the triple stage (see Fig. 9.2).

Most modern Raman systems are equipped with a confocal microscope. The use of objectives with a high numerical aperture yields spatial resolutions as low as $\sim 1\,\mu m$ as well as high optical collection powers. In the case of transparent samples, as is the case, for instance, of wide band-gap semiconductors grown on transparent substrates such as sapphire, the confocal microscope enables one to reduce the depth of focus of the measurements and thus to reduce the Raman signal from the substrate. Modern confocal micro-Raman systems, in combination with motorized piezoelectric stages, allow one to map the Raman signal from a sample with spatial resolutions as low as 200–500 nm, which is particularly interesting to investigate inhomogeneous samples. For example, confocal micro-Raman measurements have allowed probing the location and orientation of single GaN nanowires ($\sim$100 nm diameter, 1 μm length) by mapping the Raman signal of the substrate material [12].

9.4 Applications of Raman Spectroscopy in Semiconductor Physics

From the discussion of Sect. 9.2, one may anticipate the type of information that can be gained from a Raman spectrum. In general, the first-order Raman spectrum of a highly crystalline material is dominated by intense, narrow peaks. Under nonresonant conditions, the number and symmetry of Raman-active modes depend on the crystal structure of the material. Thus, Raman spectra performed in different polarization geometries may be highly relevant for structural studies. Under resonance, the enhancement of the Raman signal can be used to probe the electronic structure of the material or to study the electron–phonon interaction. Given that the frequency of the phonons is determined by the particular bonds involved in the atomic vibrations, temperature, strain/stress, or phonon confinement affects the frequency of the Raman peaks. Thus, Raman spectroscopy can be used to evaluate the strain state of an epilayer, to measure the temperature of a device, or to determine the pressure coefficient of Raman-active phonons. From the latter, the mode Grüneisen parameters can be obtained.

On the other hand, the width of first-order features is primarily determined by the finite lifetime of the phonons, which is a consequence of anharmonic decay (the spectral resolution of the spectrometer used to analyze the collected light further contributes to the broadening of the Raman peaks). In a disordered crystal and also in nanocrystals, the $q \sim 0$ selection rule is relaxed and phonons from the whole Brillouin zone can participate in the inelastic scattering phenomena. In the extreme case of an amorphous material, the first-order Raman spectra closely resemble the one-phonon density of states of the perfect crystal. With regard to higher-order Raman bands, broad features may be expected due to the contribution of different combinations of phonons with $\boldsymbol{q}_i \neq 0$ (and $\sum_i \boldsymbol{q}_i = 0$ owing to wave vector conservation). In samples with a poor lattice quality, the weak and broad features arising from higher-order Raman processes are expected to vanish. Next, we present illustrative examples about these and other issues.

9.4.1 Crystal Quality and Strain State

Raman spectroscopy is widely employed to characterize the crystal quality of as-grown semiconductor structures and processed samples. For instance, Raman scattering may provide relevant information to optimize the growth of material prepared by bulk or epitaxial methods. Raman scattering is also a highly valuable tool for the characterization of material that has been subject to mechanical, chemical, or thermal treatments. In particular, Raman spectroscopy is one of the most effective techniques to monitor the lattice damage induced by an ion beam implantation process and to determine the optimal post-implantation annealing conditions [13–19].

To illustrate the connection between the Raman spectrum of a sample and the degree of lattice disorder, we show in Fig. 9.3a $z(y'y')\bar{z}$ Raman spectra of InP samples implanted with different doses of Si^+ ions. As is well known, the ion beam bombardment displaces lattice atoms and creates point defects and disordered regions in the crystal. Thus, the $q = 0$ selection rule is relaxed in the disordered crystal, giving rise to measurable shifts and broadenings of the Raman peaks. Also, lattice damage yields internal strains that further shift the first-order peaks. While the spectrum of virgin InP (curve A) is dominated by the allowed LO mode (see Table 9.1), an increasing signal from the forbidden TO mode shows up in curves B and C (samples implanted with 10^{12} and $5 \times 10^{12}\,cm^{-2}$, respectively). Also, two weak broad structures appear around 62 and 138 cm^{-1} in curve C, associated with disorder-activated transverse and longitudinal acoustic modes (DATA and DALA modes, respectively) [21]. With increasing implantation dose, a significant loss of intensity and a broadening of the first-order peaks, as well as an enhancement of the DATA and DALA modes, are observed (curves D and E). For an implantation dose of $10^{14}\,cm^{-2}$ (curve F), the long-wavelength first-order optical phonon peaks are almost replaced by a broad band centered around 300 cm^{-1} reflecting the whole density of states. In curve G ($5 \times 10^{14}\,cm^{-2}$), the Raman signal from the first-order modes is completely lost, which confirms the complete amorphization of the crystal. The broad structure that develops at 436 cm^{-1} can be assigned to phosphorous–phosphorous vibrations arising from the formation of phosphorous clusters [22].

Figure 9.3b shows first-order Raman spectra of an InP sample implanted at a high dose of Si ions ($5 \times 10^{14}\,cm^{-2}$), as in curve G of Fig. 9.3a, after rapid thermal annealing (RTA) for 10 s at low (300°C) and high temperatures (875°C). Clearly, when the sample is annealed at 300°C, it remains amorphous, since only a broad band is detected in the optical mode region. In contrast, the Raman spectrum of the sample annealed at 875°C displays well-defined TO and LO peaks, showing that a high degree of lattice recovery has been achieved. A detailed investigation of the optimal annealing conditions in similar Si-implanted InP layers indicates that the highest degree of lattice recovery and the highest electrical activations and Hall mobilities are actually achieved at 875°C [16, 17].

The deformation of the crystal lattice (i.e., strain) plays a key role in the optical and electronic properties of semiconductors. Given that strain gives rise to a

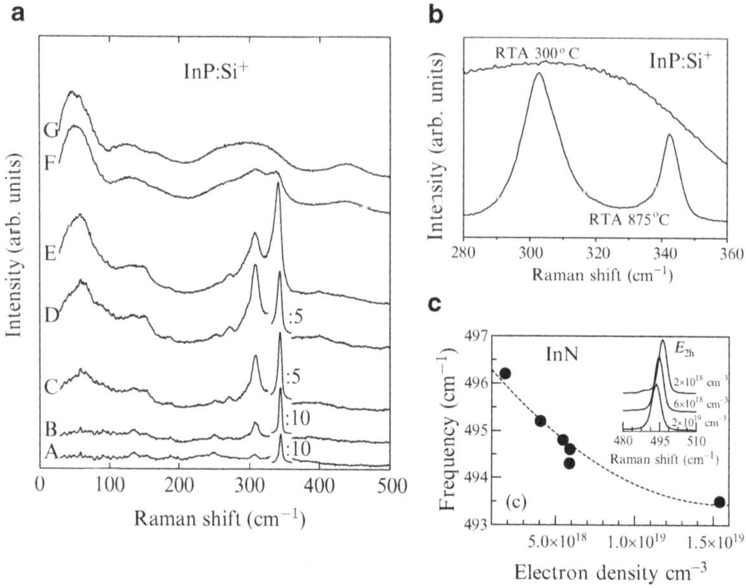

Fig. 9.3 (**a**) Room-temperature Raman spectra of InP samples implanted with 150 keV Si^+ ions at different doses [17]. A to G, respectively: virgin InP, 10^{12}, 5×10^{12}, 10^{13}, 5×10^{13}, 10^{14}, and $5 \times 10^{14}\,cm^{-2}$. (**b**) Raman spectra of InP implanted with 150 keV Si^+ ions at 150 keV and annealed at 300°C (*upper curve*) and 875°C (*lower curve*) [16]. (**c**) Frequency of the E_{2h} mode as a function of the residual electron concentration in different InN/sapphire epilayers grown by molecular beam epitaxy [20]. *Inset*: selected Raman spectra (E_{2h} mode) of InN/sapphire epilayers with different levels of residual electron densities

substantial modification of the vibrational properties of a crystal, Raman scattering is a very powerful tool to evaluate the strain state of semiconductor layers. One important source of strain in semiconductor heterostructures is the lattice mismatch between different layers, as is the case, for instance, of biaxial strain in InGaAs thin films pseudomorphically grown on GaAs substrates. Similarly, a residual strain field occurs in layers of group-III nitrides grown on hetero-epitaxial substrates like sapphire, Si(111), or SiC. Such strain fields are originated by lattice mismatch and by the different thermal expansion between the substrate and the nitride compounds. In addition, the presence of point defects in the lattice gives rise to hydrostatic strain that coexists with the uniaxial or biaxial strain arising from the mismatch between layers. The hydrostatic strain is particularly important in the III-nitrides [23].

It can be shown that up to first-order terms, the frequency of the optical phonons depends linearly on the components of the strain tensor through the so-called phonon deformation potential constants [24, 25]. Expressions for the dependence on strain of the optical phonons in cubic and hexagonal semiconductors can be found in the excellent compilation by S. Adachi [26], where values for the phonon deformation potentials for $q = 0$ optical phonons in several group-IV, III–V and II–VI semiconductors are given. Examples of the application of Raman scattering to probe the strain in semiconductor layers and structures can be found in [27] (bulk InGaAs), [28] (dilute GaAsN epilayers, see discussion in the next section), [29] (bulk InN), [30] (self-assembled InAs/GaAs quantum dots), and [31] (GaN/AlN quantum dots). A comprehensive discussion about the effect of strain on the phonons of superlattices is given by Jusserand and Cardona [32].

To illustrate the effect of strain on the Raman spectra of as-grown material, we show in the inset of Fig. 9.3c the E_{2h} peak obtained for different InN epilayers grown by molecular beam epitaxy (MBE) on sapphire. The as-grown samples contain different levels of background carrier concentration, N_e, which are originated by background donors associated with impurities or defects and also by strong electron accumulation at the InN surfaces [33]. It has been suggested that positively charged nitrogen vacancies along dislocations are responsible for the residual free charge [33]. A careful analysis of the Raman spectra reveals that the E_{2h} peak of these samples broadens with increasing N_e (see [20] for details), which indicates that the samples with higher residual electron concentrations have a poorer crystalline quality. On the other hand, a correlation between the E_{2h} frequency and N_e is also found (see Fig. 9.3c). As is well known, the nonpolar E_{2h} mode is very sensitive to biaxial strain [29]. Thus, the observed reduction of the E_{2h} frequency with increasing N_e can be attributed to the relaxation of the residual compressive strain present in the InN/sapphire layers [20]. Given that the relaxation of biaxial strain in these layers mostly occurs by forming threading dislocations, the observed correlation between the E_{2h} frequency, the E_{2h} width, and N_e seems to confirm that the background electron density is also determined by donor-type nitrogen vacancies associated with threading dislocations.

9.4.2 *Impurities and Alloys*

The ability of tuning the electronic properties of semiconductors through controlled doping is a key element in current solid state technology. Likewise, alloying of two or more pure semiconductor compounds is crucial for device applications, as it allows one to develop novel materials in which the band gap can be tailored by adjusting the composition. Raman spectroscopy is a powerful technique for the study and characterization of alloys and of intentional or unintentional impurity concentrations in semiconductors. The presence of impurities affects the vibrational properties of the host material, giving rise to different types of impurity modes which can be studied by means of Raman scattering. Similarly, the Raman spectrum of an alloy strongly depends on composition as well as on the vibrational properties of the pure end members. Thus, the Raman spectrum of an alloy can be used to assess its composition.

9.4.2.1 Impurity Modes

Let's first focus on the study of impurity modes. When an impurity atom replaces a heavier host atom, a local impurity mode, also known as local vibrational mode (LVM), may appear in the Raman spectra. The frequency of a LVM lies above the phonon frequencies of the host lattice. The amplitude of the atomic vibrations decays exponentially away from the mass defect, and the localization of the mode in the real and frequency domains (i.e., LVMs are long-lived modes) gives rise to a sharp Raman peak. This is the case, for instance, of C in Si, for which C concentrations as low as 10^{17} cm^{-3} have been detected by Raman spectroscopy [34]. Similarly, C in GaAs gives rise to a sharp LVM at 583 cm^{-1} [35]. LVM peaks arising from vacancies or other point defects not related to an impurity atom may also be observed. For example, in the case of implanted GaN, several features are usually attributed to vacancies and anti-sites induced by the ion beam implantation process [18, 19].

In order to illustrate the observation of sharp Raman peaks associated with local modes, we show in Fig. 9.4a Raman spectra of N^+-implanted and O^+-implanted ZnO. Clearly, the features that appear in the N^+-implanted sample can be attributed to N-related LVMs. The assignment of impurity modes is often controversial, and usually complex calculations are required to interpret the Raman spectra. Alternatively, the sensitivity of impurity modes to isotopic substitution can be exploited to investigate their origin. In the case of N-doped ZnO, Raman measurements carried out on ZnO samples implanted with different N isotopes indicate that N motion is not involved in the observed LVM peaks [36]. Thus, the observed LVMs do not arise from substitutional N.

In compound semiconductors with a large mass difference between constituent atoms, when a heavier impurity replaces the lighter atom, a *gap mode* may appear in the phonon gap between the acoustic and optical modes. This is the case, for instance, of As in GaP, where an As_P gap mode at 269 cm^{-1} has been reported [40]. The vibrational amplitude around the mass defect is more extended in gap modes than in LVMs. As a consequence, broader Raman lines are expected. An excellent review of impurity modes in a range of semiconductors of high technological interest has been written by McCluskey [41].

LVM spectroscopy of impurity complexes provides valuable information about the incorporation and electrical activation of donors/acceptors and non-intentional dopants in semiconductor materials. This is the case, for instance, of Mg acceptors and H impurities in group-III nitrides. In the case of Mg-doped GaN, three LVMs at 136, 262, and 656 cm^{-1} arising from to N–Mg_{Ga} vibrations are observed [42,43]. In MOCVD-grown material, it was found that the lack of *p*-type conductivity is related to the formation of Mg-N-H complexes. The LVMs associated with such complexes were detected by Raman spectroscopy, and a fingerprint LVM peak at 3,123 cm^{-1} was found to disappear after the thermal annealing treatment required to obtain *p*-type conductivity [44]. Other LVMs associated with Mg – H complexes and H-decorated vacancies are also detected in *p*-type MBE samples, but they are more stable to thermal annealing treatments [45]. Raman spectra of LVMs have also

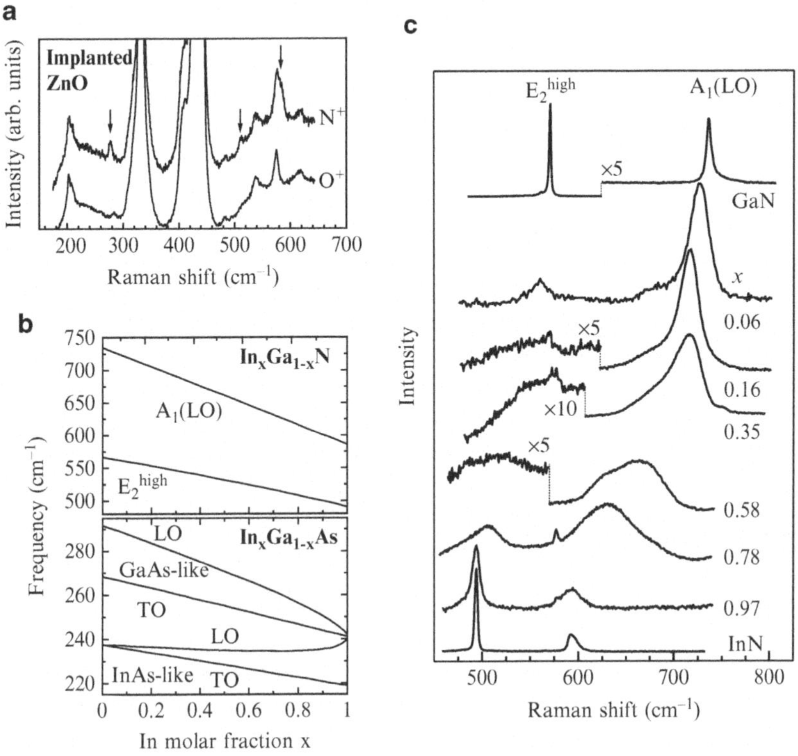

Fig. 9.4 (**a**) Raman spectra of ZnO layers implanted with N^+ and O^+ to achieve a volume impurity concentration of $\sim 5\times10^{19}\ cm^{-3}$ [36]. The *arrows* indicate the impurity modes associated with N. (**b**) Illustrative examples of one-mode behavior (InGaN, *top panel*) and two-mode behavior (InGaAs, *bottom panel*) of phonons in ternary alloys, as calculated in [37] and [38], respectively. (**c**) $z(xx)\bar{z}$ Raman spectra of InGaN layers over the whole composition range showing the A_1(LO) and E_{2h} modes evolving from one end-member binary to the other [39]

revealed the presence of C-H_n complexes in MBE-grown GaN layers due to residual contamination in the growth chamber [45].

The presence of impurities may change dramatically the optical and electronic properties of the host material. This is well illustrated by the surge of research activity in III–V dilute nitride alloys [46], where the incorporation of N impurities in small concentrations, typically below 5%, leads to strong band-gap reduction and bowing, large conduction band non-parabolicity, and anomalous changes in the electron effective mass. One should note that owing to the small N content of dilute nitrides, for lattice dynamics studies, it can be assumed that these compounds are not alloys but just heavily doped materials. In the next section, we will address the study of phonons in alloys over the entire composition range by means of Raman scattering. With regard to dilute nitrides, Raman scattering has been employed to obtain information about the incorporation of N and about the crystal quality of these compounds [28, 47–53]. A N_{As} LVM is detected at $470\,cm^{-1}$ in GaAsN [28, 47–49]. Similarly, the N_{As} LVM appears at $443\,cm^{-1}$ in dilute InAsN, although other N-related features that might arise from high-order N complexes (i.e., N_2-In-As_2 complexes and higher orders) or from non-substitutional configurations have also been reported [51, 52]. In the case of InGaAsN, LVM peaks arising from different $In_x - N - Ga_{4-x}$ substitutional N environments have been observed [50, 51, 53]. LVM frequencies obtained with first-principles density-functional calculations for dilute GaAsN and InGaAsN are in good agreement with the experimental values [54, 55].

The intensity of the LVMs can be used to determine the impurity concentration provided that previous calibrations are available. Kaschner et al. [56] proposed the intensity of impurity peaks observed in N-doped ZnO [see Fig. 9.4a] as a measure of the concentration of N incorporated to the host ZnO lattice. In the case of dilute GaAsN, it has been found that the intensity of the N_{As} LVM correlates well with the N content obtained by X-ray diffraction measurements [47].

9.4.2.2 Alloys

Raman scattering is a powerful technique to characterize the composition, crystal quality, and strain of alloys. Miscibility and phase segregation may be important issues which may also be investigated by means of Raman spectroscopy. The variation of the phonon frequencies with the molar fraction of the alloy provides a straightforward determination of the alloy composition. For a collection of polynomial fits to the composition dependency of the Raman frequency for a wide range of semiconductor alloys, see [5], p. 93. It should be emphasized that these relations apply to bulk alloys. In semiconductor heterostructures, a change in composition not only affects the intrinsic alloy frequency but also the elastic strain, and this must be taken into account in the analysis of the phonon frequency shifts (see, for instance, [30] for the case of InGaAs/GaAs self-assembled quantum dots).

In $A_xB_{1-x}C$ ternary alloys, the long wavelength optical phonons may display either a one-mode behavior, in which the frequencies vary almost linearly between the pure end members, or a two-mode behavior, in which two sets of optical modes corresponding to each sublattice (AC like and BC like) can be observed. Typical examples of frequency dependence on composition for one-mode and two-mode behavior alloys are shown in Fig. 9.4b for the case of InGaN (upper panel) and InGaAs (lower panel). In the small-x limit, the AC-like mode of the two-mode behavior alloy becomes the local or gap mode of the impurity A in the lattice BC. When the optical frequencies of the binary components differ significantly, A atoms (B atoms) are basically at rest in the BC-like (AC-like) modes. This results in a frequency downshift due to mass disorder effects. Phonon frequencies are further affected by microscopic strains associated with the distortion of AC and BC bond lengths in the alloy [38], which are important in alloys with very different lattice constants of the binary constituents. In polar lattices, a strong coupling between sublattice oscillators may occur via the macroscopic electric fields that accompany the LO modes, which results into oscillator strength transfers. Such effects give rise to a bowing in the frequency dependence on composition of the LO modes, as illustrated in the lower panel of Fig. 9.4b for the case of InGaAs (see [38] for details).

The phonon mode behavior in group-III nitride alloys has been studied theoretically by Grille et al. [37] using a modified random isodisplacement model. In contrast to the two-mode behavior of AlGaAs and InGaAs, the LO phonons in group-III nitride alloys generally display one-mode behavior on account of their light common anion. The phonon mode behavior of LO modes in group-III nitride alloys has been confirmed by Raman scattering measurements [39, 57]. While the E_{2h} modes of InGaN display one-mode behavior, there is strong experimental evidence that the E_{2l}, E_{2h}, A_1(TO), and E_1(TO) modes of AlGaN exhibit a two-mode behavior [57]. Figure 9.4c shows Raman spectra of InGaN alloys over the whole composition range. The spectra show the continuous evolution of the A_1(LO) and E_{2h} modes from InN to GaN as the In molar fraction is reduced. The substantial broadening of the Raman peaks in the alloy layers reflects the alloy disorder and composition fluctuations, which are particularly important in the mid composition range. The strong intensity displayed by the A_1(LO) mode in some of the spectra of Fig. 9.4c is due to the electronic (Fröhlich) resonance of the excitation wavelength used. This effect may not only complicate the analysis of the longitudinal polar modes, but it may also yield information about composition fluctuations in the alloy, as regions with different composition can be selectively enhanced. This topic is, however, beyond of the scope of this chapter.

9.4.3 Raman Scattering by LO Phonon–Plasmon Coupled Modes

When a significant free-carrier density N_e is present in a polar semiconductor, the longitudinal plasma oscillations couple to the LO phonon modes via their macroscopic electric fields, giving rise to LO phonon-plasmon coupled modes (LOPCMs) [58]. For high mobility plasmas, the frequency of these coupled modes is very sensitive to the free-charge concentration and can be used as a nondestructive, contactless probe to determine the charge density by optical methods.

In order to illustrate the behavior of the LOPCMs in high-mobility n-type binary semiconductors, we show in Fig. 9.5a Raman spectra of a series of n-type InP samples with electron concentrations in the $7 \times 10^{16} - 1 \times 10^{19}\,\mathrm{cm}^{-3}$ range. As can be seen in the figure, in addition to the TO and LO modes of bulk InP, two features whose frequency depends on N_e are also observed. With increasing N_e, the low-frequency coupled mode (L^-), which is visible for $N_e \gtrsim 10^{18}\,\mathrm{cm}^{-3}$, displays a phonon-like character and approaches the TO frequency. In contrast, the high-frequency mode (L^+) is plasmon like and its frequency increases as $\sim N_e^{1/2}$ approaching the plasma frequency $\omega_\mathrm{p} = (4\pi e^2 N_e/\varepsilon_\infty m^*)^{1/2}$, where ε_∞ is the high-frequency dielectric constant of the material and m^* is the effective mass of the free carriers. To understand this behavior, one should recall that owing to their longitudinal nature, the frequency of the LOPCMs basically corresponds to the zeroes of the dielectric function of the material, $\varepsilon(\omega) = \varepsilon_\infty + 4\pi(\chi_I + \chi_e)$ [7,58]. Here, $\chi_I = (\varepsilon_\infty/4\pi)(\omega_{\mathrm{LO}}^2 - \omega_{\mathrm{TO}}^2)/(\omega_{\mathrm{TO}}^2 - \omega^2 - \mathrm{i}\Gamma_i\omega)$ is the dielectric susceptibility of the ionic lattice, where the parameter Γ_i takes into account the finite lifetime of the ionic vibrations, and χ_e is the dielectric susceptibility of the free-carrier plasma [58]. As a first approximation, the free-electron contribution to the susceptibility can be included within a classical description of the plasma oscillation (Drude's model) so that $\chi_e(\omega) = -(\varepsilon_\infty/4\pi)\omega_{\mathrm{p}}^2/\omega(\omega + \mathrm{i}\Gamma_e)$ [58]. Γ_e is a phenomenological damping parameter related to the finite lifetime of the plasmons. Then, the zeroes of $\varepsilon(\omega)$ yield the LOPCM frequencies:

$$\omega_{\pm}^2 = \frac{1}{2}\left[\left(\omega_{\mathrm{LO}}^2 + \omega_{\mathrm{p}}^2\right) \pm \sqrt{\left(\omega_{\mathrm{LO}}^2 + \omega_{\mathrm{p}}^2\right)^2 - 4\omega_{\mathrm{p}}^2\omega_{\mathrm{TO}}^2}\right]. \tag{9.4}$$

This coupled mode analysis, which can be generalized to ternary [61, 62] and quaternary [63] alloys, reproduces fairly well the N_e dispersion of the L^- and L^+ branches as obtained experimentally. A more detailed dielectric model for the LOPCM Raman line shape was developed by Hon and Faust [64]. Within this model, based on fluctuation–dissipation theory, the LOPCM Raman intensity (Stokes component) can be expressed as

$$I(\omega) \propto [n(\omega) + 1]\mathcal{J}\left\{-\frac{1}{\varepsilon(\omega)}\left[\frac{\varepsilon_\infty}{4\pi} + 2A\chi_I - A^2\chi_I\left(1 + \frac{4\pi\chi_e}{\varepsilon_\infty}\right)\right]\right\}, \tag{9.5}$$

where $\mathcal{J}$ denotes the imaginary part, $n(\omega)$ is the Bose–Einstein factor, and $A = \omega_{\mathrm{TO}}^2 C/(\omega_{\mathrm{LO}}^2 - \omega_{\mathrm{TO}}^2)$, where C is the Faust–Henry coefficient [58]. The plasma contribution to the susceptibility χ_e can be treated using Drude's model or other approaches of higher complexity. One has to bear in mind that wave vector dispersion and conduction band non-parabolicity usually have an important effect on the LOPCM frequencies and should be taken into account [59]. The simple Drude's model neglects these effects and may yield significant errors [59]. In the simple, classical framework of Drude's model, wave vector dispersion can be introduced by considering a wave-vector dependent plasma frequency (hydrodynamical model) [65, 66], and the effects of band non-parabolicity can be taken into account by using the optical effective mass [66, 67]. Alternatively, one can use the Lindhard–Mermin dielectric function [58, 59, 68–70], a quantum mechanical approach that intrinsically incorporates wave vector dispersion and non-parabolicity effects. In addition, this latter approach also includes the effects of Landau damping, an energy loss mechanism by which the collective coupled mode excitations are absorbed by electronic single-particle excitations [68]. Such effects can substantially alter the coupled mode behavior, particularly in semiconductors in which LOPCMs occur at low energies [62,63].

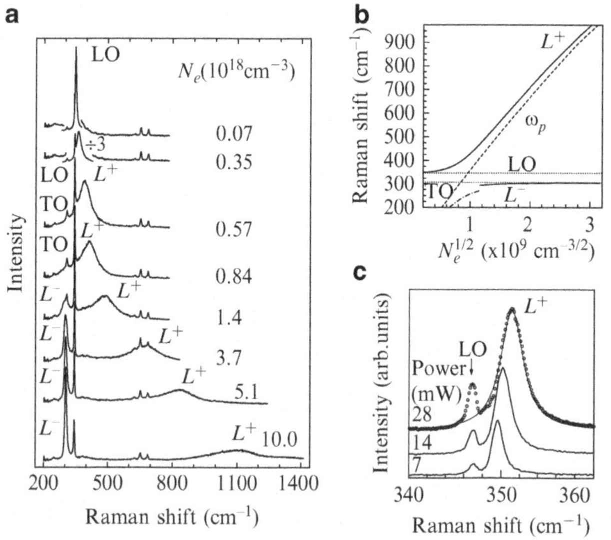

Fig. 9.5 (**a**) $z(xy)\bar{z}$ Raman spectra at 80 K of Si-doped InP samples for a wide range of free electron densities [59]. (**b**) Coupled mode dispersion calculated from (9.5) using the Lindhard–Mermin dielectric function. Note that the curvature of the plasma frequency as a function of electron concentration, $\omega_p(N_e^{1/2})$ (*dashed line*), reflects the conduction band non-parabolicity. For $N_e \lesssim 1.1 \times 10^{18}$ cm^{-3}, the L^- mode is strongly Landau damped and is not observed. The L^- dispersion for lower N_e (*dotted-dashed line*) was obtained from (9.4). (**c**) 80 K $z(xy)\bar{z}$ Raman spectra of semiconducting InP acquired at increasing excitation powers. The LO Raman peak arises from the surface depletion layer [60]. For the spectrum acquired at 28 mW, both the Raman data (*circles*) and the Lindhard–Mermin lineshape fit (*solid thin line*) are shown

We show in Fig. 9.5b the coupled mode dispersion calculated by using (9.5) (i.e., by taking the frequency corresponding to the peak intensities of calculated LOPCM lineshapes) together with the Lindhard–Mermin susceptibility. The behavior of the L^- and L^+ branches discussed above is well reproduced. It can be seen that in the strong coupling region, which in most cases includes the range of electron densities usually present in doped semiconductors, the phonon–plasmon interaction has to be suitably modeled to extract reliable information about the carrier density of the samples.

When a suitable lineshape model is used to fit the Raman spectra, the free-electron density values obtained from the Raman measurements are in good agreement with Hall data [59, 70], which validates Raman spectroscopy as a versatile contactless technique to characterize the free-carrier density in doped semiconductors. As can be seen in Fig. 9.5c for the case of a photoexcited plasma in InP (see discussion below), the Lindhard–Mermin lineshape model yields very good fits to the Raman spectra (see also [59,62,70] for the case of n-InP, n-InGaAs, and n-InN, respectively).

When the Raman spectra are excited with above-band gap radiation, photoexcited electron–hole pairs are generated. Provided that the surface recombination velocity is sufficiently low [71], the steady state free-carrier density obtained under high power-density excitation may allow the observation of LOPCMs [60, 72–75]. The low recombination rates in semiconducting InP makes this material a paradigm of photoexcitation effects in Raman spectra. Representative spectra are displayed in Fig. 9.5c, where it can be seen that the L^+ coupled mode peak is resolved from the surface depletion layer LO peak and clearly shifts to higher frequencies for moderate increases in excitation power. In this case, the lineshape analysis based on the Lindhard–Mermin model allows one to extract the photoexcited carrier density as a power of the power excitation [60]. Photoexcitation effects may be particularly relevant in micro-Raman experiments, where very high power densities are achieved [72, 75].

In semiconductors with low free-charge mobility such as p-type GaAs, only a single LOPCM peak is observed between the TO and LO phonon frequencies [76, 77]. Such different coupled mode behavior, which was shown to be due to the overdamped nature of intraheavy-hole transitions [77], has also been observed in n-type semiconductors with very low mobility as, for instance, GaAsN dilute nitride alloys [78]. Typical Raman spectra of n-GaAsN layers showing an overdamped LOPCM peak between the TO and LO frequencies are displayed in Fig. 9.6a. To illustrate the behavior of the overdamped LOPCMs, we show in Fig. 9.6b the calculated dependence of the LOPCM frequencies as a function of N_e for the case of low-mobility n-GaAsN [78]. This type of behavior is obtained when large Γ_e values ($\gtrsim 1{,}000\,\mathrm{cm}^{-1}$) are employed in order to calculate the frequency of the LOPCMs.

From the preceding discussion, it is clear that Raman scattering is a powerful, contactless tool to determine the free-charge density in doped semiconductors.

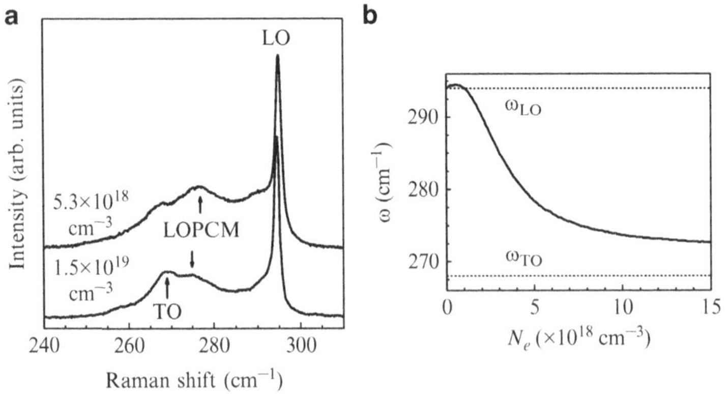

Fig. 9.6 (**a**) 80 K $z(xy)\bar{z}$ Raman spectra of Se-doped n-type $GaAs_{1-x}N_x$ epilayers with $x = 0.1\%$ ($N_e = 5.3 \times 10^{18}\,\mathrm{cm}^{-3}$) and $x = 0.36\%$ ($N_e = 1.5 \times 10^{19}\,\mathrm{cm}^{-3}$). (**b**) Calculated dependence of the coupled mode frequency on the free electron density for overdamped LOPCMs in low-mobility n-type $GaAs_{1-x}N_x$ [78]

However, the sensitivity of the Raman spectra to the presence of free charge can also be utilized in studies of other relevant physical properties of semiconductors. Thus, Raman scattering experiments have been employed to derive information about the depletion layer thickness [79], the cross-over of electronic band minima in InP under hydrostatic pressure [80], the energy separation between Γ and L electronic band minima in GaInAsSb [81], the surface recombination velocity in InN [75], or the electron effective mass in GaAsN dilute nitride alloys [78].

References

1. C.V. Raman, K.S. Krishnan, Nature **121**, 501–502 (1928)
2. G.S. Landsberg, L.I. Mandelstam, Zeitschrift für Physik **50**, 769 (1928)
3. W. Hayes, R. Loudon, *Scattering of Light by Crystals* (Wiley, New York, 1978)
4. M. Cardona, in *Light Scattering in Solids II*, Topics in Applied Physics, ed. by M. Cardona, G. Güntherodt (Springer, Berlin, 1980)
5. W.H. Weber, R. Merlin (eds.) *Raman Scattering in Materials Science*, Springer Series in Materials Science, vol. 42, (Springer, Berlin, 2000)
6. T. Beechem, A. Christensen, S. Graham, D. Green, J. Appl. Phys. **103**, 124501 (2008)
7. P.Y. Yu, M. Cardona, *Fundamentals of Semiconductors* (Springer, Berlin, 1996)
8. D.J. Harris, M.D. Bertolucci, *Symmetry and Spectroscopy: an Introduction to Vibrational and Electronic Spectroscopy* (Dover, New York, 1978)
9. M.S. Dresselhaus, G. Dresselhaus, A. Jorio, *Group Theory: Application to the Physics of Condensed Matter*(Springer, Berlin, 2008)
10. J.D. Jackson, *Classical Electrodynamics* (Wiley, New York, 1998)
11. R. Cuscó, E. Alarcón-Lladó, L. Artús, J. Ibáñez, J. Jiménez, B. Wang, M.J. Callahan, Phys. Rev. B **75**, 165202 (2007)
12. E.O. Schäfer-Nolte, T. Stoica, T. Gotschke, F. Limbach, E. Sutter, P. Sutter, R. Calarco, Appl. Phys. Lett. **96**, 091907 (2010)
13. C.S. Rao. S. Sundaram, R.L. Schmidth. J. Comas, J. Appl. Phys. **54**, 1808 (1983)
14. K. Tiong, P.M. Amirtaraj, F. Pollak. D.E. Aspnes, Appl. Phys. Lett. **44**, 122 (1984)
15. E. Bedel, G. Landa, R. Caries, J.B. Renucci, J.M. Roquais, P.N. Favennec, J. Appl. Phys. **60**, 1980 (1986)
16. L. Artús. R. Cuscó, J. Ibáñez, J.M. Martin, G. Gonzalez-Diaz, J. Appl. Phys. **82**, 3736 (1997)
17. R. Cuscó, J. Ibáñez, N. Blanco, G. González-Díaz, L. Artús, Nucl. Instrum. Methods Phys. Res. B **132**, 627 (1997)
18. W. Limmer, W. Ritter, R. Sauer, B. Mensching, C. Liu, B. Rauschenbach, Appl. Phys. Lett. **72**, 2589 (1998)
19. D. Pastor, R. Cuscó, L. Artús, G. González-Díaz, S. Fernández, E. Calleja, Semicond. Sci. Technol. **20**, 374 (2005)
20. R. Cuscó, E. Alarcón-Lladó, J. Ibáñez, T. Yamaguchi, Y. Nanishi, J. Phys. Condens. Matter **21**, 415801 (2009)
21. H. Kawamura, R. Tsu, L. Esaki, Phys. Rev. Lett. **29** 1397 (1972)
22. M. Wihl, M. Cardona, J. Taut, J. Noncryst. Solids **8–10**, 172 (1972)
23. C. Kisielowski, J. Krüger, S. Ruvimov, T. Suski, J.W. Ager III, E. Jones, Z. Liliental-Weber, M. Rubin, E.R. Weber, M.D. Bremser, R.F. Davis, Phys. Rev. B **54**, 17745 (1996)
24. F. Cerdeira, C.J. Buchenauer, F.H. Pollak, M. Cardona, Phys. Rev. B **5**, 580 (1972)
25. R. J. Briggs, A. K. Ramdas, Phys. Rev. B **13**, 5518 (1976)
26. S. Adachi, *Properties of Group-IV, III-V and II-VI Semiconductors* (Wiley, Chichester, 2005)
27. J. Groenen, G. Landa, R. Carles, P.S. Pizani, M. Gendry, J. Appl. Phys. **82**, 803 (1997)

28. T. Prokofyeva, T. Sauncy, M. Seon, M. Holtz, Y. Qiu, S. Nikishin, H. Temkin, Appl. Phys. Lett. **73**, 1409 (1998)
29. X. Wang, S. Che, Y. Ishitani, A. Yoshikawa, Appl. Phys. Lett. **89**, 171907 (2006)
30. J. Ibáñez, R. Cuscó, S. Hernández, L. Artús, M. Henini, A. Patanè, L. Eaves, M. Roy, P.A. Maksym, J. Appl. Phys. **99**, 043501 (2006)
31. A. Cros, N. Garro, A. Cantarero, J. Coraux, H. Renevier, B. Daudin, Phys. Rev. B **76**, 165403 (2007)
32. B. Jusserand, M. Cardona, in *Light Scattering in Solids V*, ed. by M. Cardona, G. Güntherodt. Topics in Applied Physics (Springer, Berlin, 1989)
33. L. Piper, T.D. Veal, C.F. McConville, H. Lu, W.J. Schaff, Appl. Phys. Lett. **88**, 252109 (2002)
34. J.C. Tsang, K. Eberl, S. Zollner, S.S. Iyer, Appl. Phys. Lett. **61**, 961 (1992)
35. J. Wagner, M. Maier, Th. Lauterbach, K.H. Bachem, A. Fischer, K. Ploog, G. Mörsch, M. Kamp, Phys. Rev. B **45**, 9120 (1992)
36. L. Artús, R. Cuscó, E. Alarcón-Lladó, G. González-Díaz, I. Mártil, J. Jiménez, B. Wang, Appl. Phys. Lett. **90**, 181911 (2007)
37. H. Grille, C. Schnittler, F. Bechstedt, Phys. Rev. B **61**, 6091 (2000)
38. J. Groenen, R. Carles, G. Landa, C. Guerret-Piecourt, C. Fontaine, Phys. Rev. B **58**, 10452 (1998)
39. S. Hernández, R. Cuscó, D. Pastor, L. Artús, K.P. O'Donnell, R.W. Martin, I.M. Watson, Y. Nanishi, J. Appl. Phys. **98**, 013511 (2005)
40. E.G. Grosche, M.J. Ashwin, R.C. Newman, D.A. Robbie, M. Sangster, T. Pletl, P. Pavone, D. Strauch, Phys. Rev. B **51**, 14758 (1995)
41. M.D. McCluskey, J. Appl. Phys. **87**, 3593 (2000)
42. A. Kaschner, H. Siegle, G. Kaczmarczyk, M. Straßburg, A. Hoffmann, C. Thomsen, U. Birkle, S. Einfeldt, D. Hommel, Appl. Phys. Lett. **74**, 3281 (1999)
43. G. Kaczmarczyk, A. Kaschner, A. Hoffmann, C. Thomsen, Phys. Rev. B **61**, 5353 (2000)
44. H. Harima, T. Inoue, S. Nakashima, M. Ishida, M. Taneya, Appl. Phys. Lett. **75**, 1383 (1999)
45. R. Cuscó, L. Artús, D. Pastor, F.B. Naranjo, E. Calleja, Appl. Phys. Lett. **84**, 897 (2004)
46. M. Henini (ed.) *Dilute Nitride Semiconductors*, (Elsevier, Amsterdam, 2005)
47. A. Mascarenhas, M.J. Seong, Semicond. Sci. Technol. **17**, 823 (2002)
48. J. Wagner, T. Geppert, K. Köhler, P. Ganser, N. Herres, J. Appl. Phys. **90**, 5027 (2001)
49. J. Ibáñez, E. Alarcón-Lladó, R. Cuscó, L. Artús, M. Hopkinson, J. Appl. Phys. **102**, 013502 (2007)
50. E.M. Pavelescu, J. Wagner, H.P. Komsa, T.T. Rantala, M. Dumitrescu, M. Pessa, J. Appl. Phys. **98**, 083524 (2005)
51. J. Wagner, K. Köhler, P. Ganser, M. Maier, Appl. Phys. Lett. **87**, 051913 (2005)
52. J. Ibáñez, R. Oliva, M. De la Mare, M. Schmidbauer, S. Hernández, P. Pellegrino, D.J. Scurr, R. Cuscó, L. Artús, M. Shafi, R.H. Mari, M. Henini, Q. Zhuang, A. Godenir, A. Krier, J. Appl. Phys. **108**, 103504 (2010)
53. S. Shirakata, M. Knodow, T. Kitatani, J. Phys. Chem. Solids **66**, 2119 (2005)
54. A.M. Teweldeberhan, S. Fahy, Phys. Rev. B **73**, 245215 (2006)
55. A.M. Teweldeberhan, S. Fahy, Phys. Rev. B **77**, 235216 (2008)
56. A. Kaschner, H. Siegle, G. Kaczamarczyk, M. Strassburg, A. Hoffmann, C. Thomsen, Appl. Phys. Lett. **74**, 3281 (1999)
57. V.Y. Davydov, I.N. Goncharuk, A.N. Smirnov, A.E. Nikolaev, W.V. Lundin, A.S. Usikov, A.A. Klochikhin, J. Aderhold, J. Graul, O. Semchinova, H. Harima, Phys. Rev. B **65**, 125203 (2002)
58. G. Abstreiter, M. Cardona, A. Pinczuk, in *Light Scattering in Solids IV*, ed. by M. Cardona, G. Güntherodt. Topics in Applied Physics (Springer, Berlin, 1984)
59. L. Artús, R. Cuscó, J. Ibáñez, N. Blanco, G. González-Díaz, Phys. Rev. B **60**, 5456 (1999)
60. R. Cuscó, J. Ibáñez, L. Artús, Phys. Rev. B **57**, 12197 (1998)
61. T. Yuasa, S. Naritsuka, M. Mannoh, K. Shinozaki, K. Yamanaka, Y. Nomura, M. Mihara, M. Ishii, Phys. Rev. B **33**, 1222 (1986)
62. R. Cuscó, L. Artús, S. Hernández, J. Ibáñez, M. Hopkinson, Phys. Rev. B **65**, 035210 (2001)

63. R. Cuscó, R.E. Alarcón-Lladó, L. Artús, J.E. Maslar, W.S. Hurst, Phys. Rev. B **81**, 195212 (2010)
64. D.T. Hon, W.L. Faust, Appl. Phys. **1**, 241 (1973)
65. U. Nowak, W. Richter, G. Sachs, Phys. Status Solidi. B **108**, 131 (1981)
66. J. Ibáñez, R. Cuscó, L. Artús, Phys. Status Solidi. B **223**, 715 (2001)
67. M. Cardona, Phys. Rev. B **121**, 752 (1961)
68. P.M. Platzman, P.A. Wolff, in *Solid State Physics*, Suppl. 13 (Academic, New York, 1973)
69. N.D. Mermin, Phys. Rev. B **1**, 2362 (1970)
70. R. Cuscó, J. Ibáñez, E. Alarcon-Lladó, L. Artús, T. Yamaguchi, Y. Nanishi, Phys. Rev. B **79**, 155210 (2009)
71. J.F. Young, K. Wan, A.J. Spring Thorpe, P. Mandeville, Phys. Rev. B **36**, 1316 (1987)
72. T. Nakamura, T. Katoda, J. Appl. Phys. **55**, 3064 (1984)
73. J.E. Kardontchik, E. Cohen, Phys. Rev. Lett. **42**, 669 (1979)
74. J.F. Young, K. Wan, Phys. Rev. B **35**, 2544 (1987)
75. R. Cuscó, J. Ibáñez, E. Alarcón-Lladó, L. Artús, T. Yamaguchi, Y. Nanishi, Phys. Rev. B **80**, 155204 (2010)
76. D. Olego, M. Cardona, Phys. Rev. B **24**, 7217 (1981)
77. K. Wan, J. F. Young, Phys. Rev. B **41**, 10772 (1990)
78. J. Ibáñez, R. Cuscó, E. Alarcón-Lladó, L. Artús, A. Patanè, D. Fowler, L. Eaves, K. Uesugi, I. Suemune, J. Appl. Phys. **103**, 103528 (2008)
79. A. Pinczuk, A.A. Ballman, R.E. Nahory, M.A. Pollack, J.M. Worlock, J. Vac. Sci. Technol. **16**, 1168 (1979)
80. S. Ernst, A. R. Goñi, K. Syassen, M. Cardona, Phys. Rev. B **53**, 1287 (1996)
81. R. Cuscó, J. Ibáñez, L. Artús, Appl. Phys. Lett. **97**, 091909 (2010)

Chapter 10
Cyclotron Resonance Spectroscopy

Oleksiy Drachenko and Manfred Helm

Abstract We give an overview of the basic theory of cyclotron resonance, discuss some experimental aspects of cyclotron resonance spectroscopy in high, mostly pulsed magnetic fields, and finally discuss some recent cyclotron resonance experiments on various semiconductor materials.

10.1 Introduction

Cyclotron resonance (CR) refers to the resonant absorption of electromagnetic radiation, usually in the microwave or far-infrared range, by electrons in a static magnetic field. It was first observed by Dresselhaus et al. [1, 2] in germanium and silicon. The resonance frequency, called cyclotron frequency, is given by $\omega_c = eB/m$, where e is the elementary charge, B the magnetic field, and m the mass of the charge carrier. In semiconductors this is the effective mass, m^*, which is usually smaller than the free-electron mass and derives from the band structure of the material. Its inverse reflects the curvature of the $E(\mathbf{k})$ dispersion relation. The above relation will be derived below in a classical and quantum mechanical framework. Since m^* is the only material parameter entering the expression for the cyclotron frequency, cyclotron resonance is a prime, direct method for the determination of the effective mass. In addition, information on scattering times can be deduced from the spectral linewidth and on the free-carrier concentration from the integrated absorption strength.

In this chapter, we will give a brief account of the basic theory of cyclotron resonance in semiconductors, from both a classical and a quantum mechanical point of view. It is not intended to provide an extensive review of the vast amount

O. Drachenko (✉) · M. Helm
Institute for Ion Beam Physics and Materials Research, Helmholtz-Zentrum Dresden-Rossendorf, P.O. Box 510119, 01314 Dresden, Germany
e-mail: o.drachenko@hzdr.de; m.helm@hzdr.de

A. Patanè and N. Balkan (eds.), *Semiconductor Research*, Springer Series in Materials Science 150, DOI 10.1007/978-3-642-23351-7_10,

of existing literature on this topic. For more extensive descriptions, the reader is referred to the book chapter by Kono [3], the books by Miura [4] and Hamaguchi [5], and chapters in the book "Landau level spectroscopy" [6].

The chapter is organized as follows: in Sect. 10.2, we start with a description of cyclotron resonance based on the classical equation of motion. This leads to the classical formula which is an extension of the Drude absorption to finite magnetic fields. Then we present the simplest quantum mechanical theory of cyclotron resonance, namely, as an absorption process between Landau levels, described by Fermi's Golden Rule. In Sect. 10.3, we address several experimental issues and describe in some detail an experimental setup for cyclotron resonance experiments in high, mostly pulsed magnetic fields. Finally in Sect. 10.4, we discuss some recent experiments obtained in our laboratory on various semiconductor material systems. This contains experiments not only on the so-called dilute nitrides, which are III–V semiconductors with a small ($<1\%$ range) fraction of nitrogen substituted for the group V element (i.e., InAsN, GaAsN), but also on p-type quantum wells (InGaAs), which show an interesting behavior due to their rather complicated valence-band structure.

10.2 Basic Theory

10.2.1 Classical Description

Let us start from the classical equation of motion of an electron in a magnetic field

$$m^*\frac{\mathrm{d}\mathbf{v}}{\mathrm{d}t} + \frac{m^*\mathbf{v}}{\tau} = -e\mathbf{E} - e(\mathbf{v}\times\mathbf{B}). \qquad (10.1)$$

Here, $\mathbf{v}$ is the electron velocity, m^* is the electron effective mass, τ a phenomenological scattering time, and e the elementary charge. The right side contains the driving terms, i.e., $\mathbf{B}$ is the static magnetic field, here assumed to point in the z-direction, and $\mathbf{E}$ is the AC electric field polarized in the xy plane, corresponding to a plane wave propagating along the z-direction (Faraday geometry). Assuming a harmonic dependence of $\mathbf{v}$ and $\mathbf{E}$, this equation can easily be solved in steady state for the velocity components. We can define the conductivity tensor σ via Ohm's law and the current density $\mathbf{j} = ne\mathbf{v} = \sigma\,\mathbf{E}$, where n is the electron density.

The non-vanishing components of σ are then given by

$$\begin{aligned}\sigma_{xx} &= \sigma_{yy} = \sigma_0\frac{1+\mathrm{i}\omega\tau}{(1+\mathrm{i}\omega\tau)^2+\omega_\mathrm{c}^2\tau^2}\\ \sigma_{xy} &= -\sigma_{xy} = \sigma_0\frac{\omega_\mathrm{c}\tau}{(1+\mathrm{i}\omega\tau)^2+\omega_\mathrm{c}^2\tau^2}\\ \sigma_{zz} &= \sigma_0\frac{1}{1+\mathrm{i}\omega\tau},\end{aligned} \qquad (10.2)$$

where $\sigma_0 = ne\mu = ne^2\tau/m^*$ is the DC Drude conductivity. Here, the cyclotron frequency ω_c has been introduced according to the above definition. The absorption of a plane wave linearly polarized in the x-direction is proportional to $\mathrm{Re}(\sigma_{xx})$, but it is more instructive to decompose the conductivity into components responding to left- and right-circularly polarized radiation:

$$\sigma_{\pm} = \sigma_{xx} \mp \mathrm{i}\sigma_{xy} = \sigma_0 \frac{1}{1 + \mathrm{i}(\omega \mp \omega_c)\tau}. \tag{10.3}$$

This represents the decomposition into a resonant and a non-resonant contribution. The resonant term, σ_+ for electrons, is called cyclotron resonance active (CRA), and the non-resonant term, σ_-, cyclotron resonance inactive (CRI). Electrons are CRA for right-circular polarization and holes for left-circular polarization, since the latter have the opposite sign of ω_c due to their positive charge (or negative effective mass). Thus the polarization of CR provides information on the nature of the charge carriers.

The real part of the resonant part is a pure Lorentzian with the half-width at half-maximum of $\gamma = 1/\tau$ (in ω-space), given by

$$\mathrm{Re}\sigma_+ = \sigma_0 \frac{1}{1 + (\omega - \omega_c)^2\tau^2}. \tag{10.4}$$

For $\omega_c\tau = \mu B \gg 1$, the non-resonant term can be neglected even for linear polarization, but one has to consider that only 50% of the radiation can be absorbed at most in this case.

The absorption coefficient can be obtained from the real part of the conductivity by

$$\alpha = \frac{\mathrm{Re}\sigma_+}{\varepsilon_0 c n_1}, \tag{10.5}$$

where n_1 is the refractive index, c the light velocity, and ε_0 the vacuum permittivity.

As an example, we plug in numbers for a semiconductor with electrons having a mobility of $\mu = 1{,}000\,\mathrm{cm^2/Vs}$ and density $n = 10^{16}\,\mathrm{cm^{-3}}$ (and refractive index of 3.5), resulting in a peak absorption coefficient of $\alpha = ne\mu/\varepsilon_0 c n_1 = 172\,\mathrm{cm^{-1}}$. This value increases linearly with both the mobility and the carrier density.

Now let us address a practical problem: What happens if the peak absorption becomes very large, due to either a narrow linewidth or a large carrier concentration? The transmission through a semiconductor slab can be approximately written as $T = (1 - \mathrm{R})^2 \mathrm{e}^{-\alpha d}$, where R is the reflectivity and d is the sample thickness. When $\alpha d \gg 1$, the transmission approaches zero and the linewidth becomes distorted; it acquires a "flat bottom" (for linear polarization at a transmission value of 0.5) and appears much broader than 2γ. For a quasi two-dimensional electron gas, the conductivity has the dimension $[\Omega^{-1}]$. As a consequence, the absorption is a dimensionless quantity and can be written as $A = \mathrm{Re}\sigma/\varepsilon_0 c n_1$, but only as long as this expression is much smaller than unity. Otherwise, the full expressions for an infinitely thin metallic slab embedded in a dielectric have to be used, for which the transmission is given by

$$T = \frac{4n_1^2}{|2n_1 + \sigma/(\varepsilon_0 c)|^2}. \tag{10.6}$$

10.2.2 Quantum Mechanical Description

In quantum mechanics, the Hamiltonian of an electron in a magnetic field is written as

$$H_0 = \frac{1}{2m^*}(\mathbf{p} + e\mathbf{A})^2, \tag{10.7}$$

where $\mathbf{A}$ is the vector potential of the magnetic field, which is given by $\mathbf{A} = (0, Bx, 0)$ in the asymmetric (Landau) gauge, when $\mathbf{B}$ has only a z-component $\mathbf{B}$=(0, 0, B). The corresponding Schrödinger equation can be written in the form of a shifted harmonic oscillator with an orbit center coordinate $x = -\hbar k_y/eB$.

This leads to energy levels of the form

$$E_N(k_z) = (N + \frac{1}{2})\hbar\omega_c + \frac{\hbar^2 k_z^2}{2m^*}. \tag{10.8}$$

For a bulk 3D semiconductor, this leads to a series of one-dimensional Landau levels (LL). The xy motion is now condensed into the Landau levels, and free motion of the electrons remains possible along the z-direction. In a two-dimensional electron system like the one that exists in a quantum well, the z-motion is quantized into electric subbands already by the quantum well potential, and application of a magnetic field results in a completely quantized system. Note, however, that the analog to an artificial atom cannot be carried too far, since each Landau level is infinitely degenerate with respect to the wavevector component k_y (or the spatial coordinate x).

In order to calculate the absorption between Landau levels according to Fermi's Golden Rule, we need the interaction Hamiltonian of the electromagnetic field acting on the unperturbed Landau levels. Note that we have to distinguish now the vector potential of the electromagnetic field $\mathbf{A}_{\mathrm{em}}$, from the static vector potential $\mathbf{A}_B$ [$\mathbf{A}$ in (10.7)] that describes the magnetic field $\mathbf{B}$. The Hamiltonian is then given by

$$H = \frac{1}{2m^*}[\mathbf{p} + e(\mathbf{A}_B + \mathbf{A}_{\mathrm{em}})]^2 \approx H_0 + \frac{e}{m^*}(\mathbf{p} + \mathbf{A}_B) \cdot \mathbf{A}_{\mathrm{em}} \tag{10.9}$$

in the one-band effective mass approximation for simplicity.

The induced transition rate from LL N to LL N', $W_{NN'}$, can then be calculated from Fermi's Golden Rule, and the absorption coefficient is related to the transition rate via $\alpha I = n\hbar\omega W_{NN'}$, where n is the electron concentration and I is the intensity.

The dipole (or momentum) matrix element contained in $W_{NN'}$ can be directly calculated from the explicit wave functions, but we can also take advantage of the known properties of a harmonic oscillator, which actually can be derived algebraically. The dimensionless oscillator strength, generally defined by

$$f_{NN'} = \frac{2m^* \omega_{N'N}}{\hbar} \left|\langle N' | \mathbf{e} \cdot \mathbf{r} | N \rangle\right|^2 \tag{10.10}$$

for a transition from LL N to LL N', where $\mathbf{e}$ is the unity polarization vector of the electric field [for CRA polarization $\mathbf{e} = (\mathbf{e}_x + i\mathbf{e}_y)/\sqrt{2}$], is given by $f_{N,N+1} = N + 1$, i.e., transitions between higher LLs become stronger. However, an electron sitting in LL N can also undergo stimulated emission to LL $N - 1$. So for any distribution (e.g., Fermi–Dirac) of electrons over the LLs, both terms have to be taken into account. In a perfectly harmonic system, $\omega_{N+1,N} = \omega_{N,N-1} = \omega_c$, thus the "extra" oscillator strength cancels because of $N + 1 - N = 1$, and there remains a net CR absorption with an oscillator strength of unity, corresponding to the classical result and consistent with the Bohr correspondence principle for high quantum numbers (and with the Thomas–Reiche–Kuhn sum rule). This is also the reason why a harmonic cyclotron resonance laser is impossible—there is always net absorption and not emission. For a perfectly harmonic system, this finally gives the same result for the absorption coefficient as in the classical calculation.

In reality, however, there is no perfect parabolic system in a semiconductor, since at high energies, all bands tend to change their curvature. Non-equidistant LLs allow the observation of different Landau level transitions and thus can provide the possibility to obtain information on the hot-electron distribution, e.g., to extract an electron temperature, or to quantify the polaron effect [7, 8]. Non-parabolicity (corresponding to an energy-dependent mass) also breaks the symmetry between absorption and emission and leads to the possibility of cyclotron resonance optical gain at certain frequencies [9]. Note that this was noticed already 60 years ago for relativistic free electrons [10], being the basis of gyrotrons [11]. A detailed account of non-parabolicity requires multiband calculations, e.g., within $\mathbf{k}\cdot\mathbf{p}$ theory, and goes beyond the scope of the present discussion [4, 5, 12]. The dilute nitrides discussed later in this chapter exhibit, of course, a very large non-parabolicity stemming from the interaction with the localized N states and described within the band anticrossing (BAC) model [13, 14]. In this situation, the observation of multiple CR lines has also been predicted [15]. Valence bands in common semiconductors, consisting of typically three bands of light, heavy, and spin–orbit split holes, respectively, always require multiband calculations, and their theoretical description [16, 17] is beyond the scope of this brief overview. Finally, we would like to mention that cyclotron resonance *emission* spectroscopy has also been performed, pioneered by Gornik et al. [6, 7, 18].

10.2.3 Further Considerations

The key material parameters that can be accessed with a CR experiment are the effective mass and the CR linewidth $\gamma(= 1/\tau)$ (half-width at half-maximum), equivalent to a characteristic scattering rate. The measured effective mass can be compared to band–structure calculations, often on the basis of **k**·**p** perturbation theory. A most valuable information is the dependence of the effective mass on energy, since this will reflect the $E(\mathbf{k})$ dispersion relation of the band. This can be gained either by investigating samples with different doping levels, and thus values of the Fermi energy, while keeping all other parameters unchanged, or by simply recording CR spectra at different magnetic field values, corresponding to different photon energies.

The linewidth appears to be related simply to the Drude scattering time τ, as seen in the classical expression (10.3). In the simple quantum mechanical expression, the linewidth is merely a phenomenological parameter as well. While the CR linewidth may be related to the mobility via $\mu = e\tau/m^* = e/\gamma m^*$ in a first approach, this does not necessarily reflect the truth in many cases. In bulk semiconductors, the CR linewidth has been found to be proportional to the square root of the impurity concentration [19] in some cases. Also, one has to consider the difference between the so-called single-particle relaxation time, τ_s, and the mobility scattering time (or transport lifetime), τ_t, the latter being weighted by $1 - \cos\theta$, where θ is the scattering angle [20, 21]. This reflects the weaker sensitivity of the mobility to forward-scattering processes. The relation of these two scattering times depends on the nature of the scattering processes; e.g., τ_s can be much smaller than τ_t for remote impurity scattering in a two-dimensional electron gas, but the two are identical for short-range scattering. The CR linewidth, on the other hand, can be different from both [22]. In a strict quantum mechanical way, one has to include the relevant scattering mechanisms in a microscopic calculation of the dynamic conductivity, via the Kubo formula or related many-body formalisms [23–28]. Phenomena such as oscillatory linewidth as a function of Landau level filling [26] or narrowing through collective effects [29] can only be obtained in such a framework. Ando has predicted a $\sqrt{B}$ dependence for electrons in a 2DEG [26]. A large amount of literature is available on many aspects of these issues, in particular concerning the CR in GaAs/AlGaAs heterojunctions [30, 31]. Some of the relevant works can be found in [5, 6].

In semiconductors, a substantial carrier concentration, necessary to observe CR, is usually achieved by doping with shallow impurities. These impurity states will also have characteristic absorption features as a function of magnetic field, which are particularly visible at not too high doping (below the Mott density) and low temperature. The most conspicuous feature is absorption due to transitions between the 1s ground state and the $2p_+$ excited state of a hydrogenic impurity. This transition is located energetically slightly above the CR, and since it exhibits [at intermediate and high fields (i.e., $\hbar\omega_c >$ impurity binding energy)] the same slope $E(B)$ as the cyclotron resonance, it is often called impurity-shifted cyclotron resonance [19, 32–34]. In the limit $B \to 0$, it extrapolates to a finite energy value, namely, the 1s–2p transition energy, whereas the CR extrapolates to zero. This behavior results from the fact that the 1s level is pinned to the lowest, $N = 0$, LL, while the $2p_+$ level is pinned to the $N = 1$ LL. Also, more general disorder-related infrared modes have been discussed [35].

By using optical excitation to generate free carriers, the complications due to impurity states can be avoided; however, for direct semiconductors, substantial optical power is needed due to the fast carrier recombination time (<1 ns). For indirect semiconductors such as Si or Ge, having a much longer recombination time (μs to ms), this method is more practical. CR can even be detected through changes in the photoluminescence [36].

10.3 Experimental Techniques

In this section, we switch to some more experimental issues. First, we will give a short overview of the methods for generation of high magnetic fields, and then we will discuss peculiarities of cyclotron resonance spectroscopy.

10.3.1 High Magnetic Fields

Today, there are several common ways to obtain high magnetic fields depending on the operating range. The first most well-known and easiest-to-operate way is based on superconducting DC coils. Here, the coils are wired using a superconducting material, which ensures dissipation-less electrical current. This method provides a relatively easy way to generate constant or slowly varying magnetic fields. In DC magnets, the best signal-to-noise ratio can be reached due to complete absence of pick-up signals, that are proportional to $\partial B/\partial t$. It is also easier to use synchronous detection, averaging and filtering, since the magnetic field can be kept constant as long as it is necessary for the experiment. Regardless of these evident advantages, there is a fundamental reason limiting the maximum operating field—the critical field of superconductor material, which breaks the electron–electron correlation and kills the superconductivity. The typical value of the critical field for commercially available coils is limited by 20–22 T (see, for example, [37]).

Another type of DC coil is the Bitter magnet (proposed by Francis Bitter [38]). This is a resistive coil where a large current is driven through the conductive plate set in a helical configuration. Typically, Bitter-type magnets are capable of delivering up to ~33 T; however, the dissipation power rises up to tens of MWatt in this case [39]. In order to dissipate this power, a huge amount of deionized water is pumped

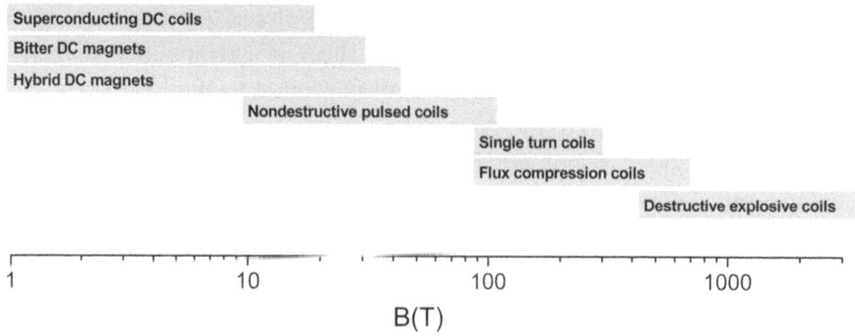

Fig. 10.1 Methods of generation of high magnetic fields arranged in order of maximum achievable field strength

through the coil.[1] The combination of an external superconducting coil with a Bitter magnet inside, a so-called hybrid magnet, allows one to obtain up to 45 T [40, 41].

Above 45 T, only pulsed magnets are available to scientists (see Fig. 10.1). Pulsed magnets fall into two general classes: non-destructive and destructive. Non-destructive coils are solenoids driven by a bank of capacitors.[2] The magnetic field in this case is limited by the ultimate tensile strength of the conductor material or the internal reinforcement and typically lies between 60 and 100 T [42–44].

Destructive pulsed magnets overstep the material's strength problem and are designed to explode within every pulse. In some cases, the sample remains intact and only the coil explodes (semi-destructive pulsed fields); in the other case, the sample explodes together with the coil. The highest magnetic fields are achieved by explosively compressing the magnetic flux around the sample. The time scale is typically fixed within the microsecond range, since it is governed by the time of shock wave propagation through the magnet. Figure 10.1 summarizes the distribution of the magnetic field types as a function of the maximum achievable field. Interested readers can refer to [45] for a detailed review.

In this chapter, we limit ourselves to non-destructive pulsed magnets available at the high magnetic field laboratory in Dresden (HLD) [42]. There are currently three coil types available for experiments characterized by their energy, pulse duration, and clear bore (see Table 10.1). Each coil is immersed into liquid nitrogen and is installed inside individual blast-protected caves. The driving circuit is basically nothing more than a typical LC circuit (Fig. 10.2a). At the time $t = 0$, fully charged capacitors discharge via stack of optically triggered thyristors into the coil. The

[1] A typical Bitter magnet at the High Field Magnet Laboratory at Radboud Universiteit Nijmegen delivers up to 33 T within 32-mm clear bore at room temperature, consumes ~17 MW, and is cooled by the 140 l/s water flow at ~30 bar [38].

[2] There exists also another very special driving circuit based on giant motor generator [43]. As usual, together with evident advantages, including the possibility to obtain arbitrary pulse shapes, there are disadvantages like a high-frequency noise in magnetic field caused by mechanical vibration that propagates into the current circuits.

Table 10.1 List of magnetic field coils available at Dresden high magnetic field laboratory (HLD)

	Type 1	Type 2	Type 3
Pulse energy	8.64 MJ	1.44 MJ	1.44 MJ
Max magnetic field	70 T	65 T	55 T
Pulse duration	150 ms	20 ms	50 ms
Clear bore at 77 K	24 mm	20 mm	24 mm
Cool down time	2 h	1 h	1 h

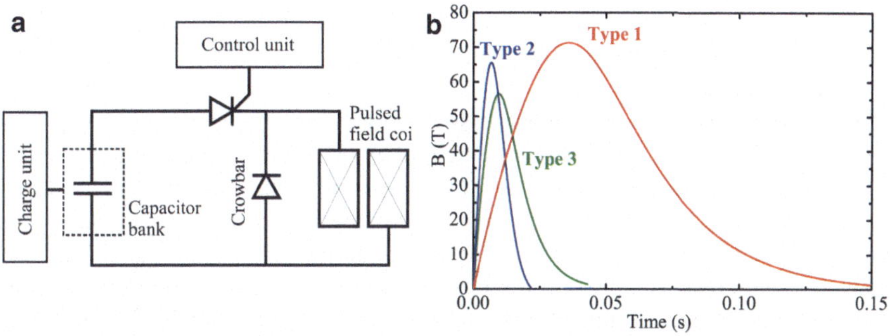

Fig. 10.2 (**a**) Schematic diagram of the driving circuit for pulsed magnets. (**b**) Typical magnetic field profiles of three coil types available at HLD

electrical current rises as a $\sin(t/\sqrt{\mathrm{LC}})$ and reaches its maximum at time $t_{\max} = \pi\sqrt{\mathrm{LC}}/2$. To avoid negative recharging of the capacitors and increased sweep-down time, a special diode (namely, *crowbar*) is installed parallel to the coil. This diode opens when the voltage on the capacitors becomes slightly negative in the vicinity of $t = t_{\max}$ and short-circuits the coil. Starting from this moment, the current is no more sine-like, but decays as $\exp(-L/R)$, where R is the coil resistance. The energy stored in the coil dissipates by Joule heat. Assuming that the coil is not deformed during the current sweep, the magnetic field is proportional to the current. Figure 10.2b shows typical magnetic field traces of the coils at HLD.

10.3.2 Cyclotron Resonance Spectroscopic Techniques

There are basically two principal ways to perform CR experiments: one is frequency-dependent magnetospectroscopy (FMS) and the other is the magnetic field-dependent or laser magnetospectroscopy (LMS). The former measures wideband transmission spectra at a given quasi-constant magnetic field, while the latter traces the transmission at a given wavelength when the magnetic field changes. The typical scheme of an FMS spectrometer includes a Fourier-transform infrared (FTIR) spectrometer connected with a DC superconducting or hybrid coil using optical waveguides (polished stainless-steel or brass pipes, see, for example, [3,4] for review). The DC magnet is an essential part of the FMS method, since the

FTIR spectrometers have mechanically movable optical components, thus spectra cannot be acquired in a millisecond timescale, which otherwise would be necessary under pulsed magnetic fields (see Fig. 10.2b for typical magnetic field profiles of pulsed magnets).[3] In turn, the static magnetic field gives rise to an important advantage of the FMS method: a high signal-to-noise ratio due to the long possible (theoretically infinite) time for integration or spectral averaging. On the other hand, FMS spectrometers are naturally limited by the highest achievable DC magnetic fields of ~45 T. If higher magnetic fields are desired for the experiment, which is usually the case when low-mobility samples are under study,[4] only pulsed magnets can be considered.

The LMS method measures transmission through the sample using a monochromatic light source while the magnetic field changes, which is naturally achieved, when pulsed magnets are used. This is, however, also an inherent disadvantage, because the sample conditions are changed during the measurement. Regarding typical values of the effective mass, the corresponding cyclotron frequency at high magnetic fields usually falls into the far-infrared or THz spectral region, which is also known as "THz gap" due to lack of compact, powerful, and reliable sources of radiation.[5] Currently, this spectral range is mainly filled with CO_2 or CO_2-pumped gas lasers, which are cumbersome and difficult to align and operate. An attractive replacement as infrared source for everyday use can be novel semiconductor quantum cascade lasers (QCLs). These lasers, based on inter-subband transitions in quantum wells, are now available in a wide spectral range from the near to the far infrared. They are compact, typically tens of microns wide and several millimeters long, and work under simple current injection (no optical alignment) and become more and more commercially available.

Another promising source of infrared radiation is the free electrons laser (FEL). It is, of course, not as portable as QCLs, or even conventional gas lasers, but offers the unique possibility of wavelength tuning over a wide spectral range. This feature can be particularly useful to probe the band dispersion of semiconductors. The FEL is based on the coherent emission of relativistic electrons while passing alternating magnetic fields (undulator or wiggler). For example, the FEL at Helmholtz-Zentrum Dresden-Rossendorf is based on a two-stage superconducting linear accelerator (ELBE) with a maximum energy of 20 MeV each, delivering 40 MeV in total. The electron beam can be directed into one of the two undulators, namely, U27 and U100, covering continuously the wavelength range from 4 up to 280 μm [47]. An important peculiarity of the FEL at ELBE (FELBE) is the so-called cw mode,

---

[3]Recently, the FMS method was also demonstrated for pulsed magnets using rapid scan time-domain spectroscopy [46].

[4]The observability condition for the cyclotron resonance is $\mu B > 1$. A magnetic field greater than 10 T is required in order to resolve the cyclotron resonance line when the carrier mobility is $\mu = 1,000\,\mathrm{cm^2/Vs}$.

[5]For an effective mass $m^* = 0.13$, which is almost twice as much as in bulk GaAs, the cyclotron frequency at 50 T would correspond to a resonance wavelength of $\lambda = 17.5\,\mu\mathrm{m}$.

when the FELBE delivers a continuous train of short Fourier-limited micro-pulses with 13-MHz repetition rate; therefore, unlike many other FELs, no additional synchronization of the FEL with the magnetic field pulses is necessary. Recently, the first CR experiment that uses FELBE radiation for the measurements under high pulsed magnetic fields was performed on p-type InGaAs/GaAs multiple quantum wells [17]. However, being very flexible (continuously tunable) and powerful sources, FELs have an important disadvantage for everyday use: the beam time is expensive and demanded for many other types of experiments.

In this section, before going to the illustration of the use of CR spectroscopy to study the basic parameters of semiconductors, we would like to introduce an example of a compact and universal LMS CR spectrometer. A typical LMS CR spectrometer is built using the following scheme: light from the externally placed excitation laser is guided inside the magnet, and then, transmitted through the sample, light is guided back toward an externally placed photodetector [3, 4]. Consequently, since external components are involved, the experimental setup requires optical alignment, and becomes susceptible to vibrations and cumbersome. On the contrary, QCLs are compact enough to be placed inside the variable temperature insert (VTI), all together with the sample and the detector. The advantages offered by this design include high vibration stability (the ensemble source–sample–detector vibrates in the same way), no external components apart from the electronics (no optical alignment), and low waveguide losses due to short optical paths. As a drawback, the performance of the QCL source and detector could be strongly perturbed by the magnetic field.

The performance of QCLs under high magnetic fields was intensively studied during the last decade [48–51]. It was found that in general, when the magnetic field is applied along the growth direction, the emission intensity exhibits giant oscillations as a function of magnetic field (see Fig. 10.3a). This effect, now known as inter-subband magneto-phonon resonance, is caused by periodic quenching of the LO phonon assisted non-radiative energy relaxation channel [49]. Some designs, especially bound-to-continuum, may also suffer from the weakening of the injection/extraction efficiency [52]. In most cases, however, in the low magnetic field limit ($B < 1$T), the emission intensity varies slowly and weakly (see Fig. 10.3a). The inset of Fig. 10.3a illustrates the axial distribution of magnetic field in the type 1 coil at HLD. It can be easily seen that when the distance from the center of the coil is more than 20 cm, the magnetic field drops below 1 T and, therefore, does not strongly affect the QCL performance. Note that the component of the magnetic field parallel to the QW plane is expected to affect strongly the injection efficiency, the radiative transition efficiency, and, finally, the emission wavelength [53]. The latter is particularly important for the LMS-type CR, because the excitation energy is expected to be constant during the sweep of magnetic field. Therefore, the in-plane component of the magnetic field has to be minimized.

On the detector side, for the mid-infrared spectral region one can use, for example, Si:B blocked impurity band (BIB) photodetectors sensitive from typically 4 up to 35 μm. The BIB structures consist of a highly doped active layer that ensures high quantum efficiency, and a blocking layer made from undoped material,

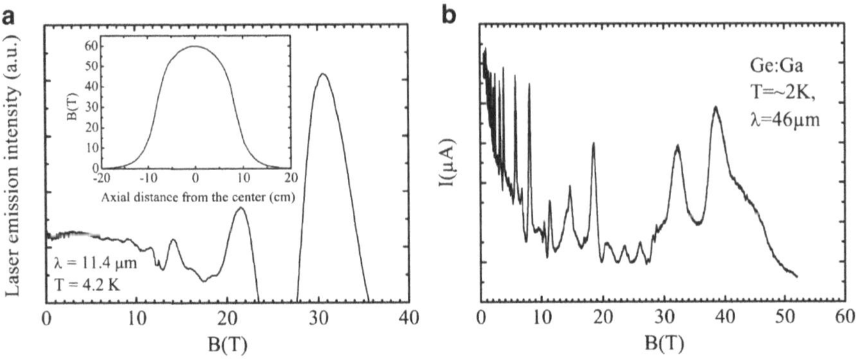

Fig. 10.3 (**a**) Typical variation of the QCL intensity as a function of magnetic field. The QCL, emitting at 11.2 μm [see P. Kruck et al., Appl. Phys. Lett. **76**, 3340 (2000) for details], is driven by 1 μs current pulses with 20-kHz repetition rate. The temperature of the heatsink is 4.2 K. The inset shows the axial distribution in the type 1 coil at HLD. (**b**) Photocurrent in a Ge:Ga photodetector as a function of magnetic field at $T = 2$ K, under FEL excitation at $\lambda = 46$ μm and constant intensity

which blocks the dark current through the impurity band of the active layer [54, 55]. The sensitivity of the Si:B BIB structures under high magnetic fields was recently studied up to 30 T [56]. It was found that the photocurrent under constant illumination monotonically drops down by a factor of ~5 (at $T = 4.2$ K and low voltage bias) when the magnetic field rises up to 30 T, remaining, however, relatively weak and slowly varying at low fields ($B < 1$T). For the far-infrared spectral region, Ge:Ga extrinsic photodetectors can be considered. In contrast to the Si:B BIB devices, these photodetectors appear to be more sensitive to the magnetic field, especially in the low-field limit. Figure 10.3b shows a typical magnetic field dependence of the photocurrent under monochrochromatic illumination of constant intensity. As can be seen, the photocurrent drops by an order of magnitude at 30 T, but the variation is not negligible even at low fields ($B < 1$T). Additionally, the photocurrent exhibits pronounced oscillations that become visible starting from 0.5 T. Consequently, the distance from the magnetic field center must be kept as long as possible.

Figure 10.4 sketches the proposed CR spectrometer consisting of several parts [57]. First, the VTI, is designed to fit the He cryostat (3) sitting in the bore of a pulsed magnet (2). The VTI integrates a QCL source (4), sample/reference holder (5), and detectors (6) with their preamplifiers (7). Outside the magnet (2) and VTI, there are a pulsed current source, a data-acquisition system, a control computer, and, if necessary, an external radiation source. The VTI is shown schematically in Fig. 10.5. If an external radiation source is used, the light is transmitted through the input window and guided down through the polished stainless-steel over-moded waveguide (light pipe) that ends with the QCL holder (note that the quality of polishing is particularly important to achieve low attenuation of the excitation laser). The QCL holder in this case is fit with a

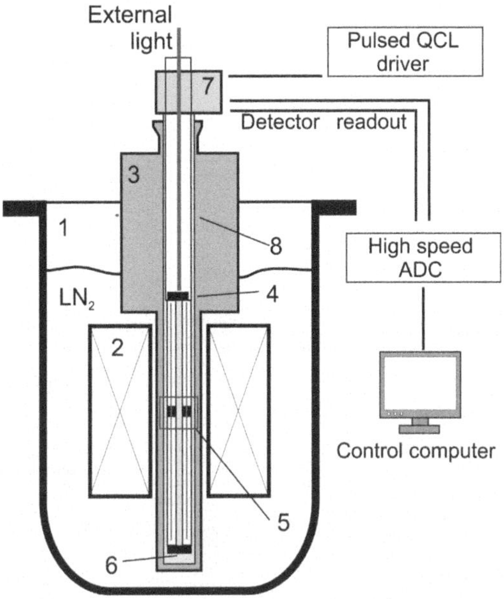

Fig. 10.4 Schematic view of the experimental setup. The pulsed magnetic field coil (2) immersed into the liquid nitrogen dewar (1) contains a ^{4}He cryostat (3) with the VTI. The VTI integrates the quantum cascade laser source (4), the sample and reference holders equipped with a heater and temperature sensor (5), the photodetector module (6), and photodetector preamplifiers (7). The VTI is made vacuum tight with a stainless-steel shield (8). The quantum cascade laser is driven by an external pulsed current source. The signals from the detectors are digitized with a fast analog-to-digital converter and transferred to the control computer for post-processing

passthrough. Otherwise, if an internal source is used, an appropriate QCL module should be plugged into the holder. Currently, QCLs are widely commercially available and can be easily adapted for the proposed LMS spectrometer. The light passing through the QCL holder (or emitted by the internal QCL source) is split into two channels/waveguides: the reference and the signal. The two-channel design is not only useful to compensate the fluctuations of the source intensity, but can also be helpful to compensate the magnetoresistance of the detectors, which is important when Ge:Ga detectors are in use (see the discussion above).

The distance from the QCL holder and the center of the coil should be chosen as a compromise between two parameters: magnetic field acting on the laser and the cooling conditions. Proper cooling of QCLs is particularly important due to high threshold currents and the quick degradation of their emission intensity when the temperature of the active zone rises. In the usual laboratory environment, the cooling of QCLs is ensured by a big copper heat sink, which is impossible to place inside pulsed magnets because of eddy currents. As a result, most of the QCLs cannot be operated in CW mode, instead they have to be driven with typically

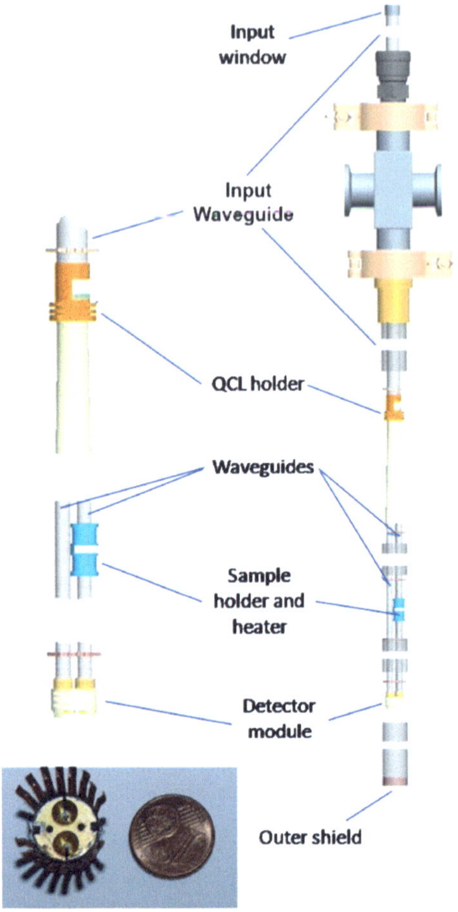

Fig. 10.5 Details of the VTI. The external light, if necessary, is transmitted through the input window and guided via the stainless-steel waveguide down to the QCL holder, which contains the passthrough for the external radiation. When an internal source is in use, the passthrough is replaced by an appropriate QCL module. The wave front is then split into two waveguides. The replaceable detector modules are installed below the sample holder. The VTI setup is made vacuum tight with an outer stainless-steel shield. The inset presents a photo of the detector module fit with two Ge:Ga crystals next to a 1 Eurocent coin

100–300-ns-long current pulses at 20–50-kHz repetition rate depending on the QCL design (although it would be highly desirable to run them in cw).

The sample holder, located in the center of the magnetic field coil, contains the temperature sensor, a magnetic field sensor (pick-up coil of Hall generator [58]), and, finally, the heater. The latter must be capable to locally heat up the sample, while the performances of QCLs and detectors should be unaffected. Good thermal decoupling of the sample holder from the detectors and QCL holder is another reason to keep the distances between them as long as possible.

The emission wavelength of QCLs may, in general, slightly depend on the injection current (and, therefore, applied voltage). Consequently, high-precision measurements may require additional calibration of QCL sources using high-precision spectrometers. Alternatively, a reference sample, exhibiting a well-defined resonance, can be installed in the reference channel.

The detectors are located *below* the sample holder. Since the magnetoresistance of the detectors is not negligible, and both Si:B BIB and Ge:Ga detectors require liquid He temperatures to be operated, the distance between the center of the coil and the detector holder should be as long as possible. It is, however, usually limited by the design of the He cryostat installed inside the pulsed magnet (see Fig. 10.4).

To measure small photocurrents, the head of the VTI probe contains a set of two high-frequency transimpedance preamplifiers, one for each of the two channels. This design, which avoids flexible cables between the detectors and the preamplifiers, helps to reduce the microphone noise greatly when measuring extra-low currents. Amplified signals are then digitized using fast analog-to-digital converters and transmitted to the control computer for post-processing. Synchronous detection locked to the repetition rate of the QCL can also be used. This, however, introduces an additional delay of the order of the integration time, which, in turn, may affect precise measurements of the resonance position.

10.4 Cyclotron Resonance Spectroscopy

In this section, we consider practical applications of cyclotron resonance spectroscopy to study basic parameters of semiconductors, which may be interesting for new researchers starting to work in this field. We will present examples from several material systems together with a very brief discussion. For the advanced readers, we suggest to refer to more detailed reviews, for example, [4, 6] or to the original papers.

10.4.1 Dilute Nitride $InAs_{1-x}N_x$

Currently, there is no other parameter of dilute nitride semiconductors reported in the literature that is scattered over such a wide range of values as the effective mass. For $InAs_{1-x}N_x$, along with a number of papers related to the optical properties (see, for example, [59–64] and references therein), only a few study the effective mass behavior as a function of the nitrogen content [65–67]. In these works, a large enhancement of the effective mass, up to $m_e^* = 0.04m_0$, even for a moderate nitrogen content of $x = 0.2\%$, has been shown, while the bulk InAs effective mass is only $m_e^* = 0.024m_0$. These changes have been, however, observed for highly doped samples, where the Fermi energy is relatively high; therefore, the band non-parabolicity may play an important role. On the other hand, little is known about the effective mass at the conduction band edge, and about the conduction band dispersion in the vicinity of the Γ-point.

Recently, an attempt of direct measurements of the effective mass was reported [68]. In this experiment, a set of nominally undoped $InAs_{1-x}N_x$ samples with x ranged between 0% (reference sample) and 1.9% has been studied under high magnetic field up to 60 T. The primary aim of the experiment was to probe the energy dispersion of the conduction band using excitation at different energies, since the conduction band of dilute nitride semiconductors is expected to be strongly non-parabolic due to the interaction with the localized nitrogen level. The samples were grown by molecular beam epitaxy (MBE) on semi-insulating (100)-oriented GaAs substrates. The epilayers have a thickness d of $\sim$1 μm and nominal N content of $x = 0$, 0.4, 0.6, 1.1, 1.9%. The N content was confirmed after growth with high-resolution X-ray diffraction measurements. Photoluminescence measurements also confirmed the systematic red shift of the emission line from 0.42 to 0.31 eV ($T = 4.2$ K) when x rises from 0 to 1.9% [59].

Figure 10.6 shows typical magneto-transmission curves, measured using the QCL emitting at $\lambda = 11.4$ μm, for samples with an N content ranging from $x = 0\%$ (Fig. 10.6a) up to $x = 1.9\%$ (Fig. 10.6c). To analyze these data, we can use the simple Drude model. As discussed in the first part of this chapter, the CR absorption coefficient in Drude model is given by a Lorentzian, where the line position yields the effective mass, the full width at half-maximum is related to the carrier mobility or carrier scattering time, and the area under the curve reflects the carrier concentration [see (10.3)–(10.5)].

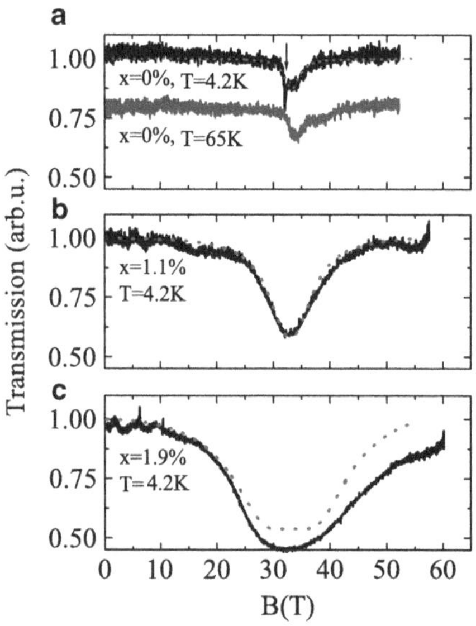

Fig. 10.6 Typical magneto-transmission curves measured for a series of $InAs_{1-x}N_x$ samples using a QCL emitting at $\lambda = 11.4$ μm (**a**) corresponds to $x = 0\%$ (reference sample) measured at $T = 4.2$ K (*upper curve*) and $T = 65$ K (curve lowered by 0.2 arb.u.), the arrow points to the impurity cyclotron resonance, (**b**) $x = 1.1\%$, (**c**) $x = 1.9\%$. *Dotted lines* show a theoretical fit using a Lorentz-like absorption coefficient with the following parameters (**a**) $m^* = 0.033$, $\mu = 7,000\,\mathrm{cm^2/Vs}$, $n = 2.3 \times 10^{16}\,\mathrm{cm^{-3}}$, (**b**) $m^* = 0.033$, $\mu = 3,500\,\mathrm{cm^2/Vs}$, $n = 3 \times 10^{17}\,\mathrm{cm^{-3}}$, (**c**) $m^* = 0.033$, $\mu = 3,500\,\mathrm{cm^2/Vs}$, $n = 2 \times 10^{18}\,\mathrm{cm^{-3}}$

The magneto-transmission curves can be then calculated using the equation $T = (1 - R)^2 e^{-\alpha d}$. Note that since the real part of the dielectric function of the sample is strongly modified by the magnetic field in the vicinity of the resonance, the CR line shape can be perturbed by interferences related to the sample thickness. In this case, the sample has to be wedged at a small angle.

Another peculiarity is the absolute CR transmission minimum, which is only 0.5 for the linearly polarized light rather than zero. This is because only one circularly polarized part of the excitation light is CR active, right or left one depending on the type of the carriers. The other part passes through the sample unaffected. When the CR absorption line falls below the 0.5 level, the excitation source most probably emits elliptically polarized light (see the discussion at the end of Sect. 10.2.1). The magneto-transmission curve corresponding to a nitrogen content $x = 1.9\%$ in Fig. 10.6c shows a good example of a CR line deformed partially by this effect ("saturation" of the CR line). In the particular case shown in Fig. 10.6c, the line is not fully reproduced by the absorption coefficient (10.5), especially in the high magnetic field part. This can be ascribed to another effect, namely, the non-parabolicity of the conduction band. The parabolic band approximation is only valid in the vicinity of band extrema. If the energy of carriers becomes large, the band dispersion deviates from the parabolic one due to the interaction with the higher bands. In general, the band starts to show sub-quadratic behavior, which would correspond to an effective mass that becomes heavier with increasing energy. In case of strong non-parabolicity and large filling factor (i.e., many Landau levels occupied), the CR line is formed by a weighted average of the transitions corresponding to different effective masses. As a result, the CR line has a clearly marked high-field tail, which cannot be modeled within the simple Drude model and requires a more sophisticated analysis (see discussion in Sect. 10.2.2). The band non-parabolicity can also result in a spin-split cyclotron resonance, since the energy and therefore the "effective mass" for the spin-down transition are higher than that for the spin-up. As a result, the spin-up transition occurs at lower fields than the spin-down at the given excitation energy (or vice versa).

The upper CR curve shown in Fig. 10.6a presents also a sharp resonance, which occurs at a magnetic field slightly lower than the main CR line. This feature can be attributed to the impurity cyclotron resonance (see Sect. 10.2.3). At low temperatures, the carriers tend to be captured at the impurity centers, which have a hydrogen-like energy spectrum. Under high magnetic fields, the 1s ground state becomes bound to the ground Landau level ($n = 0$), while the $2p_+$ state is bound to the first Landau level ($n = 1$), but the binding energy for the 1s state is slightly higher due to the smaller extension of the wave function. As a result, the transition 1s–$2p_+$ occurs at a magnetic field smaller than the free-electron cyclotron resonance at a given excitation energy. The best way to distinguish between free-electron and impurity CR lines is to perform experiments at different temperatures. When the temperature rises, the impurity states become less populated and the spectral weight of the impurity CR redistributes toward free-electron CR. According to this, the sharp line seen in Fig. 10.6a at 4.2 K disappears at higher temperatures (see the curve lowered by 0.2 arb.u.). Also, wavelength-dependent CR experiments can be performed, which would show that the impurity-shifted CR energy does not extrapolate to zero at zero magnetic fields, but rather to the 1s–2p transition energy.

Finally, Fig. 10.7 summarizes the information obtained from the experiment on $InAs_{1-x}N_x$. As can be seen, in contrast to many previous works, the effective mass shows only a moderate enhancement with increasing x, when long-wavelength (low excitation energy) lasers are used for the experiments, and practically no mass enhancement under short-wavelength excitations.

On the other hand, a huge increase in carrier density by two orders of magnitude (from $\sim 2 \times 10^{16}\,\mathrm{cm}^{-3}$ up to $\sim 2 \times 10^{18}\,\mathrm{cm}^{-3}$) was detected when the nitrogen content rises from 0 up to 1.9%, reflected in the increased area under the CR curve.

To analyze this effect, let us first calculate the reduction of the fundamental band gap due to incorporation of nitrogen according to the two-level BAC model [13, 14] (with the interaction parameter $V_N = 2.5\,\mathrm{eV}$, and the energy of the nitrogen level 1 eV above the edge of the conduction band). Then, assuming that the Fermi energy keeps its position, while the conduction band edge moves down (see inset of Fig. 10.7b), one can calculate the carrier densities for different values of nitrogen content (and, consequently, different band-gap energies and positions of the conduction band edge with respect to the pinned Fermi energy). The solid line in Fig. 10.7b shows the results of such calculation. As can be seen, the experimentally obtained values for the carrier densities (symbols) can be fit reasonably well by assuming the Fermi energy pinned at 10 meV above the conduction band edge of pure InAs. A pinning of the Fermi energy arises from strong localization of donor crystal defects, making their energy fixed relative to the vacuum level, rather than to the band structure. In this case, when the fundamental band gap reduces (conduction band edge shifts down), more and more carriers from these donors are supplied to the conduction band. Additionally, the incorporation of nitrogen itself can be a source of donor defects in the crystal.

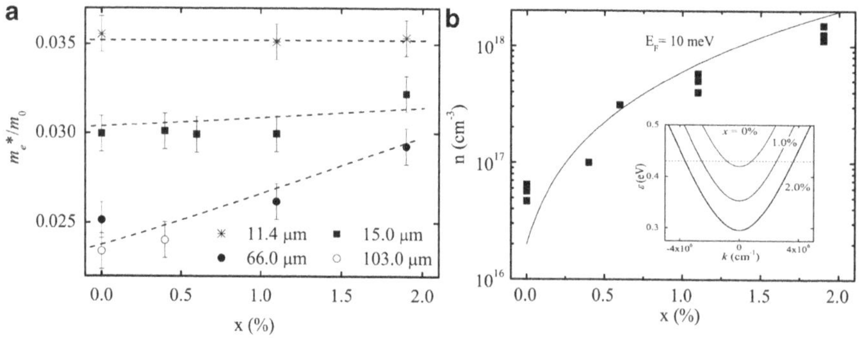

Fig. 10.7 (**a**) Dependence of the effective mass on nitrogen content obtained using different excitation lasers. (**b**) Dependence of the carrier concentration on nitrogen content, obtained experimentally (*symbols*) and calculated using the BAC model assuming the Fermi energy pinned to its level E_F at $x = 0$ (*solid lines*). The inset illustrates the Fermi energy pinning: the Fermi energy remains pinned to its position at $x = 0$ (*dashed line*), while the band moves down as a function of nitrogen content

It is also remarkable that in contrast to many previous works, a very small (or negligible, when short-wavelength lasers are used) enhancement of the effective mass was observed, even when the nitrogen content rises up to $x = 1.9\%$. For the samples with high nitrogen content, this can be explained by their high electron density. Indeed, when the nitrogen content rises to $x = 1\%$, the Fermi energy rises to ~100 meV, which is already high enough to be in the strongly non-parabolic region. In this case, the CR line contains contributions from electrons distributed over a wide range of energies, and therefore a wide range of effective masses. The CR position is then given by a weighted average of the collective CR absorption and cannot be directly related to a single effective mass. For the samples with low nitrogen content and low carrier concentration, these results confirm the negligible change in the effective mass due to nitrogen incorporation.

10.4.2 Dilute Nitride $GaAs_{1-x}N_x$

The effect of nitrogen incorporation on the effective mass of $GaAs_{1-x}N_x$ is expected to be much stronger than that in the case of $InAs_{1-x}N_x$, because the energy of the localized nitrogen level is only ~150 meV above the conduction band edge (in contrast to ~1 eV for $InAs_{1-x}N_x$). However, $GaAs_{1-x}N_x$ exhibits similar uncertainty with respect to the effective mass as $InAs_{1-x}N_x$. Some groups have reported a huge enhancement of the electron effective mass from $m_e^* = 0.067m_0$ (bulk GaAs) up to $m_e^* = 0.12m_0$ ($x = 1.2\%$) [69]; others observe only a moderate enhancement up to $m_e^* = 0.09m_0$ for $x = 1.3\%$ (see, for example, [70]). Many other values can be found in the literature, reflecting the variety of different samples under study or too large uncertainties caused by assumptions that had to be made when indirect methods are applied [71–73].

Figure 10.8 illustrates the cyclotron resonance spectra for 1-μm-thick $GaAs_{1-x}N_x$ epilayers with nitrogen content $x = 0$ (reference), 0.2, and 0.4%. For low nitrogen content ($x = 0.2\%$), the CR line is clearly resolved; however, unlike in many previous works, no significant change in the effective mass with respect to the reference sample is detected. The CR absorption quickly disappears when the nitrogen content rises due to quick degradation of the crystal quality and therefore carrier mobility, making CR experiments challenging and requiring very high magnetic fields.

10.4.3 Valence-Band Dispersion Probed by CR

An InGaAs layer grown on a GaAs substrate is a typical example of a strained system, where the strain is generated by the small lattice mismatch between the InGaAs and GaAs layers. The strain provokes significant changes in the valence-band structure, removing the Brillouin-zone center degeneracy and shifting the light-hole (lh) band down with respect to the heavy-hole (hh) band, which becomes

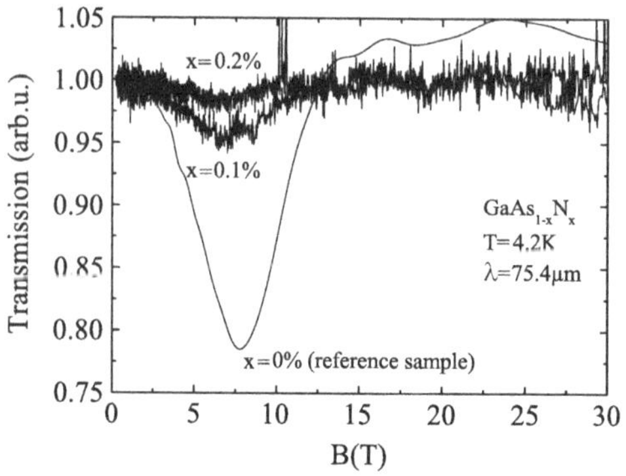

Fig. 10.8 Cyclotron resonance absorption in 1-μm-thick $GaAs_{1-x}N_x$ at different values of x

the uppermost (lowest hole energy). The dispersion of the bands also changes: the lh band becomes "heavier," while the hh band becomes "lighter"—the so-called mass reversal effect, which is also known to exist in quantum wells due to lateral confinement. The reduction of the hh band effective mass naturally leads to an enhancement of the mobility, which is particularly attractive for many applications ranging from high-speed electronics to solar cells and lasers. The valence band is also expected to be strongly non-parabolic due to the interaction between the modified lh and hh bands, which has not been fully confirmed experimentally for a long time. Only few direct measurements of the energy dispersion of the effective mass exist so far, reporting not fully consistent results [74, 75].

Recently, the strong non-parabolicity of the valence band in $In_{0.14}Ga_{0.86}As$/GaAs quantum wells (QWs) was confirmed via direct energy-dependent CR spectroscopy [17]. In this work, the sample, consisting of 50 periods of 7-nm $In_{0.14}Ga_{0.86}As$ QW and 50-nm GaAs barrier, was studied using a tunable FEL source to probe the cyclotron energy as a function of magnetic field up to 60 T. Figure 10.9a shows CR curves at various excitation energies, while Fig. 10.9b summarizes energies of each resonance as a function of the corresponding magnetic field. It can be seen that the CR energy as a function of magnetic field shows strong sub-linear dependence, confirming the strong valence-band non-parabolicity.

As was discussed in the introduction, analysis of the valence-band CR is much more complicated and always requires detailed quantum mechanical calculations of the band structure in the presence of the magnetic field. Solid lines in Fig. 10.9b show the transition energies calculated using a 4×4 Luttinger Hamiltonian that includes the deformation and QW potentials [76]. According to these calculations, the splitting of the CR line, observed when the magnetic field rises above 20 T, is related to the spin-split[6] cyclotron resonance: the first line corresponds to the transition between the ground and the first Landau levels ($n = 0$ and $n = 1$) with the projection of full momentum $J = -3/2$, and the second line corresponds to the

[6]Due to strong spin–orbit interaction in the valence band, the hole spin is always replaced by the full angular momentum.

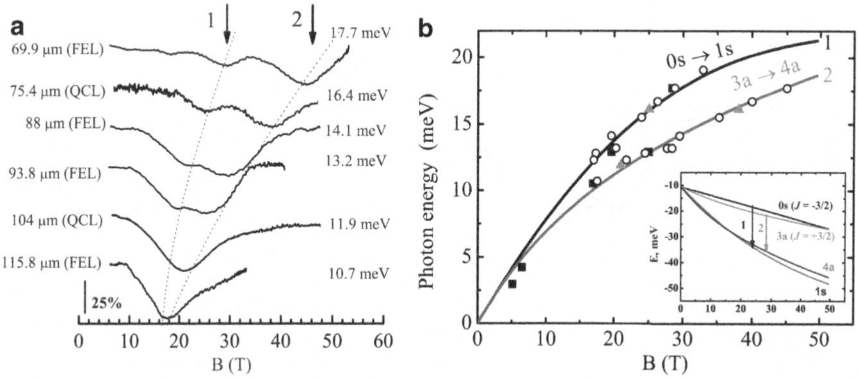

Fig. 10.9 (**a**) Selected CR curves measured in p-InGaAs/GaAs using free electron laser (FEL) and quantum cascade laser (QCL) excitations. (**b**) *Symbols* summarize the positions of the resonances as a function of magnetic field. *Solid lines* show theoretical calculations of the transition energies using a 4 × 4 Hamiltonian. The inset illustrates the transitions that correspond to lines 1 and 2 of the spin-split cyclotron resonance in (**a**)

transition between the levels with $J = 3/2$ (see inset of Fig. 10.9b). The model also reproduces very well the transition energies over the entire range of magnetic field without any fit parameters.

10.4.4 Carrier Dynamics Probed by CR

In some cases, cyclotron resonance spectroscopy can also be used to probe carrier dynamics in semiconductors [77]. Figure 10.10a shows the CR magneto-transmission curves for a sample similar to the one studied in the previous section, with the only difference being the composition of the barrier layers, which include also a thin GaAsP layer to compensate the strain in the structure. The solid line corresponds to the signal measured during sweep up of the magnetic field, and the dashed line shows the curve measured during the sweep down. The spin-split components clearly visible in the spectra basically have similar origin as in the case described in the previous section. Note that the spectral weight is distributed more toward the first line on the rising edge of the magnetic field pulse. The inset of Fig. 10.10a reflects the temporal profile of the corresponding magnetic field pulse (type-3 coil at HLD), with a sweep-up time of the order of 10 ms, while the sweep-down time is 40 ms.

A possible explanation of this unusual behavior can be a long spin relaxation time between the two ground Landau levels with projection of the full momentum $J = +3/2$ and $J = -3/2$ (levels 0s and 3a in the notation from [17], see inset of Fig. 10.9b). Indeed, when the magnetic field rises on the sweep-up edge of the magnetic field, the degeneracy of the Landau levels also rises linearly. There is a magnetic field value, when all free carriers can be accommodated by only the

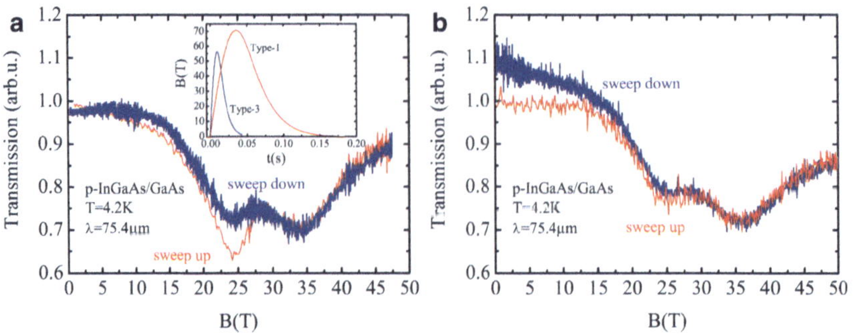

Fig. 10.10 (**a**) CR absorption curves measured for p-InGaAs/GaAs in a type-3 coil (fast). The inset shows the time dependence of the magnetic field pulses in type-1 and type-3 coils at HLD. (**b**) CR absorption curves for the same sample measured in the type-1 HLD coil (long)

two lowest Landau levels ($B \sim 10\,\mathrm{T}$ for the studied sample). When the magnetic field rises above this value, a nonequilibrium carrier distribution is formed between these levels. If the spin relaxation time is of the order or longer than the rise time of the magnetic field, the carriers do not have enough time to reach equilibrium. Consequently, the spectral weight of the CR absorption probes a nonequilibrium carrier distribution. On the falling edge of the magnetic field, which is much longer, the carriers have enough time to relax toward the lowest Landau level, which results in a different distribution of the spectral weight of the spin-split CR components. The difference between the rising and the falling edges of magnetic field tends to be smaller for longer magnetic field pulses. Accordingly, the difference between the sweeps up and down disappears when the same sample is studied using the type-1 HLD coil, which has a 35-ms rise time (Fig. 10.10b). Preliminary calculations show that the relaxation time between two lowest Landau levels is of the order of 15 ms. A similar effect was recently observed for electrons in InAs/AlSb QWs, but with characteristic lifetime of the order of microseconds [78].

Acknowledgements We would like to acknowledge collaborations with the group at the HLD Dresden led by J. Wosnitza, with our colleagues from the semiconductor spectroscopy group at HZDR (H. Schneider, S. Winnerl), with the group of V. Gavrilenko at Institute for Physics of Microstructures RAS, Nizhny Novgorod, Russia, and with Amalia Patanè (University of Nottingham, UK).

References

1. G. Dresselhaus, A.F. Kip, C. Kittel, Phys. Rev. **92**, 827 (1953)
2. G. Dresselhaus, A.F. Kip, C. Kittel, Phys. Rev. **98**, 369 (1955)
3. J. Kono, in *Methods in Materials Research*, ed. by E.N. Kaufmann, R. Abbaschian, A. Bocarsly, C.-L. Chien, D. Dollimore, B. Doyle, A. Goldman, R. Gronsky, S. Pearton, J. Sanchez (Wiley, New York, 2001), Unit 9b.2

4. N. Miura, *Physics of Semiconductors in High Magnetic Fields* (Oxford University Press, Oxford, 2008)
5. C. Hamaguchi, *Basic Semiconductor Physics* (Springer, Berlin, 2001)
6. G. Landwehr, E.I. Rashba (eds.), *Landau Level Spectroscopy* (Elsevier, Amsterdam, 1991)
7. G. Lindemann, R. Lassnig, W. Seidenbusch, E. Gornik, Phys. Rev. B **28**, 4693 (1983)
8. M. Helm, W. Knap, W. Seidenbusch, R. Lassnig, E. Gornik, R. Triboulet, L.L. Taylor, Solid State Commun. **53**, 547 (1985)
9. K. Unterrainer, C. Kremser, E. Gornik, C.R. Pidgeon, Yu.L. Ivanov, E.E. Haller, Phys. Rev. Lett. **64**, 2277 (1990)
10. J. Schneider, Phys. Rev. Lett. **2**, 504 (1959)
11. J.L. Hirshfield, J.M. Wachtel, Phys. Rev. Lett. **12**, 533 (1964)
12. P. Pfeffer, W. Zawadzki, Phys. Rev. B **74**, 115309 (2006)
13. W. Shan, W. Walukiewicz, J.W. Ager III, E.E. Haller, J.F. Geisz, D.J. Friedman, J.M. Olson, S.R. Kurtz, Phys. Rev. Lett. **82**, 1221 (1999)
14. A. Lindsay, E.P. O'Reilly, Phys. Rev. Lett. **93**, 196402 (2004)
15. P.M. Krstajic, F.M. Peeters, M. Helm, Solid State Commun. **150**, 1575 (2010)
16. J.M. Luttinger, Phys. Rev. **102** 1030 (1956)
17. O. Drachenko, D.V. Kozlov, V.Y.a. Aleshkin, V.I. Gavrilenko, K.V. Maremyanin, A.V. Ikonnikov, B.N. Zvonkov, M. Goiran, J. Leotin, G. Fasching, S. Winnerl, H. Schneider, J. Wosnitza, M. Helm, Phys. Rev. B **79**, 073301 (2009)
18. E. Gornik, Phys. Rev. Lett. **29**, 595 (1972), and in [6]
19. B.D. McCombe, R. Kaplan, R.J. Wagner, E. Gornik, W. Müller, Phys. Rev. B **13**, 2536 (1976)
20. S. Das Sarma, F. Stern, Phys. Rev. B **32**, 8442 (1985)
21. A. Gold, Phys. Rev. B **38**, 10798 (1988)
22. S. Syed, M.J. Manfra, Y.J. Wang, R.J. Molnar, H.L. Stormer, Appl. Phys. Lett. **84**, 1507 (2004)
23. P.N. Argyres, J.L. Sigel, Phys. Rev. B **10**, 1139 (1974)
24. V.K. Arora, Phys. Rev. B **14**, 679 (1976)
25. W. Götze, J. Hajdu, J. Phys. C Solid State Phys. **11**, 3993 (1978)
26. T. Ando, A.B. Fowler, F. Stern, Rev. Mod. Phys. **54**, 437 (1982)
27. Y. Shiwa, A. Ishihara, J. Phys. C Solid State Phys. **16**, 4853 (1983)
28. X. Wu, F.M. Peeters, Phys. Rev. B **41**, 3109 (1990)
29. G.M. Summers, R.J. Warburton, J.G. Michels, R.J. Nicholas, J.J. Harris, C.T. Foxon, Phys. Rev. Lett. **70**, 2150 (1993)
30. M.A. Hopkins, R.J. Nicholas, D.J. Barnes, M.A. Brummell, J.J. Harris, C.T. Foxon, Phys. Rev. B **39**, 13302 (1989)
31. H. Drexler, P. Graf, M. Besson, E. Gornik, G. Weimann, R. Lassnig, Phys. Rev. B **44**, 3105 (1991)
32. J.R. Apel, T.O. Poehler, C.R. Westgate, R.I. Joseph, Phys. Rev. B **4**, 436 (1971)
33. N.C. Jarosik, B.D. McCombe, B.V. Shanabrook, J. Comas, J. Ralston, G. Wicks, Phys. Rev. Lett. **54**, 1283 (1985)
34. W. Zawadzki, P. Pfeffer, S.P. Najda, H. Yokoi, S. Takeyama, N. Miura, Phys. Rev. B **49**, 1705 (1994), and W. Zawadzki, in [6]
35. U. Merkt, Phys. Rev. Lett. **76**, 1134 (1996)
36. J. Kono, S.T. Lee, M.S. Salib, G.S. Herold, A. Petrou, B.D. McCombe, Phys. Rev. B **52**, R8654 (1995)
37. http://www.oxford-instruments.com/
38. http://en.wikipedia.org/wiki/Francis_Bitter
39. http://www.ru.nl/hfml/
40. http://ghmfl.polycnrs-gre.fr/
41. http://www.magnet.fsu.edu
42. http://www.hzdr.de/db/Cms?pNid=580
43. http://www.lanl.gov/orgs/mpa/nhmfl/
44. http://www.toulouse.lncmi.cnrs.fr/

45. F. Herlach, N. Miura (ed.), *High Magnetic Fields: Science and Technology* (World Scientific, Singapore, 2003
46. D. Molter, F. Ellrich, T. Weinland, S. George, M. Goiran, F. Keilmann, R. Beigang, J. Léotin, Opt. Express **18**, 26163 (2010)
47. W. Seidel, E. Cizmar, O. Drachenko, M. Helm, M. Justus, U. Lehnert, P. Michel, M. Ozerov, H. Schneider, R. Schurig, D. Stehr, M. Wagner, S. Winnerl, D. Wohlfarth, S. Zvyagin, S.C. Kehr, L.M. Eng, in *Proceedings of the 30th International FEL Conference*, Gyeongju, South Korea, 2008, http://accelconf.web.cern.ch/AccelConf/FEL2008/papers/tupph060.pdf
48. D. Smirnov, C. Becker, O. Drachenko, V.V. Rylkov, H. Page, J. Leotin, C. Sirtori, Phys. Rev. B **66**, 121305 (2002)
49. D. Smirnov, O. Drachenko, J. Leotin, H. Page, C. Becker, C. Sirtori, V. Apalkov, T. Chakraborty, Phys. Rev. B **66**, 125317 (2002)
50. C. Becker, C. Sirtori, O. Drachenko, V. Rylkov, D. Smirnov, J. Leotin, Appl. Phys. Lett. **81**, 2941 (2002)
51. A. Wade, G. Fedorov, D. Smirnov, S. Kumar, B.S. Williams, Q. Hu, J. Reno, Nature Photonics **3**, 41 (2009)
52. O. Drachenko, J. Galibert, J. Leotin, J.W. Tomm, M.P. Semtsiv, M. Ziegler, S. Dressler, U. Muller, W.T. Masselink, Appl. Phys. Lett. **87**, 072104 (2005)
53. T. Chakraborty, V.M. Apalkov, Adv. Phys. **52**, 455 (2003)
54. A. Rogalski, Progr. Quant. Electron. **27**, 59 (2003)
55. J.E. Huffman, A.G. Crouse, B.L. Halleck, T.V. Downes, T.L. Herter, J. Appl. Phys. **72**, 273 (1992)
56. B.A. Aronzon, A.N. Drachenko, V.V. Ryl'kov, J. Leotin, Semiconductors **40**, 798 (2006)
57. O. Drachenko, S. Winnerl, H. Schneider, M. Helm, J. Wosnitza, J. Leotin, Rev. Sci. Instrum. **82**, 033108 (2011)
58. O.A. Mironov, M. Myronov, S. Durov, O. Drachenko, J. Leotin, Physica B **346**, 548 (2004)
59. M. de la Mare, Q. Zhuang, A. Krier, A. Patanè, S. Dhar, Appl. Phys. Lett. **95**, 031110 (2009)
60. R. Kudrawiec, J. Misiewicz, Q. Zhuang, A.M.R. Godenir, A. Krier, Appl. Phys. Lett. **94**, 151902 (2009)
61. M. Debbichi, A. Ben fredj, Y. Cuminal, J-L. Lazzari, S. Ridene, H. Bouchriha, M. Saïd, P. Christol, J. Phys. D Appl. Phys. **41**, 215106 (2008)
62. S. Ridene, M. Debbichi, A. Ben fredj, M. Said, H. Bouchriha, J. Appl. Phys. **104**, 063706 (2008)
63. Q. Zhuang, A. Godenir, A. Krier, J. Phys. D Appl. Phys. **41**, 132002 (2008)
64. S. Kuboya, Q.T. Thieua, W. Onoa, F. Nakajimaa, R. Katayamaa, K. Onabea, J. Cryst. Growth **298**, 544 (2007)
65. D.R. Hang, D.K. Shih, C.F. Huang, W.K. Hung, Y.H. Chang, Y.F. Chen, H.H. Lin, Physica E **22**, 308 (2004)
66. W.K. Hung, K.S. Cho, M.Y. Chern, Y.F. Chen, D.K. Shih, H.H. Lin, C.C. Lu, T.R. Yang, Appl. Phys. Lett. **80**, 796 (2002)
67. A. Patanè, W.H.M. Feu, O. Makarovsky, O. Drachenko, L. Eaves, A. Krier, Q.D. Zhuang, M. Helm, M. Goiran, G. Hill, Phys. Rev. B **80**, 115207 (2009)
68. O. Drachenko, A. Patanè, N.V. Kozlova, Q.D. Zhuang, A. Krier, L. Eaves, M. Helm, Appl. Phys. Lett. **98**, 162109 (2011)
69. P.N. Hai, W.M. Chen, I.A. Buyanova, H.P. Xin, C.W. Tu, Appl. Phys. Lett. **77**, 1843 (2000)
70. Y.J. Wang, X. Wei, Y. Zhang, A. Mascarenhas, H.P. Xin, Y.G. Hong, C.W. Tu, Appl. Phys. Lett. **82**, 4453 (2003)
71. F. Masia, A. Polimeni, G. Baldassarri Höger von Högersthal, M. Bissiri, M. Capizzi, P.J. Klar, W. Stolz, Appl. Phys. Lett. **82**, 4474 (2003)
72. J. Wu, W. Shan, W. Walukiewicz, K.M. Yu, J.W. Ager III, E.E. Haller, H.P. Xin, C.W. Tu, Phys. Rev. B **64**, 085320 (2001)
73. Y. Zhang, A. Mascarenhas, H.P. Xin, C.W. Tu, Phys. Rev. B **61**, 7479 (2000)
74. R.J. Warburton, R.J. Nicholas, L.K. Howard, M.T. Emeny, Phys. Rev. B **43**, 14124 (1991)

75. S.Y. Lin, H.P. Wei, D.C. Tsui, J.F. Klem, Appl. Phys. Lett. **67**, 2170 (1995)
76. V.Y. Aleshkin, V.I. Gavrilenko, D.B. Veksler, L. Reggiani, Phys. Rev. B **66**, 155336 (2002)
77. A.V. Ikonnikov, K.E. Spirin, V.I. Gavrilenko, D.V. Kozlov, O. Drachenko, H. Schneider, M. Helm, Semiconductors **44**, 1492 (2010)
78. H. Arimoto, N. Miura, R.A. Stradling, Phys. Rev. B **67**, 155319 (2003)

Chapter 11
Using High Magnetic Fields to Study the Electronic Properties of Semiconductor Materials and Nanostructures

A. Patanè and L. Eaves

Abstract This chapter examines the physics of electron motion in the presence of a magnetic field, with particular reference to recent applications in which high magnetic fields have been used to elucidate the electronic and quantum properties of some novel semiconductor materials, heterostructures, and nanostructures. We describe how magneto-tunneling spectroscopy can be used to measure the band structure of semiconductors and to investigate and manipulate the energy eigenvalues and eigenfunctions of electrons confined in low-dimensional systems.

11.1 Introduction

High magnetic fields have played a key role in elucidating the electronic properties of semiconductor materials and nanostructures. The cyclotron resonance effect, pioneered in the 1950s [1], and the Hall effect [2] are perhaps the most celebrated examples of the use of magnetic fields in semiconductor physics. The Hall effect was discovered by E.H. Hall in 1879 while investigating the magnetoresistance (MR) of metals [2]. It provides a direct means of determining the carrier concentration and mobility, and hence the scattering processes that carriers undergo as they carry electrical current down the crystal. The conventional Hall effect involves classical physics and can be understood in terms of the Lorentz force on a moving electron combined with the idea that scattering processes randomize the electron's momentum. Measurements using high magnetic fields, such as oscillatory MR effects and the quantum Hall effect, require instead a quantum mechanical description of the electron motion. Cyclotron resonance experiments undertaken on silicon and germanium in the 1950s were key milestones as they provided accurate

A. Patanè (✉)
School of Physics and Astronomy, The University of Nottingham, Nottingham NG7 2RD, UK
e-mail: amalia.patane@nottingham.ac.uk

A. Patanè and N. Balkan (eds.), *Semiconductor Research*, Springer Series in Materials Science 150, DOI 10.1007/978-3-642-23351-7_11,

measurements of the electron and hole effective masses in these important semiconductors on which the first useful transistors were based. Cyclotron resonance is the focus of Chap. 10. Thirty years before the discovery of cyclotron resonance in semiconductors, Shubnikov–de Haas and de Haas–van Alphen measurements [3] provided precise information about the nature of the Fermi surface in metals and heavily doped semiconductors. In particular, the temperature dependence of the magneto-oscillations in the electrical resistivity that characterizes the Shubnikov–de Haas effect yielded quite accurate estimates of the carrier effective mass. It is interesting to note that the Hall effect and the Shubnikov–de Haas effect have played a crucial role in three Nobel Prize discoveries in physics, namely, those concerning the integer quantum Hall effect in 1985 [4], the fractional quantum Hall effect in 1998 [5], and graphene in 2010 [6].

In Sect. 11.2, we summarize the physics of electron motion in the presence of a magnetic field, including a description of the classical Hall effect, Landau level quantization, and two of its macroscopic manifestations, the Shubnikov–de Haas and quantum Hall effects. Section 11.2 also includes a discussion of other magnetoresistance effects commonly observed in semiconductors, including the phenomenon of positive linear magnetoresistance [7–10] and of negative magnetoresistance [11]. The remainder of this chapter (Sect. 11.3) discusses the effect of high magnetic fields on the motion of electrons undergoing quantum mechanical tunneling through the barriers of a semiconductor heterostructure [12–20]. In particular, we describe how magneto-tunneling spectroscopy (MTS) [13, 16] can be used to measure the band structure and energy vs. wavevector dispersion curves of semiconductors and to investigate and manipulate the energy eigenvalues and eigenfunctions of electrons confined in low-dimensional systems.

11.2 Lorentz Force and Classical Hall Effect

The Lorentz force on a particle of charge q moving with velocity $\boldsymbol{v}$ is given by

$$\boldsymbol{F} = q\boldsymbol{v} \times \boldsymbol{B}. \tag{11.1}$$

Therefore, when current flows in a metal or a semiconductor, we can define a net Lorentz force acting on a typical charge carrier (electron or hole) with mean drift velocity $\bar{\boldsymbol{v}}$ and a mean time τ between scattering events, as shown in Fig. 11.1. Consider the case when the magnetic field is applied along the z-axis of Cartesian coordinates and the charges are electrons with $q = -e$. Provided that no current flows along the y-direction into and out of the side contacts of the Hall bar, the average motion of the electron is along the bar (parallel to the x-axis in Fig. 11.1), and the Hall force $\boldsymbol{F}$ is balanced by an induced Hall electric field $\boldsymbol{E}_{\mathrm{H}}$ along y. The Hall electric field arises from electrons which have drifted across the bar due to the Lorentz force when the current is initially switched on, thus setting up an electric field $\boldsymbol{E}_{\mathrm{H}}$ (the Hall field perpendicular to $\bar{\boldsymbol{v}}$ and $\boldsymbol{B}$). In dynamic equilibrium, we can write

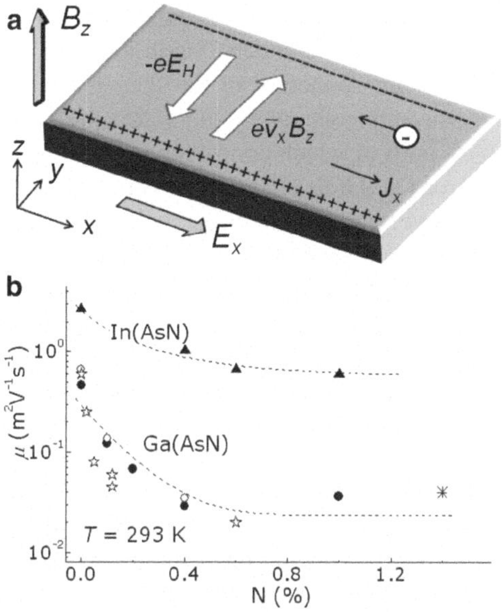

Fig. 11.1 (**a**) Sketch of the classical Hall effect. (**b**) Dependence of the electron Hall mobility on the N-content in the In(AsN) and Ga(AsN) alloys (T = 293 K) [23]. Lines are guides to the eye

$$E_{\mathrm{H}} = \bar{v}_x B = -\frac{J_x}{ne} B = \mu B E_x, \tag{11.2}$$

where $J_x = -en\bar{v}_x = en\mu E_x$ is the current density along the length of the Hall bar, n is the electron density, and μ is the electron mobility given by $\mu = e\tau/m^*$ and m^* is the electron effective mass. Hence from a measurement of B, the Hall field, $\boldsymbol{E}_{\mathrm{H}}$, and the current density, J_x, it is possible to determine both μ and n. One very important feature of the Hall effect is that the sign of the Hall field determines the type of charge of the majority carriers, i.e., electron or hole.

The Hall effect is routinely used to measure mobility and carrier densities of semiconductor materials and their temperature dependence. Studies of the temperature dependence of μ are particularly useful in separating the scattering contributions of lattice vibrations (whose quanta are phonons) and of fixed defects and impurities. An interesting example is the recent study of dilute nitride III-N-V alloys, e.g., Ga(AsN) and In(AsN) [21–23]. In these systems, which have attracted considerable interest as novel optoelectronic materials [24], the electronegativity of the N atoms combined with the stretching and compressing of neighboring bonds results in a strong perturbation of the host III–V crystal, which has significant effects on the electronic properties. In particular, the strongly localized electronic levels associated with single N-impurities and N-clusters, such as N–N pairs and higher-order clusters, admix with the delocalized band states to form the so-called amalgamated conduction band [25, 26] with localized states that limit the conductivity and mobility of the conduction electrons [27]. This unique feature of the compositional disorder exhibited by dilute nitrides has been demonstrated by detailed studies of the Hall effect (see Fig. 11.1).

Hall effect experiments can be performed at relatively modest magnetic fields (<1 T) and can be understood in terms of simple classical physics. However, care with the interpretation of the data is required in certain cases. For example, in intrinsic semiconductors, there are approximately equal numbers of electrons and holes, both of which contribute to the total current. For the special case when the electrons and holes have approximately equal mobilities, intrinsic semiconductors show a very weak Hall effect. In addition, the existence of two or more conducting channels in a semiconductor structure can obstruct the direct determination of the carrier mobility and concentration. This applies to several systems, including narrow band-gap semiconductor epilayers, such as InAs, where the pinning of the Fermi level caused by surface defects leads to the formation of a surface accumulation layer of charged carriers in addition to the bulk carriers [28]. It also applies to the so-called type II heterostructures, e.g., InAs/GaSb, in which the overlapping conduction and valence bands lead to parallel two-dimensional conducting channels of both electrons and holes [29]. In the case of InAs and the narrow gap alloy In(AsN), the conductivity is determined by two distinct parallel conduction channels, one due to a surface accumulation layer and the other to the electrons in the bulk layer [30].

It is not obvious a priori that the conventional Drude model of classical electron transport of free carriers can be applied to disordered systems in which the conduction electrons can be strongly localized, thus leading to a regime of hopping-like transport in which electrons "hop" from one localized site to a neighboring one [31]. The simple Drude model can also break down when enhanced scattering occurs for carriers whose energy resonates with that of localized states of dopant atoms, as in the case of the dilute nitrides, when the carriers are far from equilibrium (hot electron transport) or when the path length between scattering events is comparable to or longer than the size of the device (ballistic transport) [32].

11.3 Landau Level Quantization

The quantum behavior of electrons in a magnetic field gives rise to a richer variety of physical phenomena than the classical Hall effect. The Russian physicist Lev Landau laid the foundation of our understanding of the quantization of electrons in a magnetic field [33]. This phenomenon underpins two important magneto-transport effects, i.e., the Shubnikov–de Haas and the quantum Hall effect that will be discussed in Sect. 11.4. In the presence of a magnetic field, B, applied along z, the motion of the conduction electrons in the x–y plane takes the form of closed circular orbits. The (x, y) motion is, therefore, quantized into a series of discrete energy levels, called Landau levels, with an energy spacing of $\hbar\omega_c$, where $\omega_c = eB/m^*$ is the cyclotron frequency and m^* is the electron effective mass. The energy levels are, therefore, given by the relation

$$\varepsilon = \left(N + \frac{1}{2}\right)\hbar\omega_c + \frac{\hbar^2 k_z^2}{2m^*}, \tag{11.3}$$

where N is an integer [34]. The second term on the right-hand side corresponds to the kinetic energy of the free motion of charged carriers along z. This is the form of the Landau level spectrum of electrons moving in bulk semiconductors, such as GaAs, whose wavevector vs. energy dispersion curve can be described to a good approximation by a parabolic curve at k-vectors much smaller than the size of the Brillouin zone (deviations from parabolicity can be accommodated by introducing the idea of an effective mass whose value increases with ε). For the case of holes, relation (11.3) is only an approximation due to the complicated nature of valence band semiconductors. The presence of both "light" and "heavy" holes (and of a "split-off" band) combined with spin–orbit coupling leads to a nonlinear B-dependence of the hole Landau levels [35].

For low-dimensional semiconductors, such as (AlGa)As/GaAs/(AlGa)As quantum well (QW) heterostructures, the confinement provided by the (AlGa)As potential barriers of the QW quantizes the electron motion along z into a discrete series of energy levels given by ε_{iz}, where i is the quantum number for the motion along z. Hence, the energy of the quantum confined electrons is given by the relation

$$\varepsilon = \left(N + \frac{1}{2}\right)\hbar\omega_c + \varepsilon_{i,z}, \tag{11.4}$$

where i is the subband quantum number. The Landau levels of each QW subband are highly degenerate, with the degeneracy given by $g = eB/h$, for each spin orientation $m_s = \pm(1/2)$. For a field of 10 T, each of these Landau levels can contain a relatively large number of electrons; for example, when $B = 10\,\mathrm{T}$, $g = 2 \times 10^{15}\,\mathrm{m}^{-2}$, which is comparable with the sheet density of a typical two-dimensional electron gas (2DEG) in a modulation-doped (AlGa)As/GaAs heterostructure [36]. A requirement for observing the Landau level quantization is that $\omega_c\tau > 1$, which is equivalent to saying that the broadening of the Landau levels, $\hbar/\tau$, due to scattering is less than the Landau level separation, $\hbar\omega_c$. Note again that relation (11.4) is not accurate for valence band quantum wells due to admixing of the light and heavy hole states.

The picture described above applies to electron motion in most of the conventional semiconductor heterostructures, such as GaAs or (InGa)As quantum wells. However, it must be revised for systems with strongly non-parabolic energy dispersions, such as dilute nitride III-N-V alloys [14] and graphene, whose remarkable properties were discovered in 2004 [6] (see Fig. 11.2). The now celebrated energy vs. wave vector dispersion, $\varepsilon(k)$, of the two-dimensional electron system of graphene takes on a particularly interesting form, namely, that corresponding to massless Fermions with speed $v(\sim 10^6\,\mathrm{m\,s^{-1}})$ so that $\varepsilon = \pm v\hbar k$. Therefore, when a quantizing magnetic field is applied perpendicular to the plane of the graphene layer, the Landau levels are no longer equally spaced. The Landau level separation is given by $\varepsilon_N = sgn(N)\sqrt{2\hbar eBv^2|N|}$, where the integer N corresponds to electrons for $N > 0$ and holes for $N < 0$; $N = 0$ corresponds to the electron–hole degenerate level at the so-called Dirac point [37]. Since the degeneracy of each Landau level

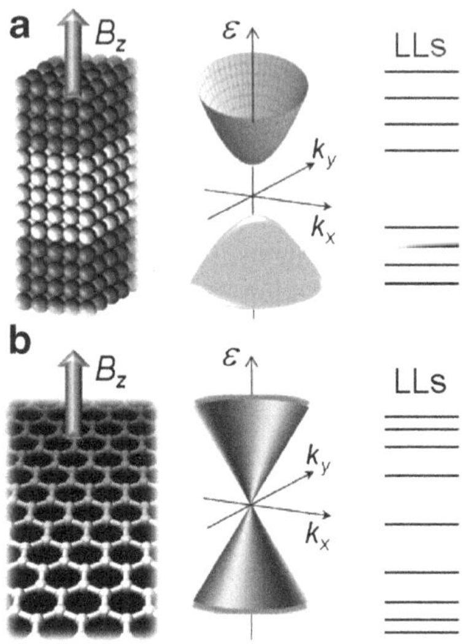

Fig. 11.2 Sketch of the energy–wavevector dispersion and Landau levels (LLs) for two-dimensional systems based on (**a**) an AlGaAs/GaAs/AlGaAs quantum well and (**b**) graphene

is given by the density of magnetic flux lines and is, therefore, independent of the dispersion curve, the relation $g = eB/h$ still holds for graphene.

The Landau levels are also non-equally spaced in QWs containing the dilute nitride alloy Ga(AsN), but for a fundamentally different reason. In Ga(AsN), the resonant interaction of the N-level with the extended conduction band states of GaAs admixes the Γ-states with states from the higher energy L- and X-bands of the III–V host crystal, thus leading to an "effective band structure" made of hybridized states with partial Γ-like character and non-parabolic $\varepsilon(k)$ dispersions [25, 26, 38].

11.4 Magnetoresistance, Shubnikov–de Haas and Quantum Hall Effects

The electrical resistance of a semiconductor usually increases with applied magnetic field, thus leading to a positive magnetoresistance effect. However, in certain cases, negative magnetoresistance effects can occur, in which the application of a small magnetic field decreases the resistance, due, for example, to the phenomenon of weak localization. We first consider the basic theory (Drude theory) of how a magnetoresistance can arise in a semiconductor where the electrons move as free particles between scattering events. Later, we will examine the effect of the Landau level quantization on the magnetoresistance. We conclude this section with the description of two interesting magnetoresistance phenomena commonly observed in semiconductors, i.e., the linear positive magnetoresistance and the negative magnetoresistance.

11.4.1 Classical Positive Magnetoresistance

To understand the nature of the magnetoresistance of a semiconductor structure or device, we need to consider the tensor relation between the current density, $\boldsymbol{J}$, and electric field, $\boldsymbol{E}$, vectors. This relation takes the form $\boldsymbol{J} = \tilde{\boldsymbol{\sigma}} \boldsymbol{E}$, where $\tilde{\boldsymbol{\sigma}}$ is the 3×3 conductivity tensor, i.e.,

$$\begin{pmatrix} J_x \\ J_y \\ J_z \end{pmatrix} = \frac{\sigma_0}{1 + (\omega_c \tau)^2} \begin{pmatrix} 1 & -\omega_c \tau & 0 \\ \omega_c \tau & 1 & 0 \\ 0 & 0 & 1 + (\omega_c \tau)^2 \end{pmatrix} \begin{pmatrix} E_x \\ E_y \\ E_z \end{pmatrix}. \tag{11.5}$$

Here, $\boldsymbol{B}$ is directed along z and τ is the scattering time. Note that σ_{xx} decreases with increasing B. The corresponding magnetoresistivity tensor is $\tilde{\boldsymbol{\rho}} = 1/\tilde{\boldsymbol{\sigma}}$. Hence, the components of the resistivity along x and z are given by the transverse magnetoresistance, $\rho_{xx} = \sigma_{xx}/(\sigma_{xx}^2 + \sigma_{xy}^2)$, and the longitudinal magnetoresistance, $\rho_{zz} = 1/\sigma_{zz}$, respectively [35].

In the simple Drude model of electrical conduction, it is assumed that the scattering of the conduction electrons can be described by a single characteristic mean scattering time, τ, independent of the electron energy. The conductivity at $B = 0$ is $\sigma_0 = ne^2\tau/m^*$. Since there is no current flow along y, from (11.5) we obtain the Hall field $E_H = -(J_x B/ne)$. The current density along x, $J_x = \sigma_0 E_x$, is independent of magnetic field, thus resulting in a resistivity $\rho_{xx} = \rho_0 = 1/\sigma_0$, which remains unchanged as the magnetic field is applied (i.e., there is "zero" magnetoresistance in this case). This result is in stark contrast with experimental observations of a positive magnetoresistance in most semiconductor samples and devices. We can resolve this inconsistency by noting that in a real semiconductor, electrons are distributed over a range of energy and hence they drift with different velocities compared to the average velocity $\bar{v}$ that determines the Hall field along y. Electrons drifting along x with lower and higher velocity than the average value tend to spread along y, resulting in an increase in ρ_{xx} with B. At low and intermediate fields, the magnetoresistance has a quadratic dependence on B. This is then often followed by a saturation to a constant value in the limit $\omega_c \tau \gg 1$ [35]. The zero-field relation, $\sigma_0 = ne^2\tau/m^*$, means that a long scattering time corresponds to a high conductivity. In contrast, when a strong magnetic field is applied perpendicular to the direction of current flow and to the applied electric field, the electron orbit centers tend to drift along the voltage equipotentials (Fig. 11.3a) and a conduction down the potential gradient is only possible due to scattering processes, so we have the interesting result that at high magnetic fields, σ_{xx} is proportional to the scattering rate, $1/\tau$. This phenomenon is nicely described by the Roth–Argyres formula [39].

The tensors $\tilde{\boldsymbol{\sigma}}$ and $\tilde{\boldsymbol{\rho}}$ can be used to understand the effect of an applied magnetic field on the three principal measurement geometries for magnetoresistance: (1) a bar-shaped sample of length L and area A, (2) a short mesa structure in which an active layer of length L and area A is sandwiched between two heavily doped layers, and (3) the circular Corbino structure, as shown in Fig. 11.3b–d. In case (1), the

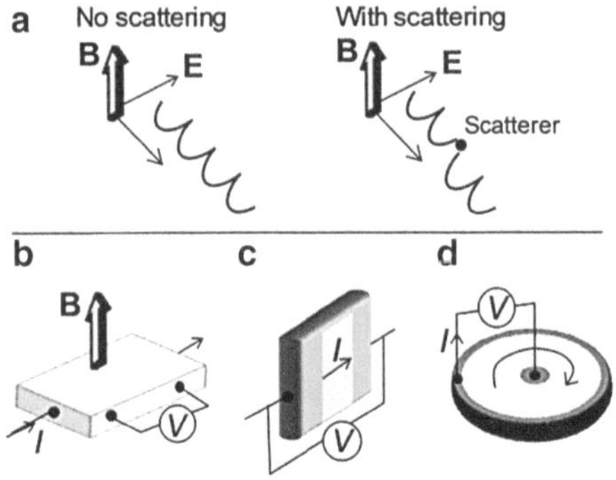

Fig. 11.3 (**a**) Electron trajectory in a magnetic field (***B***) perpendicular to the electric field (***E***) in the absence and in the presence of scattering. (**b**–**d**) Setup for measurements of the transverse magnetoresistance in (**b**) a bar-shaped sample, (**c**) a mesa structure, and (**d**) a Corbino structure

transverse magnetoresistance is $R(B) = V/I = \rho_{xx}L/A = L\sigma_{xx}/A(\sigma_{xx}^2 + \sigma_{xy}^2) \cong L\sigma_{xx}/A\sigma_{xy}^2$ when $\omega_c\tau \gg 1$. In case (2), the heavily doped contact layers which sandwich the active layer effectively act to "short-circuit" the Hall field ($\boldsymbol{E}_{\mathrm{H}} \cong 0$) so that $R(B) = V/I \cong A/L\sigma_{xx}$. A similar relation to case (2) holds for (3), the Corbino geometry, in which the inner and outer circular ring-shaped contacts "short-circuit" the Hall field. In this geometry, for higher mobility samples with small scattering rates, it can be found that when scattering is almost entirely suppressed at very high magnetic field, the current flows around the circular active layer between the inner and outer conducting rings, perpendicular to the direction of the radial Hall field, with a small dissipative current flowing radially.

11.4.2 Shubnikov–de Haas and Quantum Hall Effects

The Shubnikov–de Haas effect [3] is observed in the low temperature magnetoresistance of metals and heavily doped degenerate semiconductors with high carrier densities. In this context, "degenerate" means that the Fermi energy ε_{F} is much larger than the thermal energy $k_{\mathrm{B}}T$ and is located well above the edge of the conduction band (or well below the valence band edge in the case of p-doped semiconductors). Under these conditions, the magnetoresistivity ρ_{xx} is modulated each time a Landau level passes through the Fermi energy as the magnetic field and hence the Landau level separation, $\hbar\omega_c$, are progressively increased. The physical origins of the effect can be understood as follows: in Fig. 11.4a, the Fermi energy is located between the minima of two adjacent Landau levels, with quantum numbers N and $(N + 1)$. At this magnetic field, B_1, the density of states at the Fermi energy is relatively low because of the steep slope in the $\varepsilon(k_z)$ curve, so the number of empty states into which an electron at the Fermi energy can scatter is small: hence the elastic scattering rate is low. When the field is increased slightly to B_2, the density of states at E_{F} is much larger around the minimum of the $\varepsilon(k_z)$ curve of the $(N + 1)$ level and so the scattering rate is much higher. Note that in Fig. 11.4a, the Fermi energy at E_{F} is slightly lower than that at B_1. This is because the degeneracy of each Landau

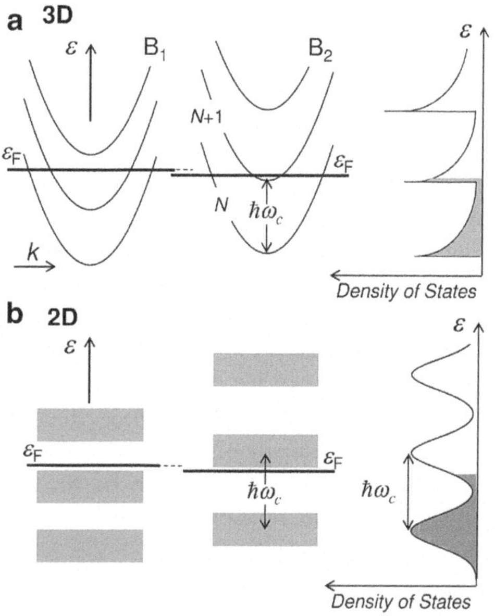

Fig. 11.4 Sketch of the Landau levels for (**a**) three- and (**b**) two-dimensional systems. Landau levels for two values of the magnetic field, B_1 and B_2, illustrating the change in the Fermi level position with B. The density of states for $B = B_2$ is shown for each system on the *right panel*

level increases with B, so the lower Landau level can accommodate more electrons. This modulation of the Fermi energy and indeed of the total electronic energy $U(B)$ also gives rise to the modulation of the magnetization μ of the conduction electrons given by $\mu = -\partial U/\partial B$. This is the famous de Haas–van Alphen effect [40].

In real crystals, the presence of electron scattering processes leads to broadening of the states in the Landau level spectrum, thus removing the singularities in the density of states at $k_z = 0$. The divergence of the conductivity tensor, σ_{xx}, which appears in the absence of scattering, is also removed so that broad maxima appear in both σ_{xx} and ρ_{xx} due to enhanced elastic intra-Landau level scattering processes when $\varepsilon_F = (N + 1/2)\hbar\omega_c$. The magneto-oscillations are periodic in $1/B$, and the periodicity, $\Delta(1/B)$, is determined by the free electron density. For a three-dimensional electron gas of density n, this periodicity is given by the relation

$$\Delta\left(\frac{1}{B}\right) = 2e\hbar^{-1}\left(3\pi^2 n\right)^{-2/3}. \tag{11.6}$$

Note that the period of the Shubnikov–de Haas oscillations in ρ_{xx} does not provide us with a measurement of the carrier effective mass, but rather the carrier density; however, m^* can be estimated by measuring the temperature dependence of the amplitude of the magneto-oscillations. As T is increased, the sharp step in the Fermi–Dirac distribution is broadened by an energy $\sim k_B T$; when this becomes comparable with $\hbar\omega_c$, the amplitude $\Delta\rho_{xx}$ of the oscillation decreases as

$$\Delta\rho_{xx} \propto \frac{\chi}{\sin h\chi}, \tag{11.7}$$

where $\chi = 2\pi^2 k_B T/\hbar\omega_c$. By measuring the dependence of $\Delta\rho_{xx}$ on T and B, it is possible to estimate m^* using the relation $\omega_c = eB/m^*$. However, for a more accurate value of m^*, cyclotron resonance is usually the best technique and it has been used extensively in elucidating the electronic properties of a wide range of semiconductor materials and heterostructures (see Chap. 10).

In a 2DEG, such as in a modulation-doped heterostructure (MDH) or a silicon MOS field-effect transistor, the Shubnikov–de Haas effect is much more pronounced than for bulk semiconductors. In MDHs, this is partly due to the much higher carrier mobility [36]. In addition, in 2DEGs, the peaks in the density of states are generally more clearly defined since the dispersion of the Landau levels arising from the k_z-dependence shown in Fig. 11.4a is absent due to the quantization of the motion along z. In this case, the discrete Landau levels give rise to δ-function-like peaks in the density of states when scattering effects are neglected (Fig. 11.4b). Each Landau level has a degeneracy of eB/h if the Zeeman spin splitting can be resolved. For a 2DEG with electron sheet density n_s, the periodicity of the magneto-oscillations in ρ_{xx} is given by the relation

$$\Delta\left(\frac{1}{B}\right) = je(hn_s)^{-1}. \tag{11.8}$$

Here, $j = 2$ when spin splitting is not resolved and $j = 1$ when spin splitting is resolved. The quantum Hall effect [4] is closely related to the Shubnikov–de Haas effect in two-dimensional systems. Over a magnetic field range corresponding to the flat steps in the Hall conduction σ_{xy}, the Fermi energy is "pinned" on the localized states in the gaps between the Landau levels. This occurs when the condition $n_s \cong NeB/h$ is satisfied ($N = 1, 2, 3\ldots$ with spin splitting resolved or $N = 2, 4, 6\ldots$ otherwise). On the flat steps, $\sigma_{xy} = Ne^2/h$ and $\rho_{xx} \cong \sigma_{xx}/\sigma_{xy^2}$ approaches zero at the low temperature limit and at high magnetic fields. Under these conditions, electrons move down the Hall bar in the free carrier states which extend along equipotentials of the Hall electric field. These extended states have energies close to the peaks of the Landau level density of states and carry current down the length of the Hall bar with almost no dissipation.

Figure 11.5 compares the magnetic field dependence of ρ_{xx} and ρ_{xy} at $T = 2$ K for a MDH based on (AlGa)As/Ga(AsN)/(AlGa)As QWs with a much higher mobility structure in which the QW is composed of undoped GaAs. The Shubnikov–de Haas oscillations and quantum Hall plateaus can be seen for the Ga(AsN) QW, although they are rather weak in QWs in which the N-content exceeds ~0.1%. These data demonstrate clearly the effect on the electronic properties of N-induced scattering and carrier trapping [41]. Shubnikov–de Haas oscillations and quantum Hall plateaus can also be seen in graphene even at room temperature due to its special electronic properties [6]. This offers the exciting prospect of producing a room-temperature resistance standard based on graphene. Before concluding this section, we make a brief mention of the fractional quantum Hall effect (FQHE) [5]. In this case, plateaux in σ_{xy} occur at values of pe^2/sh, where p and s are integers (e.g., $p/s = 1/3$, 2/3, etc.). The FQHE arises from the effect of magnetic flux

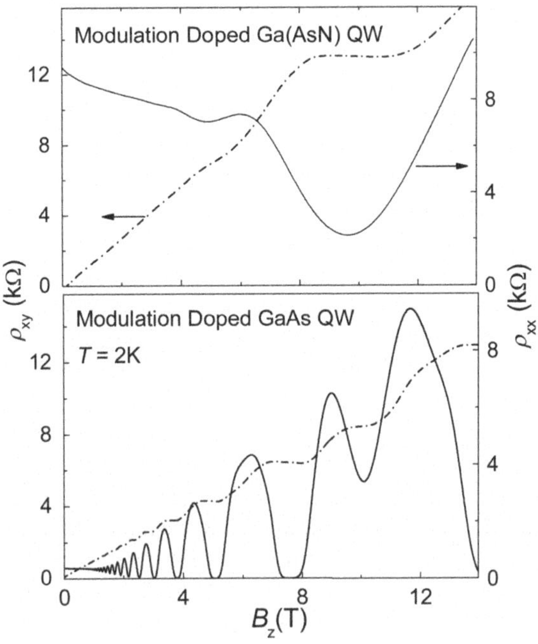

Fig. 11.5 Hall resistance (ρ_{xy}) and parallel (ρ_{xx}) resistance in a modulation-doped quantum well (QW) based on Ga(AsN) and GaAs ($T = 2$ K)

quantization on the electron–electron interaction within the 2DEG and this is best observed in high-mobility materials at low temperatures and high magnetic fields. We refer to specialized text for a detailed discussion of these quantum transport phenomena [42,43].

11.4.3 Magnetophonon Resonance

The magnetophonon resonance (MPR) effect was first predicted in a theoretical paper by Gurevich and Firsov [44] and has since been observed in the magnetoresistance of many n- and p-type semiconductors, including bulk crystals of Si, Ge, and the III–V and II–VI compounds. For a review, see [45]. It has also been observed in transverse magnetoresistance of 2DEGs in quantum wells and MDHs. The physical origin of MPR can be understood in terms of Landau level quantization of the electronic spectrum by an applied magnetic field. In semiconductors, the energy of the optical phonon modes of lattice vibration at k-vectors close to the center of the Brillouin zone is very well defined. Longitudinal optical (LO) phonons interact particularly strongly with electrons in III–V and II–VI compounds due to their polar character and hence can scatter electrons strongly; polar mode scattering is weaker in Si and Ge but nevertheless, LO phonons can scatter electrons because of the deformation potential interaction. When the energy, $\hbar\omega_{\mathrm{L}}$, of the optical phonon equals an integral number of Landau level spacings, $N\hbar\omega_{\mathrm{c}}$, i.e., when

$N\hbar\omega_c = \hbar\omega_L$, where $N = 1, 2, 3, \ldots$, it is found that the scattering is resonantly enhanced, thus giving rise to maxima in ρ_{xx} and a series of magneto-oscillations periodic in $1/B$ when the magnetic field is swept upward. Whereas in cyclotron resonance the resonant transitions occur between adjacent levels, in MPR there is no such restriction due to selection rules, and in high-quality samples, it is possible to observe resonant peaks in both ρ_{xx} and ρ_{zz} for N-values up to 20 or more. In the indirect semiconductor Si, electrons can also be scattered between the X-point conduction band minima (so-called inter-valley scattering), not only by optical phonons but also by transverse and longitudinal mode acoustic phonons, thus leading to a particularly complex and interesting series of resonances [46,47]. Since the energy of the zone-center longitudinal optical phonon $\hbar\omega_L$ can be measured very accurately by Raman spectroscopy, MPR provides us with a particularly easy and accurate way of measuring the effective masses of electrons and holes, a sort of cyclotron resonance using the phonon quanta of the crystal which dispenses with the microwave radiation source needed for cyclotron resonance. A recent example of using MPR in this way is illustrated in Fig. 11.6. These data show the effect of adding a small concentration of N to GaAs, to form the dilute nitride alloy Ga(AsN). A small concentration of N (~0.1%) leads to a significant shift of the MPR peaks which arises from the enhancement of the conduction band effective mass due to the resonant anticrossing referred to in Sect. 11.2. Note also that the N-induced scattering significantly damps out the MPR effects [48].

11.4.4 Positive Linear Magnetoresistance

A positive linear magnetoresistance, ρ_{xx}, can arise from either quantum or classical phenomena. At high enough magnetic field and at low temperatures, the electron

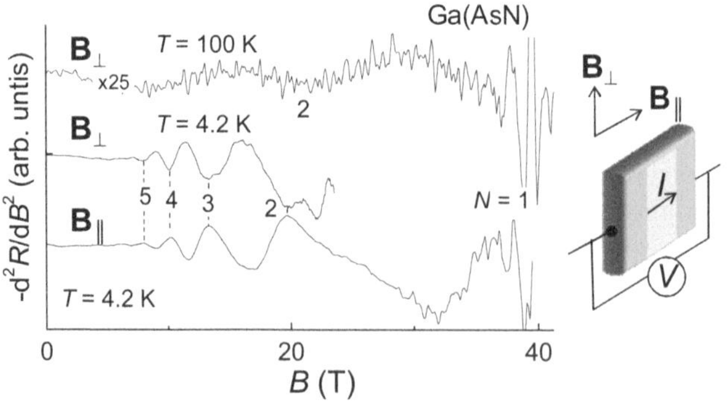

Fig. 11.6 B-dependence of $-\mathrm{d}^2R/\mathrm{d}B^2$ (where R is the magnetoresistance) for a Si-doped Ga(AsN) epilayer at $T = 4.2$ K ($\boldsymbol{B}_\perp$ and $\boldsymbol{B}_\parallel$) and $T = 100$ K ($\boldsymbol{B}_\perp$). The *inset* shows the geometry of the magneto-transport experiment

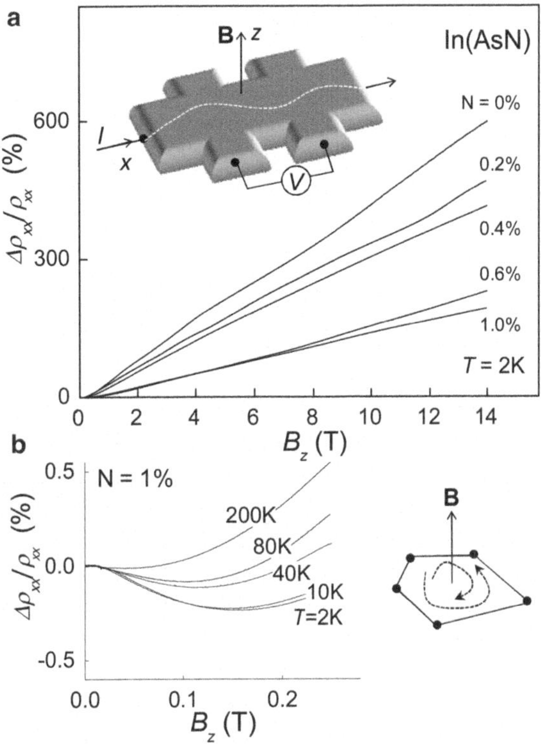

Fig. 11.7 (**a**) Magnetoresistance, $\Delta\rho_{xx}/\rho_{xx}$, as a function of magnetic field, B_z, measured in Hall bars based on In(AsN) epilayers grown on GaAs ($T = 2\,\mathrm{K}$). *Inset*: Sketch of the current deflection caused by non-homogeneties, thus causing linear positive magnetoresitance. (**b**) Negative magnetoresistance and its temperature dependence for the In(AsN) epilayer with $N = 1\%$. *Inset*: Sketch of weak localization in magnetic field

distribution enters the so-called quantum limit, in which only the lowest Landau level is occupied. This limit occurs when the cyclotron energy, $\hbar\omega_c$, exceeds the Fermi energy, E_F, and when $\hbar\omega_c$, $E_F >> k_B T$. In this regime, ρ_{xx} increases linearly with B [7, 8]. This quantum effect is generally studied at high magnetic fields and in semiconductors with a small electron effective mass. A different type of linear positive magnetoresistance can also arise from distorted current paths caused by disorder-induced macroscopic fluctuations, $\Delta\mu$, in the electron mobility [9, 10]. This effect can be explained in terms of classical physics and can be observed over a wide range of temperatures. In the classical MR model of [9, 10], the crossover of ρ_{xx} from a quadratic to a linear B-dependence occurs at a threshold magnetic field $B_t = \bar{\mu}^{-1}$ for $\Delta\mu/\bar{\mu} < 1$ or at $B_t = \Delta\mu^{-1}$ for $\Delta\mu/\bar{\mu} > 1$. Here, $\bar{\mu}$ corresponds to an average of the spatially varying mobility. Correspondingly, the magnitude of ρ_{xx} is expected to scale with $\bar{\mu}$ or $\Delta\mu/\bar{\mu} < 1$ or with $\Delta\mu$ for $\Delta\mu/\bar{\mu} > 1$. Figure 11.7a shows an example of this unusual positive magnetoresitance effect, which was observed in In(AsN) epilayers grown on GaAs [30]. This classical linear magnetoresistance effect has also been reported for other conductors, such as silver chalcogenides [49] and InSb [50], and is of topical interest as a possible route for developing new magnetic sensors.

11.4.5 Negative Magnetoresistance

Negative magnetoresistance can arise from quantum interference effects and is commonly observed in semiconductors at low temperatures [11]. An important case involves the phenomenon of weak localization. The magnetic field breaks the time reversal symmetry of the de Bröglie waves associated with the trajectories of electrons which have undergone elastic scattering processes, thus destroying the coherent backscattering of electrons due to quantum interference and leading to a quenching of the weak localization, which in most cases enhances the low temperature resistance at zero magnetic field. This process is determined by a characteristic coherent length, $L_{\varphi} = (D\tau_{\varphi})^{1/2}$. Here, $D = v_{\mathrm{F}}^2 t/3$ is the diffusion coefficient, τ_{φ} is the coherence time, τ is the elastic scattering time, and v_{F} is the electron velocity at the Fermi energy. The negative magnetoresistance effect disappears at higher temperatures when electron–phonon scattering becomes significant; this type of inelastic scattering event destroys the phase coherence of the de Bröglie waves and removes the constructive interference effect which leads to the weak localization. For weak localization of extended states in three dimensions, the negative MR, defined as $\Delta\rho_{xx}/\rho_{xx} = [\rho_{xx}(0) - \rho_{xx}(B)]/\rho_{xx}(0)$, has a characteristic dependence on B, given by B^2 or $B^{1/2}$ at low or high B, respectively. The crossover between these dependences occurs when the magnetic length, $l_{\mathrm{c}} = \sqrt{\hbar/eB}$, becomes equal to $2L_{\varphi}$ [11]. This model describes well the negative magnetoresitance effects observed in many semiconductors, including the dilute nitride alloy (InGa)(AsN) [23, 30]. In Ga(AsN), negative MR effects are observed at relative large magnetic fields ($B \sim 1$ T) corresponding to small coherence lengths ($L_{\varphi} = 15$ nm at N $= 0.1\%$) [23]. In contrast, weak localization effects in In(AsN) are observed at much smaller magnetic fields ($B \sim 0.1$ T) corresponding to coherence lengths that remain large even for high N ($L_{\varphi} = 60$ nm at N $= 1\%$), thus indicating that localization effects are significantly weaker in this small band-gap material [30] (Fig. 11.7b).

11.5 Magneto-Tunneling Spectroscopy

The combination of high magnetic fields and quantum tunneling provides a powerful tool for elucidating the nature of the quantized states of both free and bound electrons and holes in semiconductor materials, device, and nanostructures. In MTS experiments, an applied bias is used to resonantly tune the energy of a charged carrier, which tunnels from an emitter layer through a tunnel barrier into a quantum state located beyond the tunnel barrier in a resonant tunneling diode (RTD) (see Fig. 11.8). When a magnetic field is applied in the plane of the barrier, the tunneling electron or hole acquires an in-plane momentum due to the action of the Lorentz force given by

$$\hbar\mathbf{k} = qs \times \mathbf{B}, \tag{11.9}$$

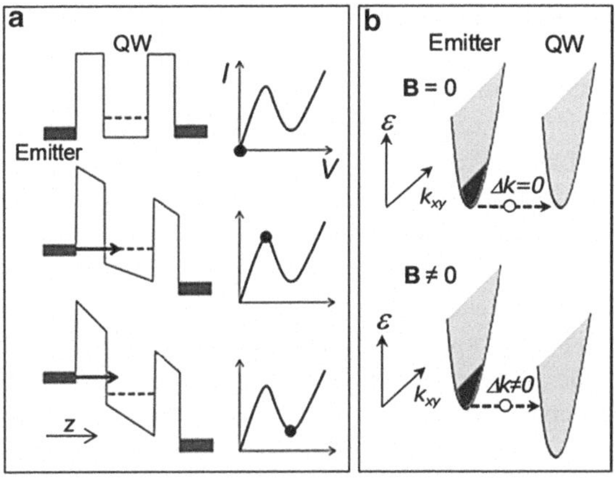

Fig. 11.8 (**a**) Sketch of electron tunneling from an emitter layer into the quasi-bound state of a quantum well (QW) under various applied biases. The *right panel* illustrates the current–voltage $I(V)$ curve. The *dot* in each $I(V)$ illustrates a particular bias condition. (**b**) Electron tunneling with and without an applied magnetic field directed in the plane of the QW

where s is the tunneling distance from the emitter to the quantum state. Thus, the combination of applied bias and magnetic field can be used to control both the in-plane wavevector ($\hbar k = qsB$) and the energy $\varepsilon = eV_b = qsF$ of a carrier when it enters the quantum state. Here, F is the electric field in the region of the emitter barrier and V_b is the voltage dropped across the region. MTS exploits the fact that molecular beam epitaxy can produce tunnel barriers which are almost atomically flat. The small thickness variations of the barriers are generally no more than an atomic monolayer, which is typically an order of magnitude smaller than the de Bröglie wavelength of the tunneling electron. This high level of translation symmetry means that in-plane momentum is conserved. The applied in-plane magnetic field then allows us to make controlled changes to the momentum.

The concept of MTS was developed from early studies of magneto-electric skipping states at a tunnel barrier interface [19] and from measurements of the change in the current–voltage characteristics, $I(V)$, of a resonant tunneling device due to a strong magnetic field applied in the plane of the quantum well [20]. It was then used to measure resonant band anticrossing (BAC) effects in the valence [13] and conduction bands [14, 15] of quantum wells, thus providing detailed plots of the energy–wavevector dispersion relations, $\varepsilon(k)$, of holes and electrons. The following two sections describe some applications of MTS to study the band structure of semiconductor materials and the eigenfunctions of low-dimensional systems.

11.5.1 Probing Band Structures

We first consider a resonant tunneling structure of the type shown in Fig. 11.8, where electrons tunnel from an emitter accumulation layer into the quasi-bound state of a quantum well. The tunneling current is along the growth direction (which is defined as the z-direction) and is affected by the presence of a magnetic field, $\boldsymbol{B}$, applied perpendicular to the z-axis. The magnetic field can be applied along different directions in the quantum well plane (x, y), but without loss of generality we can choose the direction of $\boldsymbol{B}$ as the x-axis in the following discussion.

Quantum mechanically, the effect of a magnetic field on the motion of a particle is found by replacing the canonical momentum operator p_y by $p_y - qBz$, using for the Landau gauge $\boldsymbol{A} = (0, -Bz, 0)$. The extra term $-qBz$ gives rise to a magnetic contribution to the effective potential governing the particle motion along the z-direction, normal to the barrier interfaces. However, it does not affect the plane-wave nature of the states along the y- and x-directions. For the quasi-bound states in the emitter accumulation layer and quantum well, the magnetic field can be treated as a perturbation. This is equivalent to replacing p_y by $p_y - qB < z >$, where $< z >$ is the position coordinate averaged over the quasi-bound state. Thus, the in-plane dispersion curves of the emitter and the quantum well states are displaced along k_y by an amount $\hbar k_y = qsB$, where s represents the separation $< z_{\text{well}} > - < z_{\text{emitter}} >$ between the bound state in the well and the emitter. The occupied states below the Fermi level in the emitter accumulation layer are centered around $k_y = 0$ and $k_x = 0$. In the absence of scattering, k_y, k_x, and the energy ε are conserved in the tunneling transition. The available set of states in the quantum well are shifted by an amount given by $\hbar k_y = qsB$. In turn, this produces a shift in the resonance: the peak of the resonance corresponds to carriers with $k_y = 0$ and $-k_{\text{F}} \leq k_x \leq k_{\text{F}}$ in the emitter, where k_{F} is the Fermi wavevector. The voltage shift in the resonance is determined by the energy dispersion of the particle and can be understood as follows: the magnetic field does no work on the carrier and, to maintain the resonance conditions, any gain in kinetic energy corresponding to a larger k_y in the quantum well must be compensated by an additional voltage (see Fig. 11.8b).

We now describe how this MTS technique provides a means of exploring the admixing of the extended conduction band states of GaAs with the localized states associated with N-impurities [14, 15]. The incorporation of low concentrations of N ($\sim$1%) in III–V compounds, such as GaAs, leads to a number of unusual physical properties including a large band-gap bowing with a band-gap reduction. According to a simple two-level band anticrossing model, the interaction of the extended Γ-conduction band states of GaAs with the localized N energy level causes a splitting of the conduction band into two subbands E_- and E_+, which are strongly admixed. Also, N aggregates, such as impurity N–N pairs and higher-order clusters, form strongly localized states; these states can admix to form the so-called amalgamated conduction band, thus leading to a duality of band-like and localization behaviors.

Figure 11.9 shows the current–voltage $I(V)$ characteristics of a resonant tunneling diode based on a Ga(AsN) embedded between two (AlGa)As tunnel barriers. In this structure, electrons tunnel into the N-induced subband states E_{0-} and E_{0+} in the Ga(AsN) QW layer, thus leading to two dominant resonant features in the $I(V)$. The data show a shift to higher bias of E_{0-} and E_{0+} and a general increase in current with increasing B. In particular, the B-dependence of the amplitude of the E_{0-} and E_{0+} peaks in $I(V)$ has the characteristic form of a quantum mechanical admixing effect, i.e., with increasing B, the E_{0-} feature tends to become weaker and disappears, whereas the E_{0+} peak increases significantly in amplitude. By plotting the gray-scale plot of the differential conductance vs. the applied magnetic field and applied bias, it is possible to map out the form of the in-plane dispersion curves associated with E_{0-} and E_{0+} (see inset of Fig. 11.9). The form of the measured $\varepsilon(k)$ dispersion relations of the N-induced subbands can be understood in terms of a simple two-level BAC model involving isolated N atoms. However, the Γ-character of the N-induced states depends on energy and k-vector, and manifests itself as a quenching of the resonant tunneling at an energy corresponding to that of the N-level. Similar measurements in samples containing different N-concentrations have revealed that the band structure properties change rapidly with increasing N-content. When the N-content is increased up to about 0.2%, N clustering (mostly N–N first neighbor [110] pairs and second neighbor [220] pairs) tends to reduce further the Γ-character of the electronic states and to break up the energy–wavevector dispersion relations. A further increase in N-content quenches all resonant features in $I(V)$ due to N-clustering effects. These data combined with the *L*inear *C*ombination of *I*solated *N*itrogen *S*tates (LCINS) model [15, 25] have demonstrated an unusually abrupt change in the electronic properties of Ga(AsN) with increasing N. This behavior differs significantly from the smoother variation with varying alloy composition observed in other alloy material systems, such as (InGa)As, for which the k-vector, the $\varepsilon(k)$ dispersion relations, and the effective mass values remain well defined over the whole range of In-composition. An interesting outcome of this work is that only at low N ($\sim$0.1%) there is a well-defined k-vector for the hybridized band states of Ga(AsN) over an extended range of energy, whereas at higher N-concentrations, the conduction band tends to "break up."

11.5.2 Probing and Manipulating Low-Dimensional Systems

The possibility of fine-tuning and imaging the electronic wave function in a quantum system is of fundamental interest and has potential for applications in quantum

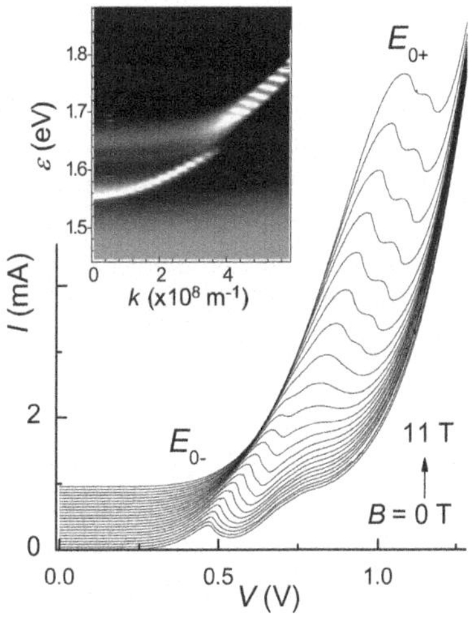

Fig. 11.9 $I(V)$ curves at $T = 4.2$ K and various B for a resonant tunneling diode based on an (AlGa)As/Ga(AsN)/(AlGa)As quantum well ($N = 0.08\%$). B is increased from 0 to 11 T by steps of 0.5 T. For clarity, the curves are displaced along the vertical axis. E_{0-} and E_{0+} indicate two dominant resonant features in $I(V)$. *Inset*: *Gray-scale* plots of the differential conductance $G = \mathrm{d}I/\mathrm{d}V$ vs. energy (ε) and wavevector (k)

information processing and other advanced technologies. This field is still in its infancy and requires the development of sensitive methods to probe and understand the nature of the quantum states. In many cases, our theoretical understanding is based on the solution of the Schrödinger equation, which gives the eigenvalues and corresponding eigenfunctions. Spectroscopic studies, on the contrary, focus traditionally on measuring transition energies and intensities and their change in response to external perturbations. The spatial form of the wave functions is generally more difficult to probe.

Various scanning probe techniques, such as scanning tunneling microscopy (STM) have been used successfully to probe atoms or electron charge distributions in a wide variety of systems [51–55], including quantum dots (QDs). QDs are artificial nanometer-sized clusters, which confine the motion of an electron in all three spatial dimensions. Some works have reported how cross-sectional STM (XSTM) of cleaved InAs/GaAs self-assembled QDs can probe the profile of the electron density along the plane of vertical confinement of the dot (i.e., along the growth direction and one in-plane direction) [51, 53, 54]. This method is both invasive and destructive as the QD is cleaved into two pieces. In addition, STM can only probe quantum states near the surface. MTS and magneto-capacitance-voltage spectroscopy (MCVS) provide alternative, non-destructive, and non-invasive methods of probing the electron wave function of a quantum state buried deep below a surface [16, 17, 56, 57]. These techniques exploit the effect of the classical Lorentz force on the motion of an electron tunneling into the quantum state and can be regarded as the momentum ($\boldsymbol{k}$-) space analog of STM imaging. In STM, a moving tip acts as a probe of the wave function in real space. In MTS and MCVS, the applied magnetic field, $\boldsymbol{B}$, acts as a variable probe in $\boldsymbol{k}$-space: the images give the probability density in $\boldsymbol{k}$-space of the electron wave function.

Figure 11.10 shows the typical structure of a resonant tunneling diode used in recent wave function imaging experiments. This consists of a GaAs QW embedded between two (AlGa)As tunnel barriers. The central plane of the QW incorporates a single layer of InAs self-assembled quantum dots. In this structure, undoped GaAs spacer layers separate the (AlGa)As barriers from an n-doped GaAs layer. Tunneling of electrons into the ground and excited states of a quantum dot generates narrow peaks in the current–voltage characteristics. The intensity of these resonances change with increasing magnetic field applied in the plane (x, y) of the QW. As in Sect. 11.5.1, we can understand this behavior semi-classically in terms of the increased momentum that is acquired by the tunneling electron due to the action of the Lorentz force. The applied voltage tunes the electron to tunnel resonantly into the energy of a particular state. Then, the variation of the tunnel current with B determines the size of the matrix element that governs the quantum transition of an electron as it tunnels from a state in the emitter layer into the quantum state. It can be shown that this matrix element represents the electron probability density in Fourier space of the quantum state [16]. In particular, by measuring the tunnel current at different orientations of B in the plane (x, y), it is possible to produce two-dimensional Fourier maps of the probability density in the (k_x, k_y) plane (see Fig. 11.10).

Recent developments have used MTS and MCVS to probe in situ the spatial compression of a quantum state induced by an applied perturbation, such as an

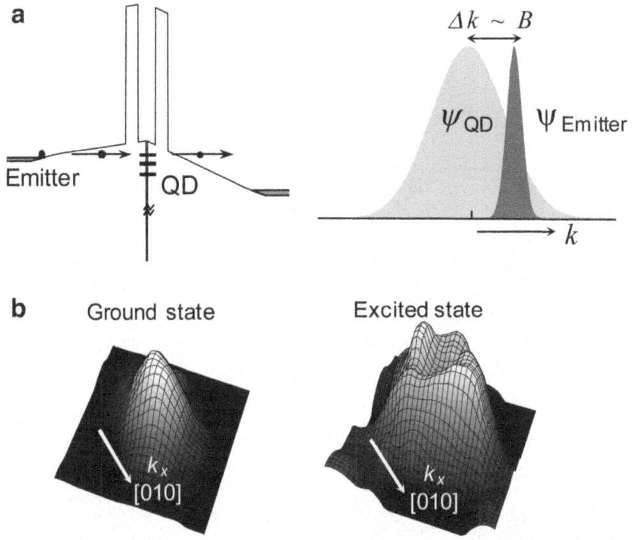

Fig. 11.10 (**a**) *Left*: Schematic conduction band profile of a single barrier GaAs/(AlGa)As tunneling device with InAs quantum dots (QDs) incorporated in GaAs. *Right*: overlap between the electron wave functions in k-space of the emitter and QD. (**b**) Distribution in the (k_x, k_y) plane of the probability density of the electron wave function for the ground and excited states of a single InAs quantum dot grown on a (100)-oriented GaAs substrate

applied magnetic field [18, 58]. To illustrate this idea, we consider the electronic wave function of a Si-donor atom in GaAs. In this compound, the donor eigenstates are well described by the hydrogenic effective mass approximation with an effective Bohr radius $a = 10$ nm. In the MTS experiment, magnetic fields of up to $B = 20$ T correspond to strengths of up to 10^6 T when scaled to the case of the hydrogen atom. The current due to electron tunneling through the ground state of Si-donors within a GaAs quantum well (QW) is studied in a magnetic field tilted at different angles to the QW plane (Fig. 11.11a). The component of $\boldsymbol{B}$ parallel to the direction of current (B_z) provides a means of increasing the donor-binding energy and hence of compressing the donor wave function in the QW plane. By measuring the current as a function of the perpendicular component, B_x, it is then possible to determine how the magneto-compression affects the spatial form of the donor wave function [18].

Figure 11.11b shows the B_x-dependence of the donor resonance peak, D, in the current–voltage $I(V)$ curves of a RTD at two values of $B_z = 4$ and 16 T. It can be seen that the B_x-dependence of the tunnel current ($\sim|\psi|^2$) becomes weaker at high B_z($\propto k$-vector) and changes from approximately polynomial at $B_z = 0$ to Gaussian at high B_z (see Fig. 11.11c). The donor wave function exhibits only very small changes when B_z is increased up to 4 T. In this regime, the confinement potential due to magnetic field is weak compared to the Coulomb potential, and the wave function retains its hydrogenic-like form. In contrast, when B_z exceeds 10 T, the tail

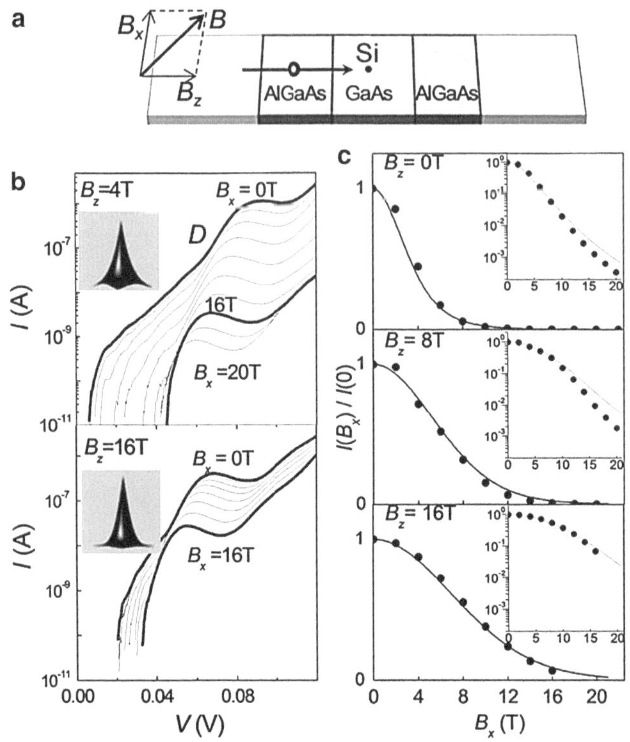

Fig. 11.11 (**a**) Sketch of electron tunneling into a Si-donor in a GaAs quantum well (QW) under a magnetic field B with components B_x and B_z. (**b**) B_x-dependence of the donor peak, D, in $I(V)$ at various B_z. The *insets* shows the magneto-compression of the donor wave function at $B_z = 4$ and 16 T. (**c**) B_x-dependence of the height of the current peak D at fixed B_z. *Dots* are the experimental data; *lines* are the calculated curves

of the wave function decays as a Gaussian, thus reflecting the parabolic confinement and magneto-compression induced by B_z. These data and analysis demonstrate that MTS is a highly sensitive probe of the change in the shape of the electronic wave function due to an externally applied perturbation. By using appropriately designed tunnel diodes, MTS could be further exploited to probe the effect of other types of applied perturbations (i.e., photo-induced charge quanta, electric field applied by a gate electrode, strain or hydrostatic pressure, etc.) on the eigenfunctions of zero-dimensional structures.

11.6 Conclusion

To date, the study of the electronic properties of semiconductor materials in magnetic field has not only provided an impetus to the improvement of growth techniques, material quality, and device performance, but has also led to unexpected advances in fundamental physics and metrology. The most recent example demonstrating the power of magnetic fields is their use in the discovery of the remarkable electronic properties of graphene [6] and in the imaging and manipulation of the wave function probability density of an electron confined in a semiconductor nanostructure [17]. Rapid advances in the technology for generating high magnetic fields will contribute to the further development of the research on semiconductor materials and nanostructures [59]. Superconducting solenoids generating fields of up to $\sim$30 T are increasingly used in semiconductor physics. Dissipative copper wire and Bitter solenoids energized by multi-megawatt power supplies are available at major facilities in Europe, Japan, and the USA: for example, the 24-MW power supply in Grenoble can generate fields of 30 T in a Bitter-type magnet. By rapidly discharging a large capacitor bank through a solenoid, very high pulsed magnetic fields ($\sim$70 T) can be generated for a short period ($\sim$10 ms). Even with these quite short pulses, it is possible to make precise magnetoresistance, magneto-tunneling, cyclotron resonance, and other magneto-optical measurements. Still higher fields can be achieved by electromagnetic flux compression ($\sim$600 T) or flux compression using high explosives ($\sim$1,000 T), though in the latter case, electrical and optical measurements in solid state physics are difficult and the user cannot expect his specimen crystal or device to survive the explosion! See [60] for a recent review of techniques for generating high magnetic fields. We can expect that high magnetic fields will find new applications in combination with optical and electrical probes of the electronic properties of semiconductors. A recent example is the direct observation using scanning tunneling spectroscopy of the unique Landau level spectrum of graphene, with non-equally spaced levels [61,62].

References

1. G. Dresselhaus, A.F. Kip, C. Kittel, Phys. Rev. **92**, 827 (1953)
2. E.H. Hall, Am. J. Math. **2**, 287 (1879)
3. L.W. Shubnikov, W.J. de Haas, Proc. R. Neth. Acad. Arts Sci. **33**, 130 (1930)
4. K. von Klitzing, G. Dorda, M. Pepper, Phys. Rev. Lett. **45**, 494 (1980)
5. D.C. Tsui, H.L. Stormer, A.C. Gossard, Phys. Rev. Lett. **48**, 1559 (1982)
6. K.S. Novoselov, A.K. Geim, S.V. Morozov, D. Jiang, Y. Zhang, S.V. Dubonos, I.V. Grigorieva, A.A. Firsov, Science **306**, 666 (2004)
7. A.A. Abrikosov, Sov. Phys. JETP **29**, 746 (1959)
8. A.A. Abrikosov, J. Phys. A Math. Gen. **36**, 9119 (2003)
9. M.M. Parish, P.B. Littlewood, Nature **426**, 162 (2003)
10. M.M. Parish, P.B. Littlewood, Phys. Rev. B **72**, 094417 (2005)
11. P.A. Lee, T.V. Ramakrishnan, Rev. Mod. Phys. **57**, 287 (1985)
12. H. Mizuta, T. Tanoue, *The Physics and Applications of Resonant Tunnelling Diodes* (Cambridge University Press, Cambridge, 1995)
13. R.K. Hayden, D.K. Maude, L. Eaves, E.C. Valadares, M. Henini, F.W. Sheard, O.H. Hughes, J.C. Portal, L. Cury, Phys. Rev. Lett. **66**, 1749 (1991)
14. J. Endicott, A. Patanè, J. Ibanez, L. Eaves, M. Bissiri, M. Hopkinson, R. Airey, G. Hill, Phys. Rev. Lett. **91**, 126802 (2003)

15. A. Patanè, J. Endicott, J. Ibáñez, L. Eaves, P.N. Brunkov, S.B. Healy, A. Lindsay, E.P. O'Reilly, M. Hopkinson, Phys. Rev. B **71**, 195307 (2005)
16. A. Patanè, R.J.A. Hill, L. Eaves, P.C. Main, M. Henini, M.L. Zambrano, A. Levin, N. Mori, C. Hamaguchi, Y.u.V. Dubrovskii, E.E. Vdovin, S. Tarucha, D.G. Austing, G. Hill, Phys. Rev. B **65**, 165308 (2002)
17. E.E. Vdovin, A. Levin, A. Patanè, L. Eaves, P.C. Main, Y.u.N. Khanin, Y.u.V. Dubrovskii, M. Henini, G. Hill, Science **290**, 124 (2000)
18. A. Patanè, N. Mori, L. Eaves, M.L. Zambrano, J.C. Arce, L. Dickinson, D.K. Maude, Phys. Rev. Lett. **105**, 236804 (2010)
19. B.R. Snell, K.S. Chan, F.W. Sheard, L. Eaves, G.A. Toombs, D.K. Maude, J.C. Portal, S.J. Bass, P. Claxton, G. Hill, M.A. Pate, Phys. Rev. Lett. **59**, 2806 (1987)
20. M.L. Leadbeater, L. Eaves, P.E. Simmonds, G.A. Toombs, F.W. Sheard, P.A. Claxton, G. Hill, M.A. Pate, Sol. St. Elect. **31**, 707 (1988)
21. C. Skierbiszewski, P. Perlin, P. Wisniewski, T. Suski, W. Walukiewicz, W. Shan, J.W. Ager, E.E. Haller, J.F. Geisz, D.J. Friedman, J.M. Olson, S.R. Kurtz, Phys. Status Solidi B **216**, 135 (1999)
22. R. Mouillet, L.A. de Vaulchier, E. Deleporte, Y. Guldner, L. Travers, J.-C. Harmand, Solid State Commun. **126**, 333 (2003)
23. A. Patanè, G. Allison, L. Eaves, N.V. Kozlova, Q.D. Zhuang, A. Krier, M. Hopkinson, G. Hill, Appl. Phys. Lett. **93**, 252109 (2008)
24. M. Henini, *Dilute Nitride Semiconductors* (Elsevier Science, Amsterdam, 2005)
25. A. Lindsay, E.P. O'Reilly, Phys. Rev. Lett. **93**, 196402 (2004)
26. V. Popescu, A. Zunger, Phys. Rev. Lett. **104**, 236403 (2010)
27. S. Fahy, A. Lindsay, H. Ouerdane, E.P. O'Reilly, Phys. Rev. B **74**, 035203 (2006)
28. W. Walukiewicz, Physica B **302**, 123 (2001)
29. H. Munekata, E.E. Mendez, Y. Iye, L. Esaki, Surf. Sci. **174**, 449–453 (1986)
30. A. Patanè, W.H.M. Feu, O. Makarovsky, O. Drachenko, L. Eaves, A. Krier, Q.D. Zhuang, M. Helm, M. Goiran, G. Hill, Phys. Rev. B **80**, 115207 (2009)
31. B.I. Shklovskii, A.L. Efros, *Electronic Properties of Doped Semiconductors*, Springer Series in Solid-State Sciences, vol. 45 (Springer, Berlin, 1984), p. 155
32. C.W.J. Beenakker, H. van Houten, in *Solid State Physics: Semiconductor Heterostructures and Nanostructures*, ed. by H. Ehrenreich, D. Turnbull (Academic, Boston, 1991), pp. 1–228
33. L.D. Landau, E.M. Lifshitz, *Quantum Mechanics: Nonrelativistic Theory* (Pergamon, Oxford, 1977)
34. J. Davies, *The Physics of Low Dimensional Semiconductors* (Cambridge University Press, Cambridge, 1998), p. 206
35. C. Hamaguchi, *Basic Semiconductor Physics* (Springer, Berlin, 2001), Chapters 2 and 7
36. R. Dingle, H.L. Stormer, A.C. Gossard, W. Wiegmann, Appl. Phys. Lett. **33**(7), 665–667 (1978)
37. K.S. Novoselov, A.K. Geim, S.M. Morozov, M.I. Katsnelson, I.V. Grigorieva, S.V. Dubonos, A.A. Firsov, Nature **438**, 197 (2005)
38. G. Allison, N. Mori, A. Patanè, J. Endicott, L. Eaves, D. Maude, M. Hopkinson, Phys. Rev. Lett. **96** 236802 (2006)
39. L.W. Roth, P.N. Argyres, *Semiconductors and Semimetals*, vol. 1 (Academic, New York, 1966), pp. 159–202
40. C. Kittel, *Solid State Physics*, 8th edn. (Wiley, New York), Chapter 6
41. D. Fowler, O. Makarovski, A. Patanè, L. Eaves, L. Geelhaar, H. Riechert, Phys. Rev. B **69**, 153305 (2004)
42. Z.F. Ezawa, *Quantum Hall Effects: Field Theoretical Effects and Related Topics* (World Scientific, Singapore, 2008)
43. S. Das Sarma, A. Pinczuk (eds.), *Perspectives in Quantum Hall Effects: Novel Quantum Liquids in Low-Dimensional Semiconductor Structures* (Wiley, New York, 1997)
44. V.L. Gurevich, Y.A. Firsov, Sov. Phys. JETP **13**, 137 (1961)
45. R.L. Peterson, in *Semiconductors and Semimetals*, vol. 10, ed. by R.K. Willardson, A.C. Beer (Academic, New York, 1975), pp. 221–289

46. L. Eaves, R.A. Hoult, R.A. Stradling, R.J. Tidey, J.C. Portal, S. Askenazy, J. Phys. C Sol. St. Phys. **8**, 1034–1053 (1975)
47. L. Eaves, R.A. Stradling, J.C. Portal, S. Askenazy, R. Barbaste, K. Hansen, Sol. St. Commun. **15**, 1281–1285 (1974)
48. G. Allison, S. Spasov, A. Patanè, L. Eaves, N.V. Kozlova, J. Freudenberger, M. Hopkinson, G. Hill, Phys. Rev. B **77**, 125210 (2008)
49. R. Xu, A. Husmann, T.F. Rosenbaum, M.-L. Saboungi, J.E. Enderby, P.B. Littlewood, Nature **390**, 57 (1997)
50. J. Hu, T.F. Rosenbaum, Nat. Mater. **7**, 697 (2008)
51. B. Grandidier, Y.M. Niquet, B. Legrand, J.P. Nys, C. Priester, D. Stie'venard, J.M. Ge'rard, V. Thierry-Mieg, Phys. Rev. Lett. **85**, 1068 (2000)
52. O. Millo, D. Katz, Y.W. Cao, U. Banin, Phys. Rev. Lett. **86**, 5751 (2001)
53. T. Maltezopoulos, A. Bolz, C. Meyer, C. Heyn, W. Hansen, M. Morgenstern, R. Wiesendanger, Phys. Rev. Lett. **91**, 196804 (2003)
54. G. Maruccio, M. Janson, A. Schramm, C. Meyer, T. Matsui, C. Heyn, W. Hansen, R. Wiesendanger, M. Rontani, E. Molinari, Nano Lett. **7**, 2701 (2007)
55. A.M. Yakunin, A.Y. Silov, P.M. Koenraad, J.H. Wolter, W. Van Roy, J. De Boeck, J.M. Tang, M.E. Flatte, Phys. Rev. Lett. **92**, 216806 (2004)
56. O.S. Wibbelhoff, A. Lorke, D. Reuter, A.D. Wieck, Appl. Phys. Lett. **86**, 092104 (2005)
57. G. Bester, D. Reuter, L. He, A. Zunger, P. Kailuweit, A.D. Wieck, U. Zeitler, J.C. Maan, O. Wibbelhoff, A. Lorke, Phys. Rev. B **76**, 075338 (2007)
58. W. Lei, C. Notthoff, J. Peng, D. Reuter, A. Wieck, G. Bester, A. Lorke, Phys. Rev. Lett. **105**, 176804 (2010)
59. J.W. Orton, *The Story of Semiconductors* (Oxford University Press, Oxford, 2004)
60. N. Miura, *Physics of Semiconductors in High Magnetic Fields* (Oxford University Press, Oxford, 2008)
61. D.L. Miller, K.D. Kubista, G.M. Rutter, M. Ruan, W.A. de Heer, P.N. First, J.A. Stroscio, Science **324**, 924 (2009)
62. G. Li, A. Luican, E.Y. Andrei, Phys. Rev. Lett. **102**, 176804 (2009)

Chapter 12
Photoconductivity and Transient Spectroscopy

Ayşe Erol and M. Çetin Arıkan

Abstract This chapter focuses on photoconductivity (PC) and transient spectroscopy techniques including photo-induced transient spectroscopy (PITS) and deep-level transient spectroscopy (DLTS). Spectral photoconductivity provides a powerful tool for measuring the band-gap energy and optical transitions in semiconductors. On the other hand, transient photoconductivity can be used to determine the time constants associated with specific recombination processes. PITS and DLTS are extensively used to characterize deep energy levels due to traps. The aim of this chapter is to discuss the fundamentals of these measurement techniques and describe typical experimental setups and methods of analysis for the determination of important trap parameters, such as the activation energy, the carrier capture cross-section, and the trap density.

12.1 Introduction

Experimental techniques based on the interaction between light and matter, such as photoconductivity (PC), can provide valuable information about semiconductors. The concept of photoconductivity emerged first in 1873 with the observation of the effect of light on selenium during the flow of an electrical current [1]. In this experiment, Willoughby Smith observed that the conductivity of selenium decreased significantly under light illumination. This discovery stimulated research in this field and soon several materials were found to be photoconductive, although the physics behind photoconductivity remained unclear. A systematic study of the measured photoconductive effects was later done by Gudden and Pohl [2], who showed that the absorption of light, luminescence, and photoconductivity processes had a similar

A. Erol (✉) · M.Ç. Arıkan
Science Faculty, Department of Physics, Istanbul University, 34134 Vezneciler, Istanbul, Turkey
e-mail: ayseerol@istanbul.edu.tr; arikan@istanbul.edu.tr

A. Patanè and N. Balkan (eds.), *Semiconductor Research*, Springer Series in Materials Science 150, DOI 10.1007/978-3-642-23351-7_12,

dependence on the photon wavelength. The findings of Pohl and Gudden also demonstrated that alkali halide crystals did not show any photoconductivity if they were illuminated by photons of energy corresponding to that of the fundamental absorption band, and that in order to observe photoconductivity, some impurities had to be incorporated into the host crystal [2, 3]. The most important outcome of these studies is the establishment of the quantum nature of the photoconductivity process, i.e., a single electric charge drifts between two electrodes per absorbed photon. To explain the experimental work on alkali halides, Mott and Gurney published a book entitled *Electronic Processes in Ionic Crystals* [4] and then an extensive study was published by Moss entitled *Photoconductivity in the Elements*, which summarizes data on photoconductivity in all non-metallic elements [5]. The discovery of photoconductivity led to the invention of photocells. Since then, photoconductive materials have been widely used for detecting a whole range of radiation from the far infrared to high-energy nuclear particles.

Photoconductivity provides a tool for investigating solids, for understanding and improving device performance, and is of great importance both as an experimental tool and as a basis for the development of new devices. The first part of this chapter describes the fundamentals of photoconductivity and typical photoconductivity experimental setups (Sects. 12.2–12.4). The chapter then focuses on transient spectroscopy techniques for the investigation of defects (Sect. 12.5).

The electronic and optical properties of semiconductors are modified by imperfections such as point defects, dislocations, and interfaces, which create localized deep levels in the band gap of a semiconductor. Some defects are beneficial and have been successfully employed either to improve existing devices or to obtain structures with new characteristics. Interesting examples include the use of gold in silicon to obtain fast recombination, oxygen and chromium in GaAs to semi-insulate GaAs, and zinc–oxygen pairs in GaP to realize red light-emitting diodes. However, many defects can have a deleterious effect on device performance, so understanding and controlling them are essential for the successful fabrication of a semiconductor device.

There are several methods to determine defect parameters but in low-dimensional structures, the characterization of defects is subject to the limitations imposed by the normally small thicknesses of the epitaxial layers. Many of the characterization techniques are inapplicable because they rely on a large number of defects in the probed volume of the semiconductor. Methods that probe both the atomic arrangement and the chemical identity of a defect, such as electron spin resonance and optically detected magnetic resonance, are in this category. Therefore, in many cases, the chemical identification of a defect must rely on either a comparison study in bulk materials or an analysis of the stoichiometric conditions of the epitaxial process. In contrast, optical characterization of defects in epitaxial layers can be more successful. For example, photoconductivity can be used to probe the presence of defects and provide information about activation energies and trapping time constants. Luminescence measurements can also give useful information about defects as long as the probability of the radiative recombination is higher than the probability of the non-radiative recombination. On the other hand, transient spectroscopy is more suitable for studying defects located in epitaxial layers and gives information about activation energies, cross-sections, and density of individual defects. In the second part of this chapter (Sect. 12.5), two techniques based on transient spectroscopy, photo-induced transient spectroscopy (PITS) and deep-level transient spectroscopy (DLTS), are presented. The chapter ends with a comparison of the DLTS and PITS techniques.

12.2 Photoconductivity

Photoconductivity measurements provide a versatile experimental tool for probing the band-gap energy of bulk semiconductors, excitonic transitions in low-dimensional semiconductors, and the presence of crystal imperfections. Although photoconductivity is one of the simplest techniques used to study excess carriers in semiconductors, the analysis of a PC spectrum is complicated as it is determined by the absorption coefficient, mobility, and carrier lifetime. As an alternative experimental tool to photoconductivity, absorption spectroscopy can measure the absorption coefficient and gives information similar to that obtained from photoconductivity measurements. However, the measurement of the absorption coefficient in epitaxially grown semiconductor structures requires the removal of the substrate. Therefore, absorption spectroscopy is a destructive technique. It is also difficult to measure the absorption intensity in very thin layers. On the other hand, photoconductivity requires only two ohmic contacts diffused through a layer and the results reflect the optical quality of the whole structure. In the following section, the basic concepts and fundamentals of photoconductivity are presented.

12.2.1 Photoconductivity: General Concepts

Photoconductivity, i.e., photon-induced conductivity, corresponds to a change in the conductivity of a material system caused by an excess of carriers generated by absorption of photons; therefore, photoconductors have the capability of converting electromagnetic radiation into an electrical current. Although different materials can exhibit photoconductive properties, here we consider only semiconductors.

Thermal excitations in semiconductors create excited electrons and holes even in dark and these carriers are free to move in the semiconductor lattice, thus contributing to the conductivity. The concentration of free carriers depends on whether the semiconductor is doped or undoped. The dark electrical conductivity of a semiconductor is given by

$$\sigma = q(n\mu_n + p\mu_p), \tag{12.1}$$

where $q = e$ is the electron charge, n and p are the free electron and hole concentrations, and μ_n and μ_p are the electron and hole mobilities, respectively.

In a homogenous material, the photoconductivity is expressed as a change in the conductivity due to excess carriers generated by the absorption of photons, i.e.,

$$\Delta\sigma = q(\Delta n\mu_n + \Delta p\mu_p), \tag{12.2}$$

where Δn and Δp are the excess electron and hole concentrations, respectively. Following the absorption of photons, the photocreated carriers are rapidly swept toward the contacts by an applied constant *dc* electric field, thus generating a *photocurrent*.

In insulators, the excess carrier concentrations can be larger than the free carrier concentrations in the dark, but in semiconductors, excess carriers are considered as only a small perturbation to the values of n and p. Also, (12.2) assumes that the carrier mobility is not changed by the photoexcitation. This is a reasonable assumption as the mechanisms that can alter the mobility can generally be neglected [6]. These mechanisms include (a) the photo-induced transfer of carriers from one of the conduction band valley to another; this process requires a momentum change, i.e., a phonon–electron scattering mechanism; (b) the photo-induced change of the scattering rate due to charged impurity atoms, whose density and/or scattering cross-section can be changed by light.

In a non-homogenous material, the free carrier concentration in the dark can be nonuniform, thus leading to high conductivity regions separated by low conductivity regions. The current passing throughout the material is limited by the low conductivity regions, which act as barriers. When light is absorbed, the barrier resistance is decreased and the current flow throughout the material becomes higher than that in the dark. The photoconductivity is controlled by the resistance of the barriers, and their effect on the photoconductivity is described in terms of an effective mobility μ_b^*, i.e.,

$$\Delta\sigma = q(n\Delta\mu_{bn}^* + p\Delta\mu_{bp}^*). \tag{12.3}$$

Hereafter, we will consider only homogenous conductors.

Let us now consider a semiconductor sample of thickness d, width w, and length l illuminated uniformly by monochromatic light of energy sufficient to generate electron–hole pairs. Let I_{L} represent the intensity of light, i.e., the incident power per unit area, and z the direction of the light propagation. The intensity of light changes along z as $I_{\mathrm{L}}(z) = I_0\mathrm{e}^{-\alpha z}$, where α is the absorption coefficient at a given photon energy. The fraction of absorbed light in a thickness d is equal to $I_0(1 - \mathrm{e}^{-\alpha d})$. If the semiconductor is sufficiently thick to absorb all the photons, the flux of photons per unit time in the semiconductor is $I_0/\hbar\omega$, where ω is the angular frequency of light. Each absorbed photon generates one electron–hole pair and the *quantum efficiency* η is defined as the fraction of those carriers that flow into the external circuit. Therefore, the total number of generated carriers in the semiconductor is $\eta I_0 wl/\hbar\omega$. The *generation rate* G is the number of carriers generated per unit volume per unit time and is given by

$$G_n = \frac{\eta I_0}{\hbar\omega d}. \tag{12.4}$$

If only some fraction of the incident light is absorbed by the semiconductor, the generation rate is instead given by

$$G = \frac{\eta\alpha I_{\mathrm{L}}}{\hbar\omega}. \tag{12.5}$$

Now we will establish the relation between the generation rate and the photocurrent using rate equations. In the presence of a uniform optical injection of carriers, the continuity (rate) equations are given by

$$\begin{aligned} \frac{\partial n}{\partial t} &= G_n - R_n + \frac{1}{q}\nabla \cdot \vec{J}_n \\ \frac{\partial p}{\partial t} &= G_p - R_p - \frac{1}{q}\nabla \cdot \vec{J}_p, \end{aligned} \tag{12.6}$$

where G_n and G_p are the electron and hole generation rates, R_n andR_p are the electron and hole recombination rates, and J_n and J_p are the electron and hole drift current densities, respectively.

For an electric field applied along the x direction, (12.6) becomes

$$\begin{aligned} \frac{\partial n}{\partial t} &= G_n - R_n + \frac{1}{q}\frac{\partial}{\partial x}J_n(x) \\ \frac{\partial p}{\partial t} &= G_p - R_p - \frac{1}{q}\frac{\partial}{\partial x}J_p(x). \end{aligned} \tag{12.7}$$

The recombination rate is defined as

$$R_n = \frac{\Delta n}{\tau_n} \quad R_p = \frac{\Delta p}{\tau_p}, \tag{12.8}$$

where τ_n and τ_p are the recombination times of the excess electrons and holes, respectively.

Two different scenarios can be envisaged corresponding to a uniform and a nonuniform excess carrier generation. When the optical illumination is uniform across the semiconductor, the generation rate has only time dependence, i.e.,

$$G(t) = G_n(t) = G_p(t), \tag{12.9}$$

and the excess carrier generation is uniform. In this case, the diffusion current density is ignored in the rate equations as in Eqs. 12.6 and 12.7. If the current passing throughout the semiconductor is kept constant during the photoconductivity measurement, the electron continuity equations become

$$\frac{\partial n}{\partial t} = G(t) - \frac{\Delta n}{\tau_n}$$
$$\frac{\partial p}{\partial t} = G(t) - \frac{\Delta p}{\tau_p}. \tag{12.10}$$

Under optical illumination, the carrier concentrations are given by $p = p_0 + \Delta p$ and $n = n_0 + \Delta n$, and the excess carrier concentrations are equal for an interband absorption. In the absence of trapping and for light with constant intensity, the generation rate is independent of time. Hence the excess carrier concentrations in steady state are given by

$$\Delta n = G\tau_n, \tag{12.11}$$

and

$$\Delta n = \Delta p = G\tau_n. \tag{12.12}$$

Using these expressions for the excess carrier concentrations, (12.2) can be rewritten as

$$\Delta\sigma = qG\tau_n(\mu_n + \mu_p). \tag{12.13}$$

Hence the total drift photocurrent density in the presence of an electric field along the x direction is

$$J = \Delta\sigma E = qG\tau_n(\mu_n + \mu_p)E = \Delta\sigma\frac{V}{l}, \tag{12.14}$$

where V is the applied voltage and l is the length of the semiconductor.

The photocurrent can be now expressed as

$$\Delta I = JA = \Delta\sigma\frac{A}{l}V = q(\mu_n + \mu_p)G\tau_n\frac{A}{l}V. \tag{12.15}$$

where A is the cross-section of the semiconductor ($A = wd$). For the case of a single type-carrier transport, the photocurrent is given by

$$\Delta I = q\mu_n G\tau_n\frac{A}{l}V. \tag{12.16}$$

Equation (12.16) shows that the photocurrent increases linearly with both the optical intensity and the applied voltage. Following the optimization of the signal-to-noise ratio, these parameters are kept constant during the experiments.

The effective quantum efficiency for photoconductivity is known as the *photoconductivity gain* G^*. This is defined as the number of charges collected in the external circuit for each absorbed photon. If F is the number of absorbed photons and the generation rate $G = F/Al$, then the gain G^* can be expressed as the ratio of the photocurrent to the generation rate, i.e.,

$$G^* = \frac{\Delta I}{q}\frac{1}{GAl}. \tag{12.17}$$

Using (12.15) in (12.17), we then obtain

$$G^* = \frac{(\mu_n + \mu_p)\tau_n}{l^2} V. \tag{12.18}$$

Alternatively, the photoconductivity gain can be expressed as the ratio of the carrier recombination lifetime to the transit time, i.e.,

$$G^* = \frac{\tau_n}{\tau}, \tag{12.19}$$

where we have used that the average drift velocity of electrons is $v_d = \mu_n E = \mu_n V/l$ and the average transit time of electrons across the semiconductor is $\tau = l/v_d$.

If the lifetime of a carrier is greater than its transit time, the carrier will make several transits throughout the semiconductor between the contacts. For example, if $\tau_n = 10^{-3}\,s$, $l = 10^{-3}$ m, and $v_d = 10^3$ m/s, the photoconductivity gain is $G^* = 1{,}000$ [7–12].

So far we have only considered the case of a uniform illumination of the semiconductor. What if a small part of the semiconductor between electrodes is illuminated? If the carrier generation is nonuniform, the diffusion of the excess carriers toward either sides of the illuminated region must be considered and the current density term in the rate equations given in (12.6) must also include the diffusion current density due to the photocreated excess carriers. This case is complicated as it requires a careful analysis of the nonuniform excess carrier distribution along the sample length [13]. Here, we will only consider the case of a uniform illumination.

12.2.2 Photoconductivity: Spectral Response

Photoconductivity induced by interband transitions is called *intrinsic photoconductivity* to distinguish it from that due to impurities, which is called *extrinsic photoconductivity*. To observe a photoconductivity signal, light must be absorbed in the semiconductor and generate free carriers by either intrinsic or extrinsic optical absorption. Therefore, there is a close relation between absorption and photoconductivity spectra, as illustrated in Fig. 12.1a.

There are three possible types of absorption transitions resulting in photoconductivity: (a) interband absorption, generating a free electron and a free hole for each photon absorbed (Fig. 12.1b (I)); (b) absorption from the valence band to an unoccupied localized trap level, producing a free hole and a bound electron (Fig. 12.1b (II)); and (c) absorption from an occupied localized trap level to the conduction band, producing a free electron and a bound hole (Fig. 12.1b (III)) [12]. Therefore, photoconductivity is a versatile experimental technique to observe optical transitions and probe the presence of traps. Due to its absorption-like character, this technique can also probe optical transitions related to the ground and excited states in low dimensional structures even at room temperature.

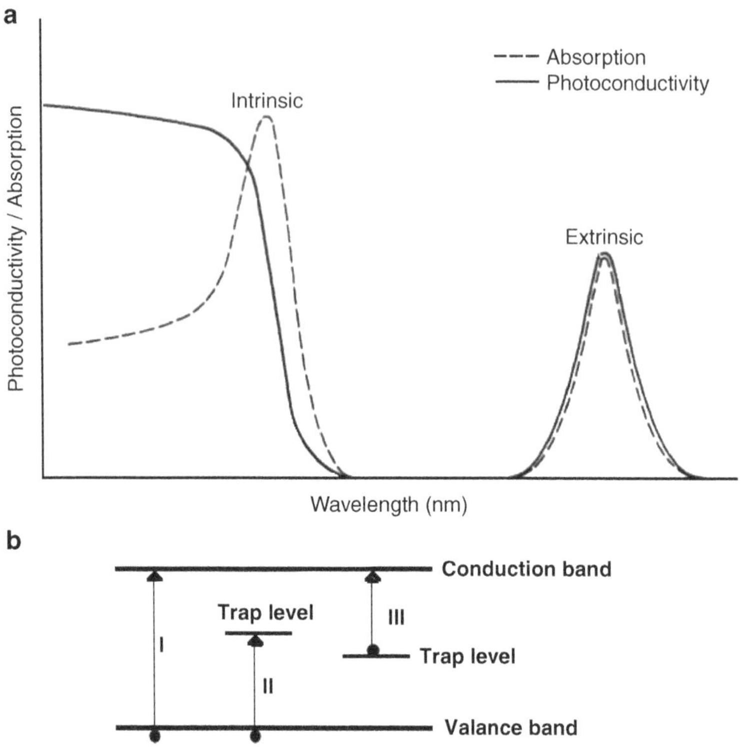

Fig. 12.1 (**a**) Comparison between photoconductivity and absorption spectra. (**b**) Three possible absorption processes involving a trap level

12.2.3 Photoconductivity Decay

Let us consider an n-type semiconductor under illumination and a steady-state photocurrent ΔI_0. At $t = t_0$, when the illumination is suddenly switched off, the photoconductivity decays exponentially to zero. The time dependence of the photocurrent ΔI can be described as

$$\Delta I = \Delta I_0[1 - \exp(-t/\tau_p)] \qquad (0 \le t \le t_0) \tag{12.20}$$
$$\Delta I = \Delta I_0[1 - \exp(-t_0/\tau_p)]\exp[-(t - t_0)/\tau_p] \qquad (t \ge t_0),$$

where τ_p is a time constant. In the absence of traps, the transient behavior of the photoconductivity can be used to determine the value of τ_p [13]. On the other hand, in the presence of traps, the transient photoconductivity cannot be described by a single exponential curve as the trapped carriers are slowly released from the trap levels. Hence, the photoconductivity decay can consist of several decay mechanisms due to trapping, each with its own characteristic time constant. In this case, the

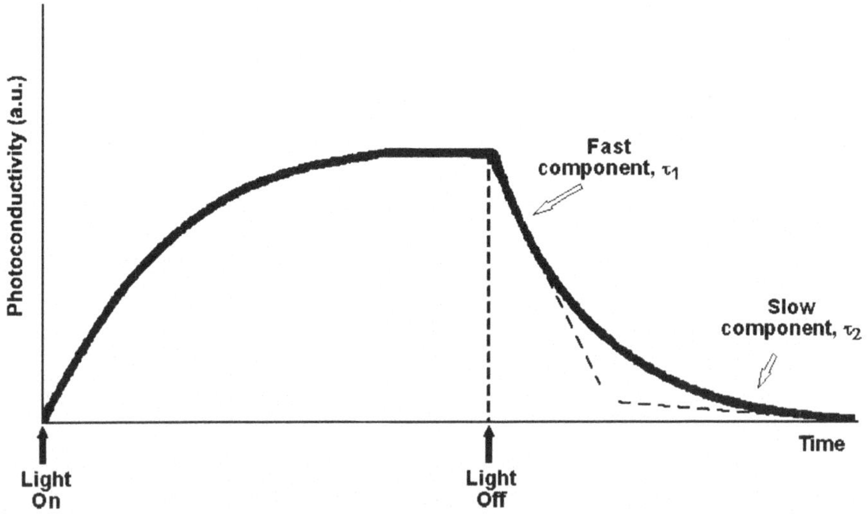

Fig. 12.2 Transient photoconductivity and exponential fits

photoconductivity decay curve is fitted using a sum of the exponential functions and the time constants for each trapping process can be obtained as illustrated in Fig. 12.2. On the decay curve, the initial rapid decrease corresponds to the recombination of the photoexcited carriers, and the slower exponential decay is due to the thermal release of electrons from the traps.

12.3 Photoconductivity Measurement Techniques

Techniques for measurements of the photoconductivity can be classified into three groups depending on the duration of the incident light, as shown in Fig. 12.3. These methods are referred to as (a) steady state, (b) modulated, and (c) pulsed illumination. Using these different excitation conditions, photoconductivity measurements are categorized as steady-state (or *dc*) photoconductivity, modulated (*ac*) photoconductivity, and transient photoconductivity.

12.3.1 Steady-State (dc) Photoconductivity

In this measurement technique, a *dc* excitation light of constant intensity and spectral distribution is used to illuminate the sample, and the photoconductivity is measured in steady state. Therefore, this technique allows the slowest trapping process to come into equilibrium with the incident radiation. Reaching the steady-state condition may take from several minutes to hours in some cases. To measure

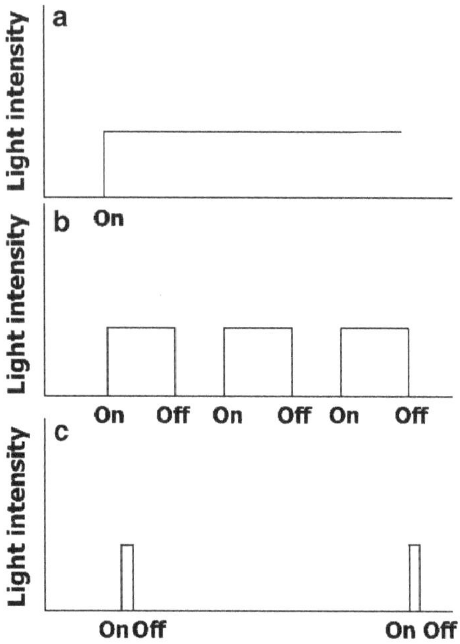

Fig. 12.3 Photoconductivity measurements under different light excitation conditions: (**a**) steady-state light, (**b**) modulated light, and (**c**) pulsed illumination

the steady-state photoconductivity, a simple circuit is used, as seen in Fig. 12.4. The current increases from its dark current value (I_0) to the value I, and the steady-state photoconductivity under uniform illumination with constant intensity is expressed as

$$\Delta\sigma_{ST} = \frac{I - I_0}{V}\frac{l}{A}, \tag{12.21}$$

where V is the applied voltage, l is the length, and A is the cross-section of the semiconductor. This circuit is useful when $\sigma_0 \ll \Delta\sigma_{ST}$, which is fulfilled in high-resistivity materials. If the semiconductor has high conductivity, then the dark conductivity will be greater than the steady-state photoconductivity. In this case, the steady-state illumination cannot be used and a modulated illumination must be chosen to avoid the contribution of dark conductivity.

12.3.2 Modulated (ac) Photoconductivity

In this technique, the photoconductivity measurement is carried out by modulating the incident light with a chopper. Modulated photoconductivity is not only useful to avoid the contribution of a high dark current, but also useful to gain information about the trapping and generation time constants. The measurement circuit is shown in Fig. 12.5. Under illumination, the current has a *dc* and an *ac* component. However, in practice, a third component is also present due to nonzero average illumination

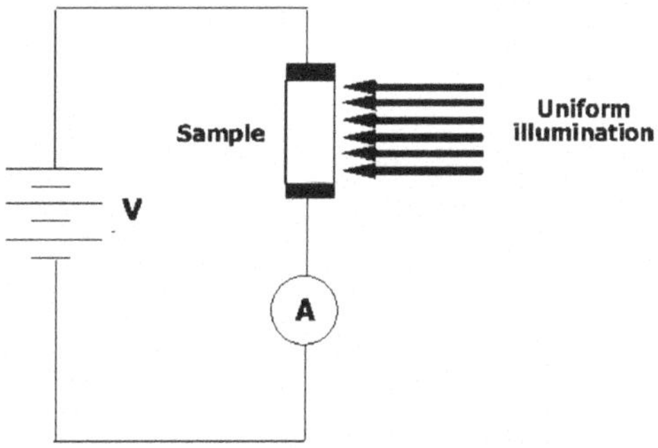

Fig. 12.4 Circuit for steady-state photoconductivity measurements

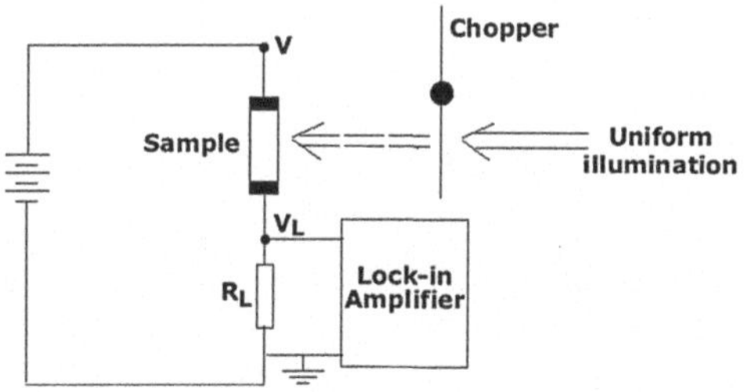

Fig. 12.5 Circuit for *ac* photoconductivity measurements

effects, which manifests through a temporal increase of the *dc* component of the photoconductivity. Therefore, we will categorize the current into three components: (1) steady dark current, (2) steady photocurrent, and (3) *ac* photocurrent. If the steady photocurrent does not change during the measurement, the *ac* photocurrent can be measured using a lock-in amplifier that is sensitive only to the *ac* component of the photocurrent and does not respond to *dc* signals.

In order to measure the *ac* photoconductivity, the circuit diagram shown in Fig. 12.5 is used. The current I is given by

$$I = \frac{V}{R_S + R_L} = \frac{V_L}{R_L}, \tag{12.22}$$

$$V_L = \frac{R_L}{R_S + R_L} V, \tag{12.23}$$

where R_S is the sample resistance, R_L is the load resistance, and V_L is the voltage across the load resistance. If $R_S \gg R_L$,

$$V_L = \frac{R_L}{R_S} V. \tag{12.24}$$

For small perturbations, the change in voltage across the load resistance due to the modulation of light is

$$\Delta V_L = -\frac{R_L}{R_S^2} V \Delta R_S, \tag{12.25}$$

where

$$R_S = \frac{1}{\sigma}\frac{l}{A}. \tag{12.26}$$

The relative change in photoconductivity is equal to the relative change in the sample resistance, i.e.,

$$\frac{\Delta\sigma}{\sigma} = -\frac{\Delta R_S}{R_S}. \tag{12.27}$$

The substitution of (12.25), (12.26) into (12.27) gives the photoconductivity

$$\Delta\sigma = \frac{l}{AR_L V}\Delta V_L. \tag{12.28}$$

12.3.3 Pulsed Photoconductivity

In this technique, a pulsed laser with short pulse width and slow repetition rate is used to illuminate the semiconductor. The short pulse duration must be sufficiently short to avoid steady-state photoconductivity. The method is useful if the pulse duration exceeds the fastest trapping rate and the repetition rate is slower than the slowest trapping rate in the semiconductor. The transient photoconductivity decay is used to determine trapping time constants [14].

12.4 Experimental Setups for Photoconductivity Measurements

12.4.1 Spectral Photoconductivity: Experimental Setup

Spectral photoconductivity provides an excellent tool to determine electronic transitions and trap levels. Fig. 12.6 shows a conventional experimental setup for measuring the *ac* spectral photoconductivity. The sample is placed in a cryostat to carry out the photoconductivity measurement at different lattice temperatures. The temperature is kept constant at any desired value during the measurements using a temperature controller. The light from a quartz halogen lamp is chopped and focused on a grating monochromator. The light intensity is controlled by a constant current source and kept constant during the measurements. The dispersed light is passed through a filter to avoid high orders of light. A small fraction of the dispersed light from the monochromator is split into two components and one of the components is diverted onto a pyroelectric detector to monitor the excitation intensity. The rest of the light is directed onto the sample. A *dc* voltage is applied to the sample with a series resistor, and the PC signal is measured by a conventional lock-in amplifier technique. The photoconductivity, $\Delta\sigma$, is the

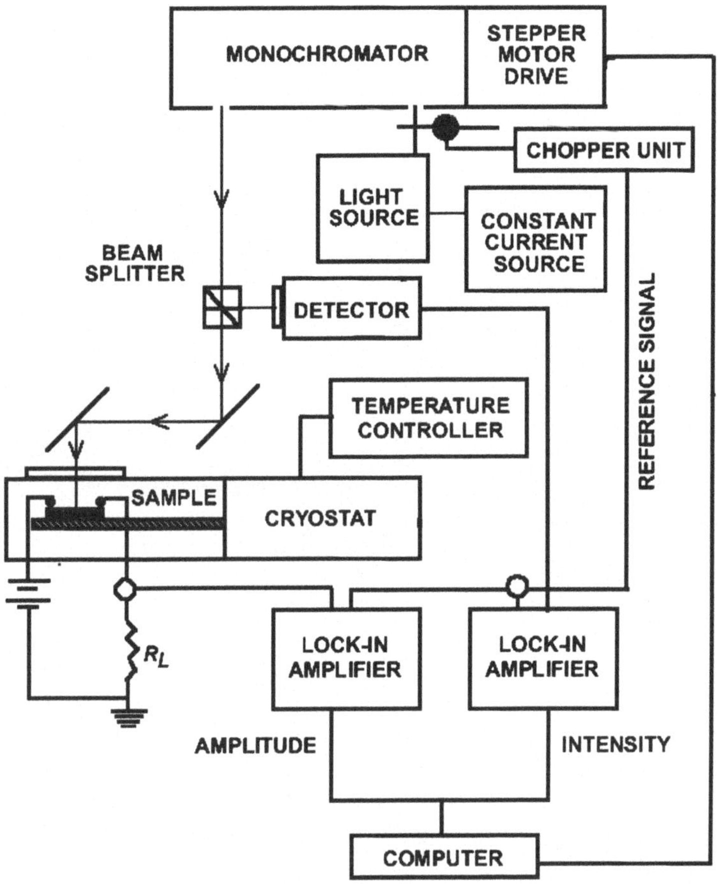

Fig. 12.6 Experimental setup of spectral photoconductivity measurements

rms value of differential *ac* photocurrent. The modulated photocurrent and the *ac* excitation intensity are simultaneously measured. The spectral photoconductivity signal is normalized using the intensity distribution of light which is the output signal of the pyroelectric detector detected by the lock-in amplifier [14, 15].

Two examples of modulated spectral PC measurements are shown in Figs. 12.7 and 12.8. Figure 12.7a, b shows, respectively, the PC and photoluminescence (PL) spectra for as-grown and annealed $Ga_{0.7}In_{0.3}N_{0.015}As_{0.985}/GaAs$ quantum well structures. The samples consist of a 6-nm $Ga_{0.7}In_{0.3}N_{0.015}As_{0.985}/GaAs$ quantum well and a 6-nm $Ga_{0.7}In_{0.3}As/GaAs$ quantum well layer on a semi-insulating GaAs substrate. In the experiments, a 100-W halogen lamp was used to create photoexcited carriers and chopped at 17 Hz. The intensity of light was kept constant during the measurements and controlled by a constant current source. The frequency of the chopper was kept low enough to have sufficiently high signal output from the pyroelectric detector as the sensitivity of the detector is much higher

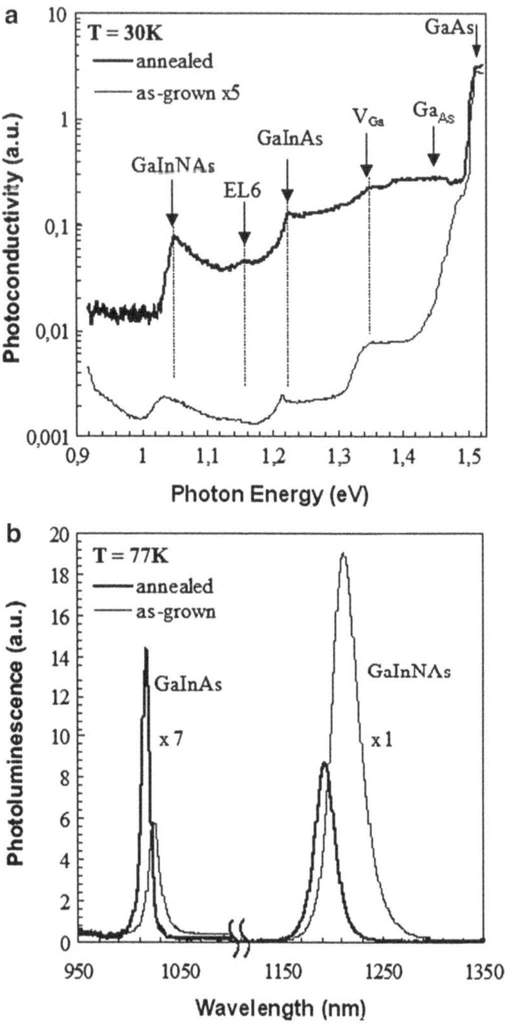

Fig. 12.7 (**a**) PC spectra and (**b**) PL spectra for as-grown and annealed GaInAs/GaAs and GaInNAs/GaAs quantum well structures

at low frequencies. The light from a halogen lamp was dispersed using a 0.5-m monochromator. A small fraction of light (15%) was split and directed onto the pyroelectric detector to monitor the excitation intensity and the rest (85%) was used to illuminate the sample uniformly. To sweep the photoexcited carriers toward the opposite electrodes, a *dc* voltage was applied to the sample with a series resistor and the PC signal was detected using a lock-in amplifier. The series resistance value was selected to give a ratio $R_S/R_L \sim 100$. The modulated spectral photoconductivity at the frequency of the chopper and the excitation intensity were simultaneously measured by a conventional lock-in technique, and the data were collected as a function of the photon energy. Samples were fabricated in the form of simple bars and ohmic contacts were alloyed by diffusing a sequence of Au/Au:Ge/Au through

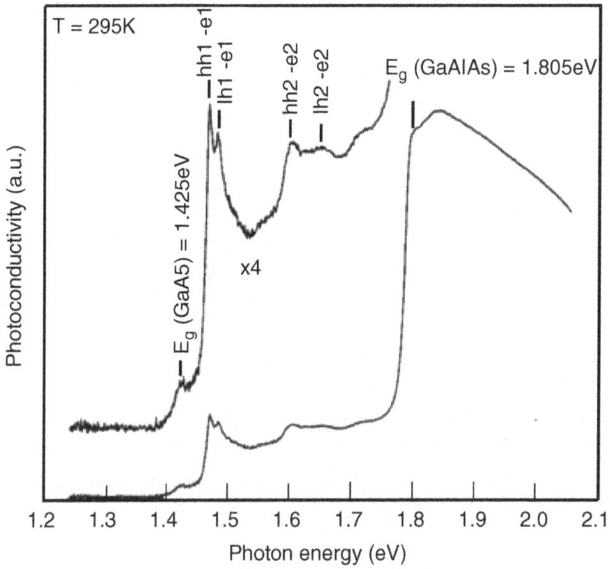

Fig. 12.8 PC spectrum for a GaAs/GaAlAs single quantum well

all the layers in the sample. The applied electric field was parallel to the layers; hence it was possible to observe contributions to the PC signal from all the layers in the PC spectra. This technique is also called *in-plane* PC.

Figure 12.7a shows that the PC spectra give information not only about interband and/or quantum well transitions, but also about transitions due to the presence of trap levels localized in the band gap (see lines labeled EL6, V_{Ga}, Ga_{As}, and e-A). In contrast, as shown in Fig. 12.7b, the PL spectra show only peaks corresponding to quantum well transitions. The energy values of the peaks labeled as GaInNAs and GaInAs in the PC and PL spectra are in good agreement when measured at the same temperature [16]. PC measurements are affected by the experimental conditions, such as excitation intensity, modulation frequency, and applied electric field. By adjusting these experimental parameters, it is possible to observe well-resolved spectra even at room temperature. As an example, a modulated spectral PC measurement is shown in Fig. 12.8 for a GaAs/AlGaAs single quantum well structure. In the PC spectrum, both the ground and higher excitonic transitions can be clearly resolved at room temperature [15].

12.4.2 *Transient Photoconductivity: Experimental Setup*

In this technique, fast-pulsed lasers and high-speed oscilloscopes are essential components of the experimental setup, as shown in Fig. 12.9. The pulsed laser light

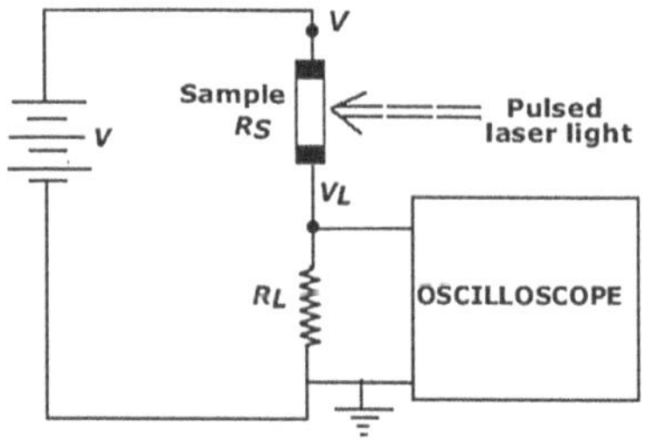

Fig. 12.9 Circuit for transient photoconductivity measurements

with higher energy than the band-gap energy of the semiconductor is directed onto the sample. A *dc* power supply is used to apply a constant voltage across the sample in series with a load resistance. When the pulsed laser light falls onto the sample, excess carriers are created, and when the light is switched off, the photoconductivity decays exponentially. To observe the transient photoconductivity signal, the voltage drop across the load resistor is recorded by a large bandwidth oscilloscope. The bandwidth of the oscilloscope is very important to observe the radiative recombination time, which is the fast component of the photoconductivity decay. In the absence of trap levels, the photoconductivity decay contains only a fast component, which corresponds to the radiative recombination from the conduction band to the valence band. In contrast, in the presence of traps, the photoconductivity decay contains the contribution of different trapping mechanisms, each characterized by a different time constant.

Figure 12.10 shows the transient PC spectrum for a GaInNAs/GaAs single quantum well. This was measured using a Q-switched Nd:YAG laser emitting at 1,064 nm, with a pulsed width of 100 ps and a repetition rate of 5 Hz. The Nd:YAG laser was chosen to excite selectively only the GaInNAs quantum well. Light pulses were directed to the sample and uniform illumination was obtained by defocusing the laser beam. A *dc* voltage was applied to the electrical circuit shown in Fig. 12.9 and the transient PC signal, which corresponds to the voltage across the load resistance, was recorded with a 300-MHz digital oscilloscope with a rise time of 4–5 ns. The transient PC decay is composed of fast and slow components and can be modeled as the sum of two exponentials, each one corresponding to a different decay process as shown in Fig. 12.10. The time constants were found to be $\tau_1 < 4$ ns and $\tau_2 \sim 80$ ns. The measured value of the time constant associated with the fast component is limited by the bandwidth of the oscilloscope [17].

12.5 Transient Spectroscopy

In real semiconductors, there are always some defects that are introduced due to either thermodynamic considerations or the presence of impurities during the crystal growth. Impurities can be intentionally introduced as dopant atoms (shallow

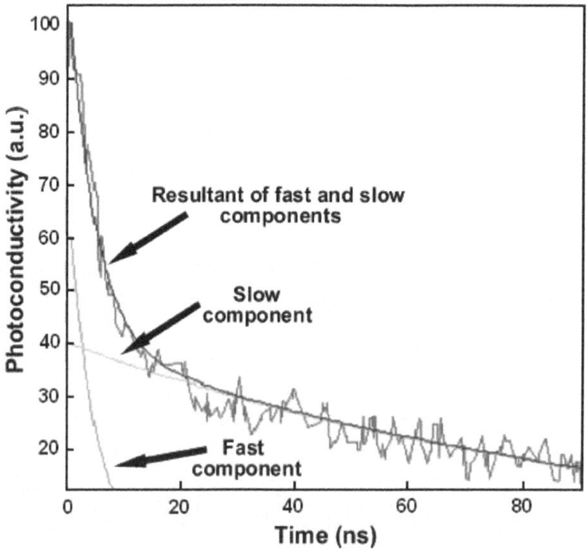

Fig. 12.10 Example of transient PC decay and its modeling by the sum of two exponentials

impurities) to control electrical properties. On the other hand, impurities are also unintentionally incorporated into a semiconductor during the crystal growth and device processing. There are many kinds of defects, which can be classified according to the dimension (D) of the defect:

- 0-D Point defects: substitutionals, vacancies, and interstitials.
- 1-D Line defects: dislocations.
- 2-D Planar defects: interfacial defects such as surfaces, grain boundaries, and phase boundaries.
- 3-D Bulk defects: voids, cracks, and pores.

Among these defects, the formation of 2-D and 3-D defects can be easily avoided in most epitaxial growth techniques. Point defects are the most common defects observed in semiconductors and are highly localized in one or a few unit cells of the semiconductor. Three common point defects are vacancies, substitutional, and interstitials. The vacancy defect is associated with the absence of an atom in the lattice. The substitutional point defect corresponds to an impurity on the lattice site of a constitutional atom. An interstitial point defect is an atom located between lattice sites and it can be self-interstitial or foreign-interstitial. An important point defect observed in compound semiconductors, such as GaAs, is the anti-site defect. In this case, one of the atoms, say Ga, sits on the As sub-lattice instead of the Ga lattice site (denoted as Ga_{As}). These simple point defects can alter electrical properties and carrier transport via diffusion, but they hardly affect thermal conductivity properties. Line defects or dislocations are not as localized as point defects and involve extended atomic sites. Their effects on the device performance can be drastic; therefore, control of dislocations during the growth process is of great importance. The concentrations of defects can be controlled, for example, by the growth temperature and pressure.

In ionic semiconductors, charged point defects, classified as Frenkel and Schottky defects, can form. The Frenkel defect is an interstitial of a charged atom and when it forms, it creates two regions of different polarity. The Schottky defect is also a pair defect, but it corresponds to the absence of both ions. Overall, the charge neutrality of the semiconductor must be maintained for the formation of a charged defect.

Traps are categorized as shallow levels and deep levels depending on the spatial extent of the wave function associated with the trapped carrier and its binding energy. Shallow levels tend to have a hydrogenic behavior and are less localized compared to deep levels. Their binding energy is in the range of a few meV. In contrast, the energy of a deep level is in the range of eV. These trap levels lead to strongly localized states in the band gap and they are usually responsible for non-radiative transitions [18–22].

To improve the optical quality and, therefore, the performance of electronic and optoelectronics devices, it is essential to identify, analyze, and eventually minimize the presence of defects in a semiconductor. Transient spectroscopy provides an experimental tool for probing the presence of defects and defect-related properties. We will now consider the fundamentals of two transient spectroscopy techniques: PITS, also called as photo-induced current transient spectroscopy (PICTS), and DLTS. PITS is based on the analysis of the change in a photoconductivity decay transient as a function of temperature in nominally undoped highly resistive semiconductors, i.e., semiconductors with a negligible dark current. On the other hand, DLTS is based on the analysis of the change in capacitance due to a change in bias condition as a function temperature and can be applied to Schottky diodes and p–n junctions.

12.5.1 Generation–Recombination Rate

The fundamental theory for understanding and analyzing PITS and DLTS is based on generation–recombination mechanisms in semiconductors. Let us consider a defect level that is located at an energy E_T, as shown in Fig. 12.11. Four possible processes are possible: (a) the capture of an electron from the conduction band to the defect level; (b) the emission of an electron from the defect level to the conduction band; (c) the capture of a hole from the valence band to the defect or the emission of an electron from the defect to the valence band; and (d) the emission of a hole from the defect to the valence band or the capture of an electron from the valence band to the defect.

A conduction electron can be captured by the defect (a) and then emitted to the valence band (c). This process is known as recombination process. Also, an electron can be captured by the defect from the valence band (d) and then emitted into the conduction band (b). This process is called generation. In both processes, a carrier

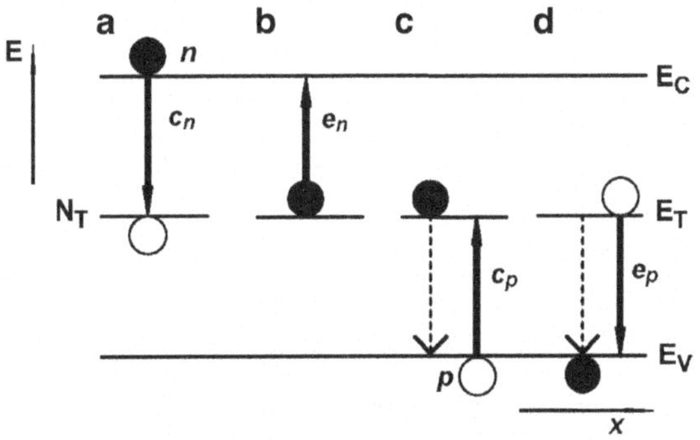

Fig. 12.11 Generation and recombination processes involving a trap level

is captured from one band and emitted into the other band. Both bands participate in these processes. Therefore, this defect is called a generation–recombination center. However, if a captured electron (a) is re-emitted to the conduction band (b) or a captured hole is emitted to the valence band, these defects are known as traps. Impurities are generally called traps even if they behave as generation–recombination centers.

Generation and recombination processes are classified as radiative or non-radiative. The radiative process involves the creation or annihilation of photons, while a non-radiative process involves a phonon–electron interaction or an exchange of energy and momentum with carriers. There are more different recombination processes, such as Auger and impact ionization, which will be not considered here [9, 19–22].

Let us assume that N_{T} is the total trap density, n and p are the free electron and hole concentrations, and n_{T} and p_{T} are the occupied and empty trap levels, respectively. The trap density is the sum of the occupied and unoccupied traps, i.e.,

$$n_{\mathrm{T}} + p_{\mathrm{T}} = N_{\mathrm{T}}. \tag{12.29}$$

The net rate of electron concentration is the difference of the emitted electrons from the trap to the conduction band (Fig. 12.11b) and the captured electrons from the conduction band to the trap (Fig. 12.11a), i.e.,

$$\frac{\mathrm{d}n}{\mathrm{d}t} = e_n n_{\mathrm{T}} - c_n n p_{\mathrm{T}}. \tag{12.30}$$

Here, e_n and c_n represent the electron emission and capture rates, respectively. Similarly, the net rate of hole concentration is

$$\frac{\mathrm{d}p}{\mathrm{d}t} = e_p p_{\mathrm{T}} - c_p p n_{\mathrm{T}}, \tag{12.31}$$

where e_p and c_p are the hole emission and capture rates, respectively.

The capture rates c_n and c_p can be expressed as

$$c_n = \sigma_n \langle \upsilon_{\text{th}} \rangle n$$
$$c_p = \sigma_p \langle \upsilon_{\text{th}} \rangle p, \tag{12.32}$$

where $\langle \upsilon_{\text{th}} \rangle$ is the average thermal velocity and $\sigma_n(\sigma_p)$ is the capture cross-section of the deep level for electrons (holes). Using (12.29)–(12.31), the net rate of electron occupancy can be expressed as

$$\frac{\mathrm{d}n_{\mathrm{T}}}{\mathrm{d}t} = \frac{\mathrm{d}p}{\mathrm{d}t} - \frac{\mathrm{d}n}{\mathrm{d}t} = (c_n n + e_p)(N_{\mathrm{T}} - n_{\mathrm{T}}) - (c_p p + e_n)n_{\mathrm{T}}. \tag{12.33}$$

For an electron trap, (12.33) can be rewritten as

$$\frac{\mathrm{d}n_{\mathrm{T}}}{\mathrm{d}t} = c_n(N_{\mathrm{T}} - n_{\mathrm{T}}) - e_n n_{\mathrm{T}}. \tag{12.34}$$

In thermal equilibrium, the net rate must be equal to zero and (12.34) gives

$$\frac{n_{\mathrm{T}}}{N_{\mathrm{T}}} = \frac{c_n}{c_n + e_n}. \tag{12.35}$$

This ratio can be expressed in terms of the Fermi–Dirac distribution function as

$$\frac{n_{\mathrm{T}}}{N_{\mathrm{T}}} \approx \frac{1}{1 + \exp(E_{\mathrm{T}} - E_{\mathrm{F}})/k_{\mathrm{B}}T}. \tag{12.36}$$

From (12.35) and (12.36), we derive

$$\frac{e_n}{c_n} \approx \exp\frac{E_{\mathrm{T}} - E_{\mathrm{F}}}{k_{\mathrm{B}}T}. \tag{12.37}$$

In a non-degenerate semiconductor, the electron concentration is given by

$$n = N_{\mathrm{C}} \exp\left(-\frac{E_{\mathrm{C}} - E_{\mathrm{F}}}{k_{\mathrm{B}}T}\right), \tag{12.38}$$

where N_{C} is the effective density of state in the conduction band and E_{C} is the energy of the conduction band edge. By substituting (12.32) and (12.38) into (12.37), we can derive the emission rate

$$e_n(T) \approx \sigma_n \langle \upsilon_{\text{th}} \rangle_n N_{\mathrm{C}} \exp\left(-\frac{E_{\mathrm{C}} - E_{\mathrm{T}}}{k_{\mathrm{B}}T}\right). \tag{12.39}$$

Similarly for a hole trap, the emission rate can be expressed as

$$e_p(T) \approx \sigma_p \langle \upsilon_{\mathrm{th}} \rangle_p N_{\mathrm{V}} \exp\left(-\frac{E_{\mathrm{T}} - E_{\mathrm{V}}}{k_{\mathrm{B}} T}\right). \tag{12.40}$$

Here, E_{V} is the energyof the valence band edge and N_{V} is the effective density of states in the valence band.

For electrons, the temperature dependence of the thermal velocity and effective density of states are given by

$$\langle \upsilon_{\mathrm{th}} \rangle_n = \left(\frac{3k_{\mathrm{B}}T}{m_n^*}\right)^{1/2}, \tag{12.41}$$

$$N_{\mathrm{C}} = 2\left(\frac{2\pi m_n^* k_{\mathrm{B}} T}{h^2}\right)^{3/2}. \tag{12.42}$$

Using (12.41)–(12.42) and (12.39), we derive

$$e_n(T) = \tau^{-1} = \gamma_n \sigma_n T^2 \exp\left(-\frac{E_{\mathrm{T}}}{k_{\mathrm{B}} T}\right), \tag{12.43}$$

where $\gamma_n = 4\sqrt{6}\pi^{3/2} m_n^* k_{\mathrm{B}}^2 / h^3$ is a material constant. Equation (12.43) illustrates that the temperature dependence of the emission rate is characterized by an exponential decay and a polynomial factor, from which the trap parameters can be derived [23–26].

12.5.2 Photo-Induced Transient Spectroscopy

Monitoring the photoconductivity transient decay provides a means of measuring the emission rate and its temperature dependence. Instead of monitoring photoconductivity transients at different temperatures, it is more convenient to monitor at different temperatures the change in the photoconductivity transient measured at two sampling times after the light is switched off. The transient decay can be described as

$$I(t) = AqE\mu_n n_T \tau_n e_n \exp(-e_n t), \tag{12.44}$$

where A is the effective cross-section of the sample, q is the electron charge, E is the applied electric field, μ_n is the electron mobility and τ_n is the lifetime of electrons. As seen in Fig. 12.12a, the difference in magnitude of the photocurrent recorded at two sampling times, t_1 and t_2, after the light is switched off, gives a PITS signal, $P(T)$, given by

$$P(T) = \Delta I(t) = I(t_1) - I(t_2). \tag{12.45}$$

The PITS signal is recorded as a function of temperature and every peak observed in the spectrum corresponds to a specific trap, as shown in Fig. 12.12b.

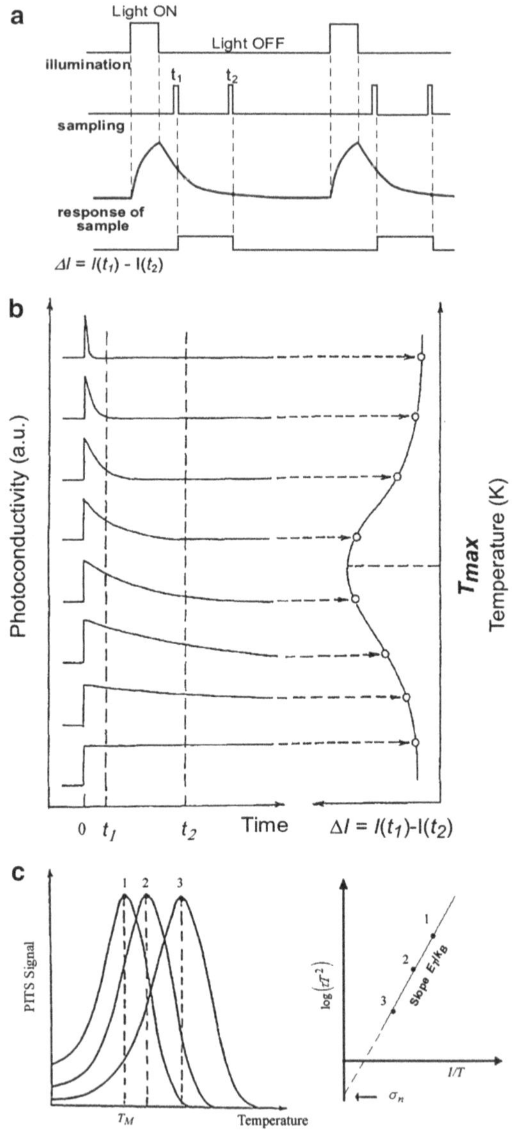

Fig. 12.12 (**a**) Transient photoconductivity response. (**b**) Transient photoconductivity responses at various temperatures and PITS signal. (**c**) Analysis of the PITS spectra to extract trap parameters

PITS peaks occur at a specific temperature, $T_{\max}$, in the spectrum, when the emission rate of trapped carriers corresponds to the rate window set by the chosen values of t_1 and t_2. Therefore, at $T = T_{\max}$, corresponding emission rate is

$$e_n = \tau_{\max}^{-1} = \gamma_n \sigma_n T_{\max}^2 \exp\left(-\frac{E_\mathrm{T}}{k_\mathrm{B} T_{\max}}\right). \tag{12.46}$$

Equation (12.43) can also be written in terms of logarithms as

$$\log\left(\tau T^2\right) = \log\left(\frac{1}{\gamma_n \sigma_n}\right) + \frac{E_{\mathrm{T}}}{k_{\mathrm{B}}}\frac{1}{T}. \tag{12.47}$$

A plot of $\log(\tau T^2)$ versus T^{-1} (*Arrhenius plot*) has a slope of E_T/k_B and an intercept on the $\log(\tau T^2)$ axis of $\log\left(\frac{1}{\gamma_n \sigma_n}\right)$ gives σ_n as shown in Fig. 12.12c.

At $T = T_{\max}$, the relation between the sampling times (t_1 and t_2) and the emission time $\tau = \tau_{\max}$ (12.46) can be found by setting the first derivative of $P(T)$ signal to zero. This gives

$$\left(1 - \frac{t_1}{\tau_{\max}}\right)\exp\left(-\frac{t_1}{\tau_{\max}}\right) = \left(1 - \frac{t_2}{\tau_{\max}}\right)\exp\left(-\frac{t_2}{\tau_{\max}}\right). \tag{12.48}$$

One of the three variables contained in this equation can be eliminated by assuming that

$$\tau_{\max} = Bt_1 \tag{12.49}$$

$$t_2 = At_1. \tag{12.50}$$

A plot of A vs. B reveals that if $A \geq 10$, $B = 1$. Therefore, if the ratio of sampling times is selected so that $t_2 \geq 10t_1$, the emission rate/time at $T = T_{\max}$ can be directly obtained from the PITS spectra as $\tau = \tau_{\max} = t_1$ [23–25].

Figure 12.13 shows a typical PITS setup and the expected waveform (left corner) at different locations from A to H on the experimental setup. The sample is placed in a cryostat and a linear temperature sweep is provided by a PID temperature controller. The light is modulated by a chopper. A *dc* voltage is applied between the sample and a load resistance connected in series. The differential photoconductivity, which is sampled between two selected sampling times, is recorded using a lock-in amplifier connected to the multivibrator/gate/sample and hold unit. The output of the lock-in amplifier, which gives the PITS signal $[P(T)]$, and the sample temperature are recorded by a computer. The PITS signal is plotted vs. the sample temperature at different sampling times keeping the ratio t_1/t_2 constant and equal to 10. Each waveform for the locations A–H shown in Fig. 12.13 can be observed using an oscilloscope.

In a PITS experiment, the following important points must be considered:

1. *Choice of the light source*: If the sample consists of materials with different band gaps, appropriate monochromatic light sources should be used to probe the presence of traps in the different materials. Both below band-gap or above band-gap excitation sources can be used.
2. *Pulse parameters*: The duration of the light pulse must be long enough to ensure that the steady-state photoconductivity condition is achieved during the illumination, i.e., all traps are filled. In contrast, the repetition rate must be kept smaller than the slowest trapping rate in the material.

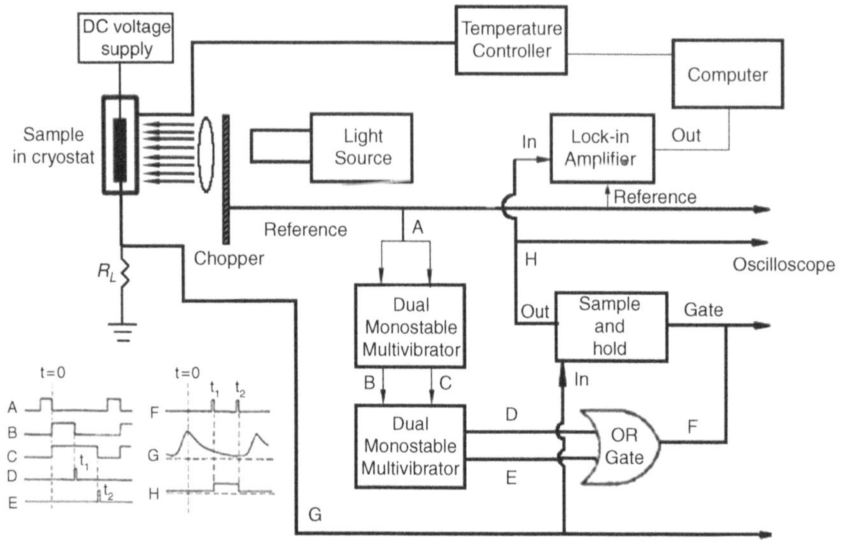

Fig. 12.13 PITS experimental setup. The signal at each location (**A**-**H**) on the setup is shown on the left corner

3. *Photoconductivity decay*: After light is switched off at $t = 0$, the fast component of the photoconductivity is related to the band-to-band recombination of the photoexcited carriers. The slow component of the decay contains information about traps. Therefore, the sampling time t_1 must be set to a value just after the fast decay transient.
4. *Sampling times*: If the ratio of sampling times is set to $t_1 \geq 10t_2$, the emission rate at $T = T_{max}$ can be directly obtained from the PITS measurements by setting $\tau_{max} = t_1$.
5. *Trap parameters*: At different sampling times and keeping the ratio $t_1/t_2 = 10$, the value of T_{max} can be changed and hence the temperature dependence of the emission rate can be obtained. Trap parameters are obtained from an Arrhenius fit of τ_{max} vs. temperature [26].

As an example, Fig. 12.14a shows the PITS spectrum for a GaInNAs/GaAs quantum well. The effective band-gap energy of this quantum well was derived from PL and was found to be equal to 1 eV at room temperature. Therefore, for the PITS experiment, a white light source and a filter with a cut-off photon energy of 1.1 eV were used to investigate the presence of the traps in the GaInNAs quantum well only, avoiding contributions to the PITS signal from the GaAs layer. The sample was placed in a liquid nitrogen-cooled cryostat. The temperature was swept from $T = 77$ K to above room temperature and a *dc* voltage was applied between the sample and the load resistance. Light pulses of width 20 ms and repetition rate of 0.8 s were used. The sampling time t_1 was set to 4, 8, 16, 32, and 64 ms and the t_2/t_1 ratio was kept equal to 10. Two peaks, A and B, were observed in the PITS

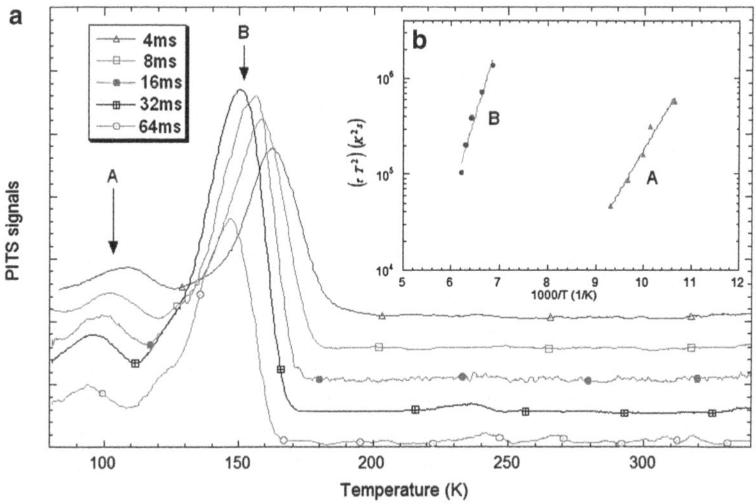

Fig. 12.14 (**a**) PITS spectra and (**b**) Arrhenius plots for a GaInNAs/GaAs quantum well

spectrum. In Fig. 12.14b, the Arrhenius plot is shown for these two peaks at five different sampling times. Using (12.47), the activation energies and capture cross-sections were found to be equal to 0.16 eV and 2×10^{-18} cm^2 for peak A, and 0.33 eV and 5×10^{-16} cm^2 for peak B [24].

12.5.3 Deep-Level Transient Spectroscopy

Both transient spectroscopy techniques, PITS and DLTS, rely on the trapping/de-trapping of carriers. The thermally induced release of the carriers from the traps can be probed by measuring transient electrical signals, such as the photocurrent (PITS) or the capacitance (DLTS). This section focuses on DLTS. DLTS was introduced by Lang in 1974 [27, 28]. Nowadays it is a standard method to study traps and the setup for DLTS experiments is also available commercially. The analysis of the emission or the capture transients is of great importance to derive trap parameters, as described in the previous section. Lang introduced the concept of rate window to characterize deep levels using the double boxcar or lock-in amplifier technique. When a capacitance transient is sampled at two different times, the largest change in capacitance between the sampling times is observed when the rate window of a boxcar average or the frequency of a lock-in amplifier is of the same order of the time rate of the capacitance transients. Monitoring changes in the repetitive capacitance transients in this rate window as a function of temperature gives a peak, which is called the DLTS spectrum.

Let us consider an n-type semiconductor with a trap level with an energy level E_T below the Fermi level, which is brought in contact with a metal to form a Schottky barrier. To reach thermodynamic equilibrium, electrons from the semiconductor diffuse into the metal leaving behind positively ionized donor atoms. This process stops when the Fermi levels of the metal and semiconductor are aligned, i.e., $E_{Fm} = E_{Fs} = E_F$. To maintain charge neutrality, the positively ionized donor atoms and the negative charged traps in the depletion layer are balanced by the electrons in the metal. At equilibrium, a depletion layer forms on the semiconductor side with a given width W. The ionized shallow donors in the space charge are positively charged and the deep levels occupied by electrons are negatively ionized. Therefore, the net charge concentrationin the depletion layer is $N_{dep} = N_D^+ - n_T^-$. If the donors are fully ionized and most of the traps are occupied, then $N_D^+ = N_D$ and $n_T^- = N_T$. If $E_T < E_F$, the traps are fully occupied, and if $E_T > E_F$, the traps are empty.

In the depletion approximation, i.e., if there is no mobile charge in the depletion region, the capacitance of a reverse biased Schottky contact is given by

$$C = \left|\frac{dQ}{dV}\right| = A\left(\frac{\varepsilon\varepsilon_0 N_{dep}}{2(V_{bi} - V)}\right)^{1/2} = A\frac{\varepsilon\varepsilon_0}{W}. \tag{12.51}$$

Here, A is the diode area, ε is the dielectric constant of the semiconductor, V is the reverse bias (< 0), V_{bi} is the built-in contact potential, N_{dep} is the net carrier density in the depletion region, and W is given by

$$W = \left(\frac{2\varepsilon\varepsilon_0}{N_{dep}}(V_{bi} - V)\right)^{1/2}. \tag{12.52}$$

At zero bias, most of the traps are occupied and the junction capacity is $C_0 = A\varepsilon\varepsilon_0/W_0$ (Fig. 12.15a). At the time $t = t_0$, the net carrier density in the depletion region is $N_{dep} = N_D - n_T$. When a pulsed reverse bias at $t_0 < t < t_1$ is applied, some of the electrons on the traps can be released and swept out of the depletion layer by the built-in electric field at the junction. This leads to a fast capacitance transient, and the width of the depletion layer increases to a new value W_1 (Fig. 12.15b). Therefore, when the Schottky diode is reverse-biased, the emission process dominates over electron capture. The electron transit time is only of a few tens of picoseconds, which is much shorter than the typical capture time constants. Therefore, the emitted electrons can escape from the depletion layer before being re-trapped. The concentration of electrons in the depletion layer during the emission process is

$$n(t) = n_T(0)\exp\left(\frac{-t}{\tau_e}\right) \approx N_T \exp\left(\frac{-t}{\tau_e}\right). \tag{12.53}$$

Here, $n_T(0)$ is the concentration of trapped electrons in thermal equilibrium and $\tau_e = 1/e_n$. The capacitance changes to $C_1 = A\varepsilon\varepsilon_0/W_1 < C_0$ and the net carrier density increases in the depletion region as electrons are emitted from the traps and the traps become neutral. As the reverse bias continues to be applied, only very

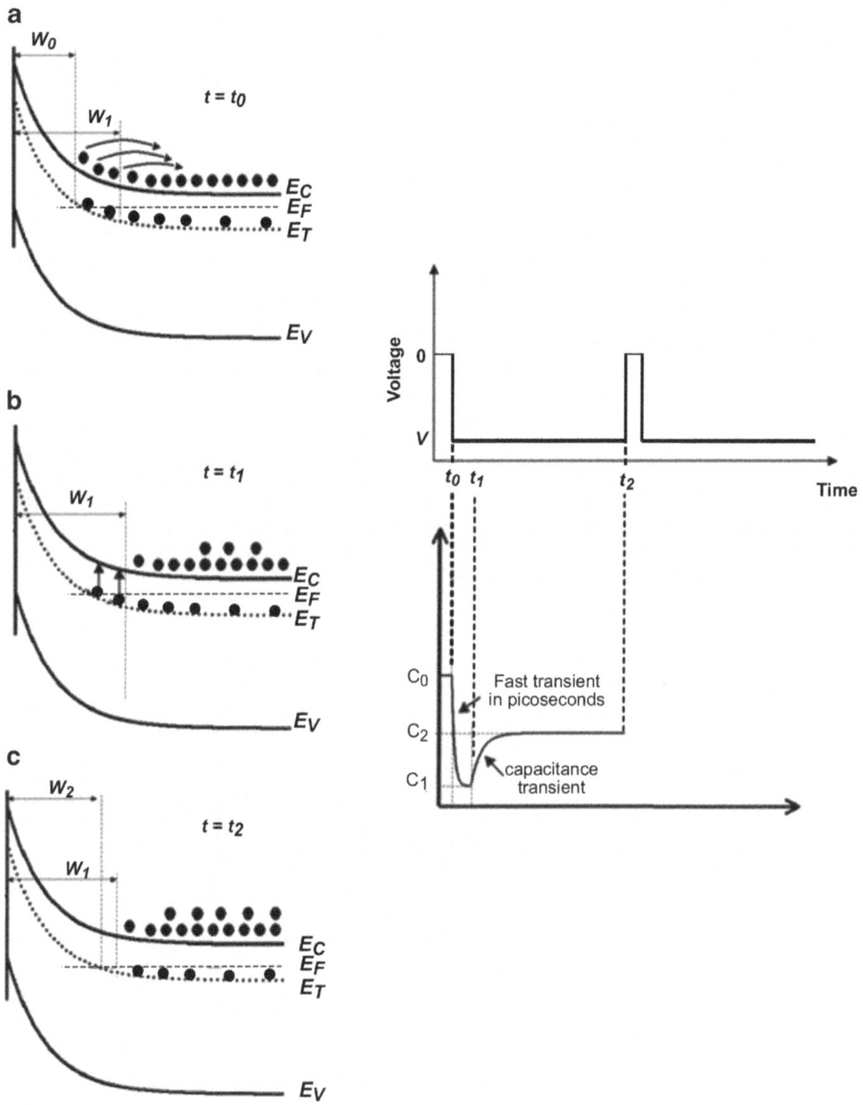

Fig. 12.15 Sketches of a reverse-biased Schottky barrier showing changes in the carrier distribution, width of depletion layer, and capacitance

few electrons are thermally excited from the deep level to the conduction band and swept out of the depletion layer. At times $t_1 < t < t_2$, the width of the depletion layer decreases to a new value W_2 as the net carrier density increases and the density of trapped electrons goes to zero. Hence the junction capacitance increases to $C_2 = A\varepsilon\varepsilon_0/W_2 > C_1$ and tends to the value C_∞ corresponding to the steady voltage V (Fig. 12.15c).

In the presence of a time-varying reverse bias, (12.51) can be rewritten as

$$C = A\left(\frac{\varepsilon\varepsilon_0 N_D}{2(V_{bi}-V)}\right)^{1/2}\left(1-\frac{n_T(t)}{N_D}\right)^{1/2} = C_0\left(1-\frac{n_T(t)}{N_D}\right)^{1/2}. \tag{12.54}$$

If $n_T \ll N_D$, (12.54) can be simplified to

$$C \approx C_0\left(1-\frac{n_T(t)}{2N_D}\right). \tag{12.55}$$

During the capacitance transient, majority carriers are emitted and swept out of the depletion layer. Using equation (12.53) in (12.55), the capacitance can be written as

$$C = C_0\left[1-\left(\frac{n_T(0)}{2N_D}\right)\exp\left(-\frac{t}{\tau_e}\right)\right]. \tag{12.56}$$

A DLTS spectrum can be explained using the capacitance transient given in (12.56) with a temperature-dependent τ_e, as described in (12.43). The emission time constant decreases with increasing temperature, as shown in Fig. 12.16a.

To obtain a DLTS spectrum, the capacitive transient is recorded at two different times, t_1 and t_2, during the transient decay, and the capacitance difference,

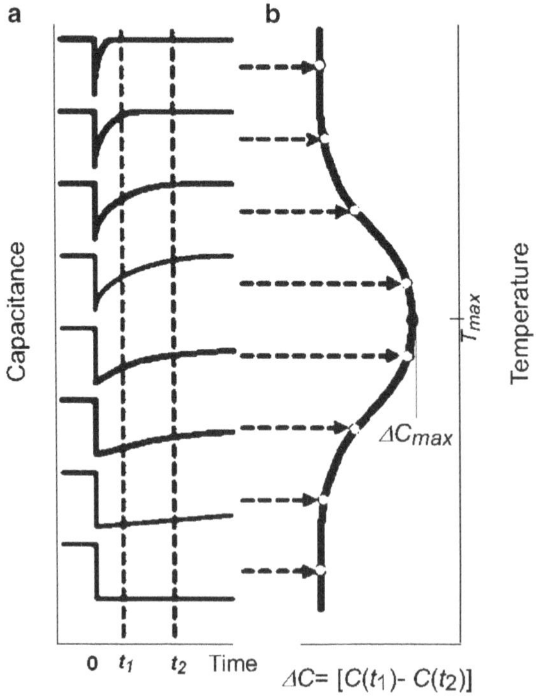

Fig. 12.16 Capacitance transients at different temperatures and DLTS spectrum

$\Delta C = C(t_1) - C(t_2)$, is then obtained as a function of temperature (see Fig. 12.16a). Since the emission rate is dependent on temperature, the capacitance decay rate decreases as temperature increases. Such a difference in a signal is a standard output feature of a double boxcar instrument and is known as *rate window* in DLTS terminology. A capacitance difference, ΔC, which is a standard output of a double boxcar average, is generated when the time constant is of the order of the gate separation $(t_2 - t_1)$. The value of ΔC has a maximum as a function of temperature, as shown in Fig. 12.16b.

It is common to describe the DLTS signal as

$$\Delta C(T) = \frac{[C(t_1) - C(t_2)]}{\Delta C(0)}, \tag{12.57}$$

where $\Delta C(0)$ is the magnitude of the capacitive transient at $T = 0\,\mathrm{K}$. Equation (12.57) can also be rewritten as

$$\Delta C(T) = \left[\exp\left(-\frac{t_1}{\tau}\right) - \exp\left(-\frac{t_2}{\tau}\right)\right], \tag{12.58}$$

where $\tau = e_n^{-1}$ or as

$$\Delta C(T) = \exp\left(-\frac{t_1}{\tau}\right)\left[1 - \exp\left(-\frac{\Delta t}{\tau}\right)\right], \tag{12.59}$$

where $\Delta t = t_2 - t_1$. The value of the emission rate at the peak of $\Delta C(T)$ is then given by

$$e_n^{-1} = \tau_{\mathrm{emax}} = \frac{t_1 - t_2}{\ln(t_1/t_2)}. \tag{12.60}$$

The choices of t_1 and t_2 determine this emission rate. Using different values for t_1 and t_2, DLTS spectra in different rate windows can be obtained. The peaks in the DLTS spectra can be used to calculate the activation energy and the emission rate of the trap, as described in Sect. 12.5 and illustrated in Fig. 12.12c. The trap concentration can be obtained from the peak value (ΔC_{max}) of the $\Delta C(T)$ curves using (12.59)–(12.60) and assuming $n_{\mathrm{T}}(0) = N_{\mathrm{T}}$, i.e.,

$$N_{\mathrm{T}} = \frac{\Delta C_{\mathrm{max}}}{C_0}\frac{2N_{\mathrm{D}}\exp\{[r/(r-1)]\ln(r)\}}{1-r} = \frac{\Delta C_{\mathrm{max}}}{C_0}\frac{2r^{r/(r-1)}}{1-r}N_{\mathrm{D}}, \tag{12.61}$$

where $r = t_2/t_1$. Note that $\Delta C < 0$ for majority carrier traps, such as electron traps in an n-type semiconductor, whereas $\Delta C > 0$ for minority traps, such as hole traps in an n-type semiconductor [20, 29–33].

The DLTS technique is capable of displaying the spectrum of traps as positive and negative peaks. The sign of the peak tells us whether the trap is near the conduction or the valence band. An electron trap is capable of trapping electrons if it is full of holes, while a hole trap can trap holes if it is full of electrons. Therefore,

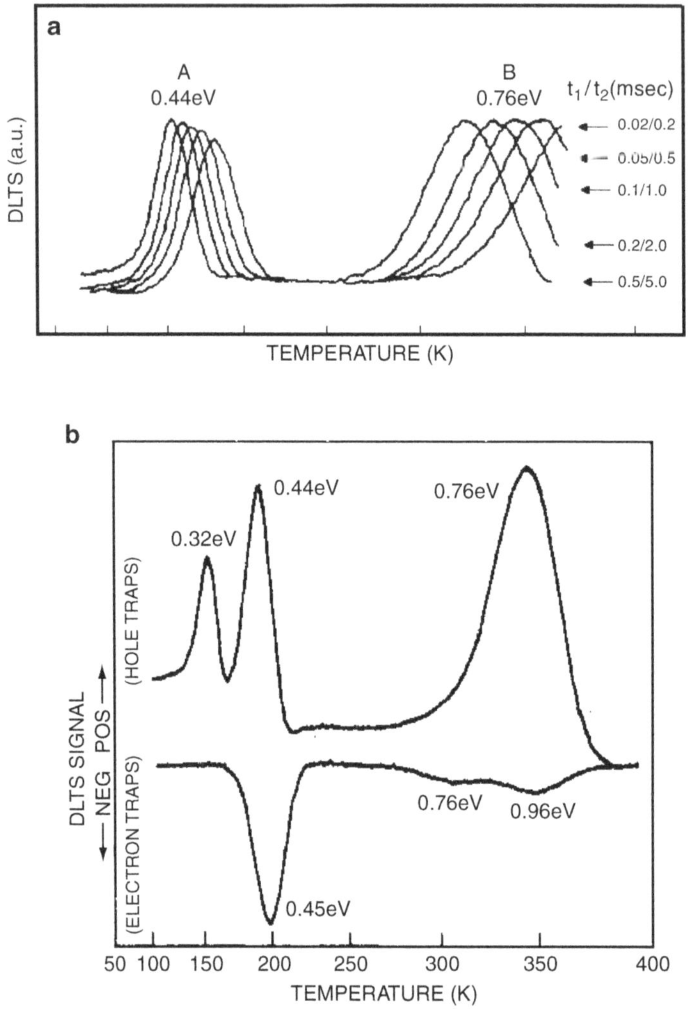

Fig. 12.17 Typical DLTS spectra for (**a**) of *n*-GaAs and (**b**) a 1-MeV electron-irradiated *n*-GaAs

electron traps tend to be closer to the conduction band, whereas hole traps tend be closer to the valence band. Examples of DLTS spectra are shown in Fig. 12.17a, b [30, 31]. The DLTS spectra were measured at five different rate windows for an *n*-type GaAs, and two minority carrier traps (hole traps) were observed and labeled as A and B. The activation energies were found to be equal to 0.44 and 0.76 eV, respectively, as measured from the valence band edge. The trap densities were found for both levels to be equal to 1.4×10^{14} cm^{-3}. In Fig. 12.17b, the DLTS spectra were taken for 1-MeV electron-irradiated *n*-GaAs. The positive signals correspond to hole traps (minority carrier traps) and the negative signals correspond to electron traps (majority carrier traps) [31].

A conventional DLTS system uses either the dual-channel boxcar averager or the lock-in amplifier for sampling of the capacitance transients, a very sensitive capacitance meter to measure the capacitance difference change as a function of temperature, a fast pulse generator to apply pulsed voltages to the sample, and a variable temperature cryostat. Because there are some disadvantages in a conventional DLTS system, such as low signal-to-noise ratio and low resolution of the time constants, several DLTS methods have been suggested. These include the Laplace DLTS (LDLTS) [34], deep-level transient Fourier spectroscopy (DLTFS) [35], optical DLTS (ODLTS) [36], and constant capacitance DLTS (CCDLTS) [37].

12.5.4 PITS versus DLTS

A trap can be generally described by three parameters: the trap density (N_T), the activation energy (E_T), and the carrier capture cross-section (σ). The density of traps cannot be measured using PITS because there is insufficient information on several parameters related to the spectral peak height. In contrast, it is possible to obtain all parameters using DLTS. Also, the nature of the trap, i.e., electron or hole trap, cannot be identified using PITS because both trap types produce a positive signal. The other weakness of conventional PITS is that the absolute concentrations of defects cannot be obtained. DLTS can be considered as the most appropriate tool to characterize defects. However, there is an important shortcoming for DLTS: it cannot be used for high resistive materials as the Debye length is longer than the width of the depletion region. Also, DLTS relies on the measurement of capacitance transients produced by the thermal emission of carriers from the traps within the depletion region of a reverse-biased p–n junction or Schottky barrier. Therefore, high electric fields are involved in DLTS measurements. The emission process is produced by a sudden change in the negative bias value toward a more negative value. Because a depletion region is required in DLTS, it cannot be used with undoped materials. In contrast, PITS is applicable to such a situation and only two ohmic contacts are necessary to observe photoconductivity decays. Finally, when the emission rates for several traps are similar, both techniques suffer from poor resolution in resolving different traps.

References

1. W. Smith, Effect of light on selenium during the passage of an electric current. Nature **7**, 303 (1873)
2. R.W. Pohl, Electron conductivity and photochemical processes in alkali-halide crystals. Proc. Phys. Soc. **49**, 3 (1937)
3. A.L. Huges, Photoconductivity in crystals. Rev. Mod. Phys. **8**, 294 (1936)
4. N.F. Mott, R.W. Gurney, *Electronic Processes in Ionic Crystals* (Oxford University Press, Oxford, 1940)

5. T.S. Moss, Photoconductivity in the elements. Proc. Phys. Soc. A **64**, 590 (1951)
6. H. Bube, *Photoelectronic Properties of Semiconductors* (Cambridge University Press, Cambridge 1992) ISBN 0521 40491 6
7. M. Fox, *Optical Properties of Solids* (Oxford University Press, Oxford, 2001) ISBN: 0 19 850613 9
8. J. Wilson, J.F.B. Hawkes, *Optoelectronics: An Introduction*, 2nd edn. (Prentice Hall, Upper Saddle River, 1989) ISBN: 0136384617
9. S.L. Chuang, *Physics of Optoelectronic Devices* (Wiley, New York, 1995) ISBN: 0-471-10939-8
10. J.R. Hook, H.E. Hall, *Solid State Physics*, 2nd edn. (Wiley, England, 1991) ISBN: 0 471 928046
11. D.A. Neamen, *Semiconductor Physics and Devices*, 3rd edn. (McGraw Hill Higher Education, New York, 2003) ISBN: 0-07-232107-5
12. R.H. Bube, *Photoconductivity of Solids* (Wiley, New York, 1960)
13. R.A. Smith (ed.), *Semiconductors*, 2nd edn. (Cambridge University Press, Cambridge, UK,. 1978) ISBN: 0521293146
14. M.Ç. Arikan, An Experimental Investigation of Photoconductivity in GaAs, PhD Thesis, Essex University, UK, 1980
15. M.Ç. Arikan, Y. Ergün, N. Balkan, B.K. Ridley, In-plane photoconductive properties of MBE-grown GaAs/GaAlAs multiple quantum wells. Semicond. Sci. Technol. **8**, 1337 (1999)
16. A. Erol, N. Akçay, M.Ç. Arikan, S. Mazzucato, N. Balkan, Spectral photoconductivity and in-plane photovoltage studies of as-grown and annealed GaInNAs/GaAs and GaInN/GaAs quantum well structures. Semicond. Sci. Technol. **19**, 1086 (2004)
17. S. Mazzucato, A. Boland-Thoms, A. Erol, N. Balkan, Transient photoconductivity and in-plane photovoltage studies in GaInNAs/GaAs quantum wells. Phys. Scripta **T114**, 236 (2004)
18. A.G. Milnes, *Deep Impurities in Semiconductors* (Wiley, New York, 1973) ISBN: 0-471-60670-7
19. U.K. Mishra, J. Singh, *Semiconductor Device Physics and Design* (Springer, The Netherlands, 2008) ISBN: 978 1-4010-6480-7
20. D.K. Schroder, *Semiconductor Material and Device Characterization*, 3rd edn. (Wiley, Canada, 2006) ISBN: 978-0-471-73-906-7
21. E.A. Irene, *Electronic Materials Science* (Wiley, New York, 2005) ISBN 9780471695974
22. E.R. Weber (ed.), *Imperfections in III/V materials, Semiconductors and Semimetals 38*, (Academic, New York, 1993) ISBN: 0-12-752138-0
23. R.E. Kremer, M.Ç. Arikan, J.C. Abele, J.S. Blakemore, Transient photoconductivity measurements in semi-insulating GaAs. I. An analogy. J. Appl. Phys. **62**, 2424 (1987)
24. A. Erol, S. Mazzucato, M.Ç. Arikan, H. Carrere, A. Arnoult, E. Bedel, N. Balkan, Photo-induced transient spectroscopy of defect levels in GaInNAs. Semicond. Sci. Technol. **18**, 968 (2003)
25. J.P. Zielinger, M. Tapiero, Assessment of deep levels in photorefractive materials by transient photoelectric methods. J. Phys. III France **3**, 1327 (1993)
26. M.Ç. Arikan, S. Cenk, N. Balkan, Photo-induced transient spectroscopy of deep levels in $GaAs/Ga_{1-x}Al_xAs$ multiple quantum wells. J. Appl. Phys. **82**, 4986 (1997)
27. D.V. Lang, Deep-level transient spectroscopy: A new method to characterize traps in semiconductors. J. Appl. Phys. **45**, 3023 (1974)
28. D.V. Lang, Recalling the origins of DLTS. Physica B **401–402**, 7 (2007)
29. A.A. Garcia, M.A.R. Baranca, Computerized DLTS system to characterize deep levels in semiconductors. Revista Mexiana De Fisica **48**, 539 (2002)
30. P.T. Devries, A.A. Khan, An efficient technique for analyzing deep level transient spectroscopy data. J. Elect. Mater. **18**, 543 (1989)
31. D.V. Lang, L.C. Kimerling, Observation of recombination-enhanced defect reactions in semiconductors. Phys. Rev. Lett. **33**, 489 (1974)
32. V.T.R. Kuoppa, Electrical characterization of Nitrogen containing III-V semiconductors. (Doctoral Dissertation TKK Dissertations 62, Helsinki University of Technology, Finland, 2007)

33. A. Dast, V.A. Singht, D.V. Lang, Deep-level transient spectroscopy (DLTS) analysis of defect levels in semiconductor alloys. Semicond. Sci. Technol. **3**, 1177 (1988)
34. L.F. Makarenko, J.H. Evans-Freeman, Application of DLTS and Laplace-DLTS to defect characterization in high-resistivity semiconductors. Phys. B: Condens. Matter **401–402**, 666–669 (2007)
35. S. Weiss, R. Kassing, Deep Level Transient Fourier Spectroscopy (DLTFS)—A technique for the analysis of deep level properties. Solid-State Electron. **31**, 1733–1742 (1988)
36. A. Blondeel, P. Clauws, Quantitative optical variants of deep level transient spectroscopy: application to high purity germanium. Mater. Sci. Eng. B **71**, 233–237 (2000)
37. M.D. Jack, R.C. Pack, J. Henriksen, A computer-controlled deep-level transient spectroscopy system for semiconductor process control. Electron Dev. IEEE Trans. **27**, 2226–2231 (1980)

Index

A. Patanè and N. Balkan (eds.), *Semiconductor Research*, Springer Series in Materials Science 150, DOI 10.1007/978-3-642-23351-7,